The organic compounds of lead

This is a volume in the series
THE CHEMISTRY OF ORGANOMETALLIC COMPOUNDS

THE CHEMISTRY OF ORGANOMETALLIC COMPOUNDS

A Series of Monographs

Dietmar Seyferth, *editor*

Department of Chemistry,
Massachusetts Institute of Technology
Cambridge, Massachusetts

PUBLISHED:

Chemistry of the iron group metallocenes

Part One: ferrocene, ruthenocene, osmocene

By Myron Rosenblum

The organic compounds of lead

HYMIN SHAPIRO

AND

F. W. FREY

Research and Development Department
Ethyl Corporation
Baton Rouge, Louisiana

Interscience Publishers
a division of
JOHN WILEY & SONS

NEW YORK · LONDON · SYDNEY · TORONTO

CHEMISTRY

Library of Congress Catalog Card Number 68-8111
SBN 470 778857

Printed in the United States of America

To Betty and Margie

Preface

Since the appearance of Krause and von Grosse's book, and particularly since the publication of Leeper, Summers, and Gilman's review of organolead chemistry, there has been a surprising growth of information on this subject. Despite this growth, there has been no comprehensive review of organolead chemistry since 1954. This volume constitutes such a comprehensive review At the same time, the volume was written in order to present also a critical review of organolead chemistry; hopefully, it describes all recent advances in critical fashion without offending old friends and co-workers in the field.

The present volume covers the organolead literature through 1966, with many references to 1967.

We are particularly indebted to Prof Dietmar Seyferth of the Massachusetts Institute of Technology, who is both the Editor of this series and the godfather of this volume. We wish to thank also Robert A. Kehoe, M.D., Professor Emeritus; Emil A. Pfitzer, Sc.D., Associate Professor of Toxicology; and George Roush, Jr., M.D., Clinical Professor of Industrial Medicine, Department of Environmental Health, College of Medicine, University of Cincinnati, for their constructive thoughts in Chapter V on Physiological and Toxicological Properties. It is a great pleasure to thank numerous associates at the Ethyl Corporation, especially Dr. J. R. Zietz for helpful suggestions, Ann Breaux and Mary Jo Pribble for bibliographic assistance, and Dorothy S. Patton, Muriel Brooks, Sue Brueck, Joyce Neal, Marie MacKenzie, and Marjorie Wilson, for the stenographic help without which there would never have been a book. Finally, we wish to thank Dr. Schrade F. Radtke, Director of Research of the International Lead Zinc Research Organization, for permission to include unpublished information obtained from ILZRO.

HYMIN SHAPIRO
F. W. FREY

Contents

1234567890

I Introduction

The first synthesis of an organolead compound, i.e., one containing a lead–carbon bond, is usually credited to Löwig (17–19), who reported the reaction of a sodium–lead alloy and ethyl iodide in 1853. Almost simultaneously, Cahours (2) reported the formation of a small amount of organolead product from the reaction of lead metal with ethyl iodide. Whether Löwig's product was tetraethyllead or hexaethyldilead has been the subject of some debate.

Nevertheless, from this inauspicious beginning, organolead chemistry has developed into one of the largest areas of organometallic chemistry. The number of known organolead compounds is exceeded only by those of tin, silicon, and mercury; it has been estimated that about 1200 organolead compounds (28) were known in 1965. Furthermore, organolead compounds enjoy a unique position among organometallic compounds as a result of Midgley's discovery of their exceptional effect in suppressing knock in the combustion of gasoline in automobile engines. In 1964, over 600 million pounds of ethyl- and methyl lead compounds were produced in the United States for use as antiknock agents in motor gasoline.

Several excellent reviews of organolead chemistry have been written. Outstanding among these are the excellent 1937 review of Krause and von Grosse (14), which covered the preparation and properties of the organolead compounds known up to that time, and the last comprehensive review, which was published in 1954 by Leeper, Summers, and Gilman (16). Recent shorter reviews by Willemsens (27) and by Willemsens and van der Kerk (28) appeared in 1964 and 1965, emphasizing the major developments in organolead chemistry that have occurred since 1953. A new *Annual Survey of Organometallic Chemistry* series by Seyferth and King (23) began to appear in 1965, and an encyclopedic review of organolead chemistry by Shapiro and Frey (25) in the *Encyclopedia of Chemical Technology* was published in 1967. Several other review-type articles—some of them excellent—also exist (see References).

Since about 1950 there has been a notable resurgence of interest in all aspects of organolead chemistry, reflecting to a certain extent a resurgence of interest in organometallic chemistry in general. This renewed interest has also been brought about by a number of special factors: an increased realization of the differences between organolead and organotin chemistry; the synthesis of new types of organolead compounds, such as, for example, Gilman's triphenylplumbyllithium types and Kocheshkov's aryllead tricarboxylates; the introduction of new methods of preparation based on alkylaluminum compounds and electrolysis; and the realization that many areas of organolead chemistry have been only lightly touched and require further elucidation.

In the last several years the horizons of organolead chemistry have been extended considerably. Many new compounds have been synthesized and characterized; these include the triorganolead hydrides, amides, and phosphides, polymetal derivatives of the type $(C_6H_5)_{12}Pb_5$, tetravinyllead and other derivatives containing olefinic groups, and organolead polymers. New methods of synthesis have been discovered, and electrolytic syntheses based on alkylaluminum and alkyl Grignard reagents have been extensively investigated. Of greater importance, however, is that increased emphasis is now being placed on a more fundamental understanding of organolead chemistry. Detailed investigations have been conducted on the kinetics of reaction of tetraorganolead compounds with various reagents, leading to a better understanding of the lead–carbon bond and reaction mechanisms. The thermal decomposition of tetraorganolead compounds has been the subject of numerous investigations, bringing a better understanding of these reactions. Presently, the coordination chemistry of organolead salts is being investigated in several laboratories. The International Lead Zinc Research Organization is currently sponsoring several programs designed to provide a better understanding of organolead chemistry. The results of the program being conducted by Willemsens and van der Kerk have been especially interesting and enlightening. Their work has been recently summarized (28).

Interest in organolead chemistry is expected to continue at a high level, with continued emphasis on the fundamental chemistry of the lead–carbon bond. Despite the extensive effort that has gone into this area in the past, many aspects of organolead chemistry have been investigated only qualitatively and are little understood. A more fundamental approach to these areas will certainly resolve some of the discrepancies and seeming anomalies which persist. Hopefully, the next several years of research in this area will prove as fruitful as have those of the recent past.

This review covers the literature through 1966, with some references to 1967, as determined from *Chemical Abstracts* and *Current Chemical Papers*. No attempt is made, however, to duplicate completely the material presented in the older comprehensive reviews, such as the review of Leeper, Summers, and Gilman (16).

REFERENCES

1. Azerbaev, I. N., and D. A. Kochkin, *Vestn. Akad. NaukKaz. SSR*, **19** (10), 18 (1963); *C.A.*, **60**, 5534 (1964).
2. Cahours, A., *Compt. Rend.*, **36,** 1001 (1853).
3. Calingaert, G., *Chem. Rev.*, **2,** 43 (1925).
4. Coates, G. E., *Organo-Metallic Compounds*, Methuen, London, 1956, pp. 133–143.
5. Dub, M., *Organometallic Compounds, Vol. II: Compounds of Germanium, Tin, and Lead, Including Biological Activity and Commercial Application*, 2nd ed., R. W. Weiss, Ed., Springer-Verlag, New York, 1967.
6. Eisch, J., and H. Gilman, "Organometallic Compounds," in *Advances in Inorganic Chemistry and Radiochemistry*, H. J. Emeleus and A. G. Sharpe, Eds., Academic Press, New York, 1960.
7. Garzuly, R., "Organometalle," in Vol. XIX of Herz, *Saumlung chem. u. chemisch-technischer Vorträge*, Ferdinand Euke, Stuttgart, 1927.
8. Gilman, H., "Organometallic Compounds," Chapter 5 of *Organic Chemistry, An Advanced Treatise*, 2nd ed., Vol. 1, Wiley, New York, 1943.
9. Goddard, A. E., and D. Goddard, *A Textbook of Inorganic Chemistry*, Vol. 11, Part 1, J. N. Friend, Ed., Charles Griffin and Co., London, 1928.
10. Klarmann, E., "Darstellung metallorganischer Verbindungen," in *Handbuch der biologischen Arbeitsmethoden*, Abt. 1, Teil 2, Heft 4, E. Abderhalden, Ed., Urban and Schwarzenberg, Berlin, 1937.

10a. Klema, F., *Mitt. Chem. Forschungsinst. Wirtsch. Oesterr.*, **20,** 1 (1966); *C.A.*, **65,** 10617 (1966).

11. Kocheshkov, K. A., *Usp. Khim.*, **3,** 83 (1934); *C.A.*, **28,** 5402 (1934).
12. Kocheshkov, K. A., *Sintet. Metody v Oblasti Metalloorg. Soedinenii, Inst. Org. Khim., Akad. Nauk SSSR*, **1947,** No. 5; *C.A.*, **46,** 11102h (1952).
13. Korshak, V. V., and G. S. Kolesnikov, *Usp. Khim.*, **15,** No. 3, 326 (1946).
14. Krause, E., and A. von Grosse, *Die Chemie der metall-organischen Verbindungen*, Borntraeger, Berlin, 1937, pp. 372–429.
15. Kumada, M., *Kagaku* (*Kyoto*), **15,** 442 (1960).
16. Leeper, R. W., L. Summers, and H. Gilman, *Chem. Rev.*, **54,** 101 (1954).
17. Löwig, C., *Ann. Chem.*, **88,** 318 (1853).
18. Löwig, C., *J. Prakt. Chem.*, **60,** 304 (1853).
19. Löwig, C., *Chem. Zentr.*, **1852,** 575.
20. Reutov, O. A., *Angew. Chem.*, **72,** 198 (1960).
21. Rochow, E. G., D. T. Hurd, and R. N. Lewis, *The Chemistry of Organometallic Compounds*, Wiley, New York, 1957, pp. 190–197.
22. Schmidt, J., *Organo-Metallverbindungen*, Wissenschaftliche Verlagsgesellschaft, Stuttgart, 1934.
23. Seyferth, D., and R. B. King, *Annual Survey of Organometallic Chemistry, Vol.* 1, *Covering the Year 1964*, Elsevier, Amsterdam, 1965.
24. Shapiro, H., "Preparation of Tetraalkyllead Compounds from Lead or Its Alloys,"

in *Metal-Organic Compounds*, Advances in Chemistry Series, Vol. 23, American Chemical Society, Washington, 1959.
25. Shapiro, H., and F. W. Frey, "Organolead Compounds," in *Encyclopedia of Chemical Technology*, Vol. 12, Interscience, New York, 1967.
26. Sidgwick, N. V., *The Chemical Elements and Their Compounds*, Vol. 1, Clarendon Press, Oxford, 1950.
27. Willemsens, L. C., *Organolead Chemistry*, Intern. Lead Zinc Res. Organ., New York, 1964.
28. Willemsens, L. C., and G. J. M. van der Kerk, *Investigations in the Field of Organolead Chemistry*, Intern. Lead Zinc Res. Organ., New York, 1965.

II Valence states and nature of bonding

The properties of organolead compounds stand in striking contrast to those of inorganic lead compounds and other compounds of lead which do not contain a lead-to-carbon bond. Thus, among the inorganic compounds, the most stable are those in which lead is in the divalent state; as a general rule, these compounds are more thermodynamically and hydrolytically stable than their analogs of tetravalent lead. On the other hand, among the organolead compounds, the tetravalent derivatives are much more stable both thermodynamically and hydrolytically than the divalent derivatives. The divalent derivatives are very unstable and decompose via intraoxidation-reduction to tetravalent lead and lead metal.

The simple divalent lead derivatives, because of their instability, are not very well characterized. Their successful isolation was reported as early as 1922 by Krause and Reissaus (5), who described the isolation of a red solid which they concluded to be diphenyllead. All subsequent attempts to confirm their results have been unsuccessful; instead it has been shown that their reaction mixture is unstable to hydrolysis and that by their procedure any divalent organolead compounds present would be hydrolyzed to lead hydroxide or halide. Alkyl derivatives of divalent lead have been reported as red oils obtained by reduction of tetravalent derivatives with active metals or by electrolysis of a ketone in aqueous sulfuric acid using a lead cathode. However, these compounds have not been characterized.

Undoubtedly, divalent organolead compounds do exist and are formed as initial intermediates in the synthesis of tetravalent compounds. These compounds probably contain covalent sigma bonds between carbon atoms and the unpaired p electrons of lead in the p_x and p_y orbitals. Stabilized derivatives of divalent lead exist as ionic compounds of the type $R_3Pb^-Li^+$

or $R_3Pb^-Na^+$, in which the third carbon atom presumably is bound to lead by an electron pair from carbon supplied to the vacant p_z orbital. These R_3PbM compounds are analogous to complexes of lead(II) halides with alkali metal halides. They have not been isolated in the pure state and are not well characterized; however, they have proved very useful for the synthesis of other organolead derivatives.

Derivatives of divalent lead containing cyclopentadienyl and methylcyclopentadienyl groups are known, namely, dicyclopentadienyllead and di(methylcyclopentadienyl)lead. These are relatively stable solids; however, in these derivatives the sigma lead-to-carbon bonds are believed to be supplemented by π bonding between the π electrons of the cyclopentadiene ring and empty inner *d* orbitals of lead. Thus, these compounds have an angular sandwich-type structure similar to that of ferrocene.

The organometallic derivatives of tetravalent lead are much better characterized. From the electronic configuration of the lead atom, one would predict bond formation of tetravalent lead to occur through hybridization of the two *s* and two *p* electrons in the outer shell to give sp^3 hybridization, resulting in a tetrahedral configuration with all four bonds directed from the center to the apices of a regular tetrahedron. All the physical properties data on the R_4Pb derivatives are consistent with this tetrahedral configuration, so that the tetraalkyl and tetraaryl derivatives are best represented as covalent molecules in which the organic groups are tetrahedrally located around the lead atom.

The lead-to-carbon bond exhibits a high degree of covalent character as a result of the small difference in the electronegativities of the two elements. Some disagreement exists concerning the preferred electronegativity value for lead. The most commonly accepted value is 1.8 (based on a value of 2.5–2.6 for carbon), a value proposed by Pritchard and Skinner (9) as a mean value derived from different methods. (C = 2.5–2.6; Si = 1.8–1.9; Ge = 1.8–1.9; Sn = 1.8–1.9; Pb = 1.8). A value of 1.13 has also been proposed for the electronegativity of lead (C = 2.55) which was calculated from force constants derived from the infrared spectrum of tetramethyllead (10). More recently, a value of 2.45 (C = 2.6) has been proposed by Allred and Rochow (1,2) on the basis of the proton chemical shifts observed in the NMR spectra of the tetramethyl derivatives of the Group IVb elements; this value would put lead next to carbon in the electronegativity scale of the Group IVb elements. However, this value has been rejected by Drago (4), who chooses to explain the proton chemical shifts in the NMR spectra on the basis of anisotropic contributions to the M—C—H bonds. Matwiyoff and Drago (6) have shown that the stability constants for coordination compounds of salts of triphenyllead and triphenyltin with various ligands indicate that the electronegativities of lead and tin are nearly equal. Furthermore, electronegativity values calculated from the C^{13} chemical shifts observed in the NMR spectra (12) of the tetramethyl derivatives of the Group IVb elements show excellent agreement with the "best" electronegativity values of Pritchard and Skinner (9). In view of these considerations the electronegativity value of 2.45 for lead must be rejected. At the same time it would seem that the value of 1.13 is too low and Pritchard and Skinner's value of 1.8 continues to be a preferred value.

Numerous references are found in the early organolead literature to "triorganolead" compounds having the composition R_3Pb. This designation arose from the faint to intense yellow color that is characteristic of the organolead compounds having this composition, and the decrease in molecular weight found with increasing dilution, as determined by cryoscopic measurements. However, the existence of such triorganolead entities as R_3Pb^0 radicals in the solid state or in solution is not supported by other physical measurements (e.g., magnetic susceptibility, electron paramagnetic resonance) and the preponderance of evidence teaches that these compositions are actually compounds of the type R_6Pb_2, or derivatives of tetravalent lead in which one of the bonds is a lead-to-lead bond. Electron diffraction measurements on hexamethyldilead show the methyl groups to be tetrahedrally located around the lead atoms at a C—Pb—Pb bond angle of 109.5° (11). A value of 2.88 Å has been calculated for the lead-to-lead bond distance from these data; this corresponds to a value of 1.44 Å for the covalent lead radius, and is in excellent agreement with Pauling's value of 1.43 Å for the covalent radius of tetravalent lead. From these same measurements, a value of 2.25 Å has been calculated for the lead-to-carbon bond distance in hexamethyldilead. Electron diffraction measurements on tetramethyllead have given values of 2.29 (3) and 2.20 Å (15) for the lead-to-carbon bond distance. Thus, the hexaorganodilead compounds are covalent compounds containing a lead-to-lead bond, with three carbon atoms located tetrahedrally around each lead atom. Because of the relatively low strength of the lead-to-lead bond, this bond is cleaved in almost all known reactions of hexaorganodilead compounds.

A large class of organometallic derivatives of tetravalent lead is that in which one or more organic groups are displaced by an anionic moiety such as chloride, oxide, sulfide, sulfate, or hydride, to mention a few. Such derivatives range from the highly ionic mono- and di-halide derivatives to the more covalent alkoxides and hydrides, the halide derivatives being the best characterized. Dipole moment measurements indicate that the degree of ionic character of the lead-to-halogen bond in di- and tri-organolead halides approaches that of the lead-to-halogen bond in lead(II) halides. Because of this ionic character, they undergo many of the metathesis reactions characteristic of inorganic lead compounds. Also, a variety of coordination compounds of organolead halides and carboxylates are known, involving such ligands as ammonia, pyridine, water, dimethylsulfoxide, and dimethylformamide.

From the infrared spectra of trimethyllead acetate, propionate, and chloroacetate, the trimethyllead moiety is proposed to be present as a planar trimethyllead ion, even in the solid compounds (7). A similar conclusion has been proposed in the interpretation of the infrared spectrum of trimethyltin fluoride (8). However, conductometric measurements of solutions of triphenyllead chloride in dimethylformamide (14), and pyridine, acetonitrile, and nitrobenzene (13) have given no evidence for the presence of triphenyllead cations in these solvents.

In recent years there has been a renewed interest in the nature of the bonding in the various types of organolead compounds. The coordination chemistry of organolead salts is of particular current interest. Several papers have appeared in recent years on the kinetics and mechanisms of reactions of organolead compounds with various electrophilic reagents which have provided valuable insight into the nature of the lead to carbon bond and the critical factors affecting its reactivity. Further investigation along these lines should be equally fruitful in elucidating the chemistry of organolead compounds.

REFERENCES

1. Allred, A. L., and E. G. Rochow, *J. Inorg. Nucl. Chem.*, **5**, 269 (1958).
2. Allred, A. L., and E. G. Rochow, *J. Inorg. Nucl. Chem.*, **20**, 167 (1961).
3. Brockway, L. O., and H. O. Jenkins, *J. Am. Chem. Soc.*, **58**, 2036 (1936).
4. Drago, R. S., *J. Inorg. Nucl. Chem.* **15**, 237 (1960).
5. Krause, E., and G. G. Reissaus, *Ber.*, **55**, 888 (1922).
6. Matwiyoff, N. A., and R. S. Drago, *Inorg. Chem.*, **3**, 337 (1964).
7. Okawara, R., and H. Sato, *J. Inorg. Nucl. Chem.*, **16**, 204 (1961).
8. Okawara, R., D. E. Webster, and E. G. Rochow, *J. Am. Chem. Soc.*, **82**, 3287 (1960).
9. Pritchard, H. O., and H. A. Skinner, *Chem. Rev.*, **55**, 745 (1955).
10. Sheline, R. K., and K. S. Pitzer, *J. Chem. Phys.*, **18**, 595 (1950).
11. Skinner, H. A., and L. E. Sutton, *Trans. Faraday Soc.*, **36**, 1209 (1940).
12. Spiesecke, H., and W. G. Schneider, *J. Chem. Phys.* **35**, 722 (1961).
13. Thomas, A. B., and E. G. Rochow, *J. Inorg. Nucl. Chem.*, **4**, 205 (1957).
14. Thomas, A. B., and E. G. Rochow, *J. Am. Chem. Soc.*, **79**, 1843 (1957).
15. Wong, C. H., and V. J. Schomaker, *J. Chem. Phys.*, **28**, 1007 (1958).

III General methods of synthesis

The various types of organolead compounds are readily synthesized by many of the methods commonly employed for the synthesis of organometallic derivatives of the other Group IVb metals. However, because of certain chemical properties peculiar to the inorganic and organic compounds of lead, as compared to the other Group IVb metals, the synthesis of organolead compounds poses special problems which are not normally encountered in the synthesis of the organometallic derivatives of these other metals.

Because of the relative instability of the inorganic salts of tetravalent lead and their susceptibility to hydrolysis, they are not generally used in organolead syntheses. On the other hand, the organolead derivatives are conveniently synthesized by the reaction of a divalent lead salt, such as $PbCl_2$, with reactive organometallic compounds such as organolithium or Grignard reagents. These reactions are usually conducted in ether solvent and produce a hexaorganodilead or tetraorganolead derivative as the major product of reaction. Probably a diorganolead moiety is formed initially, as evidenced by the formation of an intense color without precipitation of lead metal when a low reaction temperature ($\leq -10°$) is employed. When the reaction temperature is raised, precipitation of lead metal occurs and the intense color is rapidly dissipated. Krause and von Grosse (6) postulated the reaction of a divalent lead salt with the Grignard reagent to involve the following sequence.

$$6PbX_2 + 12RMgX \rightarrow 6R_2Pb + 12MgX_2$$

$$6R_2Pb \rightarrow 2R_6Pb_2 + 2Pb$$

$$2R_6Pb_2 \rightarrow 3R_4Pb + Pb$$

In the synthesis of R_6Pb_2 or R_4Pb, the overall reaction then becomes:

$$3PbX_2 + 6RMgX \rightarrow R_6Pb_2 + Pb + 6MgX_2$$

$$2PbX_2 + 4RMgX \rightarrow R_4Pb + Pb + 4MgX_2$$

so that the reaction involves an intraoxidation-reduction of the lead(II) salt. The efficacy of this sequence appears to be well documented with a number of organolead derivatives. However, the R_2Pb intermediate is too unstable to permit its isolation as a pure compound; its formation, as mentioned above, is indicated by the appearance of an intense coloration in the reaction mixture before precipitation of lead metal begins.

Inorganic or organic salts of tetravalent lead can also be used as the source of lead values in this reaction.

$$PbX_4 + 4RMgX \rightarrow R_4Pb + 4MgX_2$$

The obvious advantage for the use of a tetravalent lead salt is the possible elimination of by-product lead metal formation. Indeed, where R is aryl and X is carboxylate, the reaction does proceed without formation of any lead metal. But where R is alkyl and X is halide or carboxylate, lead metal is formed in appreciable amounts. This is attributed to the formation of an $[RPbX_3]$ derivative as the initial intermediate; the $[RPbX_3]$ is unstable (except where R is aryl and X is carboxylate, as indicated above) and decomposes according to:

$$[RPbX_3] \rightarrow RX + PbX_2$$

so that the reaction for all practical purposes becomes that of a divalent lead salt. However, lead metal formation in the divalent lead salt reaction can be eliminated by conducting it in the presence of an organic halide:

$$2PbX_2 + 6RMgX + 2RX \rightarrow 2R_4Pb + 6MgX_2$$

This reaction is discussed in a later section. Also, organolithium reagents can be used in place of the Grignard reagent in the divalent lead salt reaction; in most cases, the two reagents can be used with equivalent results. The lithium reagents are best used for organic moieties which do not readily form Grignard reagents, e.g., *p*-dimethylaminophenyl halides.

The divalent lead salt reaction represents the only truly general synthetic method for organolead compounds. By conducting the reaction at a low temperature in the presence of an excess of RLi or RMgX, compounds of the type R_3PbLi or R_3PbMgX can be formed *in situ*; these latter compounds are very useful in the synthesis of unsymmetrical tetraorganolead compounds of the type R_3PbR'.

$$PbX_2 + 3RMgX(Li) \rightarrow R_3PbMgX(Li) + 2MgX_2(LiX)$$

$$R_3PbMgX + R'X \rightarrow R_3PbR' + MgX_2$$

In fact, it has recently been shown that in a tetrahydrofuran solution containing equimolar amounts of lead(II) bromide and phenyllithium, triphenylplumbyllithium (Ph_3PbLi) is present in appreciable amounts [as evidenced by the formation of benzyltriphenyllead (28% yield) when benzyl chloride is added to the reaction mixture (3)], so that the reaction of a divalent lead salt with a Grignard reagent is not as simple as the sequence proposed by Krause and von Grosse.

Robinson (13) has found that triethyllead chloride and diethyllead dichloride (or acetate) are formed in substantial amounts when a solution of ethylmagnesium bromide or diethylzinc is forced rapidly through a column packed with a large excess of lead(II) chloride or lead(II) acetate. This procedure was used by Robinson in an attempt to prepare diethyllead, any soluble intermediate lead product being in minimal contact with unreacted organometallic reagent using this procedure. Interestingly, when triethylaluminum was reacted with $PbCl_2$ using this procedure, no ethyllead chlorides were isolated.

Organolead halides have been obtained in several instances as the major product of reaction of a lead(II) halide and a Grignard reagent. Thus, Meals (7) obtained the triorganolead chloride when he attempted to synthesize tetraorganolead derivatives of long chain alkyls (R = dodecyl, tetradecyl, hexadecyl, and octodecyl). Similarly, trineopentyllead chloride was obtained, along with hexaneopentyldilead, as the major product from the reaction of $PbCl_2$ and neopentylmagnesium chloride (15,18). The R_3PbX possibly arises by reaction of the primary product, R_6Pb_2, with by-product MgX_2, since hexaphenyldilead has been shown to react with magnesium iodide (2) or bromide (14) to form triphenyllead halide. The sequence proposed by Krause and von Grosse does not provide for the formation of an R_3PbX product in the divalent lead salt reaction.

Organolead halides and other derivatives of the type R_3PbX or R_2PbX_2 (where X is not necessarily halide) are best prepared by controlled cleavage of tetraorganolead derivatives by halogens or acids according to the equations below (where X_2 is a halogen and HX is halogen acid, carboxylic acid or inorganic acid):

$$R_4Pb + X_2 \rightarrow R_3PbX + RX$$

$$R_3PbX + X_2 \rightarrow R_2PbX_2 + RX$$

or

$$R_4Pb + HX \rightarrow R_3PbX + RH$$

$$R_3PbX + HX \rightarrow R_2PbX_2 + RH$$

Cleavage of a third R group usually results in the formation of PbX_2 and RX as a result of the instability of most $RPbX_3$ derivatives.

Organolead salts of both inorganic and organic acids are also easily prepared by metathesis reactions involving the corresponding organolead halides, hydroxides, or oxides. Typical examples are:

$$R_3PbX + NaAn \xrightarrow[\text{alcohol}]{H_2O \text{ or}} R_3PbAn + NaX$$

$$R_2PbX_2 + 2NaAn \longrightarrow R_2PbAn_2 + 2NaX$$

$$R_2PbO + 2HAn \longrightarrow R_2PbAn_2 + H_2O$$

A large number of these compounds has been prepared in this fashion.

A second important method of synthesis which has been used to prepare both tetraalkyl and tetraaryllead compounds is the reaction of an alkyl or aryl halide with an alloy of lead and a reactive metal. The best known reaction of this type is that involving sodium–lead alloy and ethyl chloride or methyl chloride, since this reaction is used for the commercial manufacture of tetramethyllead and tetraethyllead. Dimagnesium–lead alloy has also been used in this type of reaction. The main product of such reactions is the tetraorganolead compound, although some hexaorganodilead is usually obtained in trace to small amounts. However, these alloy reactions do not enjoy prominence as laboratory syntheses because of the inconvenience involved in preparing and comminuting the lead alloy. Finely divided lead metal will react directly with an alkyl halide, preferably the alkyl iodide, to form R_4Pb. This reaction has been used only to prepare tetramethyl and tetraethyllead, and more recently tetrabenzyllead, and therefore does not qualify as a general method of synthesis.

Various other reactions of a less general nature have been used to form lead–carbon bonds. The electrolysis of an organometallic electrolyte, such as RMgX, $NaAlR_4$, NaR_3AlF, and $NaBR_4$, using a lead anode, has been investigated extensively for the commercial manufacture of tetramethyl- and tetraethyllead. The electrolytic reduction of ketones in dilute sulfuric acid solution using a lead cathode also yields organolead compounds. The chemical reduction of ketones in dilute sulfuric acid solution using sodium–lead alloys also yields organolead compounds. The reaction of complexes of lead halides and phenyldiazonium chloride with zinc or copper is reported to form triphenyllead chloride or diphenyllead dichloride (5) while the reaction of phenyldiazonium tetrafluoborate with lead metal in acetone gave tetraphenyllead in low yield (8). The reaction of thiophene and 1,3-dimethoxybenzene with lead(IV) carboxylates to form di-2-thienyllead diisobutyrate (10) and 2,4-dimethoxyphenylead triacetate (11,12), respectively, has been described. The addition of the lead–hydrogen bond in triorganolead hydrides to ethylene has been used to form an ethyllead derivative (1); similarly, tributyllead hydride is reported to add to styrene, methyl acrylate, phenylacetylene, and acrylonitrile (9). These reactions of organolead hydrides are especially useful for the synthesis of organolead derivatives containing a functional group. Diazomethane has been used to effect insertion of a methylene group into trimethyllead chloride to form chloromethyltrimethyllead, Me_3PbCH_2Cl (16,17).

An intriguing method for the formation of a lead–carbon bond involves the pyrolysis of organolead salts of the type $Ar_3PbOOCCH_2(CH_2)_nCOOEt$ to yield $Ar_3PbCH_2(CH_2)_nCOOEt$ with elimination of carbon dioxide (4). This reaction represents another method of synthesis of organolead compounds containing a functional group. The formation of a lead–carbon bond in this manner is most surprising in the light of the pronounced tendency of triaryl- and trialkyllead carboxylates to undergo disproportionation reactions of the type

$$2R_3PbOOCR \rightarrow R_4Pb + R_2Pb(OOCR)_2$$

Organolead chemistry has not truly suffered from any shortage of satisfactory methods of synthesis, and new methods of synthesis are constantly being discovered. In recent years, the chemical properties of triphenylplumbyllithium, trialkyllead hydrides, trialkyllead alkoxides, and trialkyllead amides have proved amenable to the synthesis of a number of novel organolead derivatives. What is needed is a better understanding of the chemistry involved in the reactions of organolead compounds and an elucidation of the nature of the intermediates involved in the more important synthesis reactions. Some progress in this area has been made in the past few years.

REFERENCES

1. Becker, W. E., and S. E. Cook, *J. Am. Chem. Soc.*, **82,** 6264 (1960).
2. Gilman, H., and J. C. Bailie, *J. Am. Chem. Soc.*, **61,** 731 (1939).
3. Glockling, F., K. Hooton, and D. Kingston, *J. Chem. Soc.* (1961). 4405.
4. Kocheshkov, K. A., and A. P. Aleksandrov, *Ber.*, **67,** 527 (1934).
5. Kocheshkov, K. A., A. N. Nesmeyanov, and N. K. Gipp, *Zh. Obshch. Khim.*, **6,** 172 (1936); *C.A.*, **30,** 4834 (1936).
6. Krause E., and A. von Grosse, *Die Chemie der metall-organischen Verbindungen*, Borntraeger, Berlin, 1937, pp. 372–429.
7. Meals, R. N., *J. Org. Chem.*, **9,** 211 (1944).
8. Nesmeyanov, A. N., and K. A. Kocheshkov, *Bull. Acad. Sci. SSSR Classe sci. Chim.*, **1945,** 522; through *C.A.*, **42,** 5870 (1948).
9. Neumann, W. P., and K. Kühlein, *Angew. Chem.*, **77,** 808 (1965).
10. Panov, E. M., and K. A. Kocheshkov, *Dokl. Akad. Nauk SSSR*, **123,** 295 (1958); *C.A.*, **53,** 7133 (1959).
11. Preuss, F. R., and I. Janshen, *Arch. Pharm.*, **293,** 933 (1960).
12. Preuss, F. R., and I. Janshen, *Arch. Pharm.*, **295,** 284 (1962).

13. Robinson, G. C., Ethyl Corporation, Unpublished Data.
14. Setzer, W. C., R. W. Leeper, and H. Gilman, *J. Am. Chem. Soc.*, **61,** 1609 (1939).
15. Singh, G., *J. Org. Chem.*, **31,** 949 (1966).
16. Yakubovich, A. Ya., S. P. Makarov, V. A. Ginsburg, G. I. Gavrilov, and E. N. Merkulova, *Dokl. Akad. Nauk SSSR*, **72,** 69 (1950); *C.A.* **45,** 2856 (1951).
17. Yakubovich, A. Ya., E. N. Merkulova, S. P. Makarov, and G. I. Gavrilov, *Zh. Obshch. Khim.*, **22,** 2060 (1952); *C.A.*, **47,** 9257 (1953).
18. Zimmer, H., and O. A. Homberg, *J. Org. Chem.*, **31,** 947 (1966).

IV Common physical and chemical properties

The physical and chemical properties of organolead compounds continue the trend established by the lighter Group IVb metals, namely, that of lower thermal stability and greater reactivity with increasing molecular weight of the central metal atom. Hence, lead forms the least stable and most reactive organometallic derivatives of all the Group IVb metals. This trend is in keeping with the decreasing strength of the metal–carbon bond as one proceeds down the Group IVb series; for the tetraethyl derivatives, the mean bond dissociation energy of the metal carbon bond decreases as follows: germanium 56.7, tin 46.2, and lead 30.8 kcal/mole. The chemical properties of the organolead derivatives are more similar to those of their organotin analogs than to those of the organosilicon and organogermanium compounds. For example, in reactions of vinyllead or vinyltin compounds with phenyllithium vinyl group displacement occurs, while the vinylsilicon and vinylgermanium analogs react with phenyllithium via an addition of the phenyllithium across the vinyl group.

In general, organolead compounds decompose at fairly low temperatures, are very slowly oxidized in air, and are decomposed by light. Their stability is strongly dependent on the nature of the organic group bound to lead and the nature of other groups bound to lead through other than a carbon atom. Tetraorganolead compounds decompose via a free radical mechanism, while derivatives of the type R_3PbX and R_2PbX_2 decompose via a disproportionation type reaction. Hexaorganodilead compounds are less thermally stable than their tetraorganolead analogs because of the lower stability of their lead–lead bond.

With few exceptions, organolead compounds are hydrolytically stable. The main exceptions in the R_4Pb compounds are the unsymmetrical perfluoroalkyl and pentafluorophenyl derivatives, which tend to be sensitive to hydrolysis because of the high polarity of the lead-to-carbon bond resulting from the strong inductive effects of the fluoride substitutents. Certain derivatives of the type R_3PbX and R_2PbX_2 are also hydrolytically unstable, particularly where X is an anion of a very weak acid, such as alkoxide (OR^-), amide (NR_2^-), or phosphide (PR_2^-).

The physical state of organolead compounds is strongly dependent on both the nature of the organic group bound to lead and the other groups bound to the lead atom. In general, all the aryllead derivatives tend to be crystalline solids. Tetraalkyllead and hexaalkyldilead compounds tend to be liquids while alkyl derivatives of the type R_3PbX and R_2PbX_2 tend to be solids. However, the melting points of these latter compounds, i.e., R_3PbX and R_2PbX_2, decrease with increasing chain length of R and with increasing molecular weight of X, where X is the anion of an alkylcarboxylic acid.

Because of their reactivity to air and their sensitivity to light, organolead compounds are best stored in the dark under an inert atmosphere. Even under these conditions, R_3PbX and R_2PbX_2 compounds undergo slow decomposition via disproportionation. The physical and chemical properties of the different types of organolead compounds are discussed in more detail in the sections dealing with each class of compound.

V Physiological and toxicological properties

The toxic action of the organolead compounds was recognized early in the commercial use of tetraethyllead as an antiknock compound. A number of deaths occurred that dramatically pointed out this effect and led to an intensive evaluation of the toxic effects of tetraethyllead as well as of the methods for its safe handling and use. The safety and hygiene programs that were developed for the manufacture and distribution of this compound were dramatically successful in that no cases of tetraethyllead poisoning occurred when the stringent safety rules were enforced. The practical hazard related to the antiknock compound is now almost entirely confined to the cleaning of large tanks in which leaded gasoline has been stored.

1. OBSERVATIONS FROM ANIMAL STUDIES

The toxic action of many organolead compounds has been reported, but the literature on these compounds is too scanty to permit generalization, except for tetraethyllead (TEL), tetramethyllead (TML), and the related tri- and dialkyllead salts (6–10,14,21,23,25–28). Both TEL and TML are volatile, non-ionic compounds which are relatively insoluble in water, but they are completely miscible with organic solvents (see Chapter VI). The tri- and dialkyllead compounds which have been studied are generally water-soluble ionizable salts (see Chapter XI).

A. Responses to single doses

Comparative studies indicate that TEL is somewhat more toxic than TML (14,21,27). Absorption occurs readily when these compounds are ingested (7,21,26), inhaled (26), or brought into contact with the skin (26), and the symptomatology is similar following absorption via the different routes (26). Following the administration of lethal doses of TEL or TML, animals become hyper-irritable, may develop fine continuous tremors throughout the whole body, and undergo violent, intermittent convulsions, often leading to death after several days (27). Histopathological examination has revealed nonspecific degenerative changes of brain, liver, kidney, and lung tissue (21,27).

B. Responses to repeated doses

Animals exposed repetitively to sub-lethal doses of TEL or TML may develop, in time, symptomatology similar to that seen following single lethal doses (10). The quantitative nature of the response, relative to dose, may vary considerably. In general, it has been reported that TEL produces a greater toxic action than TML, although one report (10) has stated that dogs have shown a greater sensitivity to TML than to TEL following the inhalation of comparable multiple dosages. It appears that repeated doses, eventually leading to a neurotoxic syndrome and death, do not produce statistically significant changes in liver or kidney function (6), although microscopic histopathological changes are reported in the liver, pancreas, endocrine, renal, and nervous systems (10,26).

C. Mechanism of action

There is considerable evidence to support the concept that the trialkyllead salts which result from the decomposition of tetraalkyllead compounds, *in vivo*, are the ultimate toxic agents (8,14,28). Triethyllead compounds have been isolated from various body tissues in similar quantity and pattern following exposure of animals to either TEL or triethyllead compounds (28). Similar symptomatology was also noted. Diethyllead compounds were much less toxic and their effects were different (27). The microsomes of the liver cells of rats were capable of converting TEL readily to triethyllead compounds (8). Trialkyllead compounds strongly inhibit glucose oxidation of brain slices, but not kidney or liver slices, but tetraalkyllead compounds and lead acetate were without similar effect (9). This finding may suggest a relatively selective toxic action of metabolites of TEL and TML on the nervous system (8).

D. Distribution and excretion

Relatively high concentrations of lead are found in the liver, kidney, and brain (as compared with other tissues), in decreasing order as listed, following organolead intoxication (21). The concentration of lead in blood, following intoxication with organolead compounds varies from low to negligible levels (23), whereas the urinary excretion of lead is elevated following intoxication with organolead compounds, but may be quite variable in quantity (10).

E. Therapeutic measures

Therapeutic use of such compounds as British Anti-Lewisite, ethylenediaminetetraacetic acid, and penicillamine has had little effect on intoxication due to organolead compounds. This finding is supported by *in vitro* observations showing the limited chemical reactivity of tetraalkyl and trialkyllead compounds with these therapeutic agents (8). It has been hypothesized that the most useful reagent would be one that combined with trialkyllead compounds and increased their harmless elimination from the body (9). To date, no useful therapeutic agent has been demonstrated conclusively in animals, although a report by Scarinci (25) showed dramatic antidotal action when a solution was used containing sodium trithiodilactate and sodium thiolactate.

2. HUMAN EXPERIENCE

Acute poisoning by lead antiknocks has been well documented, although relatively few cases have occurred in the United States since 1926 and virtually none since 1940. Sanders (24) in 1964, in reviewing this human experience, stated that, following the cases that occurred in the early days, there had been 88 cases of tetraethyllead intoxication in the United States including 16 which terminated fatally, almost all of which occurred in connection with the cleaning of large tanks in which leaded gasoline had been stored.

When the literature is analyzed, the clinical picture of acute intoxication can be formulated (3,5,11,17,20). In general, following exposure to an atmosphere contaminated with tetraethyllead, there is a latent period of at least several hours, or as long as 7 to 10 days, before the onset of symptoms is noted. In mild cases, the patient will complain of sleeplessness, weakness, irritability, fatigue, and headache, as well, perhaps, as anorexia, nausea, and vomiting. Following more severe exposure, the clinical picture may develop within a few hours and progress rapidly to frank mania, convulsions, and coma. In the more severe intoxications, the mortality rate may approach 50%. With this clinical picture, the diagnosis in the poisoning in the seriously ill is not difficult, despite the absence of localizing physical findings. With recovery from the acute intoxication, the recovery has been complete without residua or sequellae. A laboratory test which has been invaluable in establishing the diagnosis has been the estimation of lead in the urine. The high results so obtained, when coupled with a low or normal level in the blood, point almost certainly to the absorption of an alkyllead compound.

It is important to emphasize the fact that there is a distinct difference in the laboratory findings between inorganic lead and organolead intoxication. In inorganic lead intoxication, both the blood and urinary lead concentrations are elevated, together with findings of stippling of the red blood cell, striking elevation of the urinary porphyrins, and, often, a significant degree of anemia. In organolead intoxication, the urinary lead is markedly elevated, but there is little or no significant change in the blood-lead concentration, as well as no change in the morphology of the red blood cells, no anemia, and no significant alteration of the urinary porphyrins.

Recently, Gherardi (13) has described several cases in which the symptoms were quite similar to those of mild acute tetraethyllead intoxication. He believes that these cases resulted from prolonged, intermittent exposure to TEL. The men were engaged in the production of tetraethyllead, where such exposure occurs. It has been noted, previously, that when intoxication occurs in association with prolonged exposure to the organolead compounds, it is always characterized by an irritation of the central nervous system which is revealed by the symptoms and signs associated with acute episodes.

There is little question that poisoning may occur through ingestion, as evidenced by cases of accidental and suicidal poisoning. It is also quite evident that penetration of the skin occurs when undiluted TEL comes in contact with the skin; this has been well documented by urinalysis of specimens obtained from workmen who accidentally contaminated their skin with TEL while they were protected against inhalation. The more serious hazard from exposure to TEL lies, however, in the inhalation of the vapors of this chemical. Presently the most significant occupational hazard occurs in the cleaning of leaded storage tanks, but this only emphasizes the fact that when TEL or TML, in even modest quantity, is in a confined area, a serious health hazard may be present which requires adequate precautions.

The treatment of intoxication has not been specific; that is, there is no known antidote for these compounds. Since the syndrome is self-limited, it is quite logical that supportive care with adequate sedation may be all that is required. Therefore, even in the serious cases intensive supportive care may be life-saving. Barbiturates and phenothiazines have been used for their sedative effect, and electroshock therapy, in a few cases, appeared to alter the course of the intoxication dramatically and beneficially.

The chelating agent, ethylenediaminetetraacetic acid (EDTA), is used in inorganic lead poisoning to increase the excretion of lead, and it is a generally accepted therapy in this form of lead poisoning (22). EDTA has also been used in a number of poisonings from organolead antiknock compounds but with ambiguous results (3,17). Its use has resulted in an increase in the excretion of lead in the urine, but this was to be expected and would have occurred whether or not there had been an exposure to organolead compounds. Whether chelation therapy has any place in the treatment of organolead poisoning is not clear. More specifically, whether it will hasten the excretion of metabolized tetraethyllead has not been established. As stated above, the studies of Cremer *in vitro* indicate that EDTA does not form a complex with the organolead compounds to any significant degree; this suggests that it would hardly hasten the elimination of the unmetabolized organolead compounds.

3. PREVENTION AND SAFETY MEASURES

Knowledge of the toxic chemical properties of these compounds makes it apparent that special precautions should be taken in the laboratory or wherever they are handled. These consist of working exclusively in fume hoods with good mechanical ventilation in the laboratory, thus preventing inhalation, as well as protecting the body against any type of contact with these compounds. Dispersal of dust and spilling of solution are to be avoided. The use of new rubber gloves or specially developed impervious gloves and aprons is recommended, so that there is no contact with the skin (4). Should concentrated solutions or organolead compounds be spilled on the skin, they should be rinsed off immediately with kerosene or light petroleum solvent, and the skin area should then be washed thoroughly with soap and water (12).

Accidental spills of alkyllead compounds can be decontaminated by their treatment with dilute solutions of potassium permanganate or bromine, although other decontamination agents have been suggested. For example, Beilikhis (2) recommends Chloramine-T or Dichloramine-T in a concentration of 1.5–10% in an inert hydrocarbon or chlorinated hydrocarbon solvent, as well as a 3% aqueous solution for washing the hands, but this has been shown to be unnecessary. Emelyanov et al. (11), patented the use of dilute persulfuric acid and sodium sulfide solution as decontaminants, and Taube (29) has recommended a hot 0.025% aqueous solution of potassium permanganate for washing the inside of transport containers. Sulphuryl chloride also has been used satisfactorily under appropriate conditions. Care must be exercised in the use of all decontaminants to dilute the reagent sufficiently to retard the rate of decomposition to avoid fire.

The American Conference of Governmental Industrial Hygienists (ACGIH) (1) has established 0.075 mg/m^3 as the threshold limit value (TLV) for tetraethyllead and for tetramethyllead. This means that if this average concentration of either vapor in the atmosphere is not exceeded during 8 hr of exposure, no adverse effect will be noted in workers so exposed. Sanders (24) has stated that the concentration of 1 mg/m^3 can be tolerated for 1 hr without risk, and he further states that the concentration of 100 mg/m^3 is capable of producing illness though this will not necessarily be a fatal illness.

There has been some discussion in the literature of the decontamination of waste waters (for sewage) containing TEL. For example, Kozyura et al. (18), have proposed passing the waste water successively through an acidifier, a reduction column packed with iron shavings, and a neutralyzer containing lime. In a later paper, Kozyura and his co-workers (19) recommended simply allowing the waste water or sewage to stand in a pond a sufficient length of time to allow the TEL to be decomposed naturally. Gill, Huguet, and Pearson (15) have described in detail the system for the control of pollution at a plant of the Ethyl Corporation, in Pittsburg, California.

More and more attention is being paid to air pollution, and in the recent "Survey of Lead in the Atmosphere of Three Urban Communities," alkyllead compounds were investigated (30). It was stated that the primary source of alkyllead compounds in the ambient air is the vaporization of these compounds from unburned gasoline. Samples were collected in heavy traffic, using a mobile sampler, and when these samples were analyzed, it appeared that alkyllead compounds did not reach a level of 10% of the inorganic lead. However, because of sampling and analytical errors, the precision of the results was not such as to determine how much less than 10% the actual values were.

REFERENCES

1. American Conference of Governmental Industrial Hygienists, *Threshold Limit Values for 1966* (1966).
2. Beilikhis, G. A., *Gigiena i Sanit.*, **12**, 51 (1947); *C.A.*, **43**, 1881 (1949).
3. Boyd, P. R., *Lancet*, **1**, 181 (1957).
4. Calingaert, G., and H. Shapiro, *Ind. Eng. Chem.*, **40**, 332 (1948).
5. Cassel, D. A. K., and E. C. Dodds, *Brit. Med. J.*, **2**, 68 (1946).
6. Castellino, N., G. Colicchio, B. Grieco, P. Piccoli, and A. Rossi, *Arch. Mal. Prof.*, **25**, 203 (1964).

7. Caujolle, D., and M. C. Voisin, *Ann. Pharm. Franc.*, **24,** 17 (1966).
8. Cremer, J. E., *Brit. J. Ind. Med.*, **16,** 191 (1959).
9. Cremer, J. E., *Occ. Health Rev.*, **17,** 14 (1965).
10. Davis, R. K., A. W. Horton, E. E. Larson, and K. L. Stemmer, *Arch. Environ. Health*, **6,** 473 (1963).
11. Emelyanov, B. V., Z. N. Shemyakina, and M. N. Khalyavin, USSR Pat. 128,466, May 15, 1960; *C.A.*, **54,** 25430 (1960).
12. Ethyl Corporation, *Laboratory Use of Ethyl Antiknock Compounds*, New York.
13. Gherardi, M., *Med. Lavoro*, **55,** 107 (1964).
14. Gherardi, M., and G. Salvi, *Folia Med.* (*Naples*), **45,** 1254 (1962).
15. Gill, J. M., J. H. Huguet, and E. A. Pearson, *J. Water Pollution Control Federation*, **32,** 858 (1960); *C.A.*, **55,** 860 (1961).
16. Kehoe, R. A., *J. Am. Med. Assoc.*, **85,** 108 (1925).
17. Kitzmiller, K. V., J. Cholak, and R. A. Kehoe, *Arch. Indust. Hyg. Occu. Med.*, **10,** 312 (1954).
18. Kozyura, A. S., A. P. Mirnaya, and M. D. Lad, *Tr. Obedin. Seminara po Gidrotekhn. i Vodokhoz. Stroit., Kharkov*, **1959** (2), 64 (1960), through *C.A.*, **58,** 337h (1964).
19. Kozyura, A. S., A. N. Smirnova, and A. P. Mirnaya, *Tr. Obedin. Seminara po Gidrotekhn. i. Vodokhoz. Stroit., Kharkov*, **1961** (3), 55; *C.A.*, **58,** 1233 (1964).
20. Machle, W. F., *J. Am. Med. Assoc.*, **105,** 578 (1935).
21. Magistretti, M., N. Zurlo, F. Scollo, and D. Pacillo, *Med. Lavoro*, **54,** 486 (1963).
22. Miller, L. H., *Ind. Med. and Surg.*, **28,** 144 (1959).
23. Ohmori, K., *Japan J. Hygiene*, **20,** 340 (1965).
24. Sanders, L. W., *Arch. Environ. Health*, **8,** 270 (1964).
25. Scarinci, V., *Arch. Sci. Biol.* (*Bologna*), **44,** 153 (1960).
26. Schepers, G. W. H., *Arch. Environ. Health*, **8,** 277 (1964).
27. Springman, F., E. Bingham, and K. L. Stemmer, *Arch. Environ. Health*, **6,** 469 (1963).
28. Stevens, C. D., C. J. Feldhake, and R. A. Kehoe, *J. Pharm. and Exper. Therap.*, **128,** 90 (1960).
29. Taube, P. R., *Gigiena Truda i Prof. Zabolevaniya*, **6** (8), 53 (1962); *C.A.*, **57,** 15453 (1963).
30. Working Group on Lead Contamination, *Survey of Lead in the Atmosphere of Three Urban Communities*, PHS, Div. Air Pollution, Cincinnati, Ohio, Jan. (1965).

VI Tetraalkyl- and tetraaryllead compounds, R_4Pb

1. GENERAL

Tetraorganolead compounds (R_4Pb) in which a tetravalent lead atom is bonded to four organic groups through carbon represent the most stable class of organolead compounds. Such compounds are known in which R is alkyl, aryl, cycloalkyl, aralkyl, thienyl, furyl, and substituted derivatives thereof. Compounds in which all the organic groups are identical are commonly called symmetrical, while those containing two or more different organic groups are called unsymmetrical.

By comparison to the organometallic derivatives of the metals of Groups Ia, IIa, and IIIa, R_4Pb compounds are relatively unreactive. With few exceptions they are insoluble in and stable to water and fairly unreactive to air. They can be handled in air without difficulty; however, prolonged contact with air, especially in the presence of light, causes gradual decomposition accompanied by formation of tan colored solids which have not been fully characterized. Tetraorganolead compounds are also unreactive to such materials as aldehydes, ketones, and olefins. However, they are the most reactive tetraorgano derivatives of all the Group IVb metals and react readily with such reagents as halogens, acids, metal salts, and certain active metals.

The stability and reactivity of tetraorganolead compounds are dependent on the nature of the organic group. The aryl derivatives tend to be much more thermally stable than the alkyl derivatives; on the other hand, they are more reactive with electrophilic reagents. Compounds containing primary alkyl groups are more stable than those containing *sec*-alkyl groups; *sec*-alkyl derivatives become turbid upon brief exposure to air. No derivative is known containing more than two *tert*-alkyl groups bonded to lead, although compounds containing one or two *tert*-alkyl groups have been prepared and exhibit good stability. The unsymmetrical compounds are more reactive than their symmetrical analogs, probably because of the more polar character of the unsymmetrical molecule.

Steric effects are believed to play a prominent role in the properties of tetraorganolead compounds. The failure to prepare a stable tetra-*tert*-alkyllead derivative has been attributed to steric factors. Similarly, the low yields realized in the synthesis of certain substituted aryllead derivatives have been attributed to unfavorable steric effects. Steric effects are also reflected in the relative ease of synthesis of the various R_4Pb compounds; in general, the bulkier the group, the more stringent are the conditions required to synthesize its R_4Pb derivative. However, steric effects alone are not solely responsible for these difficulties. Thus, initial attempts to prepare tetraneopentyllead were unsuccessful; only hexaneopentyldilead and trineopentyllead chloride were formed (677,803). However, tetraneopentyltin is known (802), even though the tin derivative should be more sterically crowded than the lead compound because of the smaller radius of the tin atom. Undoubtedly, side reactions of intermediates are responsible for the failure to prepare certain R_4Pb derivatives.

The low molecular weight tetraalkyllead compounds are clear, colorless liquids which are soluble in the common organic solvents such as hydrocarbons, chloroform, ether, and even absolute ethanol. The higher molecular weight alkyl derivatives are low melting solids. Tetraheptyllead (775) is a liquid; tetrakis(tetradecyl)lead (479) is reported to melt at 31°C. The tetraaryllead compounds are beautifully crystalline solids; most are white or colorless, but the more highly substituted derivatives tend to be yellow in color. The tetraaryl compounds are soluble in such solvents as chloroform, acetone, and aromatic hydrocarbons, but are largely insoluble in such solvents as ether, ethanol, and aliphatic hydrocarbons. The tetraaralkyl derivatives tend to be white to yellow solids. Tetrabenzyllead and its *o*-halo derivatives are crystalline solids, ranging in color from red to yellow (20); on the other hand, tetraphenethyllead (422) has been isolated as a colorless solid. Only two tetracycloalkyllead derivatives are known. Tetracyclohexyllead (418) is a colorless solid which is much more stable and less reactive than tetraalkyl and aryl derivatives, while tetracyclopropyllead (371) is a colorless liquid which is more reactive and less stable than its *n*-alkyl analog.

The preparation and properties of the symmetrical R_4Pb compounds are discussed in the sections which follow. Unsymmetrical compounds are discussed separately.

2. SYMMETRICAL TETRAORGANOLEAD COMPOUNDS

Many symmetrical R_4Pb derivatives are now known; they are listed in Table VI-7. Many of these were prepared by Krause and his co-workers in their pioneering research in organolead chemistry. Thus, they prepared the *n*-alkyl and *iso*-alkyl derivatives of methyl through amyl and the tetra-*sec*-propyl derivative, as well as the phenyl, *p*-tolyl, *o*-tolyl, *p*-xylyl, and certain other substituted aryl derivatives. No R_4Pb compound was known containing a trisubstituted aryl group until tetramesityllead was successfully prepared as a yellow solid by Glockling, Hooton, and Kingston (295) in 1961. Tetra-1-naphthyllead has not been prepared, the failure being ascribed to steric factors (265). For many years tetrabenzyllead defied isolation, but it and its *o*-halo derivatives were prepared in 1955 by Bähr and Zoche (20). Tetracyclohexyllead has been known for many years, and tetracyclopropyllead was prepared relatively recently by Juenge and Houser (371). Other new symmetrical compounds which have been prepared in recent years include tetraneophyllead (803) and tetrapentafluorophenyllead (234). This last compound exhibits unusual properties as a result of the presence of the five electronegative fluorine atoms on the benzene ring. Thus, it is readily hydrolyzed by aqueous caustic, but stable to acetic acid, nitric acid, and bromine, even at elevated temperatures.

A. Methods of synthesis

Several different methods are available for the synthesis of symmetrical R_4Pb compounds. These are reviewed below in the approximate order of their general utility.

(*1*) *From lead salts*

(*a*) Reactions of lead halides with organolithium and Grignard reagents

By far the most general method of synthesis of symmetrical tetraorganolead compounds is the reaction of a lead(II) halide with a Grignard reagent or an organolithium compound, according to the equations:

$$2PbX_2 + 4RMgX \rightarrow R_4Pb + Pb + 4MgX_2$$

$$2PbX_2 + 4RLi \rightarrow R_4Pb + Pb + 4LiX$$

As discussed in Chapter III, this reaction proceeds through intermediate formation of the respective R_2Pb and R_6Pb_2 compounds. The reaction is usually carried out in diethyl ether solvent, using lead chloride as the preferred halide. Syntheses have been conducted in tetrahydrofuran solvent (295,765) and the available data indicate tetrahydrofuran to be a superior solvent. In many systems, the R_6Pb_2 intermediate is too stable to be decomposed to the R_4Pb derivative at the reflux temperature of the reaction mixture, especially when diethyl ether is the solvent. This can be overcome in some cases by addition of a higher boiling co-solvent, such as benzene, to provide a higher reflux temperature (278). Alternatively, and more commonly, the intermediate R_6Pb_2 is treated with halogen at Dry Ice temperature to produce R_3PbX (and R_2PbX_2) *in situ* and the resultant R_3PbX (and R_2PbX_2) is then allowed to react with more RLi or RMgX to form the desired R_4Pb product.

In most preparations, the RLi or RMgX reactant is added to a slurry of the $PbCl_2$ in diethyl ether at room temperature and the reaction mixture is then stirred at room temperature or heated to reflux. With some derivatives, the nature of the product formed can be very dependent on the mode of addition of the reactants and the stoichiometry employed. Thus, Willemsens and van der Kerk (765) have shown that in the reaction of $PbCl_2$ and phenylmagnesium bromide, addition of the $PbCl_2$ to a stoichiometric amount of the Grignard reagent (2 moles PhMgBr/mole $PbCl_2$) yields hexaphenyldilead as the major product along with a small amount of tetraphenyllead. When the reverse mode of addition is employed, only hexaphenyldilead is formed and not a trace of tetraphenyllead is obtained. If, on the other hand, an excess of PhMgBr is employed ($\sim$2.3 moles PhMgBr/mole $PbCl_2$), a high yield of tetraphenyllead is obtained containing not a trace of hexaphenyldilead; at higher excesses of Grignard reagent ($\geq$3 PhMgBr/$PbCl_2$) low yields ($\sim$30%) of tetraphenyllead are obtained and some ($\sim$15%) hexaphenyldilead is also formed.

The formation of tetraphenyllead in these systems has been attributed to the presence of a complex of the type $Ph_3PbMgBr$, formed from diphenyllead and the excess PhMgBr (765). Formation of tetraphenyllead is postulated to occur through migration of a phenyl group according to:

$$Ph_3PbMgBr + Ph_2Pb \rightarrow Ph_3Pb—Pb(Ph)_2—MgBr$$

$$Ph_3Pb—Pb(Ph)_2—MgBr \rightarrow Ph_4Pb + Pb + PhMgBr$$

Hence, when the $PbCl_2$ is added to a stoichiometric amount of PhMgBr, excess PhMgBr is present in the initial stages of reaction so that $Ph_3PbMgBr$ is formed and tetraphenyllead is obtained as a minor product, according to the above sequence. In the absence of excess Grignard reagent, uncomplexed diphenyllead is formed; this can polymerize and undergo phenyl migration to form hexaphenyldilead, according to:

$$Ph_2Pb \longrightarrow {}^{-}\!:\!\underset{Ph}{\overset{Ph}{Pb}}-\underset{Ph}{\overset{Ph}{Pb}}-\underset{Ph}{\overset{Ph}{Pb}}{}^{+} \longrightarrow Ph_3Pb^- + Pb + Ph_3Pb^+$$

$$Ph_3Pb^+ + Ph_3Pb^- \longrightarrow Ph_6Pb_2$$

Actually, few detailed investigations have been made of the reactions of lead halides with Grignard reagents so that the chemistry of these reactions is not completely understood (see also Chapter III). With few exceptions, diethyl ether has been employed as the solvent of choice, primarily because of its routine use in the preparation of Grignard and organolithium reagents. However, recent results indicate that solvent can play a critical role in these systems. Thus Glockling and colleagues (295) obtained tetramesityllead in good yield by reaction of lead bromide and mesitylmagnesium bromide in tetrahydrofuran, whereas only hexamesityldilead is obtained when diethyl ether is used as the solvent. Since hexamesityldilead is stable to 325°, the effect of the tetrahydrofuran cannot result simply from its higher boiling point. Rather, the greater Lewis base strength of the tetrahydrofuran (versus diethyl ether) is believed to be more important. The role of the tetrahydrofuran may be to stabilize an R_3PbMgX intermediate, analogous to the $Ph_3PbMgBr$ which has been proposed to be the reactive intermediate in the formation of tetraphenyllead. Glockling and coworkers (295) have concluded that triphenylplumbyllithium is present in appreciable amounts in a reaction mixture prepared from only *equimolar* amounts of lead bromide and phenyllithium in tetrahydrofuran at −40°, since treatment of this reaction mixture with benzyl chloride gave benzyltriphenyllead in 28% yield. Similarly, Willemsens and van der Kerk (765) obtained benzyltriphenyllead in about 50% yield by addition of benzyl chloride to a reaction mixture prepared from *stoichiometric* amounts of lead chloride and phenylmagnesium bromide ($2PhMgBr/PbCl_2$) in pyridine. On the other hand, no benzyltriphenyllead was obtained when diethyl ether was used as solvent under otherwise identical conditions. Willemsens and van der Kerk postulate that the role of the tetrahydrofuran and pyridine in these systems is to stabilize the $Ph_3PbMgBr$ complex by coordination of the solvent with the magnesium atom; coordination by the solvent is proposed to increase the electron density around the magnesium so that a more covalent bond is formed between magnesium and lead in the Ph_3PbMgX intermediate.

Willemsens and van der Kerk (765) prepared tetra-*n*-butyllead in about 77% yield [based on lead(II) chloride] by the reaction of lead chloride with butyllithium in the presence of butyl bromide by the simultaneous addition of the lead chloride and butyl bromide to a cold (−20°) solution of butyllithium in tetrahydrofuran. A small amount of lead metal was also formed, but much less than is normally formed in reactions of lead(II) halides with Grignard reagents. The mode of addition (i.e., the addition of the lead chloride to the butyllithium), the low reaction temperature and the stoichiometry ($1PbCl_2/4BuLi$) would favor the formation of tributylplumbyllithium so that the reaction probably proceeded according to the sequence:

$$3BuLi + PbCl_2 \rightarrow Bu_3PbLi + 2LiCl$$

$$Bu_3PbLi + BuBr \rightarrow Bu_4Pb + LiBr$$

The formation of the Bu_3PbLi would stabilize any dibutyllead formed initially and thereby minimize its disproportionation to hexabutyldilead and lead metal, which in turn would minimize lead metal formation. In addition to decreasing lead metal formation, an additional advantage of this system for the synthesis of tetrabutyllead and possibly higher tetraalkyllead derivatives is the elimination of R_6Pb_2 formation as a by-product. This system is similar to those discussed above for the direct synthesis of benzyltriphenyllead from a lead(II) halide. The reaction of a triorganoplumbyllithium compound with an organic halide is a preferred method of preparing unsymmetrical tetraorganolead compounds of the type Ar_3PbR'; these reactions are discussed later in this section.

In view of the results obtained in tetrahydrofuran solvent, it would seem that further investigation of the role of solvent in the reaction of lead halides with organolithium and Grignard reagents is definitely warranted. By proper choice of solvent it may be possible to prepare R_4Pb derivatives which have not been previously accessible via the $PbCl_2$–RMgX route.

Organolithium reagents and Grignard reagents react with lead(II) halides with nearly equal facility. The use of organolithium reagents is especially advantageous for the synthesis of those R_4Pb derivatives wherein the organic group does not readily form a Grignard reagent, e.g., *p*-dimethylaminophenyl halides (15).

Tetravalent lead salts may be used in place of the lead(II) halides but they offer no great advantage (25,26,244,306,310,448a,458). As was mentioned earlier, the main attraction of the use of tetravalent lead salts is the possibility of avoiding by-product lead metal formation, which is at best a minor advantage in laboratory syntheses.

$$PbX_4 + 4RM \rightarrow R_4Pb + 4MX$$

Indeed, where R is phenyl and X is acetate, high yields of tetraphenyllead have been obtained without any by-product lead metal formation (244).

However, where R is alkyl, or where R is aryl and X is halide, lead metal is formed in substantial amounts because of the instabilities of the $RPbX_3$ postulated to be formed as the first intermediate. Reports in the patent literature (306,586) that R_4Pb can be prepared in good yield using lead(IV) chloride are misleading since lead(IV) chloride is unstable, especially in organic solvents; any R_4Pb formed probably arises from lead(II) chloride formed upon decomposition of the lead(IV) chloride. Lead(IV) chloride is better used in the form of its complexes with potassium chloride or tetraalkylammonium chloride. Frey (244) has prepared tetraethyllead using potassium hexachloroplumbate and Bartocha and Gray (26) have prepared tetravinyllead similarly. Ammonium hexachloroplumbate (310,458) has also been used, but the labile protons of the ammonium moiety react with the Grignard reagent to form decomposition products (hydrocarbons) and thereby decrease the amount of Grignard reagent available for reaction with the lead chloride.

Almost all of the known symmetrical tetraorganolead compounds have been successfully synthesized using the reaction of a lead(II) halide with a Grignard or an organolithium reagent. This attests to the general versatility of this route for the synthesis of R_4Pb compounds. An interesting reaction reported recently involves the synthesis of tetra(pentafluorophenyl)lead. This compound was synthesized by Fenton and Massey (234) in about 11% yield by reaction of lead(II) chloride with the pentafluorophenyl Grignard reagent. Using this reaction, Tamborski and co-workers (722) obtained the same product in only 1–3% yield and reported that the reaction of lead chloride with *pentafluorophenyllithium* was "unsuccessful"; on the other hand, the reaction of lead(IV) acetate and the lithium reagent gave the tetrakis compound in about 15% yield. About the same time, Henry and Hills (330) found that by treatment of the crude product from the reaction of lead chloride with the Grignard reagent with bromine, tetra(pentafluorophenyl)lead could be prepared in 50–60% yield; a trace of a second product was also obtained which was believed to be the tri(pentafluorophenyl)lead bromide. The role of bromine in this reaction has been attributed to its reaction with a R_3PbMgX specie to form R_3PbBr (R being pentafluorophenyl) which then reacts with more Grignard reagent to form the desired R_4Pb.

Several papers have been published on the tendency of the Radium D isotope of lead, Pb^{210}, to concentrate in the lead metal coproduced in the $PbCl_2$–RMgX reaction. Conflicting results have been reported; some investigators found that the Radium D was concentrated in the lead metal (191,212,338) while others did not detect this effect (75,76,688).

Several early patents describe the preparation of tetraethyllead via the reaction of lead(II) halide and ethyl magnesium halide (78,178,180,

415,578). In two of these, the Grignard reagent is prepared *in situ* by reaction of magnesium metal and ethyl halide in the presence of the lead halide (180,415). Another patent describes the chlorination of gasoline to form a mixture of chlorinated hydrocarbons, which is then reacted with magnesium metal and lead halide to yield a leaded gasoline (662).

(*b*) Reactions of lead salts with other organometallic reagents

A limited number of tetraalkyl- and tetraaryllead compounds has been prepared by the reaction of lead(II) salts and organometallic derivatives of zinc, aluminum, sodium, and boron. The synthesis of tetraethyllead from lead(II) chloride and diethylzinc was first reported by Buckton in 1859 (83); others (239,484,563) have also used this reaction, but only for the preparation of tetraethyllead.

$$2PbX_2 + 2Et_2Zn \longrightarrow Et_4Pb + Pb + 2ZnX_2$$

The use of an organozinc compound offers no apparent advantage over the Grignard or organolithium reagents. Tetraethyllead was prepared in 93% yield by reaction of diethylzinc and lead(II) acetate in toluene solvent, but the reaction of sodium triethylzinc and lead sulfide gave low yields of tetraethyllead (52,563). Syntheses involving lead salts other than lead(II) halides are further discussed below.

The use of alkylaluminum compounds for the synthesis of tetraorganolead and other organometallic compounds has been investigated extensively in the last decade, primarily because of the discovery of a direct synthesis of alkylaluminum compounds and their potential availability as cheap alkylating agents. Most of these investigations have been restricted almost exclusively to triethylaluminum because of the possible use of this system for the commercial manufacture of tetraethyllead.

The synthesis of tetraethyllead by reaction of triethylaluminum with divalent lead salts was investigated originally by chemists of the Ethyl Corporation (563) and later by several others (175,365,558,774). Triethylaluminum has been shown to react with various lead salts, including the acetate, formate, sulfate, sulfide, oxide, and fluoride according to the equation:

$$4R_3Al + 6PbX_2 \longrightarrow 3R_4Pb + 3Pb + 4AlX_3$$

These reactions are described in a number of patents (52–63,365). An interesting aspect of these reactions is that not only do these normally "unreactive" lead salts react with triethylaluminum, but in many cases the yields of tetraethyllead exceed that obtained from lead chloride. Generally, the highest yields are obtained using lead salts of organic acids, such as lead(II) acetate; with the latter near quantitative yields are obtained using an aromatic hydrocarbon solvent such as toluene (175,558,563,774). Near quantitative yields of tetraethyllead have also been obtained from lead(II) fluoride and triethylaluminum in pyridine solvent (61).

In the reaction of triethylaluminum and lead(II) chloride, using the stoichiometry defined in the preceding equation ($4R_3Al/6PbX_2$), the yield of tetraethyllead is always less than 50% (243,558,758). This probably results from the formation of ethylaluminum sesquichloride as a by-product of reaction; reaction ceases at this point because the sesquichloride is too weak an alkylating agent to react with lead(II) chloride.

$$8Et_3Al + 6PbCl_2 \rightarrow 3Et_4Pb + 3Pb + 4Et_3Al_2Cl_3$$

Therefore, only 50% of the ethyl groups of the triethylaluminum are utilized, despite reports to the contrary (230). The reduction can be carried beyond the sesquichloride step, however, by conducting it in the presence of an alkali metal halide to complex with the aluminum halide. Thus, Pasynkiewicz and co-workers (559) have described the preparation of tetramethyllead by reaction of methylaluminum sesquichloride and lead(II) chloride in the presence of sodium fluoride, sodium chloride, or potassium chloride. Similarly, the use of sodium fluoride to complex the aluminum chloride in the reaction of lead(II) chloride and triethylaluminum has been patented (19). In addition to tetramethyllead and tetraethyllead, tetraisobutyllead (774), and tetraphenyllead (244) have also been prepared using the PbX_2–R_3Al system.

Trialkylboranes have been used to synthesize R_4Pb compounds by reaction with lead oxide or hydroxide; dialkylmercury compounds have been prepared in similar fashion. Honeycutt and Riddle (346,617) prepared tetraethyllead by the reaction of lead oxide or hydroxide (formed *in situ* from $PbCl_2$ and aqueous caustic) with the corresponding trialkylborane in aqueous solution. These reactions are possible because trialkylboranes are stable to water and are soluble in aqueous caustic, so that these reactions are conveniently carried out in an aqueous system. Lead(II) carboxylates and oxides were also reacted with the trialkylboranes in organic solvents, but the R_4Pb yields were generally lower; tetrahexyllead was obtained in 18% yield from trihexylborane and lead(II) naphthenate in 1,2-dimethoxyethane. The highest yield of tetraethylead obtained with the aqueous system was 42%, using a stoichiometry corresponding to reacting only one of the three ethyl groups in the triethylborane.

Riddle (618) devised an improved boron system for the preparation of tetraethyllead based on the use of sodium tetraethylboron as the reactant. Since triethylborane does not react with lead(II) chloride, the reaction proceeds according to the equation:

$$4NaBEt_4 + 2PbCl_2 \rightarrow Et_4Pb + Pb + 4Et_3B + 4NaCl$$

This reaction is conveniently carried out in aqueous solution, since sodium tetraethylboron is stable to and soluble in water. Near quantitative yields of tetraethyllead were obtained, based on the above stoichiometry. The triethylborane can be recovered and treated with sodium hydride and ethylene to regenerate sodium tetraethylboron.

The boron systems enjoy little utility as laboratory syntheses because of the low reactivity of the R_3B compounds and the utilization of only one R group in the $NaBR_4$ compounds in their reactions with lead halides. Their main attraction lies in their potential utility for the commercial manufacture of tetraethyllead. With sodium tetraethylboron, only sodium, hydrogen, and ethylene would be consumed, in addition to the lead chloride, according to the overall stoichiometry:

$$4Na + 2H_2 + 4C_2H_4 + 2PbCl_2 \xrightarrow{Et_3B} Et_4Pb + Pb + 4NaCl$$

Tetraethyllead has also been prepared using sodium tetraethylaluminum (563) and lithium tetraethylaluminum (190) as the alkylating agents. However, these reagents offer no advantage over triethylaluminum. The reaction of ethylsodium with lead oxide, lead sulfide, and lead acetate has been reported by Pearson and his colleagues (563); near quantitative yields of tetraethyllead were obtained from lead acetate in heptane. This high yield is surprising since both reactants are insoluble in heptane. Nesmeyanov and Makarova (528) observed that diphenylcadmium did not react with lead(II) chloride, although it did react with tin and antimony chlorides.

(2) *Syntheses from lead metal or its alloys*

The reaction of a lead alloy with an organic halide enjoys historical significance in that this reaction was used originally to prepare the first tetraalkyl- and tetraaryllead derivatives. It also enjoys commercial significance in that it is still the major process for the commercial manufacture of tetramethyllead and tetraethyllead for use as antiknock agents. A number of lead alloys of reactive metals reacts with organic halides and other organic esters to form tetraorganolead compounds. In addition, the direct alkylation of lead metal by organic esters alone or in combination with reactive organometallic reagents has been used to prepare the low molecular weight tetralkyllead derivatives. Finally, electrolytic methods of synthesis of tetraalkyllead from lead metal, involving both anodic and cathodic processes, have been investigated in recent years, primarily as routes to the commercial manufacture of antiknock compounds. The synthesis of R_4Pb compounds from lead metal and its alloys has been reviewed by Shapiro (656).

(*a*) Reactions of lead alloys

Sodium–Lead Alloys. The reaction of a sodium–lead alloy and ethyl iodide was used in 1853 by Löwig (451) to prepare the first organolead compound; Polis (570–572) prepared the first tetraaryllead derivative by reaction of a sodium–lead alloy with bromobenzene in 1887. Although these reactions have been used in isolated instances for the laboratory synthesis of tetraalkyllead compounds, the primary interest in this system is in the commercial manufacture of tetramethyllead and tetraethyllead. Innumerable patents exist covering various modifications and variations of the reaction of sodium–lead alloys with organic halides and other organic esters. A major contribution to the successful development of the commercial alloy process for tetraethyllead was the discovery by Kraus and Callis (413,417) of conditions required for efficient reaction of sodium–lead alloy with ethyl chloride, instead of the more expensive ethyl iodide. The ethyl chloride reaction is also described in patents issued to Calcott and Daudt (91,179). The development of the alloy process has been reviewed by Edgar (213) and others (639,736,746).

The reaction proceeds according to the equation:

$$4NaPb + 4RCl \rightarrow R_4Pb + 3Pb + 4NaCl$$

Three-fourths of the lead values are converted back to lead metal and must be recycled. The reactions are carried out in pressure autoclaves under the autogenous pressure of the alkyl chloride. With ethyl chloride, the reaction proceeds at mild temperatures (80–100°) to produce tetraethyllead in 80–90% yield; no solvent is required. Tetramethyllead is manufactured similarly, except for minor modifications; a higher reaction temperature is required and an organic diluent such as toluene is used to reduce the vapor pressure of the methyl chloride and at the same time serve as a thermal stabilizer for the tetramethyllead (150).

A number of patents describes various catalysts for the methyl chloride reaction; in the absence of these catalysts, the reaction rate is extremely low. Effective catalysts include aluminum metal (636) aluminum halide, aluminum alkyls and other Lewis acids (105,741), aluminum hydrides (740), various ethers (33,35,36,225,390), ammonia–water and ammonia–alcohol (205,207) and amines (34). The addition of tetramethyllead to the alloy prior to addition of methyl chloride is reported to give better yields (364).

The reaction of sodium–lead alloy with alkyl halides is very sensitive to several reaction parameters. The mono-sodium–lead alloy, NaPb, reacts at an appreciable rate only with ethyl chloride. In fact, the presence of methyl halides or other ethyl halides in the ethyl chloride tends to poison the reaction. The reaction with ethyl chloride is characterized by an induction period. A patent describes a procedure of feeding a small amount of ethyl chloride to the NaPb alloy to initiate the reaction, and then feeding the rest of the ethyl chloride (98). The induction period can be shortened by substitution of a small amount of the sodium with an equivalent amount of potassium (on an atom basis) in the alloy (42,95, 499,658,760). At the same time, the use of potassium minimizes Wurtz-type reactions and increases the yield of tetraethyllead. Various catalysts may be used to accelerate the reaction of ethyl chloride with sodium–lead alloy. These include ketones (28,339,482), aldehydes (135), ketals (569), acid anhydrides (136), esters (340), amides (341), metal alkoxides (49), and phosphates and phosphate esters (48,50,544). A recent patent reports that catalytic amounts of iodine or inorganic iodides give improved yields (645). Iodides also act as volume reducers in that they tend to convert the by-product lead metal form to a more compact mass, and thereby increase the capacity of the autoclaves (262).

The ethyl chloride reaction is very sensitive to certain impurities in the ethyl chloride; as little as 0.0025% acetylene exerts a powerful retarding effect (656). However, the use of small amounts of acetylene as a co-catalyst has been patented (760). The rate of reaction is also affected greatly by the surface area and gross structure of the alloy. For this reason, numerous patents (22,44,74,96,97,202,203,414,454,472,655,695, 696,723,724) exist covering methods of preparing sodium–lead alloy for reaction with ethyl chloride. The reaction rate is also very sensitive to alloy composition, the most reactive alloy having the composition NaPb. The alloy becomes progressively less reactive as the sodium concentration is increased from NaPb to Na_5Pb_2, at which point reaction with ethyl chloride practically ceases. However, at these higher sodium concentrations, good yields of tetraethyllead have been obtained using ethyl bromide or ethyl iodide as the alkylating agent, but only in the presence of gross amounts of amines or hydroxyl compounds such as pyridine or water (99). Numerous patents are found in the early patent literature which describe the reaction of sodium–lead alloys and alkyl halides in the presence of small to gross amounts of protonic substances, such as water or alcohol (90,101,181,416,487,496,497,627,767,772). This so-called hydrous reaction was used for a short time in the 1920's for the commercial manufacture of tetraethyllead. At the composition of Na_9Pb_4, reaction with ethyl chloride can be effected to give tetraethyllead in excellent yield provided a catalyst is used, such as an organic ester, aldehyde, or ketone

(45). The main attraction to the use of these "high sodium" alloys is the elimination or reduction of unreacted lead metal which must be recycled in the commercial process. The alkylation of NaPb and Na_9Pb_4 by ethyl iodide in liquid ammonia solvent is described in the patent literature (397); however, the yields of tetraethyllead were low (11–20%).

In addition to tetraethyllead, hexaethyldilead is also formed in the NaPb–EtCl reaction; under certain conditions, the hexaethyldilead can be formed in appreciable amounts. A high boiling residue is also produced, which has been identified as 2,3-bis(triethylplumbyl)-butane (324).

The reaction of ethyl chloride and sodium–lead alloy undoubtedly proceeds via a free radical mechanism as evidenced by the formation of ethane, ethylene, and butane as by-products. The results of a detailed investigation of the kinetics of the NaPb–EtCl reaction have been published by Shushunov and colleagues (17,399,523,524,668–671), who found that the reaction rate is very dependent on temperature. As the reaction temperature was increased, the rate went through a maximum and then decreased (17). The "critical temperature", i.e., the temperature at which the reaction slows down greatly, is dependent on both alloy composition and ethyl chloride pressure, and is very reproducible. As the ethyl chloride pressure was increased, the critical temperature also increased, according to the formula (where K and γ are constants):

$$P_{\mathrm{EtCl}} = Ke^{-\gamma/RT_c}$$

The reaction was characterized by an induction period which decreased with an increase in ethyl chloride pressure (399). The critical temperature was found to decrease as the sodium content of the alloy was partially replaced with potassium. A similar equation was found to apply to the ternary Na–K–Pb alloy.

Shushunov and colleagues also observed that the length of the induction period was dependent on alloy composition, being shortest at the composition NaPb (524). No induction period was observed with alloy that had been allowed to stand in contact with ethyl chloride and the reaction rate of this "used" alloy was considerably faster than that of fresh alloy. Also, the addition of small amounts of potassium to the alloy greatly reduced the induction period and increased the rate.

The following reaction sequence was proposed for the ethylation of sodium–lead alloy by ethyl chloride (669):

$$\mathrm{EtCl_{(gas)}} \rightarrow \mathrm{EtCl_{(adsorbed)}}$$
$$\mathrm{NaPb} + \mathrm{EtCl_{(adsorbed)}} \rightleftharpoons \mathrm{NaPb{\cdot}EtCl}$$
$$\mathrm{2NaPb{\cdot}EtCl} \rightarrow \mathrm{Et_2Pb} + \mathrm{2NaCl} + \mathrm{Pb}$$
$$\mathrm{Et_2Pb} + \mathrm{NaPb{\cdot}EtCl} \rightarrow \mathrm{Et_3Pb} + \mathrm{Pb} + \mathrm{NaCl}$$
$$\mathrm{Et_3Pb} + \mathrm{NaPb{\cdot}EtCl} \rightarrow \mathrm{Et_4Pb} + \mathrm{Pb} + \mathrm{NaCl}$$

The formation of the intermediate compound NaPb·EtCl was postulated to be the rate determining step. A value of 12.5 kcal/mole was calculated for the apparent activation energy of the topochemical reaction occurring at the boundary surface between the sodium–lead alloy and the solid phases formed in the reaction.

Partial substitution of sodium with magnesium was found to have an effect similar to that of potassium at magnesium additions up to 0.2 wt % (193,670). Addition of magnesium up to this concentration reduced the induction period, but at higher concentrations the induction period increased in linear fashion. However, unlike potassium, magnesium did not noticeably increase the reaction rate. The break point observed at the 0.2 wt % concentration was attributed to a limiting solubility of 0.2 wt % magnesium in the alloy.

Shushunov and Sokolov (671) investigated the kinetics of the reaction of sodium–lead alloy with butyl halides. Butyl chloride was found to react fastest, then bromide and finally iodide; the butyl halides were found to react much more slowly than the ethyl halides.

The reaction of sodium–lead alloy with ethyl halides has also been investigated in some detail by Mishima (493) who observed that ethyl chloride did not react with sodium–lead alloy containing 78 wt % lead; this lead content approximates the composition Na_5Pb_2. The optimum lead content for reaction of sodium–lead alloy and ethyl bromide was reported to be 87 wt % lead, but even then the yield of tetraethyllead was low under the conditions employed. With ethyl iodide no reaction was obtained at temperatures up to 120° at any alloy composition. On the basis of the results, Mishima concluded that the alkyl halide must penetrate the alloy and break down the alloy grain in order for reaction to occur, the degree of penetration being dependent on the specific halide employed and the temperature.

Although propyl and higher halides do not react with sodium–lead alloy at an appreciable rate in an anhydrous system, these higher tetraalkyllead derivatives have been prepared by reaction of sodium–lead alloys with alkyl iodide in the presence of pyridine and water (99). Laboratory procedures for the synthesis of tetrapropyllead and tetraisobutyllead have been developed using Na_9Pb_4 in a hydrous system (322,637). Two patents describe the reaction of NaPb with gasoline that has been partially chlorinated to produce a "leaded" gasoline *in situ* (662,747), but the efficacy of these patents is questionable in the light of the data discussed above.

Sodium–lead alloys also react with diethyl sulfate to form tetraethyllead (659,704), but only one ethyl group of the diethyl sulfate is reactive in this system:

$$4NaPb + 4Et_2SO_4 \rightarrow Et_4Pb + 4NaEtSO_4 + 3Pb$$

Also, the rate is slower than with ethyl chloride and a high temperature is required. Tetraethyllead yields up to 75% have been obtained in uncatalyzed reactions, up to 80% in reactions catalyzed by iodine compounds. The reactivity of the various sodium–lead alloys with diethyl sulfate decreases in the order (656):

$$NaPb > Na_9Pb_4 > Na_5Pb_2 > Na_4Pb$$

Various sodium–lead ternary alloys have also been investigated for the synthesis of tetraethyllead. The potassium ternary alloy was discussed above, as was Shushunov's work on the magnesium ternary alloy. Calingaert and Shapiro (112) have shown that the magnesium ternary compositions which react best are those located on the sodium–lead–magnesium cross-section which contains a peritectic compound having the composition NaMgPb (116,117). Optimum yields (ca. 75%) were obtained using diethyl ether as catalyst; some hexaethyldilead was also formed. The use of ternary sodium–lead alloys containing lithium (763) and mercury (684) has been patented, and a Belgian patent describes the simultaneous production of R_4Pb and R_4Sn by the reaction of an organic halide with a ternary alloy of lead, tin and sodium (457).

The use of sodium–lead alloys in the synthesis of tetraorganolead compounds has been restricted to the alkyl derivatives, except for Polis' (570–572) original syntheses of tetraphenyllead and tetra-*p*-tolyllead. Furthermore, their use has been investigated primarily as processes for the commercial manufacture of tetraethyllead and tetramethyllead. The chore of preparing the alloy has discouraged the use of the alloy reaction in laboratory syntheses. However, NaPb alloy is now available as a stock item from the J. T. Baker Co., Chicago, Illinois; this may encourage consideration of alloy reactions as a general laboratory synthesis. In view of the pronounced activity of ethers, ammonia, and amines to catalyze the reaction of sodium–lead alloy with methyl chloride, their effect as catalysts in alloy reactions with other alkyl halides is worthy of investigation.

Other Binary Lead Alloys. Binary lead alloys other than sodium–lead alloy have been investigated for the synthesis of tetraorganolead compounds. As with sodium–lead alloy, the investigation of these other alloys has been restricted to the alkyl compounds, and primarily to the ethyl and methyl derivatives.

Dimagnesium-lead, a well known ionic compound having a fluorite structure (321), reacts with alkyl halides according to the equation:

$$Mg_2Pb + 4RX \rightarrow R_4Pb + 2MgX_2$$

An obvious advantage of this system is the possible elimination of lead metal recycle. Shapiro (654) has shown that the reaction of Mg_2Pb with alkyl iodides occurs in the absence of any catalyst; with an alkyl bromide or chloride, a catalyst is required. Tetraethyllead has been prepared in 85–90% yield from ethyl chloride in the presence of a mixed catalyst of diethyl ether and ethyl iodide; no hexaethyldilead was obtained as a by-product. Using only diethyl ether as catalyst, van der Kerk and Luijten (382) obtained tetraethyllead in only 32% yield. The high yields from the ether-iodide-catalyzed reaction could not be duplicated on a larger-than-laboratory scale (656).

The Mg_2Pb alloy is also reactive with longer chain alkyl halides. Tetra-isopropyllead and tetra-*n*-butyllead have been prepared in good yield using the ether-iodide mixed catalyst. However, the chore of preparing the alloy has discouraged its use for general laboratory synthesis.

Tetraethyllead has been prepared by Krohn and Shapiro (426,427) using calcium–lead alloys, CaPb being much more reactive than Ca_2Pb. CaPb reacted with ethyl halides, sulfate, and phosphate, usually in the presence of such catalysts as ketones, aldehydes, and esters; with ethyl chloride, tetraethyllead was prepared in 80% yield without benefit of a catalyst (426).

$$2CaPb + 4EtCl \rightarrow Et_4Pb + Pb + 2CaCl_2$$

Lithium–lead alloys have also been used to prepare R_4Pb compounds. Recent patents report the preparation of tetraphenyllead from bromo-benzene and Li_4Pb (12), and of tetraethyllead from ethyl chloride and Li_2Pb (763).

(*b*) Reactions of lead metal

With Alkyl Halides. Lead metal, like many other metals, will react directly with an alkyl halide to form an organolead derivative. Unlike most metals, especially the other Group IVb metals, this reaction forms an R_4Pb compound instead of the alkylmetal halide. Thus, while silicon and tin react with alkyl halides according to the equation

$$M + 2RX \rightarrow R_2MX_2$$

with lead, the products are R_4Pb and PbX_2, according to:

$$3Pb + 4RX \rightarrow R_4Pb + 2PbX_2$$

The reaction of lead metal with an alkyl halide was reported as early as 1853 by Cahours (88,89), who obtained a liquid product from the reaction of lead metal and ethyl iodide which was qualitatively identified as being an organolead compound.

Pearsall (561) has shown that finely divided lead metal reacts with methyl iodide, methyl bromide, and ethyl iodide without benefit of a catalyst to form R_4Pb and PbX_2, according to the equation above. Lead metal also reacts with methyl chloride, ethyl chloride, and ethyl bromide, but a catalyst is needed. Iodine and iodine-containing compounds, such as RI or PbI_2, are preferred catalysts. The reactivity of these alkyl halides decreases in the order:

$$MeI > EtI > MeBr > MeCl > EtBr > EtCl$$

Yields of tetraethyllead as high as 65% have been obtained using ethyl chloride and an iodide catalyst at reaction temperatures of 100–130°.

A reactive, oxide-free surface is required for good reactivity of the lead metal. Therefore, this system has been considered primarily for combination with the NaPb–EtCl system, since the by-product lead metal from this latter system is obtained in a finely divided, highly reactive form. The alkylation of this lead metal in a second step therefore represents a convenient way of utilizing it to produce more R_4Pb. However, this is at best a minor advantage, since only one third of the lead metal values are converted to R_4Pb.

Reaction of lead metal with a mixture of methyl chloride and ethyl chloride has also been used by Pearsall (561) to prepare a mixture of all five possible tetraalkyllead compounds. Grohn and Paudert (305) prepared tetrabenzyllead in low yield by treatment of lead metal with benzyl chloride in a ball mill; tin metal under similar conditions produced dibenzyltin dichloride. There have been no reports of reactions of lead metal with aryl halides.

With Other Alkyl Esters. Finely divided lead metal will react with dialkyl sulfates and trialkyl phosphates, but at relatively slow rates. Thus, Krohn (425) prepared tetramethyl- and tetraethyllead in 20–65% yield, using a reaction temperature of 110–125° for 20 hr; the ethyl esters require an iodide catalyst. Unlike their reactions with sodium–lead alloy all the alkyl groups are utilized, according to the equations:

$$3Pb + 2R_2SO_4 \rightarrow R_4Pb + 2PbSO_4$$

$$9Pb + 4R_3PO_4 \rightarrow 3R_4Pb + 2Pb_3(PO_4)_2$$

However, as with reactions of lead metal and alkyl halides, only one third of the lead is converted to R_4Pb. Aryl esters and higher alkyl esters have not been investigated in this system.

With Alkyl Halides and Reactive Metals. Another system which has been used to prepare R_4Pb compounds from lead metal is based on the reaction of an organic halide in combination with a reactive metal; magnesium (111), lithium (113), and zinc (99) have been employed as the reactive metals. With magnesium, the reaction proceeds according to the equation:

$$4RCl + Pb + 2Mg \rightarrow R_4Pb + 2MgCl_2$$

This reaction requires a catalyst, such as ethers, tertiary amines, or iodides, and is speculated to proceed via *in situ* formation of the Grignard reagent. Methyl, ethyl, and propyl halides (except fluorides) have been shown by Calingaert and Shapiro (111) to be reactive in the magnesium system. The stoichiometry of this reaction is identical to that of the Mg_2Pb–RX reaction. However, hexaalkyldilead (viz., hexaethyllead) is formed in appreciable amounts; this contrasts with the absence of hexaethyldilead formation in the Mg_2Pb–EtCl reaction.

The Mg–EtCl–Pb reaction can be conducted concurrently with the NaPb–EtCl process to alkylate *in situ* the by-product lead metal from the latter reaction. Tetraethyllead yields as high as 75% have been obtained in a "one-step process":

$$8EtCl + 3Mg + 2NaPb \rightarrow 2Et_4Pb + 3MgCl_2 + 2NaCl$$

The reaction of lead metal and an alkyl halide in the presence of lithium metal is vigorous and can reach violent proportions, because of the tendency of lithium metal to react directly with the alkyl halide in a Wurtz-type reaction; because of this, the yields are usually very low. The synthesis of tetraethyllead by the reaction of lead metal, zinc metal, and ethyl iodide in aqueous caustic has been mentioned by Calingaert (99). Also, Nosek (534) has recently reported the synthesis of tetrahexadecyllead via the reaction of lead(II) chloride and cetyl iodide with "activated zinc" (from reaction of zinc metal and aqueous silver nitrate) in aqueous solution. This reaction is potentially attractive as a convenient method of synthesis since aqueous solutions can be used and prior preparation of a pyrophoric organozinc compound is not required.

With Organic Halides and Organometallic Reagents. The alkylation or arylation of lead metal by certain organic halides in combination with a reactive organometallic compound proceeds with great facility. Gilman and co-workers (271,285,437) found that if the reaction of a lead(II) halide with a Grignard reagent or an organolithium compound is carried out in the presence of an organic iodide, lead metal formation is eliminated or decreased greatly and high conversions to R_4Pb are obtained. They concluded that the reaction involved the alkylation of lead metal, since a two-step operation could be used, i.e., the organic halide could be added to the reaction mixture after the lead metal had been formed in a conventional reaction of the lead(II) halide with the organometallic reagent. Tetramethyllead, tetraethyllead, and tetraphenyllead were prepared in high yield from the reaction of lead(II) chloride or iodide with the respective organolithium or Grignard reagent in the presence of the respective organic iodide. However, appreciably lower yields, accompanied by lead metal formation, were obtained when the organic bromides were used in place of the iodides; organic chlorides were not tested. On the basis of the R_4Pb yields obtained, the following order of reactivities was indicated:

$$RLi > RMgX$$

$$\text{alkyl iodide} > \text{aryl iodide}$$

$$\text{alkyl iodide} > \text{alkyl bromide}$$

The alkylation of lead metal by a reactive organometallic compound in combination with an alkyl chloride has been investigated by chemists of the Ethyl Corporation, who demonstrated that ethyl chloride was also reactive in these systems; these reactions are described in several patents. In addition to ethyl Grignard reagent (112) and ethyllithium (113) as the reactive organometallic reagents, diethylzinc (114,587), diethylcadmium (115,676), and ethylcadmium iodide (115) were shown to be reactive. Tetraethyllead was also obtained in low yield from the reaction of lead metal with sodium–naphthalene complex in the presence of ethyl chloride (657). All of these latter reactions proceed according to the general equation:

$$2RM + Pb + 2RX \rightarrow R_4Pb + 2MX$$

where RM is the reactive organometallic and RX is the organic halide. The approximate reactivities of the RM compounds decrease in the general order $RLi > RMgX > R_2Zn > R_2Cd$. These reactions were usually carried out at 80–100° in diethyl ether solvent under the autogeneous pressure of the organic halide.

Gilman and colleagues originally postulated that the PbX_2–RI–RMgX (or RLi) reaction involved the following sequence:

$$2PbX_2 + 4RM \rightarrow R_4Pb + Pb + 4MCl$$

$$2RI + Pb \rightarrow R_2PbI_2$$

$$R_2PbI_2 + 2RM \rightarrow R_4Pb + 2MI$$

wherein the formation of R_2PbI_2 would be the critical step. The efficacy of this sequence is questionable, however, since triethylaluminum is not reactive in this system (243), despite claims in the patent literature to the contrary (192). If the sequence simply involved the formation of R_2PbI_2 from lead metal and the organic iodide (or halide), one would expect triethylaluminum to be reactive, since R_3Al compounds react with organolead halides.

Although triethylaluminum will not ethylate lead metal in the presence of ethyl halides, coordination complexes of triethylaluminum, such as sodium triethylaluminum methoxide, are reactive. This reaction was investigated by Robinson and Frey and co-workers (245,623) for the synthesis of tetraethyllead. Only one ethyl group of the triethylaluminum complex is utilized:

$$2NaEt_3AlY + Pb + 2EtX \rightarrow Et_4Pb + 2NaX + 2Et_2AlY$$

where Y = Et^-, F^-, or MeO^-.

With sodium tetraethylaluminum, near quantitative conversions of lead metal were realized in the presence of certain ethers as solvents. The preferred ether solvents are the low molecular weight alkyl and cyclic ethers, such as diethyl ether, di-*n*-propyl ether, tetrahydrofuran, and tetrahydropyran (245). Steric effects from the solvent appear to play a significant role in this reaction as evidenced by the fact that aryl ethers are less effective, and di-isopropyl ether is much less effective than *n*-propyl. However, the preferred ether solvents also tend to promote the formation of hexaethyldilead. By conducting the reaction with sodium tetraethylaluminum in the presence of sodium methoxide, more than one ethyl group can be utilized, presumably because of the *in situ* formation of sodium triethylaluminum methoxide from the by-product triethylaluminum. Sodium tetraethylboron is also reactive with lead metal and ethyl chloride, but the yields are not as good as with the tetraethylaluminum; triethylborane is unreactive (245).

These reactions, especially those involving the reaction of a lead halide with an organolithium or Grignard reagent in the presence of an organic halide, are very useful as laboratory syntheses. However, they have also been considered for the ethylation of the lead metal formed in the commercial alloy process for tetraethyllead manufacture.

With Organometallic Compounds Alone. Only one synthesis of an R_4Pb compound has been reported which employs the alkylation of lead metal by an organometallic compound alone, without benefit of a second reagent. Talalaeva and Kocheshkov (718,719) have described the synthesis of tetraphenyllead by reaction of lead metal with phenyllithium in diethyl ether. This reaction is surprising since it is in the reverse direction as predicted from electromotive force considerations; phenyllithium can be prepared in good yield by reaction of tetraphenyllead and lithium metal. Tetraphenyllead was obtained from lead metal and phenyllithium in 5% yield after 185 hr; phenyl Grignard reagent was not reactive under similar conditions. With lead analgam instead of lead metal, the yield was 8.5% in 25 hr. Presumably the driving force in these reactions is the formation of lithium–lead alloy or lithium amalgam. Reaction of a ternary alloy of the composition Pb·Sn·2Hg with phenyllithium gave tetraphenyltin in 46.5% yield along with 2.6% tetraphenyllead.

In the presence of bromobenzene, tetraphenyllead was obtained in 23% yield from the reaction of lead metal and phenyllithium; tetra-*p*-tolyllead was prepared similarly (718,719). However, these latter reactions are more analogous to those discussed earlier which involved lead metal, an organic halide and a reactive metal.

(*3*) *Electrolytic syntheses*

The synthesis of an organolead compound via an electrolytic procedure dates back to 1906 when Tafel (709) obtained a brownish-red oil by electrolysis of a solution of methyl ethyl ketone in 30% sulfuric acid using a lead cathode. The product was not characterized except it was shown that it could be decomposed in ether solution to yield a lead-containing precipitate. Renger (606,607) subsequently isolated tri-*sec*-butyllead chloride and di-*sec*-butyllead dibromide upon treatment of Tafel's red oil with chlorine and bromine, respectively. A similar oil was obtained upon electrolysis of a sulfuric acid solution of diethyl ketone but none of these oily products was fully characterized. On the basis of their color and the products obtained upon halogenation, the red oils were concluded to be mixtures of dialkyllead, "trialkyllead," and possibly tetraalkyllead (711). Electrolyses of methyl isoamyl ketone (710), citral (433–435), and propionaldehyde (640) with lead electrodes have also been carried out; similar, poorly defined organolead products were obtained.

In 1925, Calingaert (100) patented a process for the synthesis of tetraethyllead by electrolysis of a solution of ethyl iodide in ethanolic caustic using a lead cathode. Except for the electrolytic synthesis of hexaethyllead (which is discussed in Chapter VII) these cathodic processes have enjoyed little utility either as laboratory methods or commercial processes for the preparation of organolead compounds. However, there is a continuing interest in these systems. Recent du Pont patents (204,208) describe an electrolytic process for the synthesis of tetraalkyllead compounds based on the electrolysis of a solution of a tetraalkylammonium halide in acetonitrile. Also, a Russian publication (737) reports the synthesis of tetra(cyanoethyl)lead by electrolysis of an acidic aqueous solution of 3-iodopropionitrile using lead electrodes.

Anodic processes for the synthesis of tetraorganolead compounds have been investigated much more extensively than cathodic processes, especially in the past 15 years. In 1924, Hein and his colleagues (327), in their classical research on the salt-like behavior of the alkyl derivatives of alkali metals, demonstrated that tetraethyllead was formed by electrolysis of an ether solution of sodium triethylzinc using a lead anode; zinc metal was deposited at the cathode. Sodium tetraethylaluminum behaved similarly, but this system was not investigated in detail. Little further interest was exhibited in anodic processes until the 1950's when Ziegler and his colleagues at the Max Planck Institut für Kohlenforschung began an extensive investigation of electrolytic syntheses based on the use of alkali metal salt complexes of triethylaluminum as fused salt electrolytes. The primary interest in this system stemmed from its possible use for the commercial manufacture of tetraethyllead.

In 1955 Ziegler published the first of a series of papers and patents on the electrolysis of the sodium fluoride complexes of triethylaluminum (785,786,798). The sodium fluoride bis(triethylaluminum) complex, $Na[Et_3Al \cdots F \cdots AlEt_3]$, melts slightly above room temperature and has a conductivity of 0.02 $ohm^{-1}cm^{-1}$ at 63°. In contrast, the one-to-one complex, $Na[Et_3AlF]$, is a solid which melts at 72–73° and has a conductivity only 5% that of the one-to-two complex. Controlled electrolysis of the liquid one-to-two complex proceeds with the formation of tetraethyllead at the lead anode and aluminum metal at the inert cathode, according to the equation:

$$4NaF{\cdot}2Et_3Al + 3Pb \xrightarrow{e^-} 3Et_4Pb + 4NaF{\cdot}Et_3Al + 4Al$$

The tetraethyllead is insoluble in the molten electrolyte and separates out as a denser liquid phase. Other alkylmetal compounds have also been prepared using this electrolyte and the respective metal as the anode. The electrolysis is conducted at elevated temperature in order to form the "by-product" $NaF{\cdot}Et_3Al$ as a melt. The electrolyte is regenerated by addition of triethylaluminum to the cell. If the electrolyte is allowed to approach the $NaF{\cdot}Et_3Al$ composition, a mixture of sodium and aluminum tends to be deposited at the cathode.

The aluminum metal deposited at the cathode can be recycled to the system by reaction with hydrogen and ethylene to form triethylaluminum. However, a major shortcoming of the system is that the aluminum tends to deposit as stringers on the cathode; within a short time these stringers become sufficiently long to short out the cell, causing the electrolytic reaction to cease.

Ziegler and his colleagues investigated a number of different electrolytes in an attempt to define a system suitable for use as a commercial process. Molten sodium tetraethylaluminum has been used as the electrolyte; sodium metal is deposited at the cathode, according to (769,781,783,800):

$$4NaAlEt_4 + Pb \xrightarrow{e^-} Et_4Pb + 4Et_3Al + 4Na$$

Sodium tetraethylaluminum melts at 124°, so that the sodium is formed as a molten deposit and drips off the cathode, thereby creating an ever-fresh cathode surface. At this temperature, however, the tetraethyllead is thermally unstable and must be distilled from the cell under reduced pressure to avoid its decomposition. Also, sodium metal reacts with tetraethyllead, so a compartmented cell must be used to separate the tetraethyllead from the sodium metal. Alternatively, Ziegler has used a pool of mercury as the cathode to amalgamate the sodium and thereby prevent its reaction with the tetraethyllead (782).

Table VI-1 Properties of Complex Organometallic Electrolytes[a]

ELECTROLYTE	M.P., °C	CONDUCTIVITY, $OHM^{-1}CM^{-1}$	TEMP., °C
$NaF{\cdot}AlEt_3$	72–73	0.002	100
$NaF{\cdot}2AlEt_3$	35	0.04	100
$NaOBu{\cdot}AlEt_3$[c]	—	0.0022	100
$KF{\cdot}AlEt_3$	56–58	0.02	100
$KF{\cdot}2AlEt_3$	129	0.07	130
$NaAlEt_4$	124	0.033 (extrapolated)	100
$2NaAlEt_4{\cdot}KCl$[d]	—	0.045	100
$KAlEt_4$	73	0.08	100
NaCl (25% in H_2O)	—	0.65	100
NaCl (molten)	—	3.66	850
$NaBEt_4$ (saturated in H_2O)[b]	—	0.009	22

[a] Ziegler and Lehmkuhl, *Chem. Ingr. Tech.*, **35**, 325 (1963).
[b] Ziegler and Steudel, *Ann. Chem.*, **652**, 1 (1962).
[c] Ziegler, Brit. Patent 864,393, April 6, 1961.
[d] Ziegler (and Lehmkuhl), Ger. Patent. 1,153,754, Sept. 5, 1963.

Addition of potassium chloride to the sodium tetraethylaluminum gives an electrolyte melting at 75° (793); this permits use of a lower cell temperature which in turn minimizes the thermal decomposition of tetraethyllead. A mixed electrolyte of sodium tetraethylaluminum and sodium triethylaluminum fluoride has also been used. Similarly, potassium tetraethylaluminum (795) and a mixture of potassium tetraethylaluminum and potassium triethylaluminum fluoride (797) have been employed as electrolytes; these potassium electrolytes are better conductors than their sodium analogs. Conductivity data on several electrolytes are given in Table VI-1.

The major problem in the use of the tetraethylaluminum electrolytes is the formation of triethylaluminum as a by-product; the triethylaluminum is soluble in the tetraethyllead and separates with the latter as a second liquid phase. Furthermore, the boiling point of triethylaluminum is very close to that of tetraethyllead and therefore they cannot be easily separated by simple physical methods. Separation is best achieved by complexation of the triethylaluminum with a Lewis base. Thus, McKay (477–478) has shown that addition of sodium hydride to the mixture forms liquid $NaH \cdot 2Et_3Al$, which is insoluble in tetraethyllead and separates as a second phase; a rapid, near-quantitative separation can be achieved in this manner. The use of other bases to complex the triethylaluminum, such as tributylamine (792), sodium azide (790), and potassium cyanide (789), is also described by Ziegler in the patent literature.

Ziegler (788) has devised a clever separation of the triethylaluminum from the tetraethyllead which is based on the reaction of the triethylaluminum–tetraethyllead mixture with sodium triethylaluminum ethoxide to form sodium tetraethylaluminum (for recycle) and diethylaluminum ethoxide. The sodium tetraethylaluminum is insoluble and separates as a second phase. The diethylaluminum ethoxide can be separated from the tetraethyllead by distillation and treated with sodium hydride and ethylene to regenerate the sodium triethylaluminum ethoxide.

$$Et_3Al + NaEt_3AlOEt \rightarrow NaAlEt_4 + Et_2AlOEt$$
$$Et_2AlOEt + NaH + C_2H_4 \rightarrow NaEt_3AlOEt$$

Taking this one step further, Ziegler and others (293,787,791,794) have used sodium triethylaluminum alkoxide as the electrolyte; in this system, diethylaluminum alkoxide is formed directly in the electrolytic process and can be separated from the tetraethyllead by distillation. However, a disadvantage with sodium triethylaluminum alkoxide as the electrolyte is that its conductivity is poorer than that of sodium tetraethylaluminum.

Tetramethyllead has been prepared by electrolysis of sodium tetramethylaluminum but, because of the high melting point of sodium tetramethylaluminum (240°), a solvent, such as tetrahydrofuran, must be used (784). Kobetz and Pinkerton (392) have electrolyzed a mixture of sodium tetramethylaluminum and sodium tetraethylaluminum as a fused melt. Only tetraethyllead was obtained; either the tetraethyllead is formed selectively in the electrolytic process, or else a mixed ethylmethyllead compound is formed initially and rapidly undergoes reaction with the trialkylaluminum by-product to form tetraethyllead. A rapid exothermic exchange reaction does occur between tetramethyllead and triethylaluminum to form ethyllead species (243,471).

Lehmkuhl, Schafer, and Ziegler (441) have suggested a combined process for the production of both tetramethyllead and tetraethyllead using a dual electrolysis system. Tetraethyllead is produced in one cell by electrolysis of molten sodium tetraethylaluminum using a mercury cathode. The sodium amalgam formed at the cathode is then treated with methyl chloride in the presence of trimethylaluminum to form sodium tetramethylaluminum, which is subsequently electrolyzed as an ether solution in a second cell to form tetramethyllead. The triethylaluminum and trimethylaluminum are recycled to regenerate the sodium tetraalkylaluminum electrolytes. However, there does not appear to be any special advantage for this dual system.

Tetraethyllead and tetramethyllead have also been prepared by electrolysis of alkylboron complexes using a lead anode in systems analogous to the aluminum systems discussed above. Tetraethyllead has been prepared by Ziegler and co-workers (796,801) by electrolysis of an aqueous solution of sodium tetraethylboron; Pinkerton (568) has prepared tetraethyllead similarly, using an ether solution of sodium tetraethylboron. A solvent is needed because pure sodium tetraethylboron has too high a melting point to be used as a fused melt. However, a mixed electrolyte of sodium tetraethylboron and sodium tetraethylaluminum has been used by Kobetz and Pinkerton (391,393). Interestingly, only triethylborane is formed as a by-product in this latter system and it can be easily distilled from the cell, thereby effecting a facile separation from the tetraethyllead. Either the sodium tetraethylboron is selectively reduced in the electrolytic process or a reaction occurs between the by-product triethylaluminum and sodium tetraethylboron to form triethylborane. Tetramethyllead has also been prepared by electrolysis of an aqueous solution of sodium tetramethylboron (801).

The breadth and depth of the research effort that have been devoted to the synthesis of tetraethyllead via electrolysis of alkylaluminum and alkylboron complexes are most impressive. Interesting and unexpected results have been obtained with the mixed electrolytes and clever solutions have been devised to some of the problems posed by the aluminum system. Obviously, these systems are of primary interest as commercial processes for the manufacture of tetraethyllead, since the overall process would involve consumption of only lead, hydrogen, ethylene and electricity. However, successful development of a commercial process has not been realized. A laboratory apparatus for the electrolytic synthesis of R_4Pb compounds has been described (799).

In addition to the aluminum and boron systems, a third anodic process which has received considerable attention in recent years is one based on the electrolysis of an ethereal solution of a Grignard reagent and an alkyl halide, using a lead anode. This system is the subject of a number of patents issued to Braithwaite and co-workers (71–73,507–513,515,517), and has been developed by Nalco Chemical Co. into a process for the commercial manufacture of tetramethyllead and tetraethyllead. The Nalco process involves the controlled electrolysis of a solution of the alkyl Grignard reagent and alkyl halide in a mixed ether solvent. The overall reaction is represented by the equation:

$$2RMgCl + 2RCl + Pb \rightarrow R_4Pb + 2MgCl_2$$

Magnesium metal is formed at the cathode, but it reacts immediately with the alkyl halide to regenerate the Grignard reagent. Close control of the voltage–current parameters is required to avoid undesirable by-products and to allow reaction of the magnesium metal with the alkyl halide as it is being deposited at the cathode. Few additional details of the process have been published (67,316). This system probably involves a chemical as well as an electrochemical process as evidenced by the fact that a current efficiency of 175% has been reported for the tetraethyllead system (73) and a current efficiency of 156% claimed for the methyl system (513).

The preparation of tetramethyllead and tetraethyllead by the electrolysis of ether solutions of alkylmetal compounds in the absence of an alkyl halide or in the presence of only a slight excess of alkyl halide has been described in several patents (221,226,446,447,517,560,625,689).

Electrochemical syntheses of organolead compounds have been reviewed recently by Marlett (470) and electrochemical syntheses based on organo-aluminum electrolytes have been reviewed by Lehmkuhl (440). Electrolytic syntheses from Grignard reagents are discussed in a recent review of organic electrode processes (701).

(4) Other synthesis methods

Various other reactions of a less general nature have been used to prepare R_4Pb compounds. Nesmeyanov and Kocheshkov (527) have prepared tetraphenyllead in low yield (~15%) by reaction of lead metal and phenyldiazonium tetrafluoborate in acetone; higher yields (30%) were obtained when sodium–lead alloy was used in place of lead metal (529). Tetra-*p*-tolyllead (16%) was also prepared by this procedure. On the other hand, reaction of lead metal and *p*-chlorophenyldiazonium chloride reportedly gave only lead(II) chloride (460). Tetraphenyllead was also prepared by the reaction of lead metal and diphenylbromonium tetrafluoborate, Ph_2BrBF_4, in acetone (530).

The preparation of tetraalkylleads by reaction of lead metal with free radicals produced by thermal decomposition of hydrocarbons has been patented by Rice (614,615) and Sullivan (702,703,705,706). Similarly, the alkylation of lead metal by free radicals produced by reaction of an "initiator" metal with an aliphatic peracid or alkyl halide has been claimed (431). Volatile organolead compounds have been obtained from the reaction of methane with lead-212 atoms generated by α-decay of polonium-216 (381). Although these reactions enjoy little utility as methods of synthesis of R_4Pb compounds, the formation of R_4Pb compounds from lead metal and free radicals has proved a useful tool for detecting free radicals (87,233,355). A technique for the analysis of minerals has been developed which is based on the synthesis of tetramethyllead via methyl radical attack on a lead mirror deposited by hydrogen reduction of the mineral and distillation of its contained lead (742,762).

The preparation of R_4Pb in very low yield by reaction of lead metal (462,500) or sodium–lead alloy (567) with an olefin and hydrogen has been patented. The addition of olefins to a trialkyllead hydride has been used to prepare symmetrical and unsymmetrical R_4Pb compounds (38,51), as well as those containing functional groups (531). The pyrolysis of benzylplumbonic acid is reported by Lesbre (443) to yield tetrabenzyllead. Pyrolysis of lead tetrabenzoate, however, did not yield a phenyllead derivative (13).

Tetraorganolead compounds are also formed in many reactions involving organolead derivatives of the type R_3PbX and R_2PbX_2. These reactions are discussed in Chapter XI.

B. Physical properties

The general physical properties of the tetraalkyl- and tetraaryllead compounds were discussed at the beginning of this chapter. Such properties as boiling point, melting point, density, and refractive index were routinely determined by Krause and his co-workers in their systematic studies on the syntheses and properties of various symmetrical and unsymmetrical tetraalkyl- and tetraaryllead compounds. Grüttner and Krause (315,420) and Jones and co-workers (369) measured and collated the *densities* and *refractive indices* of many symmetrical and unsymmetrical tetraalkyllead compounds and derived expressions for the variation of boiling points with molecular weight and for the variation of molecular volumes with the number of carbon atoms in the molecule. The *molecular refractions* of a number of tetraalkyllead compounds were calculated by Vogel and his colleagues (745) from refraction values for C—H and C—C bonds derived from the measured refractions of alkyl halides and alkanes; a mean value of 5.26 ± 0.14 was derived for the refraction of the lead–carbon bond (172). Generally good agreement was obtained between the calculated values and the observed values. Jones found that the atomic refractivity of lead in the normal alkyl derivatives increases slightly with increasing molecular weight of the alkyl group in going from methyl to butyl. Also, the atomic refractivity of an alkyl group for a normal alkyl derivative was found to be the same as that in the corresponding *iso*-alkyl derivative, but a definite increase was observed for the *sec*-alkyl group. Similar effects had been observed earlier by Grüttner and Krause (315), who calculated an average value of 18.33 (Na_D radiation) for the atomic refractivity of lead in tetraalkyllead compounds. These latter workers also derived equations for calculating the atomic refraction of lead in tetraalkyllead compounds. The boiling points, refractive indices, and densities of a large number of symmetrical and unsymmetrical, primary and secondary tetraalkyllead compounds were determined and compared.

The magnetic susceptibilities and atomic parachors of a series of symmetrical tetraalkyllead compounds have been measured by Kadomtzeff (373,374). The data are given in Table VI-2. The magnetic susceptibilities

Table VI-2 Magnetic Susceptibilities and Atomic Parachors of R_4Pb *Compounds*

COMPOUND	MAGNETIC SUSCEPTIBILITY, $\times 10^6$		ATOMIC PARACHOR	
$(CH_3)_4Pb$	−49.8		76.1	
		$\Delta = 5.8$		$\Delta = 8.0$
$(C_2H_5)_4Pb$	−44.3		68.1	
		$\Delta = 1.7$		$\Delta = 3.4$
$(C_3H_7)_4Pb$	−42.6		64.7	
		$\Delta = 8.0$		$\Delta = 7.7$
$(C_4H_9)_4Pb$	−34.6		57.0	
$(C_5H_{11})_4Pb$	—		—	
$(C_7H_{15})_4Pb$	—		66.6	

Taken from Kadomtzeff, *Compt. rend.*, **226,** 661 (1948).

were shown to decrease progressively in going from methyl to butyl and to show a definite alternation between terms containing alkyl groups with odd and even numbers of carbon atoms; a similar alternation was observed for the atomic parachors. Analagous effects were observed for the same series of tetraalkyltin compounds. The decrease in magnetic susceptibility with increasing chain length of the alkyl group was attributed to a compression of the central metal atom. Kadomtzeff's magnetic susceptibility data agree fairly well with values reported earlier by Pascal (556). However, the value for the atomic parachor of tetraethyllead, 68.1, is somewhat lower than the average value of 76 calculated by Sudgen (700). Sudgen also calculated a mean value of 456.6 for the molecular parachor of tetraethyllead and derived the following equation to express the variation of the *density* of tetraethyllead with temperature (700).

$$d_{4^0}^{t^0} = 1.673 - 0.00163t^0$$

Bothorel (65) has measured the *magnetic susceptibility* of tetraphenyllead using single crystals and also obtained Rayleigh diffusion measurements on its solutions. These data were used to determine the molecular anisotropy and orientation of the phenyl rings (66). The angle between the planes of the phenyl rings in tetraphenyllead was calculated to be 50° while the minimum distances between carbon atoms and between hydrogen atoms were calculated to be 4.34 and 4.04 Å, respectively. For tetraphenyltin, comparable values were calculated to be 42°, 3.99 Å, and 3.48 Å. Tetraphenyllead has been included in a discussion of anomalies of the diamagnetic susceptibilities of various compounds (554). A more recent

Table VI-3 Viscosities and Molecular Volumes of R_4Pb *and* R_4Sn *at 20°*

	VISCOSITY, CP		MOLECULAR VOLUME, CC	
COMPOUND	TIN	LEAD	TIN	LEAD
$(CH_3)_4M$	0.4187	0.5723	138.1	135.9
$(C_2H_5)_4M$	0.7290	0.8642	194.6	195.6
$(C_4H_9)_4M$	2.610	3.013	328.2	328.9
$(C_7H_{15})_4M$	10.44	11.26	528.3	533.5

Taken from Hügel, *Kolloid-Z.*, **131,** 4 (1953).

paper by Pascal and co-workers (557) discusses the contribution of the individual bonds to the total diamagnetic susceptibility of the aliphatic series of compounds involving lead, tin, and germanium.

The *molecular volumes* at 0°K have been calculated by Herz (334) for a series of alkyl derivatives of various metals from their densities and refractive indices; several tetraalkyllead compounds were included. For alkyl derivatives of the same metal, the molecular volume was found to decrease with increasing molecular weight of the compounds; for a series of compounds containing the same alkyl group but different metals of the same family, e.g., Group IVb, the molecular volumes were found to increase with increasing atomic weight of the central metal atom. The molecular volumes of a series of symmetrical tetraalkyllead and tin compounds have also been determined by Hügel (352); at 20°, the molecular volumes of the lead derivatives were slightly higher than those of their tin analogs, except for the tetramethyl compounds. Hügel also measured the viscosities of these same two series of R_4Sn and R_4Pb compounds and found that the viscosities increase with increasing alkyl chain length (Table VI-3). The interrelationship between viscosities (η) and molecular weights (M) for R_4Sn versus R_4Pb was found to fit the equation:

$$\frac{M_{R_4Pb}}{M_{R_4Sn}} = 1.106 \times \frac{\eta_{R_4Pb}}{\eta_{R_4Sn}}$$

The *viscosities*, *refractive indices*, and *densities* of tetramethyllead and tetramethyltin at 15–30° were determined by Bambynek (24) for use in calculation of self-diffusion coefficients. The self-diffusion coefficients (D) of the two compounds were shown to be inversely related to the square root of the molecular weights (M).

$$D_{Me_4Pb}/D_{Me_4Sn} = (M_{Me_4Sn}/M_{Me_4Pb})^{1/2}$$

The surface tension of tetraethyllead was measured by Sudgen (700); a value of 28.48 dyn/cm was obtained at 20°.

The viscosities and densities of solutions of tetraphenyllead in acetonitrile have been measured and equations derived to show the effect of concentration (679). In the very dilute solutions used (up to 3 g/100 ml), tetraphenyllead, as well as tetraphenyltin, -silane and -methane were shown to behave as ideal hydrodynamic spheres.

Tetraethyllead, as well as tetraethyltin, has been found to exhibit *polymorphism* (692,693). At least seven different polymorphs were detected for tetraethyllead based on various melting point values ranging from 135.6–142.9°K; at least ten polymorphs were observed for tetraethyltin (691). On the other hand, the tetramethyl analogs do not exhibit polymorphism; a single melting point at 242.9°K was observed for tetramethyllead by Staveley and his colleagues (691,693). This compares with a more recent value of 243 ± 2°K reported by Smith (683), who also found that the tetramethyllead had to be supercooled at least 20°K below this temperature before it solidified.

Phase diagrams were constructed by Pascal (555) for binary mixtures of tetraphenylsilane, tetraphenyltin, and tetraphenyllead. These compounds form mixed crystals in all proportions and are strictly isomorphous. For the lead–tin system, no maxima or minima were observed; for lead–silicon, an eutectic was observed which melted at 218.8°C and had the composition 34% Ph_4Si–66% Ph_4Pb.

The *dipole moments* of tetraethyllead and tetraphenyllead were determined by Strohmeier and Miltenberger (698), using solutions in heptane, benzene, and dioxane. As expected, the dipole moments of both compounds are zero, as are those of the derivatives of the other Group IVb metals.

The *heat of formation* of tetramethyllead has been determined by two groups of workers. Using rotating bomb calorimetry, Good, Scott, and co-workers (300,302) calculated the heat of formation ($\Delta H_{298°}$) of liquid tetramethyllead to be $+23.5 \pm 0.3$ kcal/mole and a value of $+32.6 \pm 0.3$ kcal/mole for its heat of formation in the gaseous state; the entropy of tetramethyllead vapor was calculated to be 100.48 ± 0.20 cal/deg/mole (302). An earlier value of $+3.2 \pm 3$ kcal/mole for the heat of formation of liquid tetramethyllead had been reported by Lippincott and Tobin (448); however, their value is considered to be the less accurate because of the method used and problems associated with measuring the heat of formation of organometallic compounds in general (301). Scott, Good, and co-workers (300,646) have also determined the heat of formation of liquid tetraethyllead by rotating bomb calorimetry; a value of $+12.8 \pm 0.6$ kcal/mole was obtained. From the rotating bomb calorimetry data, average values of 34 and 32 kcal were calculated for the dissociation energies of the lead-to-carbon bond in tetramethyllead and tetraethyllead (302,726). However, more recent values of 36.5 and 30.8 kcal have been derived for the mean dissociation energy of the lead-to-carbon bonds in tetramethyllead and tetraethyllead, respectively (450,678). Values of 837.6 (448) and 870.2 (432) kcal/mole were measured for the *heat of combustion* of tetramethyllead, while a value of 1504 kcal/mole was obtained by Lautsch (432) for the heat of combustion of tetraethyllead.

Expressions have been derived to show the variation of *vapor pressure* with temperature for both tetramethyllead and tetraethyllead. For tetramethyllead Good and co-workers (302) found:

$$\log p = 6.93767 - [1335.317/t^\circ + 219.084]$$

from which $\Delta H_{vap} = 9.075$ kcal/mole. This value for the latent heat of vaporization was used to calculate the value given above for the heat of formation of tetramethyllead in the vapor state. Vapor pressure–temperature expressions have been derived for tetraethyllead over two temperature ranges:

$$\log p = 9.4262 - [2960.0/(t^\circ + 273.1)] \text{ (from 78–150°) (369)}$$

$$\log p = 9.428 - [2938.0/(t^\circ + 273.1)] \text{ (from 0–70°) (82)}$$

The heat of vaporization of tetraethyllead has been estimated to be 13 kcal/mole (302).

Vapor pressure–temperature relationships have also been derived for tetramethyllead, tetraethyllead, and the three mixed ethylmethyllead derivatives by Calingaert and co-workers (108) who found that the dependence of the vapor pressure of all five compounds on temperature could be expressed by the general equation:

$$\log p = A - B/(t^\circ + 230)$$

The different values for A and B, as well as the boiling points of the five compounds at 50 mm, are shown in Table VI-4.

The *vapor pressure and latent heat of sublimation* of tetraphenyllead were determined using a tritium-labelled sample (128,129). At 25°, its vapor pressure was found to be 6.76×10^{-10} mm; its heat of sublimation was calculated to be 19.18 ± 0.28 kcal/mole.

Heat capacities and *heats of fusion* of tetramethyllead and tetraethyllead have been reported (693). For tetramethyllead, C_p was shown to increase from 24.2 to 36.4 cal/mole/°K over the range 100–230°K; for tetraethyllead, C_p increased from 33.65 to 62.5 cal/mol/°K over the temperature range 90–140°K. The heat capacity of tetramethyllead has also

Table VI-4 Vapor Pressure-Temperature Constants for R_4Pb

COMPOUND	A	B	BOILING POINT AT 50 MM, °C
Me_4Pb	6.9381	1378.7	33.2
Me_3PbEt	7.2760	1602.5	57.3
Me_2PbEt_2	7.5903	1810.7	77.4
$MePbEt_3$	7.8768	2000.5	93.8
Et_4Pb	8.1547	2184.6	108.4

Taken from Calingaert, Beatty, and Neal, *J. Am. Chem. Soc.*, **61,** 2755 (1939).

been calculated over the temperature range 298–1500°K (173,448). The *heats of fusion* of tetramethyllead and tetraethyllead were calculated to be 2.581 and 2.101 kcal/mole, respectively (693). Selected thermodynamic properties of tetramethyllead have been calculated by Crowder and co-workers (173).

The *solubility* of tetraphenyllead in various solvents has been determined by several investigators (502,522,699,766). Excellent solubility in chloroform and carbon tetrachloride makes these solvents useful for reactions involving tetraphenyllead. Solubility data on numerous tetraorganolead compounds were also reported by Krause and co-workers in their early publications. The solubility of tetraethyllead in water has been found to be 0.2–0.3 mg/l. at 0–38° (232). The solubility of tetramethyllead and tetraethyllead in aqueous acetic acid has also been determined (565); a 67.5% solution of acetic acid in water is reported to be useful for the separation of tetramethyllead from tetraethyllead.

The *ultraviolet, infrared, and Raman spectra* of tetramethyllead, tetraethyllead, and tetraphenyllead have been investigated extensively. The ultraviolet absorption spectra exhibit continuous absorption for both the liquid and vapor phases (383,429,442,489,608,728,729,735); this has been ascribed to photolytic decomposition of the R_4Pb compound to lead metal.

Milazzo (489) and LaPaglia (429) compared the *ultraviolet spectra* of the tetraphenyl derivatives of the Group IVb metals in the region 2100–3000 Å. The absorption spectra are similar to each other and to that of benzene; two electronic transitions are observed at 2600 and 2100 Å for tetraphenyllead which are analogous to the forbidden transitions of benzene. Rao and colleagues (588,589) have shown that the near ultraviolet spectra of the tetraphenylmetal compounds are similar to those of the Group Vb elements in the pentavalent states. The spectra show the so-called "primary" benzenoid or E band and a secondary or B band at longer wavelengths; the B bands show vibrational structure. A similar effect has been observed in the absorption spectrum of phenyltrimethyllead by Bowden and Braude (70).

The *infrared and Raman spectra* of tetramethyllead have been determined by several investigators (201,362,448,664,667,672,755,771). Jackson and Nielson (362) determined the spectra of both liquid and solid tetramethyllead at low temperature in order to inhibit photolytic decomposition during irradiation; complete assignments of the observed vibrational frequencies were made. Most recently, the infrared spectrum of tetramethyllead in the vapor state was determined by Crowder and colleagues (173). A sharp band at 120 cm^{-1} was observed, corresponding to the broader band at 130 cm^{-1} in the spectrum of liquid tetramethyllead. This revised value for this band (assigned as a f_2 C—Pb—C bending frequency) gives better agreement with the observed value of the entropy of tetramethyllead in the vapor state, $S^0_{298.15} = 100.48 \pm 0.20$ cal/deg/mole (302).

The *infrared absorption* spectrum of tetramethyllead has been compared to those of the tetramethyl derivatives of other Group IVb elements (362,448,755). The spectra exhibit a progressive shift of the symmetrical deformation frequencies to lower wave lengths with increasing molecular weight of the central metal atom. Sheppard (666) and Takenaka and Goto (717) have noted that the symmetrical methyl deformation frequencies of a number of methyl derivatives of the Groups IV–VII elements, including tetramethyllead, tend to vary regularly with the electronegativity of the element and its row in the periodic table.

Force constants and *effective methyl mass values* have been calculated for the tetramethyl derivatives of the Group IVb elements (663,673,674). From the force constants, the electronegativity values for the central metal atoms were calculated by Sheline and Pitzer (664) to be: Si = 1.76, Ge = 1.40, Sn = 1.37, and Pb = 1.13. As was mentioned in Chapter II, these values are considerably lower than those derived by other methods. Overend and Scherer (542) have calculated the repulsive force constants for the methyl groups of the tetramethyl derivatives. As expected, the constants decrease with increasing molecular weight (and diameter) of the Group IVb metal; for tetramethyllead, the force constant was calculated to be almost zero. Similarly, the energy barriers for rotation of the methyl groups in tetramethyllead and tetramethyltin have been calculated to be zero by French and Rasmussen (241); this compares with a calculated value of about 400 cal for tetramethylgermane, and measured values of 1100–1500 cal for tetramethylsilane and 4800 cal for neopentane.

The *infrared and Raman spectra of tetraethyllead* have also been investigated in detail (9,201,362,545,716,750,751), including those of the completely deuterated derivative (362), but a complete interpretation of the spectra has not been made because of the several possible conformations of the molecule. The simplicity of the spectra has been interpreted to indicate that there is little coupling between the ethyl groups as a result of the large mass of the lead atom; similar effects were found in tetraethyl- and tetrapropyltin (716).

The *infrared spectra of tetra-n-propyllead*, and tri-*n*-propyllead chloride and methoxide have been reported (9); bands in the region 376–381 and 557–568 cm^{-1} were assigned to the propyl skeletal stretching.

The *infrared spectra of tetraphenyllead* and the tetraphenyl derivatives of the other Group IVb metals have been compared (318,320,333,532,536, 588,589). The spectra exhibit mass effects similar to those observed in the spectra of the tetramethyl derivatives. Between 3–15 μ, the spectra show only minor differences. There is one sharp band between 9.0–9.5 μ (1125–1050 cm^{-1}) which shifts in a characteristic way to longer wave lengths with increasing atomic weight of the Group IVb metal (Pb = 1052 cm^{-1}). This characteristic shift was first observed by Noltes, Henry, and Janssen (532) and was ascribed to a phenyl group perturbed by the heavy metal atom; these authors have concluded that this band is useful for the identification of compounds containing a phenyl–metal bond (333,532). A similar band is not found in the spectrum of tetraphenylmethane. The spectra of compounds containing two different metals bonded to phenyl groups show the absorption bands characteristic of each metal.

In the region beyond 15 μ, the spectra of the individual tetraphenyl metal derivatives, according to Harrah, Ryan, and Tamborski (318), are more distinctive; these latter investigators conclude that this region is better for the rapid identification of new structures. The far infrared spectra of tetraphenyllead and other monosubstituted benzene derivatives have also been determined by Brown, Mohammed, and Sharp (80) and assignments made for the stronger bands.

The *singlet-triplet transitions* of the tetraphenyl metal derivatives have been determined by LaPaglia (430). The intensity of this transition was shown to increase markedly with increasing atomic weight of the central metal atom; the absorption observed for tetraphenyllead is reported to be the most intense absorption ever observed for a π electron transfer.

The infrared spectra of the organometallic derivatives of the Group IVb metals have been reviewed by Bajer (21) and by Ogawara and Sakiyama (535).

The *nuclear magnetic resonance spectrum* of tetramethyllead has been investigated extensively in recent years (7,194–197,237,644,681,687) and the significance of the observed chemical shifts and the spin-spin coupling constants has become the subject of much discussion. Allred and Rochow (7) demonstrated that a straight line relationship could be obtained by a plot of the proton chemical shift of the tetramethyl derivatives of the Group IVb metals at infinite dilution versus the electronegativity of the respective metal. From this plot, a value of 2.45 was indicated for the electronegativity of lead (compared to a value of 2.60 for carbon). This interpretation of the proton chemical shifts has been challenged by Drago (194), who chooses to explain the shifts by assuming a decreasing electronegativity of the Group IVb metal with increasing atomic number. The proton chemical shift data were ascribed by Drago to variations in the hybridization of the carbon atoms in the tetramethyl derivatives; the weaker the carbon-to-metal bond, the greater is the *s* character of the carbon–hydrogen bond of the methyl group. This causes an increase in the electronegativity of carbon and decreases the electron density around hydrogen, thereby making the hydrogen less shielded (195).

Allred and Rochow's electronegativity values as derived from proton shift data have been confirmed by Spiesecke and Schneider (687) but at the same time these latter workers showed that the C^{13} chemical shifts give electronegativity values more consistent with the "best" values calculated by Pritchard and Skinner (577). They also concluded that an approximate correlation of the observed C^{13} and H^1 chemical shifts with metal electronegativities for certain alkylmetal compounds could be obtained if contributions arising from magnetic anisotropic effects were considered. However, poor correlations were obtained with the heavier Group IVb metal (tin and lead) and a self consistent set of electronegativity values for these elements from both the H^1 and C^{13} resonances could not be established. The different electronegativity values calculated for the Group IVb elements from NMR data are shown in Table VI-5.

Drago and Matiwiyoff (196) also determined the NMR spectra of the methyl derivatives of various elements and showed that a linear relationship existed between the C^{13}—H coupling constants and the proton chemical shifts for a restricted series of these methyl compounds. However,

Table VI-5 Electronegativities of Group IVb Elements from NMR Data on Me_4M *Compounds*

ELEMENT	FROM H^1 RESONANCE		FROM C^{13} RESONANCE	
	Spiesecke and Schneider	Allred and Rochow	Spiesecke and Schneider	PRITCHARD AND SKINNER "BEST" VALUES
C	2.60	2.60	2.5	2.5–2.6
Si	2.03	1.90	1.9	1.8–1.9
Ge	2.1	2.00	1.85	1.8–1.9
Sn	2.05	1.93	1.72	1.8–1.9
Pb	2.35	2.45	1.82	1.8

Taken from Spiesecke and Schneider, *J. Chem. Phys.*, **35,** 722 (1961).

like Spiesecke and Schneider, they found that the Group IVb elements did not follow this relationship and concluded that chemical shift data cannot be used to evaluate the electronegativities of the Group IVb metals because of paramagnetic contributions to the shielding of the proton. These paramagnetic contributions were attributed to the anisotropy of the carbon atom and to inductive effects arising from the carbon to metal bond energy.

Further argument against the Allred and Rochow electronegativity scale has been offered by Drago (195,473), on the basis of the finding that the equilibrium constants for the formation of complexes of triethyllead chloride and triethyltin chloride with tetramethylenesulfoxide are nearly equal and therefore do not correlate with a gross difference in the electronegativities of the two metals.

The *NMR spectrum* of solid tetramethyllead has been measured by Smith (683). Methyl group reorientation was observed at temperatures as low as 77°K, but molecular reorientation was not observed at all. Similar effects were observed for the other Group IVb metal derivatives except that tetramethylsilane and tetramethylgermane exhibited molecular reorientation at temperatures slightly below their melting points. The energy barrier to methyl reorientation was found to vary inversely with the sixth power of the methyl group separation ($V = k/r^6$), the value of the exponent being somewhat larger than that calculated by Overend and Scherer (542) from infrared spectral data ($V = k/r^{4.8}$). The spin-spin coupling constants of tetramethyllead have been included in mathematical correlations of the variation of the coupling constants with the period of the central metal atom or its atomic number (605,682,757).

The *NMR spectrum of tetraethyllead* shows a single strong central peak arising from the ethyl groups bonded to Pb^{206} (23,516,687); the resonances of the methyl and methylene groups are coincident. The spin-spin interaction between Pb^{207} and the methyl protons is about three times greater than that with the methylene protons and the corresponding coupling constants are of opposite sign.

The *NMR spectrum of tetra-n-propyllead* has been determined by Klose (389). Analogous to tetraethyllead, the coupling between Pb^{207} and the β-methylene protons is greater than that with the α-methylene protons and presumably the coupling constants are of opposite sign. Zimmer and Homberg (803) observed no methyl proton-to-metal coupling in tetraneophyllead or hexaneophyldilead, although methyl proton-to-metal coupling was readily discernible in the analogous tin compounds. On the other hand, Singh (677) observed a very strong coupling between lead and the methyl protons in trineopentyllead halides and hexaneopentyldilead.

The spectra of tetramethyl–, tetraethyl–, and tetrapropyllead are discussed in a recent review of the NMR spectra of organometallic compounds (455).

The NMR spectrum of tetra-*p*-tolyllead and hexa-*p*-tolyldilead have been examined recently (388a). The aromatic proton chemical shifts and the coupling constants were the same for both compounds. A CH_3–Pb^{207} interaction was observed, even though this interaction had to traverse six covalent atoms. No other NMR spectra of aryllead compounds have been reported, with the exception of pentafluorophenyllead derivatives.

Electron spin resonance has been observed in gamma-irradiated tetrapropyllead (453). The spectrum exhibited hyperfine structure, but it was not sufficiently well resolved to permit determination of the exact nature of the radical responsible.

The *mass spectra* of the tetramethyl derivatives of lead and the other Group IVb metals show that Me_3M^+ is formed as the dominant species (188,189,256,337,584). The effect of temperature on the distribution of the ionized species has been determined at temperatures up to 1200° (537). Also, the appearance potentials for the formation of the various methylmetal cations have been measured. Hobrock and Kiser (337) found the observed ionization potential for tetramethyllead to be 8.0 eV versus 8.25 for tetramethyltin and 9.8 for tetramethylsilane. On the other hand, Fraser and Jewitt (240) reported a value of 11.5 eV for the ionization potential of tetramethyllead; a value of 12.5 eV was reported by the same workers for the ionization potential of tetraethyllead. Ionization potentials of the free methyl and ethyl radicals were calculated to be 11.2 and 10.6 eV, respectively. A mass spectrographic procedure for the determination of tetramethyllead and tetraethyllead has been developed (350).

The decomposition of tetraethyllead has been conducted in a glow discharge by Prilezhaeva (576) and its *emission spectrum* determined. Also, the absorption spectra arising from the flash photolysis of tetraethyllead, tetramethyllead, and deuterated tetramethyllead have been determined in the presence of various gases (138,218). No shift in the resonance line of lead was obtained, only unsymmetrical broadening (218). Two types of foreign gases were defined: (*1*) those which cause diffuse line broadening to red and (*2*) those which cause diffuse line broadening to blue.

Electron diffraction measurements have been made on tetramethyllead by Brockway and Jenkins (79). As was discussed earlier, a value of 2.29 ± 0.05 Å was calculated for the lead–carbon bond distance. However, a value of 2.203 ± 0.010 Å has been calculated more recently by Wong and Schomaker (768) and the Pb—C—H bond angle was calculated to be 109 ± 4°. Wong and Schomaker's value has become the more generally accepted preferred value for the lead–carbon bond distance in organolead compounds.

Several *X-ray investigations of the crystal structure* of tetraphenyllead have been reported (242,254,257,360,571,779,780). According to George (254), the single crystal photographs of the tetraphenyl derivatives of all the Group IVb metals are identical, except for their intensity and small dimensional variations. Some disagreement exists concerning the number (2 or 4) of tetraphenyllead molecules per unit cell (254,257). The angles of rotation of the molecules around the tetragonal axis and of the phenyl groups around the carbon-to-metal bond line have been calculated (360,779,780). The crystal structures of tetra-*m*-tolyllead and -tin have also been determined. The basic prisms of both compounds are tetragonal and are concluded to contain four molecules per unit cell (465,727).

C. Chemical properties

As was discussed earlier in this section, the tetraorganolead compounds are the most reactive of such derivatives of all the Group IVb metals. This results from the low strength of the lead-to-carbon bond relative to the strength of the metal-to-carbon bond for the other Group IVb metals. Thus, the rate of cleavage of tetraphenyllead by hydrogen chloride is some sixty times faster than that of tetraphenyltin (185); in aqueous solutions, the ratios of the rates of cleavage of aryllead bonds versus aryltin bonds by mineral acids is even greater (209). The high reactivity of tetraorganolead compounds is evidenced by their reactions with both electrophilic and nucleophilic reagents; in general, they are much more reactive with electrophilic than with nucleophilic reagents. In most cases, these reactions proceed readily under mild conditions and can be controlled to produce selective cleavage of one or two lead-to-carbon bonds. Cleavage of more than two lead–carbon bonds generally requires more stringent conditions and, because of the instability of $RPbX_3$-type compounds, usually results in the formation of the divalent lead salt, PbX_2.

Tetraorganolead compounds react with a wide variety of compounds, as is described in the following sections. Because of the relatively low strength of the lead-to-carbon bond, and the tendency of the organolead halides to undergo disproportionation type reactions, the compositions of the organolead products in these reactions are strongly dependent on reaction stoichiometry and temperature. With few exceptions, detailed investigations of the reaction parameters have not been undertaken for most of the reactions described herein. This should be considered in evaluating the products obtained; under different conditions, different lead-containing products could result.

A review of the chemical properties of tetraethyllead was published by Milde and Beatty (488) in 1959. A similar review was published in 1946 in the Russian literature by Korshak and Kolesnikov (400).

(*1*) *Pyrolysis*

The thermal decomposition of R_4Pb to lead metal and free radicals is well known from Paneth's classical experiments which demonstrated the existence of free radicals. The thermal decomposition of tetraorganolead compounds is reversible; lead metal, usually in the form of a mirror, will combine with alkyl or aryl free radicals to form the R_4Pb compound. The pyrolysis of tetraorganolead compounds is a convenient way to generate free radicals in investigations of their chemistry. Most tetraorganolead compounds decompose at relatively low temperatures; however, the thermal decomposition of only tetramethyllead, tetraethyllead, and tetraphenyllead has been investigated in any detail.

Pure tetraethyllead begins to decompose at temperatures slightly below 100°; at 100° its rate of decomposition is about 2% per hour (103). Tetramethyllead and tetraphenyllead decompose at an appreciably higher temperature; in a static system Simons and co-workers (675) found that the decomposition of tetramethyllead begins at about 265° while the thermal decomposition of pure tetraphenyllead begins at temperatures slightly above 250° (200,407,778).

Tetramethyllead is the only tetraalkyl derivative which can be distilled at atmospheric pressure. The higher alkyls are too unstable and decompose at their atmospheric boiling points so that a vacuum or steam distillation is required. Even tetramethyllead is best distilled *in vacuo* or with steam;

unexpected explosions have occurred during its distillation at atmospheric pressure possibly because of reaction with air (270,419). Rapid, autocatalytic decomposition of tetra-*n*-butyllead has occurred at a temperature of about 100° during vacuum distillation (765).

The pyrolysis of tetraethyllead has been investigated extensively by several workers (169,439,481,575,602–604,725). Decomposition proceeds by a first order process; several rate expressions have been derived:

$$k = 1.2 \times 10^{12} e^{-36,900/RT}\ \text{sec}^{-1} \qquad (439)$$

$$k = 4 \times 10^{12} e^{-37,000/RT}\ \text{sec}^{-1} \qquad (575)$$

$$k = 2.43 \times 10^{12} e^{-35,200/RT}\ \text{sec}^{-1} \qquad (602\text{–}604)$$

Lead metal, ethane, ethylene, and butane are formed as major products, along with varying amounts of higher hydrocarbons depending on conditions. The lead metal serves as a decomposition catalyst, so that the reaction is autocatalytic.

Pratt and Purnell (575) investigated the *vapor phase* decomposition of tetraethyllead at 252° using gas chromatographic techniques to identify the organic products formed at all stages of decomposition. In the initial stages, only four products were detected: ethane, ethylene, *n*-butane, and hydrogen. The amounts of ethane and butane formed relative to the fraction of tetraethyllead decomposed were shown to be constant throughout the decomposition. On the other hand, the amount of hydrogen formed relative to the fraction of tetraethyllead decomposed decreased as decomposition progressed, while the amount of ethylene formed increased at first but levelled off at about the 50% stage and then showed no further increase. Plots showing the relative amounts of these four products at progressive stages of decomposition are shown in Figure VI-1.

At later stages of decomposition, a more complex product mixture is obtained. A total of 17 chromatographic peaks, corresponding to hydrocarbons in the C_1–C_6 range, was found; the formation of higher hydrocarbons was not excluded, but they could not be detected with the equipment used. A similar complex mixture of products, including liquid hydrocarbons, was also obtained by Meinert (481) from the pyrolysis of tetraethyllead. These higher hydrocarbons probably arise from secondary reactions involving radical additions to ethylene; hence, the amount of ethylene in the product mixture reaches a constant value in the later stages of the reaction. Pratt and Purnell proposed the following reactions to account for the products formed:

(*a*) $(C_2H_5)_4Pb \rightarrow C_2H_5^0 + (C_2H_5)_3Pb^0$

(*b*) $C_2H_5^0 + (C_2H_5)_4Pb \rightarrow C_2H_6 + C_2H_4Pb(C_2H_5)_3$

(*c*) $2C_2H_5^0 \rightarrow \text{n-}C_4H_{10}$

(*d*) $2C_2H_5{}^0 \rightarrow C_2H_6 + C_2H_4$

(*e*) $C_2H_5{}^0 + (C_2H_5)_nPb \rightarrow C_2H_4 + HPb(C_2H_5)_n$

(*f*) $HPb(C_2H_5)_n \rightarrow H^0 + Pb(C_2H_5)_n$

(*g*) $H^0 + (C_2H_5)_4Pb \rightarrow H_2 + C_2H_4Pb(C_2H_5)_3$

From the observed orders of ethylene and hydrogen formation, hydrogen formation through such reactions as

$$C_2H_5{}^0 \rightarrow C_2H_4 + H^0$$

or

$$2C_2H_5{}^0 \rightarrow 2C_2H_4 + H_2$$

was ruled out.

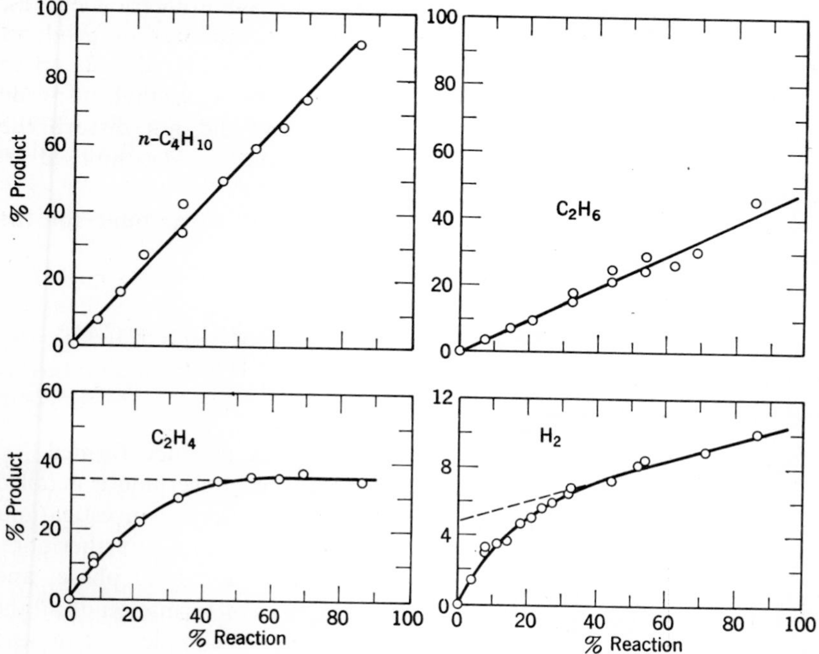

FIG. VI-1 Thermal decomposition of tetraethyllead. The variation of product yield (as % of initial TEL pressure) with extent of reaction (% TEL decomposed) for the four initial products at 252.2° for 11.0 mm Hg initial pressure of TEL. [From Pratt and Purnell, *Trans. Faraday Soc.*, **60,** 519 (1964).]

It is interesting to note that one of the organolead intermediates postulated by Pratt and Purnell has the structure $(C_2H_5)_3PbC_2H_4$ (in equation *b* above). Hedden (324) has isolated a product having a composition which corresponds to the dimer of this free radical; thus, $(C_2H_5)_3PbCH(CH_3)$-$CH(CH_3)Pb(C_2H_5)_3$ has been isolated as a heavy residue from the distillation of tetraethyllead prepared via the sodium–lead alloy–ethyl chloride reaction.

A second intermediate proposed by Pratt and Purnell is an ethyllead hydride of undefined composition, $HPb(C_2H_5)_n$. Triethyllead hydride and trimethyllead hydride have been prepared and their mode of decomposition determined. Both decompose below room temperature according to (38,198):

$$4R_3PbH \rightarrow 3R_4Pb + Pb + 2H_2$$

The *liquid phase* pyrolysis of tetraethyllead over the temperature range 105–135° was investigated by Razuvaev and colleagues (602,604) and the organolead intermediates determined by spectrophotometric methods. Hexaethyldilead and diethyllead were identified as intermediate products of decomposition. Hexaethyldilead was formed initially and built up to a maximum concentration of about 19% in the tetraethyllead; the addition of hexaethyldilead to the tetraethyllead did not disturb the reaction kinetics. The hexaethyldilead subsequently decomposed to diethyllead, which in turn decomposed to lead metal.

The following reaction sequence was proposed to account for the spectrophotometric data:

$$(a)\ (C_2H_5)_4Pb \rightarrow C_2H_5^0 + (C_2H_5)_3Pb^0$$

$$(b)\ (C_2H_5)_4Pb + (C_2H_5)_3Pb^0 \rightarrow (C_2H_5)_3PbPb(C_2H_5)_3 + C_2H_5^0$$

$$(c)\ (C_2H_5)_3PbPb(C_2H_5)_3 \rightarrow (C_2H_5)_4Pb + (C_2H_5)_2Pb$$

$$(d)\ (C_2H_5)_2Pb \rightarrow Pb + 2C_2H_5^0$$

No attempt was made to account for the organic products formed.

It is not possible to draw any firm conclusions concerning a detailed mechanism of tetraethyllead decomposition because the two investigations discussed above were conducted at different temperatures. Furthermore, one involved vapor phase decomposition, the other liquid phase, and neither investigation attempted to follow the rates of formation of all the products. Pratt and Purnell have postulated more complex organolead intermediates than Razuvaev. In any case, one point of agreement is the conclusion that the initial step involves the cleavage of one lead–carbon bond to form a triethyllead radical and a free ethyl radical. Any conclusions concerning the fate of these initial products are clouded by the complex reactions which ethyl radicals can undergo. Because of the greater simplicity of reactions of free methyl radicals, it would seem that tetramethyllead would be more suitable for investigation of the mechanism of pyrolysis of a tetraalkyllead compound. The thermal decomposition of tetramethyllead has been investigated, but not in such detail as that of tetraethyllead.

The thermal decomposition of tetraethyllead in the presence of oxygen exhibits an induction period, which decreases with increasing oxygen pressure (3). In the presence of excess oxygen, the formation of hexaethyldilead and lead metal is completely suppressed; instead, lead oxides and alkyllead hydroxides are formed. The oxidation of tetraethyllead is discussed later.

The pyrolysis of tetramethyllead has been investigated by Simons and co-workers (675) in both static and flow systems. In the static system decomposition began at a temperature of about 265°; in the flow system, much higher temperatures were required. Ethane was the major organic product, although some methane was also formed. Methane formation was ascribed to extraction of hydrogen from ethane, and not from undecomposed tetramethyllead (449). Ethylene, hydrogen, propylene, isobutylene, and acetylene were also formed in varying amounts, depending on pyrolysis temperature; carbon deposition occurred at the lower pyrolysis temperatures. Two distinct types of reactions were postulated: (*1*) a comparatively slow wall reaction at the lower temperatures and (*2*) a fast, homogeneous gas phase reaction at higher temperatures. The presence of hydrogen during pyrolysis increased methane formation. The thermal decomposition reaction obeys first-order kinetics; the rate equation has been determined by Eltenton (217) to be:

$$k = 1.5 \times 10^{10} e^{-28,000/RT} \text{ sec}^{-1}$$

whereas Romm (626) has calculated an activation energy of 23,500 cal/mole from an investigation of the polymerization of olefins initiated by methyl radicals generated by pyrolysis of tetramethyllead.

The thermal stabilities of tetramethyllead, tetraethyllead, and the unsymmetrical methyl-ethyl derivatives have been investigated in a single pulse shock tube over the temperature range 731–931° (632). The decreasing order of thermal stability was found to be: tetramethyl > trimethylethyl > dimethyldiethyl = methyltriethyl = tetraethyl; the order of stability was claimed to be relatable to the antiknock activity. The pyrolysis of tetraheptyllead has also been studied, but only in the presence of powdered nickel metal and hydrogen; at 200° and 125 atm of hydrogen tetradecane (62%) was obtained as the major product along with some heptene and heptane (775).

Rifkin and Walcutt (621), in an investigation of the mechanism of the antiknock action of organolead compounds, demonstrated that the decomposition of tetraethyllead in a fuel–air mixture in a single cylinder engine obeys first-order kinetics. Rifkin (619) also investigated the thermal decomposition of tetramethyllead, tetraisopropyllead, and isobutyltriethyllead under similar conditions. At temperatures below about 450°, the relative rates of decomposition were in the order: tetraisopropyl > tetraethyl > isobutyltriethyl > tetramethyl. A plot of the rate

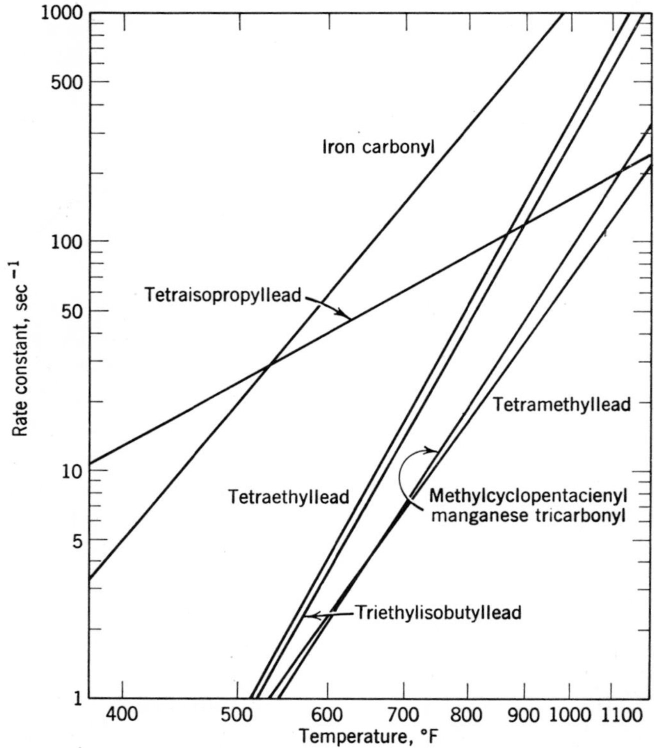

FIG. VI-2 Decomposition rates for several antiknock compounds. Data obtained from motored engines. Taken from Rifkin, 23rd Mid Year Meeting American Petroleum Industries, Los Angeles, California, May, 1958.

constants versus temperature is shown in Figure VI-2, which also includes data on the thermal decomposition of two other antiknock compounds, iron pentacarbonyl and methylcyclopentadienyl manganese tricarbonyl. It is interesting to note that at temperatures above about 875°F the rate of decomposition of tetraisopropyllead becomes less than that of the tetraethyl and isobutyltriethyl analogs.

The pyrolysis of tetra-*n*-propyllead and tetraisobutyllead was studied by Paneth and co-workers using the metal mirror technique to study the fate of the free radicals formed. The pyrolysis of tetraisobutyllead at about

300° in the presence of a lead mirror gave a yellow liquid which decomposed above 100° with separation of metallic lead and formation of minute amounts of a liquid which was not volatile at low pressures at room temperature (550,552). The product obtained from pyrolysis of tetrapropyllead in the presence of an antimony mirror was mostly tetramethyldistibine (552). Hence it was concluded that any propyl and isobutyl radicals formed underwent disproportionation to radicals of the lower alkyls. On the other hand, methyl and ethyl radicals were detected in the pyrolysis of tetramethyllead (548,549,562) and tetraethyllead (550), respectively.

The pyrolysis of tetracyclohexyllead in benzene at 160° has been investigated by Kaplin and co-workers (378). Cyclohexane was formed as the major product, along with lesser amounts of cyclohexene, dicyclohexyl, phenylcyclohexane, and condensed products containing both phenyl and cyclohexyl groups; photolysis gave similar results. Cyclohexane was also formed as the major product in the pyrolysis and photolysis of hexacyclohexyldilead. It was concluded that all of these reactions involved a free cyclohexyl radical.

The thermal decomposition of tetraphenyllead has been investigated by several workers, but no kinetic data have been obtained (200,442,552). According to Dull and Simons (200) tetraphenyllead decomposed at temperatures slightly above 250° to form lead metal and biphenyl exclusively; at higher temperatures, biphenyl was formed as the major product, along with some *p*-diphenylbenzene (at ~345°) and benzene (at ~400°). On the other hand, Leighton and Mortenson (442) obtained biphenyl as the only organic product of pyrolysis of tetraphenyllead even at 400°.

Most of the investigations of the pyrolysis of tetraphenyllead have been centered on the effect of hydrogen and finely divided metals on its thermal stability. Several finely divided metals have been shown to catalyze the decomposition of tetraphenyllead. Lead metal greatly accelerates the decomposition, so that the reaction becomes autocatalytic (407). In the presence of mercury, decomposition occurs at 215–220° with the formation of diphenylmercury; this has been cited as original evidence for the formation of free phenyl radicals in the decomposition of tetraphenyllead (199).

Koton (405,406) has shown that other finely divided metals also catalyze the thermal decomposition of tetraphenyllead at temperatures as low as 150°. The decreasing order of relative effectiveness of a series of metals was found to be Pd > Au > Ag > Ni; biphenyl was the only organic product formed. Similar results were obtained when the decomposition was conducted in ethanol.

Ipatieff and his co-workers (255,357–359) investigated the thermal decomposition of tetraphenyllead under hydrogen pressure and observed decomposition at temperatures as low as 200°; lead metal, benzene, and traces of biphenyl were formed. On the other hand, Zartman and Adkins (775) found tetraphenyllead to be stable at 200° under 125 atm of hydrogen in the absence of a metal catalyst.

Koton (406) and Razuvaev (595) also investigated the decomposition of tetraphenyllead under hydrogen pressure and in the presence of the above-mentioned metal powders. Decomposition occurred at temperatures as low as 150° and either benzene or biphenyl was formed as the major product depending on the metal powder present. Nickel powder gave mostly benzene and traces of biphenyl, while palladium gave exclusively biphenyl; gold gave a mixture of benzene and biphenyl (405). The relative order of effectiveness of these metals as catalysts for the decomposition of tetraphenyllead under hydrogen pressure was found to be Pd > Ni > Au > Ag > Cu. There does not appear to be an obvious relationship between the nature of the organic products and the effectiveness of the metal powder. Contrary to the above results, Zartman and Adkins (775) reported that the pyrolysis of tetraphenyllead at 200° under 100 atm of hydrogen pressure and in the presence of nickel metal gave biphenyl exclusively; 4,4′-dimethyl-biphenyl was obtained upon pyrolysis of tetra-*p*-tolyllead under identical conditions (775).

Ipatieff (357) has shown that hydrogen facilitates the decomposition of tetraethyllead and tetramethyllead. Tetraethyllead in benzene solution was 86% decomposed after 24 hr at 200° under 60 atm of hydrogen; in the absence of hydrogen, only 11% decomposition was obtained under identical conditions. Tetramethyllead was 89% decomposed under the above conditions under hydrogen pressure; in the absence of hydrogen, no decomposition occurred (357,358). In the absence of benzene and hydrogen, complete decomposition of the tetraethyllead would occur within a few minutes at 200°; thus is demonstrated the stabilizing effect of benzene and other aromatic compounds on the thermal decomposition of tetraethyllead. The thermal stabilization of tetraalkyllead compounds is discussed later in this chapter.

The pyrolysis of tetraethyllead under both hydrogen and ethylene pressure gave an oil which resulted from a free radical polymerization of the ethylene (169,725). The hydrogen participated in the reaction, but only to a very limited extent. Induced reactions with various paraffins and olefins were also observed in the pyrolysis of tetraethyllead at 200–300° (170); these induced reactions were concluded to be initiated by free ethyl radicals produced by the tetraethyllead. It was also concluded that the influence of the type of hydrocarbon and its molecular structure could be related to the relative amounts and reactivities of the primary, secondary, and tertiary hydrogen atoms in the molecule. No induced reactions were obtained with aromatic hydrocarbons.

The exchange of phenyl groups between benzene and tetraphenyllead or diphenylmercury has been investigated by means of C^{14} and deuterium labelling. Korshunov and Orlova (404) found that no exchange occurred with tetraphenyllead under conditions of photolysis, although appreciable exchange occurred between diphenylmercury and benzene under comparable conditions. On the other hand, Nazarova and co-workers (521) observed slight exchange between tetraphenyllead and benzene under conditions of both photolysis and pyrolysis; 0.75% exchange was obtained when tetraphenyllead and labelled benzene were heated at 200° for 20 hr. Similarly, Razuvaev and co-workers (598) found that the C^{14} activity of the tetraphenyllead was 1.7% that of the original labelled benzene after heating a mixture of the two at 240–250° for 180 hr; photolysis gave about 0.5% exchange after irradiation with ultraviolet light for 180 hr. The formation of a complex having the composition $C_6H_6 \cdot 7(C_6H_5)_4Pb$ was detected (521). If such a complex were indeed formed, only trace amounts of the complex in the recovered tetraphenyllead would lead to the conclusion that slight exchange had occurred. Similar complexes were obtained for the silicon and tin analogs.

Razuvaev and colleagues (599) also investigated the photolysis of tetraphenyllead in benzene using deuterium labelling. Photolysis in deuterated benzene gave C_6H_5D along with all three possible biphenyl derivatives: $C_6H_5 \cdot C_6D_5$ (94.7%), $C_6H_5 \cdot C_6H_5$ (4.3%), and $C_6D_5 \cdot C_6D_5$ (0.7%). Similar results were obtained using deuterium-labelled tetraphenyllead in ordinary benzene. The formation of $C_6H_5 \cdot C_6D_5$ as the major biphenyl derivative was interpreted as evidence that these decomposition reactions proceed via a homolytic process involving a complex of the type $(C_6H_5)_4Pb \cdot 2C_6H_6$ as the active intermediate. However, no evidence was given to support the existence of a complex of this composition.

The radiolysis of tetraphenyllead, and other Group IVb metal–phenyl compounds, has been investigated in benzene solution using ^{60}Co γ-radiation to determine the quenching effect of these compounds in organic scintillator solutions (428,566). The lead and tin derivatives underwent radiosensitized decomposition and gave biphenyl, phenylcyclohexadienes, and tetraquaterphenyls as decomposition products (566); this contrasts with the results from simple thermal or photolytic decomposition discussed earlier in which biphenyl is formed almost exclusively. The phenyl–metal compounds did not affect the radiolytic decomposition of the benzene solvent in which hydrogen and acetylene are formed. Unsymmetrical phenyllead derivatives, such as *p*-ethylphenyltriphenyllead, have also been investigated for their effectiveness for γ-photopeak resolution in liquid and plastic scintillators; this is discussed in Chapter XVII. Tetraphenyllead is reported to reduce slightly the radiolytic decomposition of cyclohexane by ^{60}Co γ-radiation (171).

Complete exchange between tetraphenyllead and lead metal in benzene has been demonstrated using lead metal containing Pb^{210} (Radium D) (40); a similar exchange has been shown to occur with tetraethyllead (253) and tetramethyllead (201a, 441a). With tetraethyllead, maximum exchange was obtained after 40 hr at 125°. Undoubtedly, these exchange reactions proceed via decomposition of the R_4Pb compound followed by attack of the free radicals so produced on the Radium D to reform R_4Pb. No exchange occurred between tetraethyllead and radioactive lead sulfide, sulfate, or oxide (253).

Tetramethyllead labelled with C^{14} has been prepared by heating a mixture of tetramethyllead and labelled methyl iodide in trimethylamine at temperatures up to 260° (714). Exchange was greatest at the higher temperatures but recoveries of tetramethyllead were lowest. Thus, at 260°, 14.1% exchange was obtained in 2 hr with a 60% recovery; at 200°, only 0.13% exchange was obtained in 12 hr with an 84.7% recovery. In all probability the exchange proceeds via formation of lead metal by thermal decomposition of the tetramethyllead, followed by reaction of the resultant lead metal with the labelled methyl iodide.

Perfluoromethyltrimethyllead and perfluoroethyltrimethyllead, two of the three perfluoroalkyl derivatives prepared to date, were synthesized by the reaction of tetramethyllead with perfluoroalkyl iodide at elevated temperature or under influence of ultraviolet irradiation (375,376). However, a better yield of the perfluoromethyl derivative was obtained using hexamethyldilead in place of tetramethyllead (376).

Korshunov and co-workers (401–403) found that tetraethyllead and ethyl bromide do not undergo ethyl group exchange in the absence of a catalyst or solvent; no exchange was obtained in 20 hr at 140° or under influence of ultraviolet irradiation. A decomposition reaction did occur as evidenced by the formation of ethane and ethylene containing 40–45% of the original C^{14} activity of the labelled ethyl bromide. However, ethyl group exchange was obtained when certain catalysts and solvents were used; nitromethane, dimethylformamide, ferric chloride, aluminum bromide, triethylaluminum, and aluminum chloride promoted exchange, while silica gel, cobalt(II) chloride, silver metal, and phosphorus trichloride were not effective. In dimethylformamide or nitromethane solution, both silver metal and phosphorus trichloride also promoted exchange, while cobalt(II) chloride, if it had any effect, inhibited the exchange reaction. Of the catalysts tested, the aluminum compounds were the most effective, in the relative order $AlCl_3 > (C_2H_5)_3Al > AlBr_3$. The activity of triethylaluminum is surprising since triethylaluminum and ethyl bromide do not undergo ethyl group exchange in the absence of a catalyst (27). Also, tetraethyllead and isopropyl chloride did not undergo exchange in the presence of aluminum chloride under conditions of reflux, while tetraethyltin reacted with isopropyl chloride under similar conditions to give triethyltin chloride (600).

A six-membered cyclic complex has been proposed to be the active intermediate in these exchange reactions of tetraethyllead and ethyl bromide:

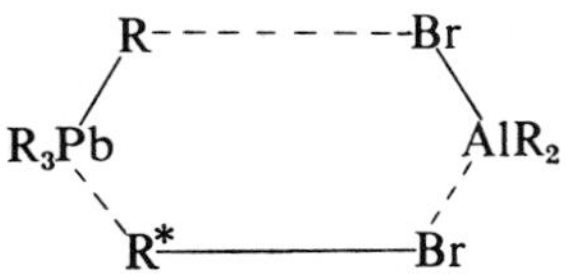

The main effect of complex formation is believed to be polarization of the ethyl bromide to weaken the carbon–bromine bond. Support for this theory rests in the finding that diethyl ether inhibits the exchange of triethylaluminum with ethyl bromide in the presence of metal halides; coordination of the ether to the triethylaluminum would lower the interaction between the triethylaluminum and ethyl bromide so that the active complex is not formed and exchange is inhibited. Polar solvents, such as nitromethane or dimethylformamide, would also effect polarization of the ethyl bromide and thereby promote exchange. The finding that halide exchange also occurs between the metal halide catalyst and ethyl bromide has been offered as additional support for the formation of a cyclic intermediate (27). However, any detailed interpretation of the mechanism of these exchange reactions is necessarily clouded by (*1*) a chemical reaction can occur between tetraethyllead and the metal halide catalyst and (*2*) under the conditions used, the tetraethyllead is pyrolyzed to form free ethyl radicals which can then undergo secondary reactions with the ethyl bromide.

Razuvaev and co-workers (591) have shown that reaction of tetraethyllead and ethyl bromide at 150° for 4 hr produced triethyllead bromide (0.8% yield) and lead(II) bromide (3.3%), along with ethane, ethylene, and butane. Mishima (493) has found that tetraethyllead reacts with ethyl chloride at 100–120° to form triethyllead chloride; irradiation of an ice-cold solution of tetraethyllead and ethyl chloride with sunlight also gave triethyllead chloride.

The reaction of tetraethyllead with benzyl bromide and various dibromoalkanes at 135–180° has been investigated by Razuvaev and co-workers (591). Dibromoalkanes were found to be more reactive than the monobromides; benzyl bromide was especially unreactive at these temperatures. The reactivity of the dibromo compounds decreased with increasing separation of the bromine atoms by carbon atoms; dibromoethane was the most reactive compound of those tested. In most cases, ethane and ethylene were formed as the major gaseous products although only propylene was evolved from 1,2-dibromopropane. Triethyllead bromide and lead(II) bromide were formed as the lead-containing products; some diethyllead dibromide may have also been formed, but it would undergo thermal disproportionation at these temperatures to form lead bromide and triethyllead bromide.

The reactions of tetraethyllead with alkyl bromides are catalyzed by light and retarded by free radical inhibitors; initial formation of a triethyllead radical is postulated, followed by abstraction of bromine from the organic bromide (690). Haloalkanes greatly enhance the thermal stability of tetraethyllead (604,749). Ethylene dibromide and chloride are added to tetraorganolead compounds marketed as antiknock fluids; the dihalides serve as thermal stabilizers during the handling, shipment, and storage of these fluids, as well as agents which help scavenge the lead from the combustion chamber of the engine.

The thermal decomposition of tetraethyllead in the presence of ethylene dibromide and gasoline has been investigated at elevated pressures in the presence of air (764). Various ratios of lead oxide, lead(II) bromide and triethyllead bromide were obtained, dependent on temperature and pressures. At temperatures up to 100° and pressures between 20–60 kg/cm^2, lead oxide was obtained as the major product; at higher temperatures, formation of lead bromide was favored. On the other hand, at temperatures below 100° and pressures above 60 kg/cm^2, triethyllead bromide was formed.

Tetraethyllead in light oil distillates does not promote poisoning of various catalysts used in the reforming reaction of natural gas and water vapor (744). Similarly, tetramethyllead is not a poison to platinum catalysts used for the hydrogenation of tetraphenylsilane to tetracyclohexylsilane (685), but it is reported to be a poison to platinum catalysts in the hydrogenation of cyclohexene (474).

The presence of tetraethyllead in the vapor phase reaction between acetylene and aniline is reported to increase the yield of indole by about 250% (348). Also, Pratt and Purnell (574) have shown that pyrolysis of a mixture of tetraethyllead and nitric oxide at 233–267° yields acetonitrile and water. In this latter system, the most important reaction is postulated to be the reaction of nitric oxide with ethyl radicals from the tetraethyllead to form acetaldoxime, which then decomposes in an entirely heterogeneous process to acetonitrile and water. Rate data calculated for the thermal decomposition of tetraethyllead, based on the rate of formation of acetonitrile, agreed well with data obtained by these same investigators for the decomposition of tetraethyllead in the absence of a second reagent (575).

The pyrolysis of tetraethyllead at 575° in the presence of carbon tetrabromide has been investigated in an attempt to gain some insight into the interaction of tetraethyllead and lead oxides with halocarbons during combustion in a gasoline engine (444). The deposition of lead-containing coatings by pyrolysis of organolead compounds is described in two patents (356,533); lead telluride has been deposited by co-pyrolysis of tetraethyllead and diethyl telluride (158). The pyrolysis of tetraethyllead in a silica tube has been used to deposit a lead glass skin on the tube (64).

The pyrolysis and photolysis of tetramethyllead have been unsuccessfully employed in an attempt to trap methyl radicals at 4°K in order to obtain their infrared spectrum (456). The pyrolysis of tetramethyllead and tetraethyllead has found wide application in investigations of free radical chemistry (490,546–552,562,616,761).

(*2*) *Photolysis*

Tetraorganolead compounds, except those containing *sec*-alkyl groups, are fairly stable to light. The compounds can be stored for extended periods in brown glass bottles without excessive decomposition if air is rigorously excluded from the system. However, irradiation with ultraviolet light causes the decomposition of tetraorganolead compounds to lead metal and free organic radicals; the continuous absorption exhibited by R_4Pb compounds in the ultraviolet region has been ascribed to photolytic decomposition.

Reports of photolytic reactions of R_4Pb derivatives are scattered through the organolead literature, but few systematic investigations have been made. Results from the photolysis of tetraethyllead, tetraphenyllead, and tetracyclohexyllead were discussed above; generally, photolysis produces much the same results as pyrolysis.

The photolysis of tetramethyllead and tetraphenyllead in the vapor and in solution has been investigated by Leighton and Mortensen (442). Photolysis of tetramethyllead in solution gave quantum yields of 0.2–0.4 versus a quantum yield of 1.11 for vapor photolysis. The high quantum yield for vapor photolysis was ascribed to involvement of short chains or an occasional second molecule in the reaction; the lower quantum yields in solution were attributed to deactivation or recombination. The presence of oxygen in the tetramethyllead vapor lowered the quantum yield, the value of unity being approached as oxygen pressure was increased; this was attributed to the reaction of oxygen with methyl radicals to retard their recombination with lead metal or their involvement in short chains. Similar results were obtained from the photolysis of tetraphenyllead in solution, the quantum yield being almost identical to that obtained with tetramethyllead in solution.

The flash photolysis of tetramethyllead and tetraethyllead at low pressures has been investigated (138,145,218). Photolysis of tetramethyllead at an energy of 2500 J gave complete decomposition in 1.2 msec and a quantum yield of 40; the products consisted of more than 90% ethane along with a trace of methane. It was concluded that a chain reaction occurred which involved attack of methyl radicals on the tetramethyllead. Similar ıesults were obtained with deuterated tetramethyllead (138).

McDonald, Blair, and co-workers (476) found that the photolysis of cumene in the presence of tetraphenyllead gave a different ratio of isomeric isopropyldiphenyls than photolysis of cumene in the presence of iodobenzene. The yields of benzene and 2,3-dimethyl-2,3-diphenylbutane were interpreted to indicate that phenyl radicals from the tetraphenyllead preferentially attacked the side chain of the cumene. Irradiation of tetramethyl-, tetraethyl-, and tetrapropyllead with gamma rays and X-rays has been conducted and their resonance spectra compared (680); it was concluded that the dominant spectra obtained from the methyl and ethyl compounds did not correspond to that of ethyl and methyl radicals. The irradiation of tetraphenyllead and the tetraphenyl derivatives of the other Group IVb metals with neutrons has been investigated (134). The yields of the different radioactive organometallic compounds were similar, indicating that the mechanism of their formation was similar.

The effect of tetraethyllead on the relative formation of the various isotopes of bromine during bombardment of ethyl bromide with slow neutrons has been investigated by Karamyan (379). The amounts of Br^{80} (4.4 hr half-life) and Br^{82} formed were not affected by the presence of tetraethyllead, but the amount of Br^{80} (18 min half life) formed in the presence of tetraethyllead was about five times greater than in the pure ethyl bromide. This latter isotope was believed to be of secondary origin.

The β decay of tetraphenyllead containing Pb^{210} (Radium D) to triphenylbismuth and of tetramethyllead to trimethylbismuth has been investigated by several workers (1,201a,214,215,502,505) in an attempt to define the mechanism of reactions involving nuclear transitions. These reactions are believed to involve an internal conversion process, and not bond rupture in the primary β decay process.

(3) *Reactions with halogens, halogen acids, and other reactive halides*

Tetraorganolead compounds react readily with halogens, even at temperatures as low as $-78°$ to yield organolead halides. Reaction with chlorine and bromine is very vigorous and must be conducted in a diluent. By variation of the temperature and solvent, either the R_3PbX or R_2PbX_2 derivative is formed as the major product. These reactions are commonly used for the synthesis of organolead halides and are discussed further in Chapter XI under Organolead Salts. Reaction can be carried beyond the R_2PbX_2 stage by use of more stringent conditions, but this results in complete decomposition to PbX_2.

The reaction of tetraethyllead with halogens represents a common method for the decomposition of tetraethyllead for its determination in gasoline; these are discussed in Chapter XVI. A convenient iodometric method has been developed for the determination of tetraethyllead in gasoline or in solution. Krause (418) reported in 1921 that the reaction of hexacyclohexyldilead with iodine proceeds quantitatively at room temperature in benzene to form tricyclohexyllead iodide and noted that titration with iodine could be used for the determination of tetracyclohexyllead.

In recent years, several papers have been published on the kinetics of reaction of various tetraalkyllead compounds with bromine and iodine in various solvents, leading to a better understanding of the mechanism of such reactions (184,258,261,611,612). The results of these investigations are presented later in a discussion of Reactions of Unsymmetrical Tetraorganolead Compounds. Reactions of tetraorganolead compounds with halogens obey second-order kinetics and are concluded to involve an electrophilic attack at a saturated carbon atom.

Tetraorganolead compounds undergo facile reaction with anhydrous halogen acids to yield triorganolead halide, diorganolead dihalide, or lead(II) halide, dependent on solvent and temperature. These reactions are very useful for the synthesis of organolead halides; they are discussed in greater detail in Chapter XI.

Tetraorganolead compounds also react with aqueous solutions of halogen acids to yield organolead halides. These reactions tend to be slower than reactions with the anhydrous acids because of the low solubility of the tetraorganolead compound in the aqueous system but nevertheless proceed at reasonable rates. The reaction of tetraethyllead with concentrated hydrochloric acid at 30–33° gave triethyllead chloride (81); reaction of phenyl-tri-*o*-tolyllead with a 40% solution of hydrobromic acid gave phenyl-di-*o*-tolyllead bromide (13). At elevated temperatures and longer reaction times, these reactions can be made to proceed to complete decomposition to PbX_2.

Tetraalkyl- and tetraaryllead compounds are cleaved readily by other reactive inorganic halides. Facile reaction occurs with disulfur dichloride (157,452), thionyl chloride (250,543), and sulfuryl chloride (543) under very mild conditions. Thus, Lutz (452) obtained triethyllead chloride from the reaction of tetraethyllead and sulfur chloride at room temperature. The reaction proceeded according to:

$$2(C_2H_5)_4Pb + S_2Cl_2 \rightarrow 2Et_3PbCl + C_2H_5SSC_2H_5$$

Gelius (250) has reported that thionyl chloride reacts at room temperature or below with equimolar amounts of tetraalkyllead compounds (methyl, ethyl, and propyl) to form the trialkyllead chloride as the major product and lesser amounts of dialkyllead dichloride and lead chloride. Padberg (543) obtained only diethyllead dichloride from the reaction of tetraethyllead and thionyl chloride in benzene, using 2 moles of thionyl chloride per mole of tetraethyllead. Diaryllead dichlorides were obtained by Gelius as the major product from the reaction of tetraaryllead and thionyl chloride in refluxing benzene.

Diethyllead dichloride and diphenyllead dichloride were formed in the reaction of sulfuryl chloride with tetraethyllead and tetraphenyllead, respectively (543); substituted sulfuryl chlorides of the type RSO_2Cl also react with tetraorganolead compounds, but with less vigor than thionyl chloride (250). Sulfuryl chloride and Dichloramine-T are reported to be effective reagents for the decontamination of wood which has been in contact with tetraethyllead (708).

Iodine trichloride reacts readily with tetraethyl- and tetraphenyllead (469); the products are dependent on stoichiometry and conditions. In refluxing chloroform, diphenyllead dichloride and triphenyllead chloride were obtained as the major products along with iodobenzene; benzene was also formed in trace amounts. A method has been developed for the determination of tetraethyllead in gasoline which is based on its decomposition with iodine trichloride to lead chloride (219). Tetraethyllead also undergoes facile reaction with iodine monochloride to from diethyllead dichloride (363,503,504). This reaction has been employed as a means of determining tetraethyllead in air (503).

Tetraalkyl- and tetraaryllead compounds also undergo reaction with ammonium and alkyl ammonium halides at elevated temperatures. Krause and Schlöttig (421) isolated triphenyllead chloride from the reaction of tetraphenyllead and ammonium chloride at 180°. Triphenyllead chloride was also obtained by Koton (409) from the reaction of tetraphenyllead with trimethylammonium chloride at 130°. The use of trialkylammonium halide for the removal of tetraethyllead from gasoline has been patented (386).

(4) *Reactions with other inorganic acids and anhydrides*

Tetraorganolead compounds react with most inorganic acids and their anhydrides. Jones and co-workers (369) prepared a number of dialkyllead dinitrate derivatives as the dihydrates, $R_2Pb(NO_3)_2{\cdot}2H_2O$, by reaction of the tetraalkyllead derivatives with cold, concentrated nitric acid. With hot, concentrated nitric acid complete decomposition to lead(II) nitrate can be effected; however, this reaction can occur with explosive violence, even at low temperatures (99,369). The vapor phase reaction of nitric acid and tetraethyllead at 150° in a heated tube has been shown to result in complete conversion of the ethyl radicals to nitromethane and ethyl nitrate (475).

Tetraphenyllead (570,651), phenyl-tri-*o*-tolyllead (13), tetra-*p*-tolyllead (572), and phenyltriethyllead (353) react with hot concentrated nitric acid to yield the dinitrate salts of diphenyllead, *o*-tolylphenyllead, di-*p*-tolyllead, and diethyllead, respectively. Diphenyllead dinitrate will react with nitric acid (in the presence of sulfuric acid) to produce di-(*m*-nitrophenyl)lead dinitrate in low yield (132,748). This is one of the few reactions of an organolead compound in which cleavage of a lead-to-carbon bond is not involved.

Hetnarski and Urbanski (335) have reported that tetraethyllead and tetrapropyllead react with dinitrogen tetroxide at 0° in diethyl ether to yield crystalline products having the structure $[R_4Pb(NO)_2](NO_3)_2$. The ethyl and propyl analogs are soluble in a number of polar organic solvents and useful in the synthesis of other dialkyllead salt derivatives by reaction with other acids. The reaction of tetraethyllead and dinitrogen tetroxide has been patented by Rifkin and Ewen (620) as a method for removing tetraethyllead from gasoline; at room temperature, the tetraethyllead is precipitated from gasoline as diethyllead dinitrate. Gilman and Melstrom (273) obtained impure diphenyllead dinitrate from the reaction of dinitrogen tetroxide with tetraphenyllead in chloroform. The reaction of tetraphenyllead with a mixture of dinitrogen trioxide, N_2O_3, and nitric oxide was investigated by Makarova and Nesmeyanov (459); benzenediazonium nitrate was formed in quantitative yield. Triphenyllead chloride gave a 50% yield of the same product but diphenyllead dichloride did not react (459).

Tetraethyllead did not undergo reaction at room temperature with dilute (1:1) sulfuric acid (99); however, reaction did occur upon mild heating to form triethyllead sulfate and ethane. Tetra-2-furyllead is reported to undergo a vigorous decomposition in concentrated sulfuric acid (288) but tetraphenyllead in hot chloroform did not react with 35% sulfuric acid

(581). Tetraethyllead and other tetraalkyl derivatives react with sulfur dioxide under mild conditions. The reaction of tetraethyllead with sulfur dioxide was reported originally in 1879 by Frankland and Lawrence (239) who obtained a product having the composition $(C_2H_5)_2PbS_2O_3$. Heap, Saunders, and Stacey (323) later prepared a dialkyllead sulfite, R_2PbSO_3, by the reaction of tetraalkyllead compounds with sulfur dioxide in wet ether; similar results were obtained by Padberg (543).

The reaction of tetraalkyllead compounds with sulfur dioxide in anhydrous media has been investigated more recently by both Padberg (543) and Gelius (251). Reaction of tetraethyllead with sulfur dioxide gas in dry benzene or hexane gave an amorphous white solid having the composition $C_8H_{20}PbS_2O_4$, which was concluded to be diethyllead bis(ethanesulfinate), $(C_2H_5)_2Pb(SO_2C_2H_5)_2$, based on its infrared spectrum and the products obtained upon reaction of the product with 1,2-dichloroethane. The product slowly decomposed on standing via a typical disproportionation to form tetraethyllead, lead(II) ethanesulfinate, and diethylsulfane (251); at elevated temperature, decomposition occurred with evolution of sulfur dioxide to form lead(II) ethanesulfinate (543). Triethyllead ethanesulfinate was obtained in small amounts as an unstable intermediate when the reaction was conducted in diethyl ether solution at 0°. It was concluded that the reaction of tetraethyllead with sulfur dioxide proceeded with step-wise cleavage of ethyl groups leading to the formation of ethyllead tris(ethane sulfinate), which disproportionated into the lead(II) ethanesulfinate and diethylsulfane (251). The reaction of tetraethyllead with sulfur dioxide has been patented as a method for its removal from gasoline (653,738).

Tetramethyllead (251,351a,543) and tetraphenyllead (351a,543) reacted similarly with sulfur dioxide at room temperature to yield $R_2Pb(SO_2R)_2$, except that tetramethyllead reacted less rapidly than the ethyl analog (251). Dimethyllead bis(methanesulfinate) was less stable to air and was easily oxidized to the sulfonate. Trimethyllead methanesulfinate was obtained in nearly quantitative yield when tetramethyllead was reacted with sulfur dioxide in cold ether. At room temperature in hexane, tetrabutyllead gave a product having the approximate composition $(C_4H_9)_6Pb_2(SO_2)_3$ and unsymmetrical tetraorganolead compounds, such as allyltriethyllead and ethyltriphenyllead, gave products having the overall composition $R_3R'PbS_2O_4$ (251). Benzyltriethyllead gave white needles having a chemical composition which did not fit any likely compound, the carbon-to-lead atomic ratio being 5.31.

The reaction of tetraethyllead and sulfur trioxide was investigated by Padberg (351a,543). In benzene solution at room temperature, bis(triethyllead) sulfate precipitated from the reaction mixture as white needles. A gel-like product was isolated by evaporation of the benzene filtrate; this residue contained lead, sulfur, carbon, hydrogen, and oxygen, but not in a ratio corresponding to any likely compound. On the other hand, Gelius and Müller (252a) obtained trimethyllead methanesulfonate, dimethyllead bis(methanesulfonate), and diethyllead bis(ethanesulfonate) from the reaction of sulfur trioxide with tetramethyl- and tetraethyllead in methylene dichloride; the trimethyllead derivative was obtained using equimolar amounts of the two reactants. Some alkyllead sulfate was obtained as by-product.

Triethyllead dihydrogenphosphate has been obtained from the reaction of tetraethyllead and phosphoric acid at room temperature in the presence of silica gel (279). Similarly, the reaction of dialkyl hydrogenphosphates with tetraethyllead yields the corresponding triethyllead dialkylphosphate (319). The reactions of R_4Pb with *m*-nitrophenylarsonic (328) and selenic acid (281) have been shown to yield the corresponding R_3PbX derivatives.

(5) *Reactions with carboxylic acids*

Tetraorganolead compounds undergo reaction with carboxylic acids under relatively mild conditions to yield the mono- or di-carboxylate, dependent on conditions. These reactions are commonly used for the preparation of trialkyllead- and dialkyllead carboxylates (81,297,322,350a); the synthesis of organolead carboxylates is discussed in Chapter XI. The conditions required to effect reaction with the weaker carboxylic acids are such that the dicarboxylate, R_2PbX_2, tends to be the major product and little or no R_3PbX is isolated (637). However, Browne and Reid (81) have shown that the reaction of R_4Pb with carboxylic acids is promoted by silica gel and catalysis with silica gel has been used for the synthesis of R_3PbX derivatives of the weaker carboxylic acids.

Robinson (622) has shown that tetraethyllead slowly reacts with acetic acid at room temperature, and despite reports to the contrary (370), triethyllead acetate is formed. At elevated temperatures, complete decomposition of tetraethyllead to lead(II) acetate occurs. Jones and Werner (368) obtained lead(II) acetate, ethane and ethyl acetate from the reaction of tetraethyllead and acetic acid at 250–260°; these products can be explained by cleavage of three ethyl groups to yield unstable ethyllead triacetate which then disproportionates at elevated temperatures to lead(II) acetate and ethyl acetate. Acetolysis of tetraphenyllead at elevated temperatures is reported to yield lead(II) acetate, benzene, and a tar (368).

(6) Reactions with metal salts

Tetraorganolead compounds undergo reaction with a wide variety of metal salts to cleave one or more lead–carbon bonds to form organolead salts. These reactions involve simple metathesis of the type

$$R_4Pb + MX_n \rightarrow R_{4-x}PbX_x + RMX_{n-x}$$

and have been employed as a convenient synthesis of organometallic derivatives of several metals, e.g., arsenic, antimony, phosphorus, mercury. With those metals which do not form stable sigma-bonded organometallic derivatives, the reaction is accompanied by reduction of the metal halide reactant to a lower oxidation state. Most of these reactions involve the utilization of one or two of the organic groups bonded to lead, although at higher temperatures, three of the four R groups in the R_4Pb compound can be utilized.

Tetraethyllead and aluminum chloride undergo vigorous reaction even at room temperature; triethyllead chloride, diethyllead dichloride, and/or lead(II) chloride are formed, dependent on stoichiometry, temperature, and/or solvent (81). This system was studied in some detail by Gilman and Apperson (264). In refluxing hexane, lead(II) chloride was formed as the major product, along with some triethyllead chloride and ethyl chloride; the ethyl chloride undoubtedly arose from decomposition of ethyllead trichloride. In chloroform at room temperature, Manulkin (466) obtained lead(II) chloride in 81% yield. Tetraethyllead also reacted with ethylaluminum sesquichloride; triethyllead chloride was formed in 60% yield (264). Reaction of tetraphenyllead with aluminum chloride yielded a mixture of triphenyllead chloride and diphenyllead dichloride. Ethyl or phenyl aluminum chlorides were the other products formed in these reactions. The general stoichiometry is represented by the equations:

$$R_4Pb + AlCl_3 \rightarrow R_3PbCl + RAlCl_2$$

$$R_3PbCl + RAlCl_2 \rightarrow R_2PbCl_2 + R_2AlCl$$

Diethylaluminum chloride did not react with tetraethyllead to any appreciable extent; triethyllead chloride was formed in only 3–5% yield (264). The triethyllead chloride may have actually resulted from ethylaluminum sesquichloride in the diethylaluminum chloride.

Tetramethyllead and tetraethyllead also undergo reaction with boron halides. The synthesis of trimethylborane and triethylborane by the reaction of boron trichloride with tetramethyllead or tetraethyllead has been described in the patent literature (342a,697). Boron trifluoride (104) and its diethyl etherate (206,752) also undergo reaction with tetraalkyllead compounds as evidenced by their activity as catalysts for the exchange of alkyl groups between two different R_4Pb compounds.

Tetraalkyl- and tetraaryllead compounds react with the halides of bismuth, arsenic, phosphorus, and antimony under mild conditions; these reactions represent preferred methods of synthesis of the mono- and di-alkyl derivatives of these metals (264,297,387,467,638). The reaction of tetraethyllead or tetramethyllead with phosphorus trichloride is an excellent method for synthesizing ethyldichlorophosphine and diethylchlorophosphine (39,86,182,183,387). These reactions proceed in distinct stages. With tetraethyllead at 25°, two ethyl groups are used, yielding diethyllead dichloride; at 100°, three ethyl groups are used and lead(II) chloride is formed, according to the stoichiometry:

$$Et_4Pb + 3MCl_3 \xrightarrow{100^\circ} 3EtMCl_2 + EtCl + PbCl_2$$

where M = As, P, or Sb. Tetraethyllead has been reacted with ethyldichloroarsine at 120° to form diethylchloroarsine and diethyllead dichloride. The reaction of tetraphenyllead with arsenic and bismuth halides gave the diphenylchloro derivatives of these metals (297,580). The reaction of ethyldichlorophosphine with tetramethyllead gave methylethylchlorophosphine; ethyldichlorophosphine oxide and methylethylchlorophosphine oxide were obtained when the reaction of tetraethyllead with the corresponding chlorophosphine was conducted in the presence of air (804). The preparation of various tertiary phosphines by reaction of a tetraalkyllead compound with a diorganochlorophosphine is described in a Monsanto patent (498).

Diphenylantimony trichloride has been prepared by the reaction of tetraphenyllead and antimony pentachloride (297); diphenyllead dichloride was formed as coproduct. The reaction of phosphorus pentachloride and tetraethyllead gave a syrup from which triethyllead chloride was isolated (81). The reaction of tetraethyllead and 10-chloro-9,10-dihydrophenarsazine has been shown to yield triethyllead chloride and the corresponding 10-ethyl derivative (590) while the reaction of tetraethyllead and 10-formyl-9,10-dihydrophenarsazine gave lead(II) formate and the 10-ethyl derivative. Phenylarsine, $PhAsH_2$, and tetraethyllead underwent reaction at 150–170° to yield lead metal and tetraphenyldiarsine (525). The following sequence was postulated to account for the products:

$$(C_2H_5)_4Pb + 2C_6H_5AsH_2 \rightarrow Pb + 4C_2H_6 + C_6H_5As{=}AsC_6H_5$$

$$2C_6H_5As{=}AsC_6H_5 \rightarrow (C_6H_5)_2As{-}As(C_6H_5)_2 + 2As$$

Tetraorganolead compounds undergo reaction with the Group IVb metal halides, the order of reactivity increasing with increasing atomic weight of the Group IVb metal, Silicon tetrachloride is reported to react with tetraethyllead to form triethyllead chloride, but reaction conditions were not defined (81). On the other hand, tetraphenyllead reacts sluggishly, if at all, with silicon tetrachloride on refluxing chloroform. Puchinyan (581) obtained triphenyllead chloride from this reaction, but also recovered 80% of the silicon tetrachloride unchanged; the yield of triphenyllead chloride was 71%, based on the reacted silicon tetrachloride. Conversely, Manulkin (468) recovered 98.5% of the silicon tetrachloride unchanged after it had been heated with tetraphenyllead in chloroform solution; benzene was formed in 90.5% yield (the basis on which the benzene yield was calculated was not defined) and a small amount of tetrachloroethylene was also formed. Manulkin's results suggest that the tetraphenyllead reacted directly with chloroform and not with the silicon tetrachloride.

Mironov and Kravchenko (492) prepared ethyldichlorogermane, $C_2H_5Ge(H)Cl_2$, by reaction of trichlorogermane with tetraethyllead or tetraethyltin; some ethyltrichlorogermane was also formed. Ethyltrichlorogermane was prepared in 90% yield by these same workers (491) by the reaction of germanium tetrachloride with tetraethyllead; tetramethyltin and germanium tetrachloride did not react under comparable conditions.

Both tin(II) chloride and tin(IV) chloride react with R_4Pb compounds. Calingaert and co-workers (104,121) have shown that tin(IV) chloride is an effective catalyst for the redistribution of organic groups between two different R_4Pb compounds. Also, the use of tin(IV) chloride for the removal of tetraethyllead from gasoline is reported in the patent literature (246,385). Goddard and colleagues (297) obtained diphenyllead dichloride from the reaction of tetraphenyllead and tin(II) chloride dihydrate. The reaction was postulated to proceed according to the equation:

$$Ph_4Pb + 2SnCl_2{\cdot}2H_2O \rightarrow Ph_2SnCl_2 + Ph_2PbCl_2 + Sn + 2H_2O$$

The diphenyltin dichloride was isolated as $Ph_2Sn(OH)Cl$; hydrolysis of the diphenyltin dichloride by the water in the system was postulated.

Tetraorganolead compounds react with tetravalent lead salts under relatively mild conditions, but not with divalent lead salts, even at elevated temperature (336,571). In the absence of a solvent, lead(IV) chloride reacts violently with tetraethyllead even at temperatures below 20° (243). Lead(IV) acetate and R_4Pb compounds undergo a smooth reaction in the presence of a solvent to form organolead acetates. Kocheskov and Freidlina (394,395) obtained diphenyllead diacetate in nearly quantitative yield from the reaction of tetraphenyllead with lead(IV) acetate in refluxing chloroform. On the other hand, a similar reaction of tetraethyllead and lead(IV) acetate gave triethyllead acetate and lead(II) acetate (394,395). The formation of the lead(II) acetate undoubtedly results from the instability of the ethyllead triacetate formed as the initial product.

Tetraorganolead compounds also react with the Group IVa metal halides, such as zirconium(IV) chloride (104) and titanium(IV) chloride (29,81,231,380,461,581). Ethyltitanium trichloride was prepared by Bawn and Gladstone (29) by the reaction of tetraethyllead and titanium tetrachloride at $-80°$ using a stoichiometry of $2TiCl_4/Et_4Pb$; this reaction is also described in the patent literature (231). At higher temperatures, Malatesta (461) obtained ethane and butane as the major products, along with substantial amounts of ethylene and ethyl chloride; these products undoubtedly arise from decomposition of the ethyltitanium chlorides formed. Kashireninov and co-workers (380,665) have reported that tetraethyllead and tetraethyltin form complexes with titanium(IV) chloride or bromide of the type $Et_4Pb{\cdot}2TiX_4$, and that these complexes catalyze the polymerization of ethylene and styrene; the polymerization of olefins catalyzed by tetraorganolead compounds in a Ziegler-type system is discussed in a later chapter. Finally, Puchinyan (581) obtained triphenyllead chloride and diphenyllead dichloride in nearly quantitative yield from the reaction of equimolar amounts of tetraphenyllead and titanium(IV) chloride in refluxing chloroform. Obviously, the composition of the products obtained from the reaction of R_4Pb compounds with titanium(IV) halides is strongly dependent on reaction temperature and stoichiometry.

Tetraorganolead compounds undergo facile reaction with silver, copper, and gold salts even at $-78°$. These reactions are convenient for the preparation of organometallic derivatives of these latter metals. The reaction of tetramethyllead or tetraethyllead with an alcoholic solution of copper-(II) nitrate has been extensively investigated by Bawn (30–32) and Costa (159–164,167,168) and their co-workers. The reaction is postulated to involve the initial reduction of the copper(II) salt to copper(I), followed by further reaction of the latter to form alkylcopper according to:

$$Cu(NO_3)_2 + R_4Pb \xrightarrow{ROH} Cu^+ + R^0 + R_3Pb^+$$

$$Cu^+ + R_4Pb \longrightarrow RCu + R_3Pb^+$$

These reactions of tetraalkyllead and copper(II) nitrate involve bimolecular processes. For the reaction with tetraethyllead in ethanol, Bawn and Johnson (30) calculated an activation energy of 15.2 kcal/mole. At $-30°$, a brown precipitate of ethylcopper is obtained, along with ethane and ethylene as gaseous by-products. At higher temperatures the ethylcopper decomposes to form butane and ethane, without formation of any ethylene. The following reaction sequence was proposed to explain these observations and products (30):

$$\le -30^\circ \left\{ \begin{array}{l} Cu^{++} + Et_4Pb \rightarrow Cu^{+} + Et_3Pb^{+} + Et^0 \\ 2Et^0 \rightarrow C_2H_6 + C_2H_4 \\ Et^0 + EtOH \rightarrow C_2H_6 + EtO^0 \\ Cu^{+} + Et_4Pb \rightarrow EtCu + Et_3Pb^{+} \end{array} \right.$$

$$> -30^\circ \left\{ \begin{array}{l} CuEt + EtOH \rightarrow CuOEt + C_2H_6 \\ 2CuEt \rightarrow 2Cu + C_4H_{10} \end{array} \right.$$

A similar reaction occurs between tetramethyllead and copper(II) nitrate; however, methylcopper is less stable than the ethyl analog and decomposes above -50°. The decomposition of methylcopper in the absence of a solvent produces ethane exclusively (168). However, in the presence of methanol, methane is formed exclusively in a mole ratio of $1CH_4/1CuMe$ (168); in the presence of tetramethyllead and methanol, methane is formed in a mole ratio greater than one. Methane formation was attributed to extraction of a proton from the methanol by the methylcopper to yield methane and copper(I) methoxide; the copper(I) methoxide then extracts a methyl group from the tetramethyllead to reform methylcopper along with trimethyllead methoxide (31,168). The following sequence was proposed to account for the products formed:

$$Cu^{++} + Me_4Pb \rightarrow Cu^{+} + Me_3Pb^{+} + Me^0$$

$$2Me^0 \rightarrow C_2H_6$$

$$Cu^{+} + Me_4Pb \rightarrow MeCu + Me_3Pb^{+}$$

$$MeCu + CH_3OH \rightarrow CH_4 + CuOCH_3$$

$$CuOCH_3 + Me_4Pb \rightarrow MeCu + Me_3PbOMe$$

Small amounts of formaldehyde were detected in the reaction mixture; its formation was ascribed to the disproportionation of methoxy radicals according to:

$$2CH_3O^0 \rightarrow CH_3OH + CH_2O \qquad (159,160,168)$$

The reaction of copper(II) sulfate with tetramethyllead proceeded similarly to that of copper(II) nitrate; on the other hand, the reaction of copper(II) chloride with tetramethyllead gave only copper(I) chloride and methyl chloride; no methylcopper was formed (32).

Gilman and Woods (291) also investigated the reaction of tetramethyllead and copper(II) nitrate and concluded that methylcopper is formed as an intermediate. They also obtained triphenyllead nitrate, benzene, and copper metal from the reaction of tetraphenyllead and copper(II) nitrate in hot ethanol.

Tetraalkyllead compounds and alcoholic silver nitrate undergo facile reaction at −78° to yield an alkylsilver derivative which decomposes via a free radical mechanism at higher temperatures. This reaction was demonstrated originally by Krause and co-workers in their early investigations of organolead chemistry; more detailed investigations of these reactions have been made by Costa (165,166), by Semerano and Riccoboni (647–650) and by Bawn and Whitby (31). The reaction of tetramethyllead and silver nitrate at −78° in methanol produces a yellow precipitate of methylsilver (647), which decomposes at temperatures above −40° to give ethane exclusively (31,647). Tetraethyllead and silver nitrate react similarly; Semerano and Riccoboni (647) obtained a red-brown precipitate while Gilman and Woods (291) reported their product to be light yellow. At higher temperatures, the ethylsilver decomposes to silver metal and a mixture of ethane, ethylene, and butane. Tetrapropyllead gave similar results (647); the resultant propylsilver was less stable than the ethylsilver and decomposed to form hexane, propylene, and propane (647,650).

Symmetrical tetraaryllead compounds do not react with alcoholic silver nitrate solution at −78°; however, the unsymmetrical aryltriphenyllead derivatives do react, presumably to form an arylsilver compound (307,423). Gilman and Woods (291) isolated biphenyl as the major product from the reaction of tetraphenyllead and ethanolic silver nitrate at room temperature. On the other hand, Spice and Twist (686) obtained benzene as the major product and only a small amount of biphenyl when the reaction was conducted in ethanol–benzene at 60°; benzene formation was postulated to result from abstraction of a proton from the ethanol by a phenyl radical. The reaction was shown to involve a bimolecular process having a temperature coefficient of 18.5 kcal/mole.

Tetraalkyllead compounds and gold(III) chloride undergo reaction at low temperature to form an alkylgold derivative which decomposes at higher temperatures via a free radical mechanism (41,609,610,713,715). With tetramethyllead in methanol, a yellow-orange color was formed; at −65° the orange-yellow color slowly changed over to green. At −30° the green color disappeared and a reddish-brown precipitate was formed which decomposed at higher temperatures to methane and ethane. Methane formation was attributed to proton extraction from the methanol by a methyl radical as evidenced by the formation of ethylene glycol (via combination of two $\cdot CH_2OH$ radicals). Ethylene glycol was also formed in the reaction of tetraethyllead with $HAuCl_4$, along with ethane and ethylene.

The following reaction sequence was proposed to account for the color changes and products from tetramethyllead:

$$Me_4Pb + 2AuCl_3 \xrightarrow{MeOH} 2Me_3Au$$
$$2Me_3Au \rightarrow 2MeAu + 2C_2H_6$$
$$2MeAu \rightarrow 2Me^0 + 2Au$$
$$2Me^0 + 2CH_3OH \rightarrow 2CH_4 + 2{\cdot}CH_2OH$$
$$2{\cdot}CH_2OH \rightarrow HOCH_2CH_2OH$$

Trimethyllead chloride, dimethyllead dichloride, or lead(II) chloride was also formed, dependent on the ratio of tetramethyllead to gold chloride.

The reactions of tetraorganolead compounds and mercury(II) salts have been investigated extensively. These reactions represent a convenient method for synthesizing organolead salts; the reaction of tetraethyllead with mercury(II) salts is also used for the commercial manufacture of organomercury fungicides (345,384,759) (see Chapter XVII). Mercury(II) halides and carboxylates react with tetraalkyllead compounds to form the RHgX derivative; the lead product is either R_3PbX or R_2PbX_2, depending on stoichiometry. With aryllead compounds and mercury carboxylates, step-wise cleavage of R_4Pb all the way to the $RPbX_3$ compound can be achieved (553). This is a preferred synthesis of aryllead tricarboxylates (although one usually starts with a diaryllead dicarboxylate). The reaction of both tetraalkyl- and tetraaryllead compounds with mercury(II) halides proceeds beyond the R_2PbX_2 stage only with great difficulty.

Tetraphenyllead and mercury(II) oxide do not react to any appreciable extent. However, reaction of triphenyllead chloride or triphenyllead hydroxide with mercury(II) oxide, according to Nesmeyanov and Kocheshkov (526), gave diphenylmercury; diphenyllead dichloride gave phenylmercuric chloride. Ethylmercuric dihydrogen phosphate has been prepared by the reaction of tetraethyllead with mercury(II) oxide in the presence of phosphoric acid; diethyllead bis(dihydrogen phosphate) was also formed (2).

The reaction of mercury(I) nitrate with tetramethyllead and tetraethyllead in methanol has been demonstrated by Tagliavini and Belluco (712) to yield both diphenylmercury and mercury metal along with the di- and trialkyllead nitrates.

Iron(III) chloride is reduced by tetraethyllead at room temperature to iron(II) chloride (264,445); this reaction is claimed to be useful for the removal of tetraethyllead from gasoline (133). The reaction of tetrabutyllead and tetraethyllead with iron(III) chloride in hot chloroform gave tributyllead chloride and lead(II) chloride, respectively (466). Thallium(III) chloride and tetraphenyllead gave diphenylthallium chloride and diphenyllead dichloride (298); on the other hand, triethyllead chloride and thallium(III) chloride gave diethyllead dichloride, thallium(I) chloride and ethyl chloride (298).

Tetraethyllead did not react with iron(II) chloride, iron(II) iodide, cobalt(II) bromide, or nickel(II) bromide at room temperature (77,264). However, Razuvaev and colleagues (583) found that the dihalides of cobalt, nickel, manganese, and copper were reduced to the respective metals upon reaction with tetraphenyllead in a hot solution of monoethyl ether of ethylene glycol. Except with copper(II) chloride, benzene was formed, presumably by abstraction of a proton from the solvent by phenyl radicals produced upon decomposition of the unstable phenylmetal intermediates; copper(II) chloride gave chlorobenzene in 83% yield. Tetraethyllead reduced platinum(IV) chloride to platinum metal (264).

Setzer and co-workers (651) have shown that tetraphenyllead does not react with magnesium bromide in diethyl ether, nor did tetraphenyllead and tetra-*p*-methoxyphenyllead undergo reaction with a mixture of magnesium metal and magnesium iodide or bromide. These reactions were considered to be a possible source of triorganolead halides in reactions of lead halides with Grignard reagents. However, Komarov and co-workers (398) claim that some triethyllead bromide was obtained from the reaction of tetraethyllead with basic magnesium bromide.

Tellurium tetrachloride and tetraphenyllead undergo reaction in refluxing toluene–xylene to form diphenyldichlorotelluride and diphenyllead dichloride (297).

(7) *Reactions with metals*

Tetraorganolead compounds undergo facile reaction with the alkali and alkaline earth metals in liquid ammonia or ammonia–ether to form compounds of the type R_3PbM (46,267). These reactions proceed according to the equation:

$$R_4Pb + 2M + NH_3 \xrightarrow{NH_3} R_3PbM + RH + MNH_2$$

The reaction of tetraphenyllead with sodium in liquid ammonia has been used to synthesize triphenylplumbylsodium, which in turn is useful for the synthesis of unsymmetrical derivatives of the type Ph_3PbR. The reactivity of the various metals with tetraphenyllead decreases in the approximate order K, Na > Li, Ca, Sr > Ba; with tetraethyllead, calcium, lithium, and sodium are about equally reactive. Magnesium metal in diethyl ether does not react with tetraphenyllead or tetraethyllead (265,651); its reactivity in liquid ammonia has not been tested.

The reaction of tetramethyllead with sodium, lithium, and potassium in liquid ammonia has been investigated in some detail by Holliday and Pass (343), who showed that all three metals react in the initial stages according to the equation shown above. Methane is formed as the principal organic product, along with some ethane. On the basis of the types and amounts of products obtained, the following reaction sequence was proposed for the initial stages of the reaction; the stoichiometry of the reaction was followed by observing the disappearance of the characteristic blue color of the alkali metal in ammonia as tetramethyllead was added to the ammonia solution.

$$Me_4Pb + e^- \rightarrow Me_3Pb^- + Me^0$$
$$Me^0 + e^- \rightarrow Me^-$$
$$Me^- + NH_3 \rightarrow CH_4 + NH_2^-$$
$$Me^0 + Me^0 \rightarrow C_2H_6$$

Gas evolution continued after the disappearance of the characteristic blue color of ammoniacal alkali metal solutions, with methane being formed exclusively. This was attributed to a solvolysis reaction according to:

$$Me_3Pb^- + NH_3 \rightarrow Me_2PbNH_2^- + CH_4$$

With sodium and lithium, the reaction stopped at this point; termination was attributed to a disproportionation of the $Me_2PbNH_2^-$ to Me_2Pb + NH_2^- as a result of precipitation of *insoluble* lithium or sodium amide.

In the potassium system solvolysis proceeded beyond the formation of $Me_2PbNH_2^-$, presumably because potassium amide is soluble in liquid ammonia. Thus,

$$Me_2PbNH_2^- + NH_3 \rightarrow MePb(NH_2)_2^- + MeH$$
$$MePb(NH_2)_2^- + NH_3 \rightarrow Pb(NH_2)_3^- + MeH$$
$$Pb(NH_2)_3^- \rightarrow PbNH + NH_3 + NH_2^-$$

The final product from the potassium system was an orange solid which, like lead imide, tended to decompose explosively.

Holliday and Pendlebury (344) have concluded that the reaction of tetravinyllead with sodium metal in liquid is similar to that of tetramethyllead.

Schmitz-Dumont and Ross (643) have shown that tetraphenyllead is completely solvolyzed by liquid ammonia in the presence of potassium amide, even when less than an equimolar amount of potassium amide was used. The phenyl groups were converted to benzene and an ammonia-soluble product was formed which was shown to be $KPb(NH_2)_3$. In addition, a black, ammonia-insoluble solid was formed which decomposed explosively when dried; this latter solid was deduced to be a nitrogen derivative of tetravalent lead, since lead dioxide was formed in appreciable amounts when the black solid was hydrolyzed. The solvolysis reaction was proposed to involve a stepwise sequence, according to:

$$Ph_4Pb + NH_2^- \longrightarrow [Ph_4Pb(NH_2)]^-$$
$$[Ph_4Pb(NH_2)]^- \longrightarrow Ph_3PbNH_2 + Ph^-$$
$$Ph^- + NH_3 \longrightarrow C_6H_6 + NH_2^-$$
$$Ph_3PbNH_2 + NH_2^- \longrightarrow [Ph_3Pb(NH_2)_2]^-$$
$$[Ph_3Pb(NH_2)_2]^- \longrightarrow Ph_2Pb(NH_2)_2 + Ph^-$$
$$Ph^- + NH_3 \longrightarrow C_6H_6 + NH_2^-$$
etc.

Finally

$$PhPb(NH_2)_3 + NH_2^- \longrightarrow [PhPb(NH_2)_4]^-$$
$$[PhPb(NH_2)_4]^- \longrightarrow Pb(NH_2)_4 + Ph^-$$
$$Ph^- + NH_3 \longrightarrow C_6H_6 + NH_2^-$$
$$Pb(NH_2)_4 + NH_2^- \longrightarrow Pb(NH_5)^- \qquad [\text{or} \quad KPb(NH_2)_5]$$

The hypothetical $KPb(NH_2)_5$ might then decompose according to:

$$Pb(NH_2)_5^- \longrightarrow Pb(NH_2)_3^- + N_2H_4$$
$$N_2H_4 + 2Pb(NH_2)_5^- \longrightarrow N_2 - 2NH_3 + Pb(NH_2)_3^-$$

The efficacy of this latter sequence is supported by the identification of nitrogen as a reaction product.

Holliday and Pass (343) also investigated the reaction of trimethyllead chloride with potassium and lithium and concluded that this reaction is clearly different from that of tetramethyllead. The initial step is the formation of trimethyllead (or hexamethyldilead):

$$2Me_3PbCl + 2M \longrightarrow Me_6Pb_2 + 2MCl$$

According to Holliday and Pass, the resultant hexamethyldilead could then react further by different paths. One possible path could involve:

$$Me_6Pb_2 + 2e^- \longrightarrow 2Me_3Pb^-$$

and the resultant Me_3Pb^- could undergo solvolysis as described above for the tetramethyllead reaction. With potassium, the overall stoichiometry by this path would then become (on the basis of the tetramethyllead system):

$$Me_3PbCl + 2K + 2NH_3 \longrightarrow PbNH + 3MeH + KNH_2 + KCl$$

A second path considered was:

$$Me_3Pb + 3K + 3NH_3 \longrightarrow Pb + 3MeH + 3KNH_2$$

or $$Me_3PbCl + 4K + 3NH_3 \longrightarrow Pb + 3MeH + 3KNH_2 + KCl$$

Methane formation in this system would arise by the sequence:

$$Me_3Pb + e^- \longrightarrow Me_2Pb + Me^-$$
$$Me_2Pb + e^- \longrightarrow MePb + Me^-$$
$$MePb + e^- \longrightarrow Pb + Me^-$$
$$3Me^- + 3NH_3 \longrightarrow 3MeH + 3NH_2^-$$

The third possible path considered was:

$$4Me_3Pb \longrightarrow 3Me_4Pb + Pb$$

or

$$4Me_3PbCl + 4K \longrightarrow 3Me_4Pb + Pb + 4KCl$$

From the order and amounts of the products obtained, it was concluded that all three of the above possible paths were involved.

In the reaction of trimethyllead chloride with lithium metal, lesser amounts of methane were formed. This was attributed to cessation of the solvolysis reaction at the $Me_2PbNH_2^-$ stage as a result of precipitation of $LiNH_2$, as was postulated for the reaction of tetramethyllead with lithium or sodium metal.

$$Me_3PbCl + Li \longrightarrow Me_3Pb + LiCl$$
$$Me_3Pb + Li + NH_3 \longrightarrow Me_2Pb + LiNH_2 + MeH$$

Tetraethyllead and sodium metal undergo reaction in the absence of ammonia to give ethylsodium; sodium–potassium alloy behaves similarly (292). However, these systems are less suitable for the synthesis of ethylsodium than is the reaction of sodium metal with diethylmercury (641). Phenyllithium has been prepared by the reaction of lithium metal with tetraphenyllead in diethylether or xylene (281) and vinyllithium has been prepared in similar fashion from tetravinyllead (372). Sodium metal did not react with tetraphenyllead in benzene or petroleum ether at temperatures up to 100° (274,275). The reactions of organolead compounds with sodium metal have been reviewed (238,756).

Gilman and Apperson (264) were not able to obtain any reaction between tetraphenyllead and finely divided bismuth metal; Goddard and Goddard (298) also failed in an attempt to prepare a phenylthallium derivative by reaction of tetraphenyllead with thallium metal. Oils containing tetraethyllead are reported to be highly corrosive to lead metal; the corrosive action has been attributed to attack on the lead metal by free radicals generated by the tetraethyllead (131). Irradiation of the oils with ultraviolet radiation markedly reduced the corrosive action.

(8) *Reactions with organometallic compounds*

Tetraorganolead compounds will undergo organic group exchange with other organolead compounds as well as with other organometallic compounds. The exchange of organic groups between organometallic derivatives of two similar or dissimilar metals was discovered by Calingaert and co-workers (104,107,120), who called such reactions "redistribution reactions" (106).

The redistribution of organic groups between two different tetraorganolead compounds occurs in a random manner to produce a statistical distribution of all five possible tetraorganolead derivatives based on the relative concentrations of the two or more original R_4Pb reagents (108).

$$R_4Pb + R_4'Pb \rightleftharpoons R_3PbR' + R_2PbR_2' + RPbR_3'$$

Since the enthalpy change for reactions involving transfer of different organic groups to the same atom is nearly zero, the entropy of the system becomes the driving force; hence a statistical distribution is obtained (694). Similar distribution reactions will also occur between tetraorganolead and triorganolead halide to yield a random mixture.

A catalyst is required to effect the redistribution reaction between tetraorganolead compounds. Effective catalysts are Lewis acid compounds, such as zinc chloride, zinc fluoride, mercury(II) chloride, boron trifluoride, aluminum chloride, aluminum bromide, and dimethylaluminum chloride as well as zirconium(IV) chloride, tin(IV) chloride, triethyllead chloride, triethyllead iodide, phosphorus trichloride, arsenic trichloride, bismuth trichloride, iron(III) chloride, and platinum(IV) chloride (37,104, 106,109,119,206,752). Activated alumina and other activated metal oxides are also effective redistribution catalysts (137,224). Slow exchange occurs between R_4Pb and triorganolead halides in the absence of any catalyst.

Redistribution reactions involving the Group IVb metals have been reviewed by Moedritzer (494) who calculated equilibrium constants for the reactions:

$$2Me_3PbEt \leftrightarrows Me_2PbEt_2 + Me_4Pb \qquad K_1 = 0.356$$

$$2Me_2PbEt_2 \leftrightarrows Me_3PbEt + MePbEt_3 \qquad K_2 = 0.426$$

$$2MePbEt_3 \leftrightarrows Me_2PbEt_2 + Et_4Pb \qquad K_3 = 0.317$$

It was concluded that these values show good agreement with the value required for ideal randomness.

Tetraorganolead compounds also exchange organic groups with organometallic derivatives of other metals. Thus, Calingaert and co-workers (110,122) obtained exchange of methyl and ethyl groups between tetraethyllead–dimethylmercury and tetramethyllead–diethylmercury. About two thirds of the methyl groups ended up preferentially bound to mercury, but a random distribution of the ethyl and methyl groups was obtained within the R_4Pb compounds and within the R_2Hg compounds. This exchange reaction required a catalyst.

Nazarova (518) investigated the exchange of ethyl groups between two different ethylmetal derivatives by means of C^{14} labelling. No exchange was detected between tetraethyllead and diethylmercury after 20 hr at 100° in the absence of a catalyst. However, triethylaluminum and tetraethyllead gave 41% exchange in 5 hr at 100°, ethylmagnesium bromide–tetraethyllead in diethyl ether gave about 5% exchange in 5 hr at 100° and tetraethyllead–ethylsodium gave less than 1% exchange in 2.5 months at room temperature. Diethylmercury underwent a similar exchange reaction with these same ethylmetal compounds, and with greater facility.

Using similar techniques, Nazarova (519) also investigated phenyl group exchange between the phenyl derivatives of these same metals. Tetraphenyllead and phenylsodium gave 4% exchange in 5 hr at 100°, triphenylaluminum and tetraphenyllead gave about 11% exchange under similar conditions, but no exchange occurred between tetraphenyllead and phenylmagnesium bromide. Tetraphenyllead and diphenylmercury gave 84–92% exchange in 20 hr at 150°; the high degree of exchange in this latter system is most surprising since no exchange was obtained in the tetraethyllead–diethylmercury system. Nazarova (519) established the following order of decreasing ease of exchange of phenyl groups: $Ph_2Hg > Ph_4Pb \geq Ph_3Al > Ph_4Sn > PhNa > Ph_4Si$, and related the ability to undergo phenyl group exchange to reactivity as catalysts in the polymerization of styrene (520). By similar techniques, Razuvaev and co-workers (597) have shown that tetraphenyllead and diphenylzinc undergo appreciable phenyl group exchange in 15–20 hr at 180–200°; tetraphenylsilane and tetraphenylgermane gave little or no exchange with diphenylzinc.

Tetramethyllead and triethylaluminum undergo rapid exchange at room temperature without benefit of a catalyst; the methyl groups become preferentially bound to the aluminum (471). The driving force for this exchange is attributed to the formation of a methyl–aluminum bridge bond. When a molar ratio of $4Et_3Al/Me_4Pb$ was employed in the absence of a solvent, only tetraethyllead was formed and all the methyl groups were transferred to the aluminum as methyl–aluminum bridge bonds (243); in the presence of ethers, a mixture of methyl–ethyllead compounds was obtained. Tetramethyllead and triethylborane did not undergo exchange (471).

The exchange of alkyl groups between two different Group IVb metals was investigated by Pollard and co-workers (573) using gas chromatographic techniques (see Table VI-6). It was shown that exchange between tetraethyllead and tetramethyllead was complete within 2 min at 160° in the presence of aluminum chloride as catalyst. Slower exchange occurred between tetraethyllead and tetra-*n*-propyltin, while tetraethyllead did not undergo alkyl group exchange with tetra-*n*-butylgermane. Exchange occurred only between adjacent members of the Group IVb metals; this was interpreted as further evidence in support of the mechanism proposed by Russell (630,631) for alkyl exchange between silanes. This mechanism is based on a four-centered state for the alkylsilane involving electrophilic

Tcble VI-6 Alkyl Group Rearrangements Between Two Different Group IVb Atoms[a]

STARTING MATERIAL	PROPORTIONS $AlCl_3$ CATALYST, MOLE %	TEMP., °C	TIME, MIN	DEGREE OF REDISTRIBUTION
$Si(C_3H_7^n)_4 + Pb(C_2H_5)_4$	2	150	30	None
	5	170	120	None
				Some decomposition of $Pb(C_2H_5)_4$
$Ge(C_4H_9^n) + Pb(C_2H_5)_4$	2	145	3	None
	10	160	60	None
$Sn(C_3H_7^n)_4 + Pb(C_2H_5)_4$	2	140	20	40%
	5	140	30	Complete
$Sn(C_3H_7^n)_4 + Pb(C_2H_5)_4$	2	170	5	Complete for tin, decomposition of higher lead alkyls

[a] Taken from Pollard, Nickless, and Uden, *J. Chromatogr.*, **19**, 28 (1965).

substitution on carbon and nuclear substitution on silicon. A similar mechanism has been postulated for exchange reactions between different Group IVb metals (573).

$$
\begin{array}{ccc}
 & R_1 & \\
R_3M & & M'R_3 \\
 & R_2 & \\
 & \vdots & \\
 & AlX_3 &
\end{array}
$$

The role of the catalyst then is to help achieve the transition state by polarization of the R_4M compound. Failure of non-adjacent Group IVb metals to undergo alkyl exchange results from the greater energy required to achieve the transition state. It was also speculated that unfavorable steric effects may also result from the larger difference in size between two non-adjacent Group IVb metal atoms.

Alkyllithium compounds undergo reaction with tetraaryllead compounds in diethyl ether solution; Talalaeva and Kocheshkov (720) employed this reaction for the synthesis of aryllithium compounds. Gilman's group (274,275) has also demonstrated that *n*-butyllithium reacts with tetraaryllead compounds; tetrabutyllead and benzoic acid (from carboxylation of the phenyllithium product) were isolated as products from the reaction of *n*-butyllithium and tetraphenyllead. The relative reactivity of the various alkyllithium compounds with tetraphenyllead, based on the amount of benzoic acid formed, was shown to be: ethyl > *n*-propyl > *n*-butyl > methyl > phenylethynyl (275). The relative ease of cleavage of aryl groups by *n*-butyllithium was found to be: *p*-tolyl > phenyl > *o*-tolyl (274,275). *n*-Butyllithium preferentially cleaved the *p*-chlorophenyl group from di-*p*-chlorophenyldiphenyllead and *p*-chlorophenyltriphenyllead; *p*-bromophenyltriphenyllead did not react with *sec*- or *tert*-butyllithium under comparable conditions (275).

Zasosov and Kocheshkov (777) obtained *p*-bromophenyllithium and tetraphenyllead in about equivalent yields (24–27%) from the reaction of *p*-bromophenyltriphenyllead with phenyllithium; actually, metallation of the *p*-bromophenyl group was expected according to:

$$p\text{-BrC}_6\text{H}_4\text{Pb(C}_6\text{H}_5)_3 + \text{C}_6\text{H}_5\text{Li} \not\rightarrow p\text{-LiC}_6\text{H}_4\text{Pb(C}_6\text{H}_5)_3 + \text{C}_6\text{H}_5\text{Br}$$

However, Gilman and co-workers (269) have effected the successful metallation of *p*-bromophenyltriphenyllead using diethylbarium; *p*-carbomethoxyphenyltriphenyllead was isolated in low yield from the reaction of *p*-bromophenyltriphenyllead with diethylbarium, followed by carboxylation of the reaction mixture and reaction with diazomethane.

Vinyllithium has been prepared by Juenge and Seyferth (372) by the reaction of phenyllithium with a vinyllead compound. Vinyltin compounds react similarly (652).

The ability of organolead compounds to undergo facile exchange with certain organolithium compounds has posed problems in the synthesis of unsymmetrical tetraorganolead compounds because of the occurrence of further reaction of the desired unsymmetrical organolead product with the organolithium reagent. These reactions are undoubtedly equilibrium reactions, the more electronegative organic group being preferably bound to the more electropositive lithium atom than to lead. Hence, in a system containing tetraaryllead and alkyllithium, formation of aryllithium is favored.

Gilman and colleagues (269) isolated benzoic acid after carboxylation of the reaction mixture from diethylbarium and ethyltriphenyllead; presumably its formation resulted from a metal–metal interconversion reaction to form a phenylbarium derivative. The reaction of benzylsodium with tetraphenyllead in refluxing toluene gave a tarry residue; no metal–metal interconversion was detected (274). However, butylsodium did undergo reaction with tetraphenyllead in petroleum ether or benzene at temperatures below 100° since benzoic acid was formed upon carboxylation of the reaction mixture (275). No reaction occurred between tetraethyllead and tetraphenyldiphosphine in refluxing benzene (361).

(*9*) *Miscellaneous reactions*

Conflicting reports exist concerning the feasibility of electrolytic reduction of tetraethyllead. Saikina (633) has reported that tetraethyllead does not undergo polarographic reduction. In contrast, Vertyulina and Korshunov (743) obtained polarographic reduction of tetraethyllead in absolute alcohol at a half-wave potential of 0.65–0.70 V, compared to a half-wave potential of 1.8–2.0 V observed for the reduction of hexaethyldilead. Dessy and co-workers (186) have observed the electrochemical reduction of tetraphenyllead at a half-wave potential of -3.2 V (versus a 10^{-3} MAg^{+}/Ag electrode); this compares with an observed value of -2.0 V for the half-wave potential of hexaphenyldilead under identical conditions. Polarographic methods for the determination of tetraethyllead have been developed which are based on its prior conversion to a triethyllead halide or a lead(II) halide (11,579,805). Both triethyllead halide and hexaethyldilead can be determined in the presence of tetraethyllead by polarographic procedures (187,743).

Tetraorganolead compounds are slowly oxidized by air to form ill-defined solids which undoubtedly are organolead oxides, hydroxides, or carbonates. Oxidation by air is catalyzed by light and can be prevented by the addition of various compounds to the tetraorganolead or its solutions to prevent undesirable sludge formation. The stabilization of tetramethyllead and tetraethyllead is discussed later in this chapter.

The reaction of tetraethyllead with oxygen under conditions of pyrolysis and photolysis has been investigated by Aleksandrov and co-workers (3,4). At 90°, oxidation occurred at an appreciable rate (3). However, the reaction was characterized by an induction period which was shortened by an increase in oxygen pressure. The reaction was shown to be autocatalytic; an activation energy of 35.0 kcal/mole was calculated. Ethane, ethylene, butane, acetaldehyde, water, triethyllead hydroxide, diethyllead hydroxide, lead oxide, and lead peroxide were identified as reaction products.

The reaction of tetraethyllead with oxygen in the presence of ultraviolet light was concluded to consist of both light and dark reactions with formation of (triethylplumbyl) ethyl peroxide, $Et_3PbOOEt$, as an initial intermediate (4). Triethyllead ethoxide was obtained as the major end product, along with lesser amounts of the above peroxide, triethyllead hydroxide, and lead monoxide; acetaldehyde and a trace of tar were also formed. The triethyllead ethoxide was postulated to arise from reaction of the peroxide with tetraethyllead; the other lead-containing products were attributed to a free radical dissociation of the peroxide into $Et_3PbO\cdot$ and $EtO\cdot$ radicals followed by subsequent reaction and decomposition.

Tetramethyllead and tetraethyllead will burn in air when heated in the presence of an ignition source (216). Tetramethyllead burns steadily at first but the combustion becomes violent when the liquid becomes hotter and an orange flash of considerable height is produced. Tetraethyllead behaves similarly, but its combustion is less vigorous than that of tetramethyllead. The combustion of tetraethyllead vapors in admixture with heptane was investigated by Gelius and Franke (252); the major product was lead carbonate (52–86% meq %), along with lead monoxide (13–45% meq %) and lesser amounts of lead dioxide and lead nitrite. Tetramethyllead and tetrapropyllead gave similar products. The reaction of tetramethyllead with oxygen in the upper atmosphere is reported to produce a chemiluminescent gas trail which can be used at night to measure winds and other properties of the atmosphere at altitudes above 90 km (347). The combustion of tetraethyllead has been studied extensively relative to its effect as an antiknock agent.

Tetraethyllead is reported to react readily with ozone at $-68°$; acetaldehyde, water, triethyllead hydroxide, diethyllead oxide, triethyllead ethoxide, and ethanol were identified as reaction products (5). Tetraethyltin reacted similarly; in addition to acetaldehyde and water, an organotin oligomer was formed which contained diethyltin and —$CH(CH_3)$— groupings bound together by oxygen or peroxide linkages. Despite the facile reaction of ozone with tetraethyllead, a recent patent describes the use of ozone to remove organobismuth and hexaorganodilead compounds from tetramethyllead and tetraethyllead (139,140). Presumably, the ozone reacts selectively with the organobismuth compound without causing gross decomposition of the tetraalkyllead.

The large majority of tetraorganolead compounds is stable to water; the few exceptions are those derivatives containing an alkenyl or alkynyl group, or a perfluoroalkyl, perfluorovinyl, or pentafluorophenyl group. Although tetrakis(pentafluorophenyl)lead is remarkably stable and is unreactive to acids and halogens (330,331), it is readily hydrolyzed by aqueous caustic and sodium carbonate (329,722); unsymmetrical pentafluorophenyl derivatives, such as pentafluorophenyltriphenyllead and pentafluorophenyltrimethyllead, are hydrolyzed even by neutral water (234). Similarly, trimethylperfluoroethyllead (375) is hydrolyzed by aqueous caustic. The simple tetraalkyllead derivatives are sufficiently stable to hydrolysis so that they are commonly purified by steam distillation. However, Ipatieff (357) observed the formation of a white solid when tetramethyllead and water were heated at 200° in a sealed tube; undoubtedly hydrolysis of the tetramethyllead occurred, since tetramethyllead is stable to about 265°.

Pedinelli and co-workers (564) have reported the synthesis of trialkyllead alkoxides by the reaction of tetramethyllead or tetraethyllead with alcohols at 100°. In view of their excellent stability to steam distillation, the reaction of these tetraalkyllead compounds with alcohols at this temperature is most surprising. Phenols and thiophenols undergo reaction with tetraorganolead compounds at elevated temperatures. The reaction of tetraphenyllead with *p*-nitrophenol and *p*-cresol at 150° gave the lead(II) phenoxide; on the other hand, 2,4,6-trichlorophenol gave only lead(II) chloride (408,410). The reaction of thiophenol and *p*-methylthiophenol with tetraphenyllead at 130° gave the respective lead(II) thiophenoxides (411); similar products were obtained from the reaction of tetraethyllead with thiophenol and *p*-thiocresol at 100° (276).

Henry and Krebs (332) obtained triphenyllead thiophenoxide in about 2% yield from the reaction of tetraphenyllead with thiophenol in refluxing chloroform; most of the tetraphenyllead was recovered unchanged. The preparation of several triethyllead phenoxides is described in the patent literature (8,127); triethyllead picrate was prepared by the reaction of tetraethyllead with picric acid at room temperature in the presence of silica gel (127).

Padberg (351,543) has shown that tetraethyllead and sulfur undergo reaction at 50° in benzene; a greenish-gold solution was formed from which bis(triethyllead) sulfate precipitated upon addition of oxygen. The greenish-gold color was attributed to the presence of bis(triethyllead) sulfide. No organolead products were obtained when a mixture of tetraethyllead and sulfur was irradiated with ultraviolet light at room temperature; only lead metal, lead sulfide, diethyl sulfide and diethyl disulfide were formed (543).

Tetramethyllead and sulfur underwent reaction at room temperature in benzene solution and in the presence of air to form bis(trimethyllead) sulfate; tetra-*n*-butyllead under similar conditions gave dibutyllead sulfate (543). Contrariwise, Schmidt and Schumann (642) obtained no reaction between tetrabutyllead and sulfur at 125° in the absence of a solvent; however, an explosive decomposition was obtained as the temperature was slowly raised above 125°. Lead metal, lead sulfide, and dibutyl sulfide were formed.

The reaction of sulfur with allyltriethyllead in the presence of air gave a yellow oil upon evaporation of the benzene solution (543); oxidation of the oil with air produced a white solid which was postulated to be bis(allyldiethyllead) sulfate since the infrared absorption spectrum of the solid contained an absorption band characteristic of the allyl group. The reaction of benzyltriethyllead with sulfur in benzene in the absence of air gave lead metal, lead sulfide, and a greenish yellow solution, but the composition of the benzene-soluble organolead products could not be established by elemental analyses (543).

The reaction of tetraphenyllead with sulfur at elevated temperatures did not yield a phenyllead sulfide (349,596,642); at 150°, lead sulfide, diphenyl sulfide, and diphenyl disulfide were formed (596,642).

Tetraethyllead dissolved in carbon disulfide to produce a red coloration within a short time from which a black, flocculent solid subsequently precipitated (81). This solid was insoluble in ether, alcohol, and petroleum ether and did not melt at temperatures up to 350°, but it burned readily in air; it was not further characterized. According to Corli (157), the reaction of tetraethyllead with hot carbon disulfide gave lead sulfide.

Tetraorganolead compounds are reported to react with selenium selenocyanate, $Se(SeCN)_2$, under relatively mild conditions to yield triorganolead selenocyanates, according to the equation:

$$R_4Pb + Se(SeCN)_2 \rightarrow R_3PbSeCN + RSeCN + Se$$

Such reactions have been carried out with tetraethyllead, tetramethyllead, and tetraphenyllead in refluxing chloroform to form the corresponding $R_3PbSeCN$ compounds (18). Prolonged refluxing of tetraphenyllead and selenium selenocyanate in chloroform solution also gave some diphenyllead diselenocyanate. Tetraphenyllead and selenium dicyanide underwent slow reaction in boiling benzene; triphenyllead cyanide and phenyl selenocyanate were identified as the products (18).

Tetramethyllead and tetraethyllead undergo a slow reaction with carbon tetrachloride at room temperature; triethyllead chloride was isolated in 90% yield after 30 days (220,486). Hein and co-workers (326) obtained a vigorous reaction between tetraethyllead and carbon tetrabromide or hexabromomethane in the presence of air; diethyllead dibromide and bromophosgene were obtained. Diphenyllead dichloride, hexachloroethane, and chlorobenzene were obtained when a solution of tetraphenyllead in carbon tetrachloride was irradiated with ultraviolet light; these products are indicative of a free radical mechanism (594). Tetraphenyllead also appears to undergo reaction with hot chloroform since benzene and small amounts of perchloroethylene were formed when a mixture of tetraphenyllead and silicon tetrachloride was refluxed in chloroform and 98.9% of the silicon tetrachloride was recovered unchanged (468). The reaction of tetraethyllead with ethyl bromide and other haloalkanes was discussed earlier in this section under Pyrolysis.

Gilman and Apperson (264) obtained no reaction between tetraphenyllead and benzoyl chloride in the absence of a catalyst. However, Browne and Reid (81) report that tetraethyllead reacted with benzoyl chloride in the presence of silica gel to form lead(II) chloride; acetyl chloride, under similar conditions, gave triethyllead chloride. Malinovskii (464) reported that hot acetyl chloride in the presence of aluminum chloride cleaved one phenyl group from tetraphenyllead and proposed that additional phenyl groups can be cleaved successively. Tetraethyllead reacted with chloral at 100° in the absence of a catalyst (263,480); triethyllead chloride was formed.

Reactions of R_4Pb compounds with organic peroxides are described in several Russian publications. Razuvaev and colleagues (592,601) report that tetraethyllead and dibenzoyl peroxide underwent reaction at 80° to form triethyllead benzoate, ethane, ethylene, and butane, along with a small amount of carbon dioxide; tetraethyltin gave similar products. On the basis of the products formed, the reactions were postulated to proceed via a cyclic intermediate of the type

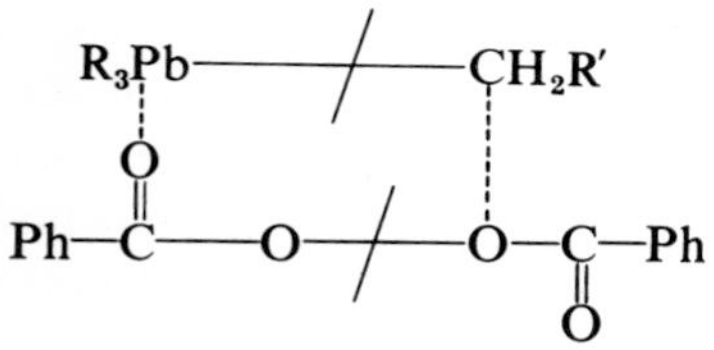

```
R3Pb——/——CH2R'
 :          :
 O          :
 ||         :
Ph—C——O—/—O—C—Ph
             ||
             O
```

as well as a free radical mechanism. On the other hand, the reaction of tetraethylsilane with dibenzoyl peroxide gave $Et_3SiCH(CH_3)CH(CH_3)$-$SiEt_3$. The formation of this silicon derivative from tetraethylsilane suggests that the reaction proceeded exclusively via a free radical mechanism; benzene, benzoic acid, and carbon dioxide were also formed. Razuvaev (601) has also shown that acetylbenzoyl peroxide reacts with tetraethyllead; both triethyllead acetate and diethyllead diacetate were formed, in about a two-to-one mole ratio, respectively.

Diphenyllead dichloride was obtained as the major product from the reaction of tetraphenyllead and dibenzoyl peroxide in refluxing chloroform (539); chlorobenzene, hexachloroethane, benzoic acid, and diphenyllead dibenzoate were also formed. These products are indicative of a free radical reaction initiated by radicals produced from the dibenzoyl peroxide. No reaction was obtained using diacetyl peroxide under similar conditions (538).

A recent patent describes the use of peroxycarboxylic acids to decompose and remove methylbismuth compounds which are formed in the manufacture of tetramethyllead from sodium–lead alloy (141).

No organolead peroxides have been isolated from these reactions of R_4Pb compounds with organic peroxides. However, organolead peroxides have been synthesized by reactions involving triorganolead hydroxides or halides; their synthesis is discussed in Chapter XI.

Tetraethyllead is decomposed at room temperature by an aqueous solution of potassium permanganate. The product has not been characterized but probably is diethyllead oxide or hydroxide. This reaction is a recommended method for decontaminating outdoor areas where tetraethyllead spillage has occurred. According to Austin (14), tetraaryllead compounds do not react with potassium permanganate in acetone solution.

Gilman and Nelson (276) obtained no reaction of tetraethyllead with various nitro, acetylene, azo, or imine compounds. Also, ketene and benzaldehyde did not react with tetraethyllead (354).

Tetraethyllead, in admixture with enough tetranitromethane to form carbon monoxide, water, and lead oxide, is reported to give an explosive composition of unusual strength and rate of detonation (770). The mixture is sensitive to both friction and shock.

Tetramethyllead is reported to undergo facile reaction with aluminum borohydride even at $-60°$ and below (342). The resultant product was speculated to be unstable trimethyllead borohydride. Tetramethyllead also reacted with diborane in dimethoxyethane solution, but only at room temperature or above (342b). The products were postulated to be unstable trimethyllead hydride and methylboranes. Tetravinyllead reacted with diborane in similar fashion with even greater facility; at room temperature reaction occurred with explosive violence (344a).

D. Stabilization of tetraorganolead compounds

As discussed earlier in this Chapter, most tetraalkyllead compounds undergo an exothermic, thermal decomposition reaction which, if not controlled, can give rise to a potentially catastrophic chain reaction. This property becomes an important consideration in the commercial manufacture and use of tetramethyllead, tetraethyllead, and unsymmetrical methylethyllead derivatives which, although very stable at room temperature, can decompose rapidly at elevated temperatures. To minimize this problem, various compounds are added to these antiknock compounds to increase their thermal stability and guard against uncontrolled decomposition reactions during their manufacture, handling, shipment, and storage.

A number of different types of compounds has been shown to be effective as thermal stabilizers for tetraethyllead and tetramethyllead. As mentioned earlier, alkyl halides greatly enhance the thermal stability of tetraethyllead, as does a small amount of air. Numerous patents exist describing the use of various compounds as stabilizers (92,102,142,143,146–156,222,227–229, 235,294,366,367,486,624,628,660,661,730–734); effective compounds comprise aromatic hydrocarbons, including alkylbenzenes and fused ring aromatics, aliphatic hydrocarbons, phenols, carboxylic acids, aldehydes, ethers, alcohols, olefins, cyclic dienes, substituted hexadienes, phosphate esters, alkyl nitrites, aqueous solutions of metal nitrites, nitroalkanes, haloalkanes, lecithin, glycerides, and various other free radical inhibitors and mixtures thereof. As an example of their effectiveness, by the addition of as little as 0.01% styrene, the decomposition of liquid tetraethyllead can be prevented for over 10 hr at 130°. Tetramethyllead is usually manufactured as a 70% solution in toluene, one effect of the toluene being to enhance the thermal stability of the tetramethyllead. The most important class of thermal stabilizers is that of the haloalkanes, since these serve not only the secondary function of stabilizing the tetraalkyllead compound but also the primary function of facilitating the removal of lead from the combustion chamber of the engine. Surprisingly, synergistic effects have been observed for various combinations of the above thermal stabilizers, particularly for combinations containing the haloalkanes (see for example references 151, 223, and 229).

In addition to their effect as thermal stabilizers, many of these compounds are also useful for preventing the air oxidation and photochemical decomposition of tetraalkyllead compounds. For example, mixtures of aromatic hydrocarbons with 1,1-dichloroethane, 1,2-dichloroethane, or dichloropropanes have been found synergistic for this purpose. Numerous patents describe other compounds which are effective for enhancing the light or air stability of tetraalkyllead compounds. These include hydroxythiophenols (629), alkylol amines (92,94), diaryl and dialkyl amines (93,124), aqueous sodium fluoride (485), furalaniline (123), lecithin (540), cinnamylideneimine (125), benzalaniline (126), 2,4-dimethyl-6-*tert*-butylphenol and other phenols (304,396,754), aminophenol hydrohalides (483), *p*-hydroxy-diphenylamine (303,304,412), phenylenediamines (74), and isovalerylactone and certain other pentanediones (144).

Except for the investigations of the Russian workers on the reaction of tetraethyllead with alkyl halides and with oxygen at elevated temperatures discussed earlier, no detailed investigations of the mechanism of the stabilizing activity of these compounds have been reported.

Table VI-7 Symmetrical Tetraorganolead Compounds, R_4Pb

COMPOUND	M.P., °C	B.P., °C	n_D^{20}	REFERENCES
Tetraamyllead		170^{1}	1.4966	352,369,438
Tetra-*dl*-amyllead		$150^{0.5}$	1.4957	369,438
Tetrabenzyllead	66			20,317,422,438,443
Tetra-*o*-bromobenzyllead	122			20
Tetra-*p*-bromophenyllead	187			438
Tetra-*n*-butyllead		$92^{0.0007}$ $108–10^{0.01}$ 140^{1}	1.5119	9,177,352,369,373,374,438,466,637, 654,765
Tetra-*o*-chlorobenzyllead	99.5			20
Tetracyclohexyllead	d. 160			307,317,418,438
Tetra-*p*-dimethylaminophenyllead	187–9; 197–8			15,285,438,585,707
Tetra-*p*-dimethylaminophenyllead tetramethiodide	187–9			284,753
Tetra-*o*-ethoxyphenyllead	219–20			265,438
Tetra-*p*-ethoxyphenyllead	110			265,438
Tetraethyllead		78^{10}; 82^{13}		9,43,131,299,325,363,431,438,462, 541,751,758,774
Tetra-*o*-fluorobenzyllead	74–5			20
Tetraheptyllead				352,373,438,775
Tetrakis(hexadecyl)lead	40.5–1.0; 42			438,479,534
Tetrahexyllead				346

Tetraisoamyllead			1.4947	313,315,420,438,622
Tetraisobutyllead	−23	$101–2^{1.5}$	1.5079; 1.5043	313,315,322,438,550,551,774
Tetraisopropyllead		120^{14}	1.5223	314,315,420,438
Tetramesityllead	242			295
Tetra-*o*-methoxyphenyllead	148–9			265,438
Tetra-*p*-methoxyphenyllead	145–6			265,289,438
Tetramethyllead	−27.5; −30.2	108^{720}; 110^{760};		9,248,438,506,541,546,698
Tetraneopentyllead	139–41			677
Tetraneophyllead	88–9			803
Tetrakis(pentafluorophenyl)lead	199–200; 204–6			234,330,722
Tetraphenethyllead				422,438
Tetraphenyllead	223			15,317,438,765
Tetrapropyllead		$75–80^{0.01}$; 126^{13}	1.5094	9,109,310,315,352,373,420,438,453, 550,551,622,637,650
Tetrakis(tetradecyl)lead	31			438,479
Tetra-*m*-tolyllead	122–3			265,438
Tetra-*o*-tolyllead	201–2			13,14,265,438
Tetra-*p*-tolyllead	240			14,104,109,250,265,274,388a,438, 570,572,719,775,
Tetra-2,5-xylyllead	255			424,438

3. UNSYMMETRICAL TETRAORGANOLEAD COMPOUNDS

The unsymmetrical tetraorganolead compounds, in which two or more of the groups attached to lead are different, are very similar in properties to the symmetrical derivatives; they are listed in Table VI-8. They merit separate discussion primarily because of the special methods employed in their synthesis and because of their utility in establishing the order of reactivity of organic groups bound to lead. Also, certain organic groups form stable organolead derivatives only as unsymmetrical compounds; symmetrical derivatives of these organic groups have not been isolated. For example, no symmetrical *tert*-alkyl derivative has been prepared. Such derivatives have been isolated only as the unsymmetrical derivatives of the type $RPbR_3'$ and R_2PbR_2' where R is *tert*-butyl and R′ is methyl, ethyl, or phenyl (109,266,277,286,287). The only known symmetrical *sec*-alkyl lead derivative is tetra-isopropyllead; *sec*-butyl derivatives are known only as unsymmetrical compounds (266,277). However, the symmetrical tetra-isobutyl and tetra-isoamyl derivatives have been prepared.

Tetra-2-pyridyllead has not been prepared; 2-pyridyltriphenyllead has been prepared, but an attempt to synthesize di-2-pyridyldiphenyllead was unsuccessful (268). On the other hand, tetra-2-thienyllead and tetra-2-furyllead are relatively stable compounds.

Indenyl and fluorenyl derivatives of lead are known only as the triphenyllead and diphenyllead derivatives; attempts to prepare the tetra derivatives via a lead halide reaction gave reduction to lead metal (10). Similarly, tetracyclopentadienyllead has not been prepared, although a number of unsymmetrical cyclopentadienyl derivatives has been described (247). These compounds and the heterocyclic derivatives described above are discussed in later chapters.

Tetra-1-naphthyllead has escaped preparation, although hexa-1-naphthyldilead has been prepared and unsymmetrical 1-naphthyllead derivatives are known (290,299,308,424). Until recently, tetraneopentyllead had not been prepared and an attempt to prepare phenyltrineopentyllead from trineopentyllead iodide and phenylmagnesium chloride gave only hexaneopentyldilead (677,803). Tetraneopentyllead and several alkyltrineopentyllead compounds have now been prepared and their NMR spectra examined (677).

The first synthesis of an unsymmetrical tetraorganolead compound is usually credited to Grüttner, who reported the preparation of diphenyldicyclohexyllead in 1914 (307). In the years that followed, Grüttner, Krause, and colleagues synthesized a large number of unsymmetrical compounds. By far the largest number of known unsymmetrical derivatives are those of the type R_3PbR' or R_2PbR_2', in which only two different organic groups are present. A small number of compounds is known in which three different organic groups are present, $R_2PbR'R''$, and a few compounds are known in which four different organic groups are bonded to lead, $RR'PbR''R'''$. The large majority of these latter compounds was prepared by Krause and his co-workers some 50 years ago.

The physical and chemical properties of the unsymmetrical R_4Pb compounds are very similar to those of the symmetrical derivatives. In general they are more reactive and less thermally stable than the symmetrical derivatives. The tetraalkyl and aryltrialkyl derivatives are usually liquids, while the tetraaryl compounds are usually solids. Generally, the melting point of the unsymmetrical tetraaryl compound is lower than that of either of the respective symmetrical tetraorganolead derivatives. Unsymmetrical dialkyldiaryl and alkyltriaryl derivatives may be liquids or solids. Thus, for example, methyltriphenyllead is a liquid, ethyltriphenyllead is a low melting solid.

Some of the newer unsymmetrical tetraorganolead compounds to be prepared in recent years include di-2-thienyl-di-2-selenieyllead (773), *p*-cyclohexylphenyltriphenyllead and di-*p*-cyclohexylphenyldiphenyllead (582), and a series of substituted aryltricyclohexyllead compounds (210). Bis(4-biphenyl)dimethyllead, instead of 4-biphenyltrimethyllead, was obtained from the reaction of trimethyllead chloride and 4-lithiobiphenyl (174); its formation was attributed to disproportionation of the expected 4-biphenyltrimethyllead. Attempts to prepare 4,4′-bis(trimethylplumbyl)-biphenyl gave a yellow oil which crystallized to ill-defined yellow solids; disproportionation of the desired product was suspected (174). However, 2,2′-biphenyldiphenyllead has been prepared; in this latter compound the lead is bound to the biphenyl moiety as part of a heterocyclic ring system (249). A number of unsymmetrical compounds is known in which one of the organic groups contains a functional group. These are discussed in Chapter XII.

A. Synthesis

Only two methods are generally suitable for the preparation of unsymmetrical R_4Pb compounds and one of these is restricted to the synthesis of compounds of the type R_3PbR'. These two methods are: (*1*) the reaction of a triorganolead halide or diorganolead dihalide with a Grignard or organolithium reagent:

$$R_3PbX + R'MgX \text{ (or } R'Li) \rightarrow R_3PbR' + MgX_2 \text{ (or } LiX)$$

$$R_2PbX_2 + 2R'MgX \text{ (or } 2R'Li) \rightarrow R_2PbR_2' + 2MgX_2 \text{ (or } 2LiX)$$

and (*2*) reaction of an organic halide with a triorganoplumbylmetal compound:

$$R_3PbM + R'X \rightarrow R_3PbR' + MX$$

(*1*) *Syntheses from R_3PbX and R_2PbX_2*

The reaction of an organolead halide with a Grignard reagent or an organolithium reagent represents by far the most general method of synthesis of unsymmetrical tetraorganolead compounds. These reactions are commonly carried out in diethyl ether solvent. By using R_3PbX or R_2PbX_2 as the reactants, an R_3PbR' or R_2PbR_2' compound is formed. Alkyl, aryl, and cycloalkyl derivatives are readily synthesized in this manner. The resultant reaction mixture can be treated with halogen at $-78°$ to produce an unsymmetrical $R_2R'PbX$ compound *in situ*, which in turn can be treated with a different Grignard or lithium reagent to yield an unsymmetrical compound containing three different organic groups, $R_2PbR'R''$. Repetition of this cycle then yields a tetraorganolead derivative containing four different alkyl groups.

$$R_3PbR' + X_2 \rightarrow R_2PbR'X \xrightarrow{R''MgX} R_2PbR'R''$$

$$R_2PbR'R'' + X_2 \rightarrow RR'PbR''X \xrightarrow{R'''MgX} RR'PbR''R'''$$

Using this scheme, Grüttner, Krause, and co-workers prepared a large number of unsymmetrical compounds containing two, three, and four different groups. They found that the cleavage of an organic group by halogen or acid is not random and observed that as a general rule: (*1*) aryl groups are cleaved preferentially to alkyl; (*2*) shorter chain alkyl groups are cleaved more easily than longer chain groups; and (*3*) the more highly substituted the aryl group, the more readily it is cleaved. Over the past 50 years this qualitative order of reactivity has been expanded greatly and a better understanding of this observed order developed from kinetic and theoretical considerations. These data are discussed later in this section.

In general, the reaction of an organolead halide proceeds better with a Grignard reagent than with an organolithium reagent, mainly because in some systems the organolithium reagents present the possibility of the occurrence of a transmetallation reaction with the desired product, leading to undesired by-products (268,282).

$$R_3PbX + R'Li \rightarrow R_3PbR' \xrightarrow{R'Li} R_2PbR_2' + RLi$$

Gilman and co-workers have devised a simple, but clever, solution to this problem. With those organic groups which do not readily form a

Grignard reagent, the lithium reagent is prepared initially and treated with magnesium chloride to form the Grignard reagent and thereby minimize the possibility of a transmetallation reaction. Thus, while the reaction of *p*-dimethylaminophenyllithium and triphenyllead chloride gave a complex mixture of products, *p*-dimethylaminophenyltriphenyllead was obtained in 66% yield when the same reaction was conducted in the presence of magnesium chloride, (282).

A few systems have been encountered in which the Grignard reaction is dependent on either the organolead halide used or the mode of addition of the reactants. Thus, in the synthesis of *tert*-butyltrimethyllead, satisfactory results were obtained using trimethyllead chloride and *tert*-butylmagnesium chloride; on the other hand, trimethyllead bromide gave tetramethyllead and lead metal (118). Similarly, the addition of triphenyllead bromide (or chloride) to *tert*-butylmagnesium chloride is reported to yield only hexaphenyldilead; inverse addition gave the desired *tert*-butyltriphenyllead (286). Fortunately, however, such systems are the exception and not the rule.

The use of organometallic reagents other than the Grignard or organolithium reagent generally is unsatisfactory for the preparation of unsymmetrical tetraorganolead compounds because these other reagents either undergo a transmetallation reaction with the organolead product or they catalyze a redistribution-type reaction in which all five possible compounds are formed (269).

(2) *Syntheses from $R_3PbM + R'X$*

The second method commonly employed for the synthesis of unsymmetrical R_4Pb compounds is the reaction of an organic halide with triorganoplumbylmetal derivatives of the type R_3PbM, where M is usually lithium or sodium:

$$R_3PbM + R'X \rightarrow R_3PbR' + MX$$

However, this reaction is not particularly suitable for the synthesis of R_2PbR_2' compounds from R_2PbM_2 and $R'X$. The R_2PbM_2 compounds are not well characterized; they appear to be unstable and tend to give a complex mixture of products. Conversely, the R_3PbM compounds are relatively stable, at least in solution, and can be prepared by any of several methods. Their synthesis is discussed in Chapter XI.8.

The reaction of R_3PbM with an organic halide has been used mostly for the synthesis of unsymmetrical organo-triaryllead derivatives of the type Ar_3PbR (267), but a few unsymmetrical organo-trialkyllead compounds have also been prepared in this manner (296).

The synthesis of an unsymmetrical derivative of the type $R_2PbR'R''$ was attempted by Gilman and co-workers (285) according to the following scheme:

$$2o\text{-}CH_3C_6H_4Li + PbCl_2 \xrightarrow{-10^\circ} (o\text{-}CH_3C_6H_4)_2Pb$$

$$(o\text{-}CH_3C_6H_4)_2Pb + C_6H_5Li \rightarrow (o\text{-}CH_3C_6H_4)_2Pb(C_6H_5)Li$$

$$(o\text{-}CH_3C_6H_4)_2Pb(C_6H_5)Li + nC_3H_7Br \rightarrow (o\text{-}CH_3C_6H_4)_2Pb(C_6H_5)(C_3H_7)$$

However, a complex mixture of products was obtained. Compounds of the type R_2PbR_2' or $R_2PbR'R''$ might be synthesized by the reaction of an unsymmetrical triorganoplumbylmetal compound with an organic halide according to:

$$R_2R'PbM + R'X \rightarrow R_2PbR_2' + MX$$

$$R_2R'PbM + R''X \rightarrow R_2PbR'R'' + MX$$

Evidence for the existence of phenyldiethylplumbylsodium has been reported but its reactions with organic halides were not investigated (46).

The reaction of triorganoplumbylmetal compounds suffers certain limitations in the synthesis of specific unsymmetrical organolead compounds (47). *sec*-Butyltriethyllead has been prepared in 80% yield by reaction of triethylplumbylsodium in liquid ammonia with *sec*-butyl bromide; however, reaction with *tert*-butyl bromide gave only tetraethyllead. Similarly, benzyltriethyllead was formed only in low yield (~10%) in the reaction of triethylplumbylsodium and benzyl chloride. Phenyltriethyllead was formed in only 15% yield from triethylplumbylsodium and iodobenzene; with bromobenzene, the yield was 77%. Triphenylplumbylsodium is reported to be unreactive with *p*-chlorotoluene, *p*-bromotoluene, *p*-iodonitrobenzene, and 2,4-dinitrochlorobenzene, while *p*-iodotoluene gave *p*-tolyltriphenyllead in less than 5% yield (47). The reaction of triphenylplumbyllithium with *p*-halobenzenesulfonamides or sulfonanilides gave poorer results than the reaction of triphenyllead chloride with the *p*-lithio derivative of the benzenesulfonamide or anilide (232).

Reactions of triorganoplumbyllithium with organic halides have been carried out mostly in diethyl ether solution, although recent results indicate that tetrahydrofuran may be a better solvent because of the greater solubility of triorganoplumbyllithium compounds in tetrahydrofuran (765). The reactions of triorganoplumbylsodium compounds are usually carried out in liquid ammonia or ammonia–ether. The use of ammonia–ether mixed solvent is preferred for reactions with organic halides, possibly because of the increased solubility of the organic halide in the ether.

(3) *Other synthesis methods*

Few other methods of synthesis of unsymmetrical tetraorganolead compound are known which enjoy general utility. The reaction of sodium–lead alloy with a mixture of ethyl chloride and methyl chloride yields a mixture of all five possible ethylmethyl compounds (105). The preparation of a mixture of unsymmetrical methyl-3-pentyllead compounds from NaPb, methyl iodide, and 3-chloropentane (or from magnesium metal and lead chloride) is claimed in the patent literature (6). The redistribution of two different symmetrical R_4Pb derivatives produces unsymmetrical R_4Pb derivatives, but here again the product consists of all five possible R_4Pb compounds. Redistribution reactions of R_4Pb compounds were discussed earlier.

The reaction of lead metal with phenyllithium and ethyl iodide is reported to yield diethyldiphenyllead; with methyl iodide or *p*-iodotoluene, tetraphenyllead was formed in high yield (438,707). Ethyltrimethyllead has been prepared by addition of ethylene to trimethyllead hydride (38).

A Nalco patent (513) claims the synthesis of mixtures of unsymmetrical methyl-ethyl and methyl-*tert*-butyl derivatives of lead by electrolysis, in the presence of a lead anode, of an ethereal Grignard solution containing a mixture of methyl chloride and ethyl chloride, and methyl chloride and *tert*-butyl chloride, respectively.

B. Physical properties

The physical properties of unsymmetrical tetraorganolead compounds are similar to those of the symmetrical derivatives and therefore do not require a detailed discussion here. Most of the available physical properties data on the unsymmetrical derivatives were included in the earlier discussion of the physical properties of the symmetrical derivatives. Boiling points, molecular refractivities, and molecular volumes have been determined for a large number of unsymmetrical tetraalkyl derivatives containing alkyl groups ranging from methyl to amyl, including normal, *iso*- and *sec*-alkyl derivatives (315,369,420). Bond refractions have been calculated for a number of compounds and compared to measured values with generally good agreement (745). The dipole moment of phenyltriethyllead has been reported to be 0.86 μ (463); this value, small as expected, compares with values of 4.66 and 4.88 μ units for the dipole moments of triethyllead chloride and triethyllead bromide, respectively.

The infrared and Raman spectra of the three unsymmetrical methyl-ethyl R_4Pb derivatives have been measured (362). A band occurs very close to 1015 cm^{-1}, the intensity of which increases as the number of ethyl groups in the molecule increases. Therefore, it is concluded to be a fundamental vibration involving largely carbon–carbon stretching. The ultraviolet absorption spectrum of phenyltriethyllead is reported to be very similar to that of tetraphenyllead (70).

Physical properties data are scattered throughout the literature in papers dealing with the synthesis and characterization of specific unsymmetrical tetraorganolead compounds but, except for the references discussed above, few systematic physical properties measurements have been made.

C. Chemical properties

As mentioned earlier, the chemical properties of unsymmetrical tetraorganolead compounds are very similar to those of the symmetrical derivatives except they tend to be more reactive (130,280). Thus, ethyltriphenyllead has been shown to react with trichloroacetic acid about a thousand times faster than does tetraphenyllead and about forty times faster than does tetraethyllead (130). This higher degree of reactivity apparently results from the greater polarity of the molecule from the presence of different organic groups.

No comprehensive studies have been made of the pyrolysis of unsymmetrical tetraorganolead compounds. An investigation of the relative thermal stabilities of the five methyl-ethyl compounds has been reported; the stabilities were found to decrease in the order tetramethyl > trimethylethyl > dimethyldiethyl = methyltriethyl = tetraethyl (632). It was concluded that this order of thermal stability could be related to the antiknock activities.

The major interest in the chemical properties of the unsymmetrical derivatives has been directed at defining the relative order of reactivity of organic groups bound to lead by the determination of which group is preferentially cleaved by electrophilic reagents. As was discussed earlier, Krause and co-workers established early in their investigations in organolead chemistry that cleavage of an organic group bound to lead is not a random process. Some groups are preferentially cleaved over others and general rules concerning the selective cleavage of an alkyl or aryl group from lead were established. As new unsymmetrical organolead derivatives were synthesized, it became common practice to determine which group was preferentially cleaved by acid or halogen. From these numerous investigations, the following orders of relative reactivities with halogens or halogen acids have been established for various organic groups bound to lead.

1. α-naphthyl > *p*-xylyl > *p*-tolyl > phenyl > methyl > ethyl > *n*-propyl > isobutyl > isoamyl > cyclohexyl (422)
2. 2-furyl > 2-thienyl ∼ 2-selenyl > phenyl (288,773)
3. *p*-cyclohexylphenyl > phenyl (582)
4. vinyl > cyclopropyl > ethyl (370)

A similar order applies to the cleavage of organic groups in diorganomercury compounds.

Originally, the relative ease of cleavage of an organic group bound to lead or mercury was considered to be a measure of its relative electronegativity or aromaticity, the more electronegative or aromatic group being preferentially cleaved. However, in 1959 Kaesz, Phillips, and Stone (375,376) reported that an alkyl group is preferentially cleaved in reactions of a perfluoroalkyltrialkyl derivative of lead or tin with hydrogen chloride and boron trichloride. Thus, the reaction of trimethylperfluoromethyllead and trimethylperfluoroethyllead with hydrogen chloride yields methane and not the perfluoroalkane although, from electronegativity considerations alone, cleavage of the perfluoroalkyl group would be expected. This result was interpreted to indicate that the order of cleavage of lead–carbon bonds by electrophilic reagents is actually one based on attack at the carbon atom having the highest electron density. In alkylperfluoroalkyllead compounds, the carbon atom having no carbon to fluorine bonds has a higher electron density than a carbon atom bonded to one or more fluorines because of the strong inductive effect of the fluorine; therefore, the nonfluorinated alkyl group is preferentially cleaved by electrophilic reagents. Conversely, in reactions of these compounds with nucleophilic reagents such as sodium hydroxide, a nucleophilic attack on the lead atom should occur and the perfluoroalkyl group should be cleaved preferentially because of its high polarity. This indeed is what occurs. In fact, the unsymmetrical perfluoroalkyl- and pentafluorophenyllead compounds are readily hydrolyzed by water alone (234).

For alkenyllead derivatives, Gilman, Towne and Jones (290) established the following order of reactivity with electrophilic reagents:

$$CH_2{=}CHCH_2{-} \simeq C_6H_5CH{=}CH{-} > C_6H_5{-} > C_2H_5{-} > CH_2{=}CHCH_2CH_2{-}$$

Also, Juenge and Cook (370) have shown that the vinyl group is cleaved by acid more readily from vinylethyllead compounds than is the ethyl group. [The perfluorovinyl group is cleaved preferentially from alkylperfluorovinyltin compounds (377); in this case the inductive effect of the fluorine atoms in the perfluorovinyl group apparently is not sufficient to overcome the sp_2 hybridization of the carbon atom bound to tin so that the presence of a p-π orbital in the perfluorovinyl group is sufficient to attract the electrophilic reagent.] As expected, the reactivity of an alkenyl group is dependent on the location of its double bond with respect to the lead-to-carbon bond, the ethyl group being cleaved more readily by hydrogen chloride than the 3-butenyl group. This result is consistent with a decreasing inductive effect with increasing displacement of the double bond from the lead-to-carbon bond.

The high reactivity of the allyl group in the above series suggested to Gilman's group (290) that its structure might actually be $CH_3CH{=}CH{-}Pb$. The remarkable reactivity of the allyl group has been noted by others in allyltin and allylmercury compounds. Thus, tetraallyltin has been shown to be about 3×10^8 times more reactive to iodine than is tetraisopropyltin (258). An SE2′ mechanism, free from steric effects, was assumed to explain this high reactivity. On the other hand, Kharasch and Swartz (388) have suggested the following explanation for the high reactivity of allylmercury compounds:

$$CH_2{=}CHCH_2{-}hg + H^+Cl^- \rightarrow \left[CH_3\overset{+}{C}HCH_2Hg\right]Cl^-$$
$$\rightarrow CH_3CH{=}CH_2 + hgCl$$

The reaction of silver nitrate with various unsymmetrical triethyllead derivatives in ethanol solution has been investigated by Glockling and Kingston (296). With all the compounds investigated, cleavage of the unsymmetrical group in Et_3PbR' occurred, where R′ was *p*-trifluoromethylphenyl, benzyl, styryl, *p*-dimethylaminophenyl, and 2-methylprop-2-enyl. It was speculated that the course of these reactions may have been governed by steric factors.

Prior to 1960, little data were available on kinetics of reactions of halogens and halogen acids with tetraorgano derivatives of lead and the other Group IVb metals. However, many papers have been published on this subject in recent years, with Eaborn and Pande and co-workers at the University of Leicester and Gielen and Nasielski at the University of Brussels being particularly active in this area. Eaborn and Pande (209) found that the approximate relative rates of reaction of perchloric acid in water–ethanol with the phenyltriethyl derivatives of the Group IVb metals are in the order: silicon = 1, germanium = 36, tin = 3.5×10^5, lead = 2×10^8; the phenyl group is selectively cleaved. The reactivities of a series of aryltricyclohexyllead compounds in this same system were shown by these same investigators to increase in the order: *m*-chlorophenyl < *p*-chlorophenyl < phenyl < *p*-tolyl < *p*-methoxyphenyl. Eaborn and Pande (210) pointed out that this order of reactivity of the substituted aryllead compounds is consistent with an electrophilic attack at the carbon atom of the aryl–metal bond. The effect of the various substituents on the benzene ring on the reaction rates can be satisfactorily related to substituent effects in common electrophilic reactions, although their effect was found to be smaller than in most other electrophilic substitutions. A plot of the relative reaction rates of the lead compounds (relative to the rate of phenyltricyclohexyllead) versus the relative rates of the analogous tin compounds gives a straight line, indicating that a linear free energy relationship exists between substituent effects in these dearylation reactions of tin and lead. A similar straight line relationship could not be constructed between the lead (or tin) analogs and the germanium or silicon analogs.

These reactions of arylmetal compounds with aqueous acids were also found to show a definite solvent isotope effect (68). Thus, in the reaction of phenyltriethyllead with hydrochloric acid in aqueous dioxane, the rate of cleavage of the phenyl group is appreciably slower in deuterium oxide solution than in water; a similar but smaller difference is observed for the same reaction of the other Group IVb metals. The slower rate in deuterium oxide has been interpreted to indicate that a proton transfer from medium to substrate is involved in the rate determining step. The simplest mechanism consistent with this and the other data on dearylation reactions of Group IVb metals was proposed to be:

$$H_3O^+ + C_6H_5MR_3 \underset{(b)}{\overset{(a)}{\rightleftharpoons}} H_2O + [C_6H_5(H)(MR_3)]^+ \longrightarrow C_6H_6 + R_3M\overset{+}{O}H_2$$

The $k_{H_2O}:k_{D_2O}$ ratios for the lead and tin compounds (3.05 and ~2.5, respectively) approach the theoretical maximum for reactions in which proton transfer is rate controlling (85,86) and have been interpreted to mean that the O—H bond of the oxonium ion is broken in the transition state of step (*a*) above to about the same extent as the C—H bond is formed (68). The lower ratios found for the solvent isotope effects for silicon and germanium suggest that the transition state of step (*a*) in their reactions is closer in structure to the intermediate shown, so that the forming C—H bond is stronger and this compensates for the breaking of the O—H bond. Eaborn and co-workers have not ruled out the possibility of a four-center process or of a more complex cyclic process. However, they consider the four-center process to be unlikely, since it was shown that cleavage by bromine of the *p*-methoxyphenylsilicon bond in *p*-methoxyphenylmethyl-1-naphthylphenylsilane occurs with inversion of configuration of the silicon atom (211). A four-center process would be much more likely for cleavage by bromine than for cleavage by acid.

The increase in the reactivity of the arylmetal compounds with hydrochloric acid in going from silicon to lead is attributed to increasing stabilization of the transition state (68). Any $d\pi$-$p\pi$ bonding between the aromatic ring and the MR_3 moiety probably decreases in the sequence Si > Ge > Sn > Pb; increased bonding of this type would stabilize the initial $ArMR_3$ compound and thereby decrease the rate of its reaction with electrophilic reagents.

Several investigations of the kinetics of the reactions of symmetrical tetraalkyllead compounds with halogens and halogen acids have also been reported which shed further light on the mechanism of reaction of lead–carbon bonds and the reason for the observed order of reactivity of various organic groups bound to lead (184,258–261,611–613). These reactions of symmetrical tetraorganolead compounds have been shown to obey second-order kinetics ($r = k[MR_4][X_2]$) and are concluded to be SE_2-type reactions involving an electrophilic substitution at a saturated carbon atom. Gielen and Nasielski (258) have shown that the reactivity of a series of symmetrical tetraalkyl*tin* compounds with iodine in methanol decreased in the sequence: methyl > ethyl > *n*-propyl > *n*-butyl; they attributed this order of reactivity to an increasing steric repulsion between the incoming and leaving groups. The same investigators (261) also showed that the relative reactivities of tetramethyllead and tetraethyllead with bromine in methanol were found to be in the ratio 1.2 (Me_4Pb/Et_4Pb), versus a (calculated) ratio of 2.4 for the corresponding tin compounds. The lower ratio for the lead derivatives was interpreted to indicate that the difference in the lengths of the metal-to-carbon bonds for tin versus lead is more important than the increased volume of the alkyl group being cleaved. The same results were obtained in other solvents such as acetic acid and dimethyl sulfoxide, the ratio of reactivity of tetramethyltin to tetraethyltin always being greater than that of the analogous lead compounds; however, the magnitude of these ratios is dependent on solvent. A transitory complex of the type has been proposed to be the reactive

M ← $O^{\delta+}$(H)(CH_3); C; Br ······ $Br^{\delta-}$

intermediate, in which M is tin or lead. Support for this structure is found in the observation that the reaction rate is dependent on the ionic strength of the reaction medium (261,613). The very similar sensitivity of the tetraalkyltin and lead compounds to solvent effects in iodometallation reactions also suggested that the mechanisms of cleavage of the carbon–metal bonds are very similar, with bond breaking being the rate determining factor (260).

The iodination of tetramethyllead in various solvents has also been investigated by Riccoboni and co-workers (611,612) using a coulombic–amperometric technique to generate the iodine *in situ.* The relative rates of reaction in different solvents were found to increase in the order: ethanol $= 1 <$ methanol $= 3.4 <$ acetonitrile $= 260$ (612).

The rates of reaction of phenyltrimethyllead and phenyltrimethyltin with iodine in methanol were also investigated by Nasielski's group (184). Both reactions were found to obey second-order kinetics with selective cleavage of the phenyl group; the lead compound reacted at a faster rate. A transitory complex of the type

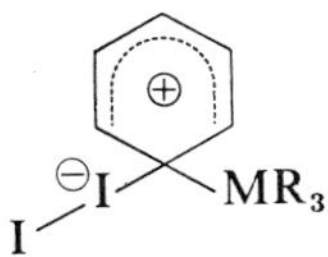

was proposed as the reactive intermediate in these reactions with iodine. This intermediate is similar to that proposed by Eaborn and co-workers (68) for reactions of aryllead and aryltin compounds with aqueous acids. The dependence of reaction rates on the ionic strength of the reaction medium was interpreted as evidence in support of the proposed intermediate.

Finally, Robinson (622) has shown that the rates of acetolysis of a series of tetraalkyllead compounds decrease in the order methyl > ethyl > propyl $\simeq$ butyl $\simeq$ isoamyl; the entropies and enthalpies of activation were very similar for all five compounds. A similar reactivity sequence was found for the reactions of these compounds with perchloric acid in acetic acid, except the reaction rates were much higher with perchloric acid.

The reactivity of organic groups bound to lead with sodium metal in liquid ammonia is in the order (438): allyl > benzyl $\simeq$ *sec*-butyl > *n*-butyl > ethyl > methyl > phenyl > *p*-dimethylaminophenyl. This order is almost the exact reverse of that for cleavage by halogens or acids and therefore is consistent with a mechanism involving a nucleophilic attack of lead at the carbon atom having the lowest electron density.

The kinetic data on the reaction of R_4Pb compounds with halogens and halogen acids teach that all of these reactions involve electrophilic substitution at a saturated carbon atom. The effect of substituents on the organic group is similar to that involved in common electrophilic substitutions. Inductive effects play an important role in these reactions of organolead compounds, although less so than in reactions involving the lighter Group IVb metals or in common electrophilic organic reactions. Solvent and

Table VI-8 *Unsymmetrical Tetraorganolead Compounds*

COMPOUND	M.P., °C	B.P., °C	n_D^{20}	REFERENCES
	A. R_3PbR' *Compounds*			
o-Aminophenyltriphenyllead	164–5			283,438
p-Aminophenyltriphenyllead	166–7			282,283,438
Amyltriethyllead		121^{15}	1.5097	315,438
sec-Amyltriethyllead		118^{13}	1.5158	314,315,438
Amyltriphenyllead	16–7			16,438
(Benz[*a*]anthracen-7-yl)triphenyllead	204–5			438
Benzohydryltriphenyllead	122; d. 130			265,438
Benzyltricyclohexyllead	228			265,438
Benzyltri-*p*-ethoxyphenyllead	76–7			265,438
Benzyltriethyllead		$80^{0.05}$; 149–50.5 (d.)^{13}	$1.5843^{21.4}$	267,271a,296,308,438
Benzyltri-*o*-methoxyphenyllead	80–1			265,438
Benzyltrimethyllead		124^{13}		308,438
Benzyltriphenyllead	91–3; 95–6			265,267,272,285,290,295, 422,438,707,721,765
Benzyltris(pentafluorophenyl)lead	108			330
Benzyltri-*p*-tolyllead	81–2			265,285,438
p-Bromobenzyltriphenyllead	66–8			284,438
[5-(*p*-Bromophenylazo)-2-dimethylamino-phenyl]triphenyllead				283,438
[5-(*p*-Bromophenylazo)-2-hydroxyphenyl]-triphenyllead				283,438
p-Bromophenyltriethyllead		$143^{0.002}$	1.5968	273,438

p-Bromophenyltriphenyllead	115; 121			269,275,438,776
Butyltriethyllead		108^{13}	1.5123	267,287,312,315,438
sec-Butyltriethyllead		103^{13}	1.5186	267,287,314,315,438
tert-Butyltriethyllead		76^{5}		287,438
Butyltriisobutyllead		$145–6^{10}$		177,438
Butyltrimethyllead		64.5^{14}	1.5046	309,315,438
sec-Butyltrimethyllead		59^{13}	1.5133	118,312,438
tert-Butyltrimethyllead	5.7	47^{13}	1.5089	104,109,118,438
tert-Butyltrineopentyllead	75–7			677
Butyltriphenyllead	47			286,438
sec-Butyltriphenyllead	84			286,438
tert-Butyltriphenyllead	150–50.5			286,438
(Carbethoxymethyl)triphenyllead	59–60			393a,438
(α-Carbethoxyphenethyl)triphenyllead	82–4			393a,438
(*p*-Carbomethoxyphenthyl)triphenyllead	125–7			269,273,438
(Carboxymethyl)triethyllead	137 (d.)			438
(Carboxymethyl)triphenyllead	194			438
[5-(*p*-Carboxyphenylazo)-2-dimethylamino-phenyl]triphenyllead				283,438
[5-(*p*-Carboxyphenylazo)-2-hydroxyphenyl]-triphenyllead				283,438
(*p*-Carboxyphenyl)triphenyllead	256–8			269,273,438
[5-(*p*-Chlorophenylazo)-2-dimethylamino-phenyl]triphenyllead				283,483
[5-(*p*-Chlorophenylazo)-2-hydroxyphenyl]-triphenyllead				283,438
m-Chlorophenyltricyclohexyllead	73–4			210

(*continued*)

Table VI-7 (Continued)

COMPOUND	M.P., °C	B.P., °C	n_D^{20}	REFERENCES
p-Chlorophenyltricyclohexyllead	102–3			210
p-Chlorophenyltriphenyllead				275,438
Cyclohexyltriphenyllead	119			421,423,438
p-Cyclohexylphenyltriphenyllead	138			582
[2-Dimethylamino-5-(*p*-iodophenylazo)-phenyl]triphenyllead				283,438
[2-Dimethylamino-5-(*p*-nitrophenylazo)-phenyl]triphenyllead				283,438
p-Dimethylaminophenyltriethyllead		130^{1}; $150^{0.06}$	1.5442^{25}	283,296,438
p-Dimethylaminophenyltriethyllead methiodide	131–2			296
o-Dimethylaminophenyltriphenyllead	101–2			283,438
p-Dimethylaminophenyltriphenyllead	124–5			15,267,438
p-Ethoxyphenyltriphenyllead	119–20			423,438
p-Ethylphenyltriphenyllead				634,635,739
Ethyltriisoamyllead			1.4981	313,315,438
Ethyltriisobutyllead			1.5064	313,315,438
Ethyltrimethyllead		$27–8^{11}$; $128–30^{751}$	1.5154; 1.5132	104,108,109,120,309,312, 315,398,438
Ethyltrineopentyllead			1.5030^{25}	677
Ethyltriphenyllead	42; 49–50			130,267,269,285,421,423, 438,585,707
Ethyltripropyllead		118.2^{14}	1.5120	109,313,315,438
[2-Hydroxy-3,5-bis(*p*-nitrophenylazo)phenyl]-triphenyllead				283,438

(*p*-1-Hydroxyethylphenyl)triphenyllead	68–70			273,438
(*p*-2-Hydroxyethylphenyl)triphenyllead	87–8			273,438
[2-Hydroxy-5(*p*-iodophenylazo)phenyl]-triphenyllead				283,438
(*o*-Hydroxymethylphenyl)triphenyllead	134–6			273,438
(*m*-Hydroxymethylphenyl)triphenyllead	113–4			273,438
(*p*-Hydroxymethylphenyl)triphenyllead	98–100			273,438
[*o*-(2-Hydroxy-1-naphthylazophenyl]-triphenyllead				283,438
[*p*-2-Hydroxy-1-naphthylazophenyl]-triphenyllead	d. 135			282,438
o-Hydroxyphenyltriphenyllead	216–8			271a,283,438
Isoamyltriisobutyllead			1.5012	313,315,438
Isoamyltrimethyllead		69–71[12]	$1.4926^{20.3}$	309,438
Isoamyltripropyllead			1.5051	313,315,438
Isobutyltrimethyllead		58–9[15]; 165–6[769]	1.5050	309,315,438
Isobutyltriphenyllead	68–8.5			286,438
Isobutyltripropyllead			1.5079	313,315,438
Isopropyltrimethyllead		75[60]	1.5095	104,109,314,315,438
Isopropyltrineopentyllead			1.5065^{25}	677
[2-Methoxy-5-(*p*-nitrophenylazo)phenyl]-triphenyllead				283,438
[4-Methoxy-3-(*p*-nitrophenylazo)phenyl]-triphenyllead				283,438
o-Methoxyphenyltriphenyllead	128–9			273,438
Methyltrineopentyllead			1.4980^{25}	677
p-Methoxyphenyltriphenyllead	150–1 (d.)			273,289,438

(*continued*)

Table VI-7 (Continued)

COMPOUND	M.P., °C	B.P., °C	n_D^{20}	REFERENCES
Methyltriphenyllead	60; 61–2			109,421,438,765
Methyltripropyllead		106^{13}	1.5101	109,313,315,438
p-Methylaminophenyltriphenyllead	97–8			283,438
1-Naphthyltriphenyllead	101; 131			290,423,438
9-Phenanthryltriphenyllead	169–71			271a,438
Phenethyltriphenyllead	116–7			422,438
p-Phenoxyphenyltriphenyllead	127			421,438
Phenyltri-*o*-tolyllead	161–2			13,438
Tricyclohexylphenyllead	146–7			210
Tricyclohexyl(*p*-dimethylaminophenyl)lead	125			210
Tricyclohexyl(*p*-dimethylaminophenyl)lead methiodide	185–7			210
Tricyclohexyl-*p*-methoxyphenyllead	78			210
Tricyclohexylphenyllead	146–7			210
Tricyclohexyl-*p*-tolyllead	97–8			210
Triethylisoamyllead		114.5^{13}	1.5099	309,312,315,438
Triethylisobutyllead		102^{10}; 108.2^{16}	$1.5120^{21.5}$	287,309,312,315,438
Triethylisopropyllead		90^{13}	1.5181	314,315,438
Triethylmethyllead		$70\text{–}70.5^{16}$	1.5183	104,108,109,120,299,309, 315,438
Triethyl-1-naphthyllead		176^{13} (d.)		308,438
Triethyloctadecyllead	73–4			438
Triethylphenyllead		$126\text{–}30^{566}$; 135^{12}	1.5732; 1.5762	68,209,267,308,393a,398, 438,463,585
Triethyl-*p*-phthalimidophenyllead		$100^{0.0001}$		283,438

Triethylpropyllead		97^{12-3}; 99.5^{16}	1.5168	109,309,312,315,438
Triethyl-*o*-tolyllead		153.5^{13}	$1.5740^{21.5}$	308,438
Triethyl-*p*-tolyllead		154^{13}	1.5686	308,438
Triethyl-*p*-trifluoromethylphenyllead		$80^{0.06}$		296
Triisoamylisobutyllead			1.4960	313,315,438
Triisoamylmethyllead			1.4971	313,315,438
Triisoamylpropyllead			1.4979	313,315,438
Triisobutylmethyllead		121^{12} (d.)	1.5030	313,315,438
Triisobutylpropyllead			1.5054	313,315,438
Triisopropylmethyllead				109,438
Tri-*o*-methoxyphenyl(triphenylmethyl)lead	145–6			265,438
Trimethylneopentyllead		32.5^{21}	1.4990^{25}	677
Trimethylphenyllead		$77–8^{3}$; 104^{13}	1.5837	70,184,308,438
Trimethylpropyllead		$48–9^{16}$; $151–2^{755}$	$1.5095^{21.4}$	309,312,315,438
Trimethyl-*o*-tolyllead		$117.5–8.0^{13}$	1.5793	308,438
Trimethyl-*p*-tolyllead		$118–9^{13}$	1.5732	308,438
Triphenyl(triphenylmethyl)lead	196–7			265,438
Triphenylpropyllead	69–70			16,285,438,707
Triphenyllead *p*-benzene sulfonanilide	156–61			232
Triphenyllead-*p*-*N*,*N*-diethylbenzene sulfonamide	106–8			232
Triphenyl-*p*-tolyllead	125.5			14,15,267,421,423,424,438, 707
Triphenyl-2,4-xylyllead	111.5–2.0			176,423,438
Triphenyl-2,5-xylyllead	104.5			421,423,438
B. R_2PbR_2' Compounds				
Diamyldi-*dl*-amyllead		160^{1}	1.4923	369,438
Diamyldibutyllead		170^{1}	1.4984	369,438

(continued)

Table VI-7 (Continued)

COMPOUND	M.P., °C	B.P., °C	n_D^{20}	REFERENCES
Di-*dl*-amyldibutyllead		$135^{0.5}$	1.4993	369,438
Diamyldiethyllead		$116–118^{0.5}$	1.5038	369,438
Diamyldihexyllead		$180^{0.5}$		369,438
Diamyldiisoamyllead		$160^{0.5}$	1.4959	369,438
Diamyldiisobutyllead		$140^{0.5}$	1.4994	369,438
Diamyldimethyllead		113^{4}	1.5009	369,438
Diamyldipropyllead		150^{1}	1.5019	369,438
Di(benz[*a*]anthracen-7-yl)diphenyllead	295–6 (d.)			271a,438
Dibenzyldiphenyllead	d. 127; m. 135			422,438
Di(biphenylyl)diphenyllead	134–5			438
Di(*p*-bromophenyl)diphenyllead	123			438
Dibutyldiethyllead		90^{1}	1.5093	323,369,438
Di-*tert*-butyldiethyllead		80^{10}		438
Dibutyldiisoamyllead		$135^{0.5}$	1.4982	369,438
Dibutyldiisobutyllead		135^{1}	1.5021	369,438
Dibutyldimethyllead		108^{10}	1.5049	369,438
Di-*sec*-butyldimethyllead				118,438
Dibutyldiphenyllead				266,438
Di-*sec*-butyldiphenyllead	Unstable			266,438
Di-*tert*-butyldiphenyllead	177			266,438
Dibutyldipropyllead		120^{1}	1.5062	369,438
Di-(*p*-chlorophenyl)diphenyllead				275,438
Dicyclohexyldiphenyllead	178–80 (d.)			271a,307,438
Di(*p*-cyclohexylphenyl)diphenyllead	170			582

Bis(*p*-dimethylaminophenyl)diphenyllead	134–5			15,438
Diethyldiisoamyllead		142^{13}	1.5041	311,315,438
Diethyldiisobutyllead		124^{13}	1.5086	311,315,438
Diethyldiisopropyllead		95.5^{14}	1.5169	311,314,315,438
Diethyldimethyllead		51^{13}	1.5177	104,108,109,120,311,315, 438,495
Diethyldi-1-naphthyllead	116			423,438
Diethyldineopentyllead			1.4994^{25}	677
Diethyldioctadecyllead	58–60			438
Diethyldiphenyllead		176^{8}	1.5959^{18}	280,299,438,495,707
Diethyldipropyllead		99^{10}; 105^{13}	1.5149	104,109,311,315,323,369, 438
Diisoamyldimethyllead		$122–3^{13}$	1.5005	311,315,438
Diisobutyldimethyllead		$95.5–6^{13}$	1.5024	104,109,311,315,438
Diisobutyldiphenyllead	Unstable			266,438
Diisopropyldimethyllead				109,314,438
Dimethyldineopentyllead			1.4973^{25}	677
Dimethyldiphenyllead		$151–2^{2}$	1.6263	12,109,438
Dimethyldipropyllead		72^{10}; $77–8^{13}$	1.5086	109,311,315,369,438
Di-1-naphthyldiphenyllead	197			290,299,423,438
Bis(pentafluorophenyl)diphenyllead	91			330
Di-9-phenanthryldiphenyllead	208–10			271a,438
Diphenyldi-*o*-tolyllead	134–5			436,438
Diphenyldi-*p*-tolyllead	121–2			15,274,438
Diphenyldi-2,5-xylyllead	94			421,438

(continued

Table VI-7 (Continued)

COMPOUND	M.P., °C	B.P., °C	n_D^{20}	REFERENCES
	C. $R_2R'PbR''$ Compounds			
Butyldiethylmethyllead		67^{5}	1.5125	267,438
sec-Butyldiethylmethyllead		87^{10}	1.5180	267,438
Cyclohexyldiphenylmethyllead				421,438
Diethyl-*sec*-amylpropyllead		121^{12}	1.5137	314,315,438
Diethylbutylpropyllead		116^{13}	1.5094	312,315,438
Diethyl-*sec*-butylpropyllead		115^{13}	1.5170	314,315,338
Diethylisoamylisobutyllead		131^{14}	1.5050	312,315,438
Diethylisoamylmethyllead		103^{13}	1.5082	312,315,438
Diethylisoamylpropyllead		127.5^{15}	1.5075	312,315,438
Diethylisobutylmethyllead		87^{13}	1.5117	312,315,438
Diethylisobutylpropyllead		110^{13}	1.5115	312,315,438
Diethylisopropylpropyllead		107^{17}	1.5162	314,315,438
Diethylmethylphenyllead		132^{15}		267,438
Diethylmethylpropyllead		80.5^{15}	1.5150	109,267,312,315,396,438
Diisoamylethylpropyllead		145.5^{13}	1.5034	312,315,438
Dimethyl-*sec*-amylethyllead		90^{15}	1.5114	314,315,438
Dimethyl-*sec*-butylethyllead		74^{13}	1.5128	314,315,438
Dimethyl-*sec*-butylisoamyllead		$111.5–2.5^{14}$	1.5066	314,315,438
Dimethylethylisoamyllead		92^{14}	1.5059	312,315,438

Dimethylethylisobutyllead		74^{13}	1.5081	312,315,438
Dimethylethylisopropyllead		61.2^{15}	1.5135	314,315,438
Dimethylethylpropyllead		65^{15}	1.5118	109,312,315,438
Dimethylisoamylpropyllead		105^{15}	1.5020	312,315,438
Dimethylphenylpropyllead		93^{3}		267,438
Dipropylethylmethyllead				109,438
Di-*o*-tolylphenylpropyllead				16,438
	D. RR′R‴Pb Compounds 49–50			
Butylethylisoamylpropyllead		144^{14}	1.5035	312,315,438
Butylethylmethylpropyllead		103^{13}	1.5083	312,315,438
Ethylisoamylmethylpropyllead		115^{15}	1.5068	312,315,438
(*p-sec*-Octyloxyphenyl)phenylpropyl-*o*-tolyllead	oil			16,438

ionic strength also greatly influence the reaction rate; in general the more polar the solvent, the faster the rate.

The kinetic data for the reactions of tetraorganometals with halogens and halogen acids have advanced greatly our understanding of the mechanism of such reactions and the nature of the metal–carbon bond. More information is needed to provide further elucidation of the nature of the transitory intermediates involved in these reactions and the critical parameters. Undoubtedly, this area of organolead chemistry, and of organometallic chemistry in general, will receive continued investigation.

REFERENCES

1. Adloff, M., and J. P. Adloff, *Compt. rend.*, **259,** 141 (1964).
2. Ainley, A. D., L. A. Elson, and W. A. Sexton, *J. Chem. Soc.*, **1946,** 776.
3. Aleksandrov, Yu. A., T. G. Brilkina, and V. A. Shushunov, *Tr. po Khim. i Khim. Tekhnol.*, **4,** 3 (1961); through *C.A.*, **56,** 492 (1962).
4. Aleksandrov, Yu. A., B. A. Radbil, and V. A. Shushunov, *Zh. Obshch. Khim.*, **37,** 208 (1967); through *C.A.*, **66,** 104991q (1967).
5. Aleksandrov, Yu. A., and N. G. Sheyanov, *Zh. Obshch. Khim.*, **36,** 953 (1966); through *C.A.*, **65,** 8955 (1966).
6. Alleman, G., U.S. Pats. 1,949,948–9 (to Sun Oil Co.), March 6, 1934; through *C.A.*, **28,** 3229 (1934).
7. Allred, A. L., and E. G. Rochow, *J. Inorg. Nucl. Chem.*, **5,** 269 (1958); **20,** 167 (1961).
8. Altamura, M. S., U.S. Pat. 2,171,423 (to Socony Vacuum Oil Co.), Aug. 29, 1940; *C.A.*, **34,** 116 (1940).
9. Amberger, E., and R. Hönigschmid-Grossich, *Chem Ber.*, **98,** 3795 (1965).
10. d'Ans, J., H. Zimmer, and M. v. Brauchitsch, *Chem Ber.*, **88,** 1507 (1955).
11. Anzaldi, O., *Bol. inform. petrol.* (*Buenos Aires*), **312,** 170 (1959); *C.A.*, **53,** 19361 (1959).
12. Associated Lead Manufacturers, Ltd., and F. B. Lewis, Brit. Pat. 854,776, Nov. 23, 1960; through *C.A.*, **55,** 11362 (1961).
13. Austin, P. R., *J. Am. Chem. Soc.*, **53,** 1548 (1931).
14. Austin, P. R., *J. Am. Chem. Soc.*, **53,** 3514 (1931).
15. Austin, P. R., *J. Am. Chem. Soc.*, **54,** 3726 (1932).
16. Austin, P. R., *J. Am. Chem. Soc.*, **55,** 2948 (1933).
17. Avramenko, L. I., M. I. Gerber, M. B. Neiman, and V. A. Shushunov, *Zh. Fiz. Khim. SSSR*, **20,** 1347 (1946); through *C.A.*, **41,** 2969 (1947).
18. Aynsley, E. E., N. N. Greenwood, G. Hunter, and M. J. Sprague, *J. Chem. Soc.* (*A*), **1966,** 1344.
19. Badische Anilin and Soda Fabrik. A.-G. (by H. Müeller), Brit. Pat. 840,619, July 6, 1960; *C.A.*, **55,** 4363 (1961).
20. Bähr, G., and G. Zoche, *Chem Ber.*, **88,** 542 (1955).
21. Bajer, F. J., *Progress in Infrared Spectroscopy*, H. A. Syzmanski, Ed., Plenum Press, N.Y., 1964, pp. 151–76.
22. Bake, L. S., U.S. Pat. 2,029,301 (to E. I. du Pont de Nemours & Co.), Feb. 4, 1936; *C.A.*, **30,** 1811 (1936).
23. Baker, E. B., *J. Chem. Phys.*, **26,** 960 (1957).

24. Bambynek, W., *Z. Physik. Chem.* (*Frankfurt*), **25**, 403 (1960).
25. Bartocha, B., U.S. Pat. 3,100,217, Aug. 6, 1963; *C.A.*, **60**, 551 (1964).
26. Bartocha, B., and M. Y. Gray, *Z. Naturforsch.*, **14b**, 350 (1959).
27. Batalov, A. P., and I. A. Korshunov, *Zh. Obshch. Khim.*, **31**, 1649 (1961); *C.A.*, **55**, 27024 (1961).
28. Baumgartner, W. E., and N. O. Brace, U.S. Pat. 2,917,527 (to E. I. du Pont de Nemours & Co.), Dec. 15, 1959; through *C.A.*, **54**, 6550 (1960).
29. Bawn, C. E. H., and J. Gladstone, *Proc. Chem. Soc.*, **1959**, 227.
30. Bawn, C. E. H., and R. Johnson, *J. Chem. Soc.*, **1960**, 4162.
31. Bawn, C. E. H., and F. J. Whitby, *Discussions Faraday Soc.*, **1947**, 228.
32. Bawn, C. E. H., and F. J. Whitby, *J. Chem. Soc.*, **1960**, 3926.
33. Beaird, F. M., and P. Kobetz, U.S. Pat. 3,188,333 (to Ethyl Corp.), June 8, 1965; *C.A.*, **63**, 13316 (1965).
34. Beaird, F. M., and P. Kobetz, U.S. Pat. 3,188,334 (to Ethyl Corp.), June 8, 1965; *C.A.*, **63**, 13316 (1965).
35. Beaird, F. M., and P. Kobetz, U.S. Pat. 3,226,408 (to Ethyl Corp.), Dec. 28, 1965; *C.A.*, **64**, 9768 (1966).
36. Beaird, F. M., and P. Kobetz, U.S. Pat. 3,226,409 (to Ethyl Corp.), Dec. 28, 1965; *C.A.*, **64**, 9768 (1966).
37. Beatty, H. A., and G. Calingaert, U.S. Pat. 2,270,108 (to Ethyl Corp.), Jan. 13, 1942; *C.A.*, **36**, 3190 (1942).
38. Becker, W. E., and S. E. Cook, *J. Am. Chem. Soc.*, **82**, 6264 (1960).
39. Beeby, M. H., and F. G. Mann, *J. Chem. Soc.*, **1951**, 411.
40. Belluco, U., L. Cattalini, and G. Tagliavini, *Ric. sci. Rend.*, *Sez A*, **32**, 110 (1962); *C.A.*, **57**, 13786 (1962).
41. Belluco, U., L. Riccoboni, and G. Tagliavini, *Ric. sci.*, **30**, 1255 (1960); through *C.A.*, **55**, 7273 (1961).
42. Benning, A. F., and C. A. Sandy, U.S. Pat. 3,239,548 (to E. I. du Pont de Nemours & Co.), March 8, 1966; through *C.A.*, **64**, 15926 (1966).
43. Berezovskaya, F. I., E. K. Varfolomeeva, and V. G. Stefanovskaya, *Zh. Fiz. Khim. SSSR*, **18**, 321 (1944); through *C.A.*, **39**, 2024 (1945).
44. Beste, G. W., U.S. Pat. 2,664,605 (to Ethyl Corp.), Jan. 5, 1954.
45. Beste, G. W., H. Tanner, and H. Shapiro, U.S. Pat. 2,653,159 (to Ethyl Corp.), Sept. 22, 1953; *C.A.*, **48**, 10057 (1954).
46. Bindschadler, E., *Iowa St. Coll. J. Sci.*, **16**, 33 (1941); through *C.A.*, **36**, 4476 (1942).
47. Bindschadler, E., and H. Gilman, *Proc. Iowa Acad. Sci.*, **48**, 273 (1941); through *C.A.*, **36**, 1595 (1942).
48. Biritz, L. F., U.S. Pat. 3,048,611 (to Houston Chemical Corp.), Aug. 7, 1962; *C.A.*, **57**, 16655 (1962).
49. Biritz, L. F., U.S. Pat. 3,057,898 (to Houston Chemical Corp.), Oct. 9, 1962; *C.A.*, **58**, 5723 (1963).
50. Biritz, L. F., U.S. Pat. 3,108,127 (to Houston Chemical Corp.), Oct. 22, 1963; *C.A.*, **60**, 3007 (1964).
51. Blitzer, S. M., M. W. Farrar, T. H. Pearson, and J. R. Zietz, Jr., U.S. Pat. 3,136,795 (to Ethyl Corp.), June 9, 1964; *C.A.*, **61**, 5691 (1964).
52. Blitzer, S. M., and T. H. Pearson, U.S. Pat. 2,859,225 (to Ethyl Corp.), Nov. 4, 1958; *C.A.*, **53**, 9149 (1959); Brit. Pat. 824,848; *C.A.*, **54**, 9847 (1960).
53. Blitzer, S. M., and T. H. Pearson, U.S. Pat. 2,859,226 (to Ethyl Corp.), Nov. 4, 1958; *C.A.*, **53**, 9149 (1959).

54. Blitzer, S. M., and T. H. Pearson, U.S. Pat. 2,859,227 (to Ethyl Corp.), Nov. 4, 1958; *C.A.*, **53,** 9149 (1959).
55. Blitzer, S. M., and T. H. Pearson, U.S. Pat. 2,859,228 (to Ethyl Corp.), Nov. 4, 1958; *C.A.*, **53,** 9149 (1959); Brit. Pat. 826,475; *C.A.*, **54,** 22497 (1960).
56. Blitzer, S. M., and T. H. Pearson, U.S. Pat. 2,859,229 (to Ethyl Corp.), Nov. 4, 1958; *C.A.*, **53,** 9149 (1959).
57. Blitzer, S. M., and T. H. Pearson, U.S. Pat. 2,859,230 (to Ethyl Corp.), Nov. 4, 1958; *C.A.*, **53,** 9149 (1959).
58. Blitzer, S. M., and T. H. Pearson, U.S. Pat. 2,859,231 (to Ethyl Corp.), Nov. 4, 1958; *C.A.*, **53,** 9149 (1959); Brit. Pat. 824,849; *C.A.*, **54,** 5571 (1960).
59. Blitzer, S. M., and T. H. Pearson, U.S. Pat. 2,859,232 (to Ethyl Corp.), Nov. 4, 1958; *C.A.*, **53,** 9149 (1959).
60. Blitzer, S. M., and T. H. Pearson, U.S. Pat. 2,955,124 (to Ethyl Corp.), Oct. 4, 1960; *C.A.*, **55,** 11303 (1961).
61. Blitzer, S. M., and T. H. Pearson, U.S. Pat. 2,985,675 (to Ethyl Corp.), May 23, 1961; *C.A.*, **55,** 25757 (1961).
62. Blitzer, S. M., and T. H. Pearson, U.S. Pat. 2,989,558 (to Ethyl Corp.), June 20, 1961; *C.A.*, **55,** 2345 (1961).
63. Blitzer, S. M., and T. H. Pearson, U.S. Pat. 3,007,955 (to Ethyl Corp.), Jan. 7, 1962; *C.A.*, **56,** 8744 (1962).
64. Bloom, M., *J. Am. Chem. Soc.*, **48,** 649 (1965).
65. Bothorel, P., *Ann. chim. (Paris)*, **4,** 669 (1959).
66. Bothorel, P., and A. Unanue, *Compt. rend.*, **255,** 901 (1962).
67. Bott, L. L., *Hydrocarbon Process Petrol. Refiner*, **44,** 115 (1965); *C.A.*, **62,** 7788 (1965).
68. Bott, R. W., C. Eaborn, and P. M. Greasley, *J. Chem. Soc.*, **1964,** 4804.
69. Bourn, A. J. R., D. G. Gillies, and E. W. Randall, *Proc. Chem. Soc.*, **1963,** 200.
70. Bowden, K., and E. A. Braude, *J. Chem. Soc.*, **1952,** 1068.
71. Braithwaite, D. G. U.S. Pat. 3,007,857 (to Nalco Chemical Co.), Appl. July 31, 1957; *C.A.*, **56,** 3280 (1962).
72. Braithwaite, D. G., U.S. Pat. 3,007,858 (to Nalco Chemical Co.), Appl. May 6, 1959; *C.A.*, **56,** 4526 (1962).
73. Braithwaite, D. G., J. S. D'Amico, P. L. Gross, and W. Hanzel, U.S. Pat. 3,141,841 (to Nalco Chemical Co.), July 21, 1964; *C.A.*, **61,** 10323 (1964).
74. Bramer, H. von, and A. C. Ruggles, U.S. Pat. 2,436,838 (to Eastman Kodak Co.), March 2, 1948; *C.A.*, **42,** 3950 (1948).
75. Brennen, H., *Compt. rend.*, **180,** 282 (1925).
76. Brennen, H., *Ann. Chim. et Phys.*, **1925,** 128.
77. Breynan, Th., *Angew. Chem.*, **62,** 430 (1950).
78. Britton, E. C., U.S. Pat. 1,805,756 (to Dow Chemical Co.), May 19, 1931; through *C.A.*, **25,** 3667 (1931).
79. Brockway, L. O., and H. O. Jenkins, *J. Am. Chem. Soc.*, **58,** 2036 (1936).
80. Brown, D. H., A. Mohammed, and D. W. A. Sharp, *Spectrochim. Acta*, **21,** 659 (1965).
81. Browne, O. H., and E. E. Reid, *J. Am. Chem. Soc.*, **49,** 830 (1927).
82. Buckler, E. J., and R. G. Norrish, *J. Chem. Soc.*, **1936,** 1567.
83. Buckton, G. B., *Ber.*, **109,** 218 (1859).
84. Bunton, C. A., and V. J. Shiner, Jr., *J. Am. Chem. Soc.*, **83,** 3207 (1961).
85. Bunton, C. A., and V. J. Shiner, Jr., *J. Am. Chem. Soc.*, **83,** 3214 (1961).
86. Burg, A., *Studies, on Boron Hydrides*, Univ. of Southern California, 9th Report to U.S. Office of Naval Research, Nov. 1, 1955.

87. Burton, M., J. E. Ricci, and T. W. Davis, *J. Am. Chem. Soc.*, **62,** 265 (1940).
88. Cahours, A., *Compt. rend.*, **36,** 1001 (1853).
89. Cahours, A., *Ann. Chem.*, **62,** 257 (1861).
90. Calcott, W. S., U.S. Pat. 1,559,405 (to E. I. du Pont de Nemours & Co.), Oct. 27, 1925; through *C.A.*, **20,** 209 (1926).
91. Calcott, W. S., and H. W. Daudt, U.S. Pat. 1,692,926 (to E. I. du Pont de Nemours & Co.), Nov. 27, 1928; through *C.A.*, **23,** 608 (1929).
92. Calcott, W. S., and A. E. Parmelee, U.S. Pat. 1,724,640 (to E. I. du Pont de Nemours & Co.), Aug. 13, 1929; *C.A.*, **23,** 4816 (1929).
93. Calcott, W. S., and A. E. Parmelee, U.S. Pat. 1,835,140–1 (to E. I. du Pont de Nemours & Co.), Dec. 8, 1931; through *C.A.*, **26,** 997 (1932).
94. Calcott, W. S., and A. E. Parmelee, U.S. Pat. 1,843,942 (to E. I. du Pont de Nemours & Co.), Feb. 9, 1932; through *C.A.*, **26,** 1945 (1932).
95. Calcott, W. S., A. E. Parmelee, and F. B. Lorriman, U.S. Pat. 1,664,021, March 27, 1928; through *C.A.*, **22,** 1677 (1928).
96. Calcott, W. S., A. E. Parmelee, and H. F. Meschter, U.S. Pat. 1,944,167 (to E. I. du Pont de Nemours & Co.), Jan. 23, 1934; *C.A.*, **28,** 1982 (1934).
97. Calcott, W. S., A. E. Parmelee, and H. F. Meschter, U.S. Pat. 1,962,173 (to E. I. du Pont de Nemours & Co.), June 12, 1934; through *C.A.*, **28,** 4847 (1934).
98. Calcott, W. S., A. E. Parmelee, and J. L. Stecher, U.S. Pat. 1,983,535 (to E. I. du Pont de Nemours & Co.), Dec. 11, 1934; through *C.A.*, **29,** 817 (1935); Brit. Pat. 453,271, Sept. 1, 1936; through *C.A.*, **31,** 1043 (1937).
99. Calingaert, G., *Chem. Rev.*, **2,** 43 (1925).
100. Calingaert, G., U.S. Pat. 1,539,297, May 26, 1925; *C.A.*, **19,** 2210 (1925).
101. Calingaert, G., U.S. Pat. 1,622,233, March 22, 1927; through *C.A.*, **21,** 1546 (1927).
102. Calingaert, G., U.S. Pat. 2,660,591–5 (to Ethyl Corp.), Nov. 24, 1953; *C.A.*, **48,** 2085 (1954).
103. Calingaert, G., U.S. Pat. 2,660,596 (to Ethyl Corp.), Nov. 24, 1953; *C.A.*, **48,** 2085 (1954).
104. Calingaert, G., and H. A. Beatty, *J. Am. Chem. Soc.*, **61,** 2748 (1939).
105. Calingaert, G., and H. A. Beatty, U.S. Pat. 2,270,109 (to Ethyl Corp.), Jan. 13, 1942; *C.A.*, **36,** 3190 (1942).
106. Calingaert, G., and H. A. Beatty, "The Redistribution Reaction," in *Organic Chemistry, An Advanced Treatise*, Vol. II, 2nd edition, H. Gilman, Ed., Wiley, N.Y., 1943, Chapter 24, pp. 1806–20.
107. Calingaert, G., H. A. Beatty, and L. Hess, *J. Am. Chem. Soc.*, **61,** 3300 (1939).
108. Calingaert, G., H. A. Beatty, and H. R. Neal, *J. Am. Chem. Soc.*, **61,** 2755 (1939).
109. Calingaert, G., H. A. Beatty, and H. Soroos, *J. Am. Chem. Soc.*, **62,** 1099 (1940).
110. Calingaert, G., and H. Shapiro, *J. Am. Chem. Soc.*, **63,** 947 (1941).
111. Calingaert, G., and H. Shapiro, U.S. Pat. 2,535,190 (to Ethyl Corp.), Dec. 26, 1950; *C.A.*, **45,** 3864 (1951).
112. Calingaert, G., and H. Shapiro, U.S. Pat. 2,535,191–3 (to Ethyl Corp.), Dec. 26, 1950; *C.A.*, **45,** 3865 (1951).
113. Calingaert, G., and H. Shapiro, U.S. Pat. 2,558,207 (to Ethyl Corp.), June 26, 1951; *C.A.*, **46,** 131 (1952).
114. Calingaert, G., and H. Shapiro, U.S. Pat. 2,562,856 (to Ethyl Corp.), July 31, 1951; *C.A.*, **46,** 1581 (1952).
115. Calingaert, G., and H. Shapiro, U.S. Pat. 2,591,509 (to Ethyl Corp.), April 1, 1952; *C.A.*, **46,** 11229 (1952).
116. Calingaert, G., H. Shapiro, and I. T. Krohn, *J. Am. Chem. Soc.*, **68,** 520 (1946).

117. Calingaert, G., H. Shapiro, and I. T. Krohn, *J. Am. Chem. Soc.*, **70,** 270 (1948).
118. Calingaert, G., and H. Soroos, *J. Org. Chem.*, **2,** 535 (1938).
119. Calingaert, G., and H. Soroos, *J. Am. Chem. Soc.*, **61,** 2758 (1939).
120. Calingaert, G., H. Soroos, and H. Shapiro, *J. Am. Chem. Soc.*, **62,** 1104 (1940).
121. Calingaert, G., H. Soroos, and H. Shapiro, U.S. Pat. 2,390,988 (to Ethyl Corp.), Dec. 18, 1945; *C.A.*, **40,** 1025 (1946).
122. Calingaert, G., H. Soroos, and G. W. Thomson, *J. Am. Chem. Soc.*, **62,** 1542 (1940).
123. Cantrell, T. L., and C. L. Suplee, U.S. Pat. 2,296,199 (to Gulf Oil Corp.); through *C.A.*, **37,** 1595 (1943).
124. Cantrell, T. L., and C. L. Suplee, U.S. Pat. 2,296,200 (to Gulf Oil Corp.); through *C.A.*, **37,** 1595 (1943).
125. Cantrell, T. L., and C. L. Suplee, U.S. Pat. 2,303,818 (to Gulf Oil Corp.), Dec. 1, 1942; through *C.A.*, **37,** 3263 (1943).
126. Cantrell, T. L., and C. L. Suplee, U.S. Pat. 2,303,819 (to Gulf Oil Corp.), Dec. 1, 1942; through *C.A.*, **37,** 3263 (1943).
127. Carothers, W. H., U.S. Pat. 2,008,003 (to E. I. du Pont de Nemours & Co.), July 16, 1935; *C.A.*, **29,** 5862 (1935); Brit. Pat. 408,967, April 18, 1934.
128. Carson, A. S., R. Cooper, and D. R. Stranks, *Radioisotopes Phys. Sci. Ind., Proc. Conf. Use, Copenhagen*, **3,** 495 (1960); *C.A.*, **57,** 9300 (1962).
129. Carson, A. S., R. Cooper, and D. R. Stranks, *Trans. Faraday Soc.*, **58,** 2125 (1962).
130. Catlin, W. E., *Iowa St. Coll. J. Sci.*, **10,** 65 (1935); through *C.A.*, **30,** 935 (1936).
131. Chakravarty, N. K., *Indian J. Technol.*, **1,** 181 (1963); through *C.A.*, **59,** 3611 (1963).
132. Challenger, F., and E. Rothstein, *J. Chem. Soc.*, **1934,** 1258.
133. Choudhuri, B. K., C. R. Viswanathan, S. S. Vats, and A. R. Aijar, *Defence Sci. J.*, **2,** 34 (1961); through *C.A.*, **56,** 628 (1962).
134. Claridge, R. F. C., E. Merz, and J. H. Riedel, *Nukleonik*, **7,** 53 (1965); through *C.A.*, **62,** 16293 (1965).
135. Clem, W. J., and R. J. Plunkett, U.S. Pat. 2,515,821 (to E. I. du Pont de Nemours & Co.), July 18, 1950; *C.A.*, **44,** 9978 (1950).
136. Clem, W. J., and H. Podolsky, U.S. Pat. 2,426,598 (to E. I. du Pont de Nemours & Co.), Sept. 2, 1947; *C.A.*, **42,** 203 (1948).
137. Closson, R. D., U.S. Pat. 3,231,511 (to Ethyl Corp.), Jan. 25, 1966; *C.A.*, **64,** 11251 (1966).
138. Clouston, J. G., and C. L. Cook, *Nature*, **179,** 1240 (1957); *Trans. Faraday Soc.*, **54,** 10001 (1958).
139. Collier, H. E. Jr., U.S. Pat. 3,270,042 (to E. I. du Pont de Nemours & Co.), Aug. 30, 1966; through *C.A.*, **65,** 15428 (1966).
140. Collier, H. E., Jr., J. W. Eberlin, and W. S. Hillman, U.S. Pat. 3,277,134 (to E. I. du Pont de Nemours & Co.), Oct. 4, 1966; *C.A.*, **66,** 95197u (1967).
141. Collier, H. E., Jr., and G. S. Hammond, U.S. Pat. 3,187,028 (to E. I. du Pont de Nemours & Co.), June 1, 1965; through *C.A.*, **63,** 17770 (1965).
142. Collier, H. E., Jr., and J. D. Sterling, Jr., U.S. Pat. 3,274,224 (to E. I. du Pont de Nemours & Co.), Sept. 20, 1966; *C.A.*, **65,** 19910 (1966).
143. Compagnia Italiana Petrolio S.p.A., Ital. Pat. 546,378, July 18, 1956; *C.A.*, **53,** 4135 (1959).
144. Condo, A. C., Jr., U.S. Pat. 3,261,674 (to Atlantic Refining Co.), July 19, 1966; through *C.A.*, **65,** 10402 (1966).
145. Cook, C. L., and J. G. Clouston, *Nature*, **177,** 1178 (1956).

146. Cook, S. E., U.S. Pat. 3,081,326 (to Ethyl Corp.), March 12, 1963; *C.A.*, **58,** 13688 (1963).
147. Cook, S. E., U.S. Pat. 3,147,294 (to Ethyl Corp.), Sept. 1, 1964; *C.A.*, **61,** 13345 (1964).
148. Cook, S. E., and H. Shapiro, U.S. Pat. 3,021,350 (to Ethyl Corp.), Feb. 13, 1962; *C.A.*, **56,** 11899 (1962).
149. Cook, S. E., and H. Shapiro, U.S. Pat. 3,038,916–9 (to Ethyl Corp.), June 12, 1962; *C.A.*, **57,** 8810 (1962).
150. Cook, S. E., and T. O. Sistrunk, U.S. Pat. 3,049,558 (to Ethyl Corp.), Aug. 14, 1962; *C.A.*, **57,** 16656 (1962).
151. Cook, S. E., and T. O. Sistrunk, U.S. Pat. 3,221,037 (to Ethyl Corp.), Nov. 30, 1965; *C.A.*, **64,** 6378 (1966).
152. Cook, S. E., and T. O. Sistrunk, U.S. Pat. 3,221,039 (to Ethyl Corp.), Nov. 30, 1965; *C.A.*, **64,** 4841 (1966).
153. Cook, S. E., and T. O. Sistrunk, French Pat. 1,367,570 (to Ethyl Corp.), July 24, 1964; *C.A.*, **62,** 10276 (1965).
154. Cook, S. E., and W. H. Thomas, U.S. Pat. 3,098,089 (to Ethyl Corp.), July 16, 1963; *C.A.*, **59,** 7289 (1963).
155. Cook, S. E., and W. H. Thomas, U.S. Pat. 3,133,097–9 (to Ethyl Corp.), May 12, 1964; *C.A.*, **61,** 1695 (1964).
156. Cook, S. E., and W. H. Thomas, U.S. Pat. 3,133,103 (to Ethyl Corp.), May 12, 1964; *C.A.*, **61,** 4393 (1964).
157. Corli, B., *Ann. Chim. Applicata*, **25,** 634 (1935); through *C.A.*, **30,** 5389 (1936).
158. Cornish, E. H., *J. Applied Chem.* (*London*), **11,** 41 (1961).
159. Costa, G., A. Camus, and N. Marsich, *J. Inorg. Nucl. Chem.*, **27,** 281 (1965).
160. Costa, G., A. M. Camus, and E. Pauluzzi, *Gazz. chim. ital.*, **86,** 997 (1956); through *C.A.*, **53,** 1122 (1959).
161. Costa, G., and G. De Alti, *Gazz. chim. ital.*, **87,** 1273 (1957); through *C.A.* **52,** 19452 (1958).
162. Costa, G., and G. De Alti, *Atti. Accad. Naz. Lincei, Rend., Classe Sci. Fis. Mat. Nat.*, **28,** 1 (1960).
163. Costa, G., and G. De Alti, *Atti Accad. Naz. Lincei. Rend., Classe Sci. Fis. Mat. Nat.*, **28,** 845 (1960); through *C.A.*, **55,** 10305, (1961).
164. Costa, G., G. De Alti, and S. Lin, *Atti. Accad. Naz. Lincei, Rend., Classe Sci Fis. Mat. Nat.*, **31,** 265 (1961); through *C.A.*, **58,** 1336 (1963).
165. Costa, G., G. De Alti, and N. Sillani, *Gazz. chim. ital.*, **86,** 77 (1956); through *C.A.*, **50,** 12719 (1956).
166. Costa, G., G. De Alti, and N. Sillani, *Univ. studi Trieste Fac. sci. Ist. chim.*, **No. 26** (1958); through *C.A.*, **53,** 13776 (1959).
167. Costa, G., G. De Alti, and L. Stefani, *Atti Accad. Naz. Lincei, Rend., Classe Sci. Fis. Mat. Nat.*, **31,** 267 (1961); through *C.A.*, **57,** 8599 (1962).
168. Costa, G., G. De Alti, L. Stefani, and G. Boscarato, *Ann. Chim.* (*Rome*), **52,** 289 (1962); through *C.A.*, **57,** 12523 (1962).
169. Cramer, P. L., *J. Am. Chem. Soc.*, **56,** 1234 (1934).
170. Cramer, P. L., *J. Am. Chem. Soc.*, **60,** 1406 (1938).
171. Cramer, W. A., *Dissertation, University of Amsterdam*, 1961.
172. Cresswell, W. T., J. Leicester, and A. I. Vogel, *Chem. Ind.* (*London*), **1953,** 19.
173. Crowder, G. A., G. Gorin, F. H. Kruse, and D. W. Scott, *J. Mol. Spectry.*, **16,** 115 (1965).
174. Curtis, M. D., and A. L. Allred, *J. Am. Chem. Soc.*, **87,** 2554 (1965).

175. Dahlig, W., S. Pasynkiewicz, and K. Wazynski, *Przemysl Chem.*, **39,** 436 (1960); through *C.A.*, **55,** 15335 (1961).
176. Dannin, J., S. R. Sandler, and B. Baum, *Intern. J. Appl. Radiation Isotopes*, **16,** 589 (1965); *C.A.*, **64,** 2979 (1966).
177. Danzer, R., *Monatsh.*, **46,** 241 (1925).
178. Daudt, H. W., U.S. Pat. 1,705,723 (to E. I. du Pont de Nemours & Co.), March 19, 1929; through *C.A.*, **23,** 2192 (1929).
179. Daudt, H. W., U.S. Pat. 1,749,567 (to E. I. du Pont de Nemours & Co.), March 4, 1930; through *C.A.*, **24,** 2138 (1930).
180. Daudt, H. W., U.S. Pat. 1,798,593 (to E. I. du Pont de Nemours & Co.), March 31, 1931; through *C.A.*, **25,** 3016 (1931); Brit. Pat. 283,913, Jan. 20, 1927; through *C.A.*, **22,** 4134 (1928); French Pat. 642,120, Oct. 8, 1927; through *C.A.*, **23,** 1143 (1929).
181. Daudt, H. W., A. E. Parmelee, and K. P. Monroe, U.S. Pat. 1,717,961 (to E. I. du Pont de Nemours & Co.), June 18, 1929; through *C.A.*, **23,** 3931 (1929).
182. Davis, M., and F. G. Mann, *J. Chem. Soc.*, **1964,** 3770.
183. Davis, M., and F. G. Mann, *J. Chem. Soc.*, **1964,** 3786.
184. Delhaye, A., J. Nasielski, and M. Planchon, *Bull. Soc. Chim. Belges*, **69,** 134 (1960).
185. Dessy, R. E., and J. Y. Kim, *J. Am. Chem. Soc.*, **83,** 1167 (1961).
186. Dessy, R. E., and P. M. Weissman, *J. Am. Chem. Soc.*, **88,** 5124 (1966).
187. De Vries, J. E., A. Lauw.-Zecha, and A. Pellecer, *Anal. Chem.*, **31,** 1995 (1959).
188. Dibeler, V. H., *J. Res. Natl. Bur. Stand.*, **49,** 235 (1952).
189. Dibeler, V. H., and F. L. Mohler, *J. Res. Natl. Bur. Stand.*, **47,** 337 (1950).
190. Dickson, R. S., and D. O. West, *Australian J. Chem.*, **15,** 710 (1962).
191. Dillon, T., R. Clarke, and V. Hinchy, *Sci. Proc. Roy. Dublin Soc.*, **17,** 53 (1922); through *C.A.*, **16,** 4125 (1922).
192. Dobratz, E. H., U.S. Pat. 2,816,123 (to Koppers Co.), Dec. 10, 1957; *C.A.*, **52,** 7344 (1958).
193. Downing, F. B., and L. S. Bake, U.S. Pat. 2,000,069 (to E. I. du Pont de Nemours, & Co.), May 7, 1935; through *C.A.*, **29,** 4026 (1935).
194. Drago, R. S., *J. Inorg. Nucl. Chem.*, **15,** 237 (1960).
195. Drago, R. S., *Record Chemical Prog.*, **26,** 157 (1965).
196. Drago, R. S., and N. A. Matwiyoff, *J. Organometal. Chem.*, **3,** 62 (1965).
197. Dresskamp, H., *Z. Physik Chem.* (*Frankfurt*), **38,** 121 (1963); *Z. Naturforsch.*, **19a,** 139 (1964).
198. Duffy, R., J. Feeney, and A. K. Holliday, *J. Chem. Soc.*, **1962,** 1144.
199. Dull, M. F., and J. H. Simons, *J. Am. Chem. Soc.*, **55,** 3898 (1933).
200. Dull, M. F., and J. H. Simons, *J. Am. Chem. Soc.*, **55,** 4328 (1933).
201. Duncan, A. B. F., and J. W. Murray, *J. Chem. Phys.*, **2,** 146 (1934); **2,** 636 (1934).
201a. Duncan, J. F., and F. G. Thomas, *J. Inorg. Nucl. Chem.*, **29,** 869 (1967).
202. E. I. du Pont de Nemours & Co., Brit. Pat. 469,244, July 19, 1937; through *C.A.*, **32,** 479 (1938).
203. E. I. du Pont de Nemours & Co., Brit. Pat. 481,498, March 7, 1938; through *C.A.*, **32,** 6612 (1938).
204. E. I. du Pont de Nemours & Co., Brit. Pat. 949,925, Feb. 19, 1964; through *C.A.*, **61,** 3935 (1964).
205. E. I. du Pont de Nemours & Co., Brit. Pat. 1,015,227, Dec. 31, 1965; through *C.A.*, **64,** 8240 (1966)
206. E. I. du Pont de Nemours & Co., (by F. S. Arimoto), French Pat. 1,328,932,

June 7, 1963; through *C.A.*, **60**, 550 (1964); Ger. Pat. 1,168,430, April 23, 1964; through *C.A.*, **61**, 1893 (1964).
207. E. I. du Pont de Nemours & Co. (by R. L. Pedrotti and C. A. Sandy), French Pat. 1,406,132, July 16, 1965; *C.A.*, **63**, 14904 (1965).
208. E. I. du Pont de Nemours & Co., Neth. Appl. 6,508,049, Dec. 24, 1965; through *C.A.*, **64**, 17048 (1966).
209. Eaborn, C., and K. C. Pande, *J. Chem. Soc.*, **1960**, 1566.
210. Eaborn, C., and K. C. Pande, *J. Chem. Soc.*, **1961**, 3715.
211. Eaborn, C., and O. M. Steward, *J. Chem. Soc.*, **1965**, 521.
212. Ebert, G., *Ion*, **2**, 177 (1911); through *C.A.*, **5**, 829 (1912).
213. Edgar, G., *J. Chem. Ed.*, **31**, 560 (1954).
214. Edwards, R. R., and C. D. Coryell, *Conf. Chem. Effects of Nuclear Transformations, AECU 50* (*BNL-C-7*), 63–73 (1948); through *C.A.*, **45**, 4148 (1951).
215. Edwards, R. R., J. M. Day, and R. F. Overman, *J. Chem. Phys.*, **21**, 1555 (1953).
216. Egerton, A. C., and S. Rudrakanchana, *Proc. Royal Soc.* (*London*), **A225**, 427 (1954).
217. Eltenton, G. C., *J. Chem. Phys.*, **15**, 465 (1947).
218. Erhard, K., *Naturwiss.*, **49**, 417 (1962).
219. Escolar, L. G., and E. N. Martinez, *Anales real. soc. espan. fis. y quim.*, **53B**, 161 (1957); through *C.A.*, **53**, 12647 (1959).
220. Ethyl Corporation, Unpublished Data.
221. Ethyl Corp., Brit. Pat. 842,090, July 20, 1960; *C.A.*, **55**, 5199 (1961).
222. Ethyl Corp., Brit. Pat. 949,268, Feb. 12, 1964; *C.A.*, **60**, 14311 (1964).
223. Ethyl Corp., Brit. Pat. 976,972, Dec. 2, 1964; *C.A.*, **62**, 6326 (1965).
224. Ethyl Corp. (by R. D. Closson), French Pat. 1,362,696, June 5, 1964; *C.A.*, **62**, 4052 (1965).
225. Ethyl Corp. (by P. Kobetz and F. M. Beaird), French Pat. 1,372,724, Sept. 18, 1964; through *C.A.*, **62**, 586 (1965).
226. Ethyl Corp. (by A. P. Giraitis), Ger. Pat. 1,046,617, Dec. 18, 1958; *C.A.*, **55**, 383 (1961).
227. Ethyl Corp., Neth. Appl. 6,402,472, Sept. 14, 1964; *C.A.*, **62**, 10277 (1965).
228. Ethyl Corp., Neth. Appl. 6,403,049, Sept. 21, 1965; *C.A.*, **64**, 6694 (1966).
229. Ethyl Corp., Neth. Appl. 6,412,633, May 3, 1965; *C.A.*, **63**, 11223 (1965).
230. Farbwerke Hoechst A.-G., Brit. 839,370, June 29, 1960; *C.A.*, **55**, 3435 (1961), Belg. Pat. 547,962.
231. Farbwerke Hoechst, A.-G., Ger. Pat. 1,089,382, Sept. 22, 1961; through *C.A.*, **56**, 1480 (1962); Belg. Pat. 553,477.
232. Feldhake, C. J., and C. D. Stevens, *J. Chem. Eng. Data*, **9**, 241 (1964); **8**, 196 (1963).
233. Feldman, M. H., J. E. Ricci, and M. Burton, *J. Chem. Phys.*, **10**, 618 (1942).
234. Fenton, D. E., and A. G. Massey, *J. Inorg. Nucl. Chem.*, **27**, 329 (1965).
235. Fields, E. K., and A. E. Brehm, U.S. Pat. 2,901,335 (to Standard Oil Co. of Indiana), Aug. 25, 1959; *C.A.*, **53**, 22889 (1959).
236. Fischer, H. H., A. E. Parmelee, and J. L. Stecher, U.S. Pat. 2,276,031 (to E. I. du Pont de Nemours & Co.), March 10, 1942; through *C.A.*, **36**, 4472 (1942).
237. Flitcroft, N., and H. D. Kaesz, *J. Am. Chem. Soc.*, **85**, 1377 (1963).
238. Fowles, G. W. A., and D. Nicholls, *Quart. Rev.*, **16**, 19 (1962).
239. Frankland, E., and A. Lawrence, *J. Chem. Soc.*, **35**, 244 (1879).
240. Fraser, R. G. J., and T. N. Jewitt, *Proc. Royal Soc.* (*London*), **A160**, 563 (1937); *Phys. Rev.*, **50**, 1091 (1936).

241. French, F. A., and R. S. Rasmussen, *J. Chem. Phys.*, **14,** 389 (1946).
242. Frevel, L. K., H. W. Rinn, and H. C. Anderson, *Ind. Eng. Chem., Anal. Ed.*, **18,** 83 (1946).
243. Frey, F. W., Unpublished Data.
244. Frey, F. W., and S. E. Cook, *J. Am. Chem. Soc.*, **82,** 530 (1960).
245. Frey, F. W., P. Kobetz, G. C. Robinson, and T. O. Sistrunk, *J. Org. Chem.*, **26,** 2950 (1961).
246. Friedman, M., U.S. Pat. 2,392,846 (to United States of America), Jan. 15, 1946; *C.A.*, **40,** 1038 (1946).
247. Fritz, H. P., and K. E. Schwarzhans, *Chem. Ber.*, **97,** 1390 (1964).
248. Garzuly-Janke, R., *J. prakt. Chem.*, **142,** 141 (1935).
249. Gelius, R. von, *Angew. Chem.*, **72,** 322 (1960).
250. Gelius, R. von, *Z. anorg. allgem. Chem.*, **334,** 72 (1964).
251. Gelius, R. von, *Z. anorg. allgem. Chem.*, **349,** 22 (1967).
252. Gelius, R. von, and W. Franke, *Brennstoff-Chem.*, **47** (9), 280 (1966).
252a. Gelius, R. von and R. Müller, *Z. anorg. allgem. Chem.*, **351,** 42 (1967).
253. Genta, V., and A. Ansaloni, *Gazz. chim. ital.*, **84,** 921 (1954); through *C.A.*, **49,** 5993 (1955).
254. George, W. H., *Proc. Royal Soc.* (*London*), **113A,** 585 (1927).
255. Gershbein, L. L., and V. N. Ipatieff, *J. Am. Chem. Soc.*, **74,** 1540 (1952).
256. Ghate, M. R., and K. N. Bhide, *Indian J. Chem.*, **2,** 243 (1964); *C.A.*, **61,** 9019 (1964).
257. Giacomello, G., *Gazz. chim. ital.*, **68,** 422 (1938); through *C.A.*, **32,** 9046 (1938).
258. Gielen, M., and J. Nasielski, *Bull. Soc. Chim. Belges*, **71,** 32 (1962).
259. Gielen, M., and J. Nasielski, *Bull. Soc. Chim. Belges*, **71,** 601 (1962).
260. Gielen, M., and J. Nasielski, *J. Organometal. Chem.*, **7,** 273 (1967).
261. Gielen, M., J. Nasielski, J. E. Dubois, and P. Fresnet, *Bull. Soc. Chim. Belges*, **73,** 293 (1964).
262. Gilbert, O. G., U.S. Pat. 2,621,199 (to Ethyl Corp.), Dec. 9, 1952; *C.A.*, **47,** 9996 (1953).
263. Gilman, G., and R. K. Abbott, *J. Org. Chem.*, **8,** 224 (1943).
264. Gilman, H., and L. D. Apperson, *J. Org. Chem.*, **4,** 162 (1939).
265. Gilman, H., and J. C. Bailie, *J. Am. Chem. Soc.*, **61,** 731 (1939).
266. Gilman, H., and L. Balassa, *Iowa St. Coll. J. Sci.*, **3,** 105 (1929); through *C.A.*, **23,** 4201 (1929).
267. Gilman, H., and E. Bindschadler, *J. Org. Chem.*, **18,** 1675 (1953).
268. Gilman, H., W. A. Gregory, and S. M. Spatz, *J. Org. Chem.*, **16,** 1788 (1951).
269. Gilman, H., A. H. Haubein, G. O'Donnell, and L. A. Woods, *J. Am. Chem. Soc.*, **67,** 922 (1945).
270. Gilman, H., and R. G. Jones, *J. Am. Chem. Soc.*, **68,** 517 (1946).
271. Gilman, H., and R. G. Jones, *J. Am. Chem. Soc.*, **72,** 1760 (1950).
271a. Gilman, H., and R. W. Leeper, *J. Org. Chem.*, **16,** 466 (1951).
272. Gilman, H., O. L. Marrs, and S. Y. Sim, *J. Org. Chem.*, **27,** 4232 (1962).
273. Gilman, H., and D. S. Melstrom, *J. Am. Chem. Soc.*, **72,** 2953 (1950).
274. Gilman, H., and F. W. Moore, *J. Am. Chem. Soc.*, **62,** 3206 (1940).
275. Gilman, H., F. W. Moore, and R. G. Jones, *J. Am. Chem. Soc.*, **63,** 2482 (1941).
276. Gilman, H., and J. Nelson, *J. Am. Chem. Soc.*, **59,** 935 (1937).
277. Gilman, H., and J. Robinson, Unpublished Data, Reference 122 in *Chem. Rev.*, **54,** 101 (1954) by R. W. Leeper, L. Summers, and H. Gilman.
278. Gilman, H., and J. Robinson, *J. Am. Chem. Soc.*, **49,** 2315 (1927).

279. Gilman, H., and J. Robinson, *Rec. trav. chim.*, **49,** 766 (1930).
280. Gilman, H., E. L. St. John, N. B. St. John, and M. Lichtenwalter, *Rec. trav. chim.*, **55,** 577 (1936).
281. Gilman, H., S. M. Spatz, and M. J. Kolbezen, *J. Org. Chem.*, **18,** 1341 (1953).
282. Gilman, H., and C. G. Stuckwisch, *J. Am. Chem. Soc.*, **64,** 1007 (1942).
283. Gilman, H., and C. G. Stuckwisch, *J. Am. Chem. Soc.*, **72,** 4553 (1950).
284. Gilman, H., and L. Summers, *J. Am. Chem. Soc.*, **74,** 5924 (1952).
285. Gilman, H., L. Summers, and R. W. Leeper, *J. Org. Chem.*, **17,** 630 (1952).
286. Gilman, H., O. R. Sweeney, and J. E. Kirby, *Iowa St. Coll. J. Sci.*, **3,** 1 (1928); through *C.A.*, **23,** 1888 (1929).
287. Gilman, H., O. R. Sweeney, and J. Robinson, *Rec. trav. chim.*, **49,** 205 (1930).
288. Gilman, H., and E. B. Towne, *Rec. trav. chim.*, **51,** 1054 (1932).
289. Gilman, H., and E. B. Towne, *J. Am. Chem. Soc.*, **61,** 739 (1939).
290. Gilman, H., E. B. Towne, and H. L. Jones, *J. Am. Chem. Soc.*, **55,** 4689 (1933).
291. Gilman, H., and L. A. Woods, *J. Am. Chem. Soc.*, **65,** 435 (1943).
292. Gilman, H., and R. V. Young, *J. Am. Chem. Soc.*, **57,** 1121 (1935).
293. Giraitis, A. P., U.S. Pat. 3,177,130 (to Ethyl Corp.), April 6, 1965; *C.A.*, **63,** 1816 (1965).
294. Giraitis, A. P., H. Shapiro, and J. D. Johnston, U.S. Pat. 3,133,105 (to Ethyl Corp.), May 12, 1964; *C.A.*, **61,** 2894 (1964).
295. Glockling, F., K. Hooton, and D. Kingston, *J. Chem. Soc.*, **1961,** 4405.
296. Glockling, F., and D. Kingston, *J. Chem. Soc.*, **1959,** 3001.
297. Goddard, A. E., J. N. Ashley, and R. B. Evans, *J. Chem. Soc.*, **121,** 978 (1922).
298. Goddard, A. E., and D. Goddard, *J. Chem. Soc.*, **121,** 256 (1922).
299. Goddard, A. E., and D. Goddard, *J. Chem. Soc.*, **121,** 482 (1922).
300. Good, W. D., and D. W. Scott, *Pure Appl. Chem.*, **2,** 77 (1961).
301. Good, W. D., and D. W. Scott, "Combustion in a Bomb of Organometallic Compounds" in *Experimental Thermochemistry*, Vol. 2, H. A. Skinner, Ed., Interscience, N.Y., 1962, pp. 57–76
302. Good, W. D., D. W. Scott, J. L. Lacina, and J. P. McCullough, *J. Phys. Chem.*, **63,** 1139 (1959).
303. Goppoldova, M., and Z. Smrz, *Ropa Uhlie*, **6,** 107 (1964); through *C.A.*, **61,** 5424 (1964).
304. Goppoldova, M., and Z. Smrz, *Sb. Praci Vyzkumu Chem. Vyuziti Uhli, Dehtu, Ropy*, **2,** 141 (1962); through *C.A.*, **62,** 5117 (1965).
305. Grohn, H., and R. Paudert, *Chem. Tech.* (*Berlin*), **12,** 430 (1960); *C.A.*, **55,** 4086 (1961).
306. Groll, H. P. A., U.S. Pat. 1,938,180 (to Shell Development Co.), Dec. 5, 1934; *C.A.*, **28,** 1053 (1934).
307. Grüttner, G., *Ber.*, **47,** 3257 (1914).
308. Grüttner, G., and G. Grüttner, *Ber.*, **51,** 1293 (1918).
309. Grüttner, G., and E. Krause, *Ber.*, **49,** 1125 (1916).
310. Grüttner, G., and E. Krause, *Ber.*, **49,** 1415 (1916).
311. Grüttner, G., and E. Krause, *Ber.*, **49,** 1546 (1916).
312. Grüttner, G., and E. Krause, *Ber.*, **50,** 202 (1917).
313. Grüttner, G., and E. Krause, *Ber.*, **50,** 278 (1917).
314. Grüttner, G., and E. Krause, *Ber.*, **50,** 574 (1917).
315. Grüttner, G., and E. Krause, *Ann.*, **415,** 338 (1918).
316. Guccione, E., *Chem. Eng.*, **72,** 102 (1965).

317. Hardmann, M., and P. Backes, Ger. Pat. 508,667 (1926); Chem. Zentr., **1930,** II, 3195.
318. Harrah, L. A., M. T. Ryan, and C. Tamborski, *Spectrochim. Acta*, **16,** 12 (1962).
319. Hartle, R. J., U.S. Pat. 3,055,925 (to Gulf Research and Dev. Corp.), Sept. 25, 1962; through *C.A.*, **59,** 1681 (1963).
320. Harvey, M. C., and W. H. Nebergall, *Appl. Spectry.*, **18,** 21 (1962).
321. Hausen, M., *Der Aufbau der Zweistofflegierugen*, Julius Springer, Berlin, 1936.
322. Heap, R., and B. C. Saunders, *J. Chem. Soc.*, **1949,** 2983.
323. Heap, R., B. C. Saunders, and G. J. Stacey, *J. Chem. Soc.*, **1951,** 658.
324. Hedden, G. D., U.S. Pat. 3,110,719 (to E. I. du Pont de Nemours & Co.), Nov. 12, 1963; *C.A.*, **60,** 3006 (1964).
325. Hein, F., A. Klein, and H. J. Mesee, *Z. anal. Chem.*, **115,** 177 (1939).
326. Hein, F., E. Nebe, and W. Reimann, *Z. anorg. allgem. Chem.*, **251,** 125 (1943).
327. Hein, F., E. Petzchner, K. Wagler, and F. A. Segitz, *Z. anorg. allgem. Chem.*, **141,** 161 (1924).
328. Henry, M. C., *Inorg. Chem.*, **1,** 917 (1962).
329. Henry, M. C., and H. Gorth, *Quarterly Report No. 15, to Intern. Lead Zinc Res. Organ.*, Project No. LC-28, Jan.–March, 1965.
330. Henry, M. C., and K. Hills, *Quarterly Report No. 14, Project LC-28, to Intern. Lead Zinc Res. Organ.*, New York, Oct.–Dec., 1964.
331. Henry, M. C., and K. Hills, *Summary Report No. 14A to Intern. Lead Zinc Res. Organ.*, Proj. LC-28, Dec. 31, 1964.
332. Henry, M. C., and A. W. Krebs, *J. Org. Chem.* **28,** 225 (1963).
333. Henry, M. C., and J. G. Noltes, *J. Am. Chem. Soc.*, **82,** 555 (1960).
334. Herz, W., *Z. anorg. allgem. Chem.*, **182,** 173 (1929).
335. Hetnarski, B., and T. Urbanski, *Tetrahedron*, **19,** 1319 (1963).
336. Hevesy, G. von, and L. Zechmeister, *Ber.*, **53,** 410 (1920).
337. Hobrock, B. G., and R. W. Kiser, *J. Phys. Chem.*, **65,** 2186 (1961).
338. Hoffmann, K. A., and V. Wolfe, *Ber.*, **40,** 2425 (1907).
339. Holbrook, G. E., U.S. Pat. 2,464,397 (to E. I. du Pont de Nemours & Co.), March 15, 1949; *C.A.*, **43,** 4287 (1949).
340. Holbrook, G. E., U.S. Pat. 2,464,398 (to E. I. du Pont de Nemours & Co.), March 15, 1949; *C.A.*, **43,** 4287 (1949).
341. Holbrook, G. E., U.S. Pat. 2,464,399 (to E. I. du Pont de Nemours & Co.), March 15, 1949; *C.A.*, **43,** 4287 (1949).
342. Holliday, A. K., and W. Jeffers, *J. Inorg. Nucl. Chem.*, **6,** 134 (1958).
342a. Holliday, A. K., and G. N. Jessop, *J. Chem. Soc. A*, **1967,** 889.
342b. Holliday, A. K., and G. N. Jessop, *J. Organometal. Chem.*, **10,** 291 (1967).
343. Holliday, A. K., and G. Pass, *J. Chem. Soc.*, **1958,** 3485.
344. Holliday, A. K., and R. E. Pendlebury, *J. Chem. Soc.*, **1965,** 6659.
344a. Holliday, A. K., and R. E. Pendlebury, *J. Organometal. Chem.*, **10,** 295 (1967).
345. Holt, L. C., U.S. Pat. 2,344,872 (to E. I. du Pont de Nemours & Co.), March 21, 1944; *C.A.*, **38,** 3776 (1944).
346. Honeycutt, J. B., and J. M. Riddle, *J. Am. Chem. Soc.*, **82,** 3051 (1960).
347. Hord, R. A., and H. B. Tolefson, *Virginia J. Sci.*, **16,** 105 (1965); *C.A.*, **63,** 16086 (1965).
348. Horie, S., *Nippon Kagaku Zasshi*, **78,** 1795 (1957); through *C.A.*, **53,** 21868 (1959).
349. Horn, H., *Dissertation, Univ. of Aachen*, 1964.
350. Howard, H. E., W. C. Ferguson, and L. R. Snyder, *Anal. Chem.*, **32,** 1814 (1960).

351. Huber, F., and F. J. Padberg, *19th Internat. Congr. Pure Appl. Chem.*, (*London*), *Div. A* (1963), 196.
351a. Huber, F., and F. J. Padberg, *Z. anorg. allgem. Chem.*, **351,** 1 (1967).
352. Hügel, G., *Kolloid-Z.*, **131,** 4 (1953); *C.A.*, **47,** 9087 (1953).
353. Hurd, C. D., and P. R. Austin, *J. Am. Chem. Soc.*, **53,** 1543 (1931).
354. Hurd, C. D., and A. S. Roe, *J. Am. Chem. Soc.*, **61,** 3355 (1939).
355. Hurd, D. T., and E. G. Rochow, *J. Am. Chem. Soc.*, **67,** 1057 (1945).
356. Imhausen, A., Ger. Pat. 362,814, Sept. 29, 1923; *J. Soc. Chem. Ind.* (*London*), **42,** 315 (1923).
357. Ipatieff, V. N., in *Catalytic Reactions at High Pressures and Temperatures*, Macmillan, N.Y., 1937, Chapter VII, pp. 350–9.
358. Ipatieff, V. N., G. A. Razuvaev, and I. F. Bogdanov, *Ber.*, **63,** 335 (1930).
359. Ipatieff, V. N., G. A. Razuvaev, and I. F. Bogdanov, *J. Russ. Phys. Chem. Soc.*, **61,** 1791 (1929); through *C.A.*, **24,** 2660 (1930).
360. Ismailzade, I. G., and G. S. Zhdanov, *Zhur. Fiz. Khim. SSSR*, **26,** 1619 (1952); *C.A.*, **49,** 2811 (1955); **26,** 1139 (1952); *C.A.*, **49,** 2147 (1955).
361. Isslieb, K., and F. Krech, *Z. anorg. allgem. Chem.*, **328,** 21 (1964).
362. Jackson, J. A., and J. R. Nielsen, *J. Mol. Spectry.*, **14,** 320 (1964).
363. Jahr., K. F., *Chem. Tech.* (*Berlin*), **2,** 88 (1950).
364. Jarvie, J. M. S., M. J. Schuler, and J. D. Sterling Jr., U.S. Pat. 3,048,610 (to E. I. du Pont de Nemours & Co.), Aug. 7, 1962; *C.A.*, **58,** 550 (1963).
365. Jenkner, H. (to Kali-Chemie, A.-G.), U.S. Pat. 3,072,697, May 21, 1962; U.S. Pat. 3,061,647, Oct. 30, 1962; Brit. Pat. 768,765, Feb. 10, 1957; *C.A.*, **52,** 421 (1958).
366. Johnston, J. D., H. Shapiro, and A. P. Giraitis, U.S. Pat. 3,097,222 (to Ethyl Corp.), July 9, 1963; *C.A.*, **59,** 7289 (1963).
367. Johnston, J. D., H. Shapiro, and A. P. Giraitis, U.S. Pat. 3,133,104 (to Ethyl Corp.), May 12, 1964; *C.A.*, **61,** 2894 (1964).
368. Jones, L. W., and L. P. Werner, *J. Am. Chem. Soc.*, **40,** 1257 (1918).
369. Jones, W. J., D. P. Evans, T. Gulwell, and D. C. Griffiths, *J. Chem. Soc.*, **1935,** 39.
370. Juenge, E. C., and S. E. Cook, *J. Am. Chem. Soc.*, **81,** 3578 (1959).
371. Juenge, E. C., and R. D. Houser, *J. Org. Chem.*, **29,** 2040 (1964).
372. Juenge, E. C., and D. Seyferth, *J. Org. Chem.*, **26,** 563 (1961).
373. Kadomtzeff, I., *Compt. rend.*, **226,** 661 (1948).
374. Kadomtzeff, I., *Bull. soc. chim. France*, **1949,** D394.
375. Kaesz, H. D., J. R. Phillips, and F. G. A. Stone, *Chem. Ind.* (*London*), **1959,** 1409.
376. Kaesz, H. D., J. R. Phillips, and F. G. A. Stone, *J. Am. Chem. Soc.*, **82,** 6228 (1960).
377. Kaesz, H. D., S. L. Stafford, and F. G. A. Stone, *J. Am. Chem. Soc.*, **81,** 6336 (1959).
378. Kaplin, Yu. A., L. F. Kudryavtsev, and G. Petukhov, *Zh. Obshch. Khim.*, **36,** 1061 (1966); through *C.A.*, **65,** 10617 (1966).
379. Karamyan, A. S., *Dokl. Akad. Nauk SSSR*, **68,** 269 (1949); *C.A.*, **44,** 3368 (1950).
380. Kashireninov, O. E., *Materialy 4-oi* (*Chetvertoi*) *Nauchn. Konf. Aspirantov* (*Rostov-on-Don: Rostovsk. Univ.*) *Sb.*, **1962,** 144; through *C.A.*, **60,** 10792 (1964).
381. Kay, J., and F. S. Rowland, *J. Am. Chem. Soc.*, **80,** 3165 (1958).

382. Kerk, G. J. M. van der, and S. G. A. Luijten, *J. Applied Chem.* (*London*), **4,** 307 (1954).
383. Kettering, C. F., and W. W. Sleator, *Physics*, **4,** 39 (1933).
384. Kharasch, M. S., U.S. Pat. 1,987,685 (to E. I. du Pont de Nemours & Co.), Jan. 15, 1935; *C.A.*, **29,** 1436 (1935).
385. Kharasch, M. S., U.S. Pat. 2,453,138 (to United States of America), Nov. 9, 1948; *C.A.*, **44,** 2746 (1950).
386. Kharasch, M. S., U.S. Pat. 2,504,134 (to United States of America), April 18, 1950; *C.A.*, **44,** 5581 (1950).
387. Kharasch, M. S., E. V. Jensen, and S. Weinhouse, *J. Org. Chem.*, **14,** 429 (1949).
388. Kharasch, M. S., and S. Swartz, *J. Org. Chem.*, **3,** 405 (1938).
388a. Kitching, W., V. G. Kumar Das, and P. R. Wells, *Chem. Commun.*, **1967,** 356.
389. Klose, G., *Ann. Physik*, **10,** 391 (1963).
390. Kobetz, P., and F. M. Beaird, U.S. Pat. 3,188,332 (to Ethyl Corp.), June 8, 1965; *C.A.*, **63,** 13316 (1965).
391. Kobetz, P., and R. C. Pinkerton, U.S. Pat. 3,028,321 (to Ethyl Corp.), April 3, 1962; *C.A.*, **58,** 3457 (1963).
392. Kobetz, P., and R. C. Pinkerton, U.S. Pat. 3,028,322 (to Ethyl Corp.), April 3, 1962; *C.A.*, **57,** 11235 (1962).
393. Kobetz, P., and R. C. Pinkerton, U.S. Pat. 3,028,323 (to Ethyl Corp.), April 3, 1962; *C.A.*, **58,** 3457 (1963).
393a. Kocheshkov, K. A., and A. P. Aleksandrov, *Ber.*, **67,** 527 (1934).
394. Kocheshkov, K. A., and R. Kh. Freidlina, *Izvest. Akad. Nauk SSSR, Otdel Khim. Nauk* **1950,** 203; through *C.A.*, **44,** 9342 (1950).
395. Kocheshkov, K. A., and R. Kh. Freidlina, *Uchenye Zapiski Moskov. Gosudarst. Univ. M. V. Lomonosova*, No. 132, *Org. Khim.*, **7,** 144 (1950); through *C.A.*, **50,** 7728 (1956).
396. Kolka, A. J., and G. G. Ecke, U.S. Pat. 2,836,609–10 (to Ethyl Corp.), May 27, 1958; *C.A.*, **52,** 16735 (1958); Brit. Pat. 828,910, Feb. 24, 1960; through *C.A.*, **54,** 10302 (1960).
397. Kolka, A. J., and I. T. Krohn, U.S. Pat. 2,621,200 (to Ethyl Corp.), Dec. 9, 1952; *C.A.*, **47,** 10550 (1953); Brit. Pat. 719,913, Dec. 8, 1954.
398. Komarov, N. V., T. I. Ermolova, and N. F. Chernov, *Izv. Akad. Nauk SSSR. Ser. Khim.*, **1966,** 1679; through *C.A.*, **66,** 65605h (1967).
399. Kornilov, I. I., *Compt. rend. reunion ann. avec. comm. thermodynam. Union intern. phys.* (*Paris*), *1952, Changements de phases 291;* through *C.A.*, **47,** 7304 (1953).
400. Korshak, V. V., and G. S. Kolesnikov, *Usp. Khim.*, **15,** 326 (1946).
401. Korshunov, I. A., R. V. Amenitskaya, A. A. Orlova, and A. P. Batalov, *Zh. Obshch. Khim.*, **29,** 1992 (1959); *C.A.*, **54,** 8248 (1960).
402. Korshunov, I. A., and A. P. Batalov, *Zh. Obshch. Khim.*, **29,** 3135 (1959); *C.A.*, **54,** 12981 (1960).
403. Korshunov, I. A., A. P. Batalov, and A. A. Orlova, *Radiokhimiya*, **1,** 679 (1959); through *C.A.*, **54,** 18033 (1960).
404. Korshunov, I. A., and A. A. Orlova, *Zh. Obshch. Khim.*, **28,** 45 (1958); *C.A.*, **52,** 11769 (1958).
405. Koton, M. M., *Ber.*, **66,** 1213 (1933).
406. Koton, M. M., *Zh. Obshch. Khim.*, **4,** 653 (1934); through *C.A.*, **29,** 3662 (1935).
407. Koton, M. M., *J. Am. Chem. Soc.*, **56,** 1118 (1934).
408. Koton, M. M., *Zh. Obshch. Khim.*, **17,** 1307 (1947); through *C.A.*, **42,** 1903 (1948).

409. Koton, M. M., *Zh. Obshch. Khim.*, **18,** 936 (1948); *C.A.*, **43,** 559 (1949).
410. Koton, M. M., E. P. Moskvina, and F. S. Florinskii, *Zh. Obshch. Khim.*, **19,** 1675 (1949); through *C.A.*, **44,** 1436 (1950).
411. Koton, M. M., E. P. Moskvina, and F. S. Florinskii, *Zh. Obshch. Khim.*, **20,** 2093 (1950); *C.A.*, **45,** 5644 (1951).
412. Kozyreva, L. S., *Trudy Tsentral. Nauch.-Issledovatel. Lab., Glavnoe Upravlenie Gosudarst. Material. Reservov pri Sovete Ministrov SSSR,* **1956,** No. 4, 167 (1956); through *C.A.*, **56,** 1669 (1962).
413. Kraus, C. A., and C. C. Callis, U.S. Pat. 1,612,131 (to Standard Oil Dev. Co.), Dec. 28, 1926; *C.A.*, **21,** 593 (1927).
414. Kraus, C. A., and C. C. Callis, U.S. Pat. 1,655,908 (to Standard Oil Development Co.), Jan. 10, 1928; *C.A.*, **22,** 940 (1928).
415. Kraus, C. A., and C. C. Callis, U.S. Pat. 1,690,075 (to Standard Oil Development Co.), Oct. 30, 1928; *C.A.*, **23,** 245 (1929).
416. Kraus, C. A., and C. C. Callis, U.S. Pat. 1,694,268 (to Standard Oil Development Co.), Dec. 4, 1928; through *C.A.*, **23,** 970 (1929).
417. Kraus, C. A., and C. C. Callis, U.S. Pat. 1,697,245 (to Standard Oil Dev. Co.); *C.A.*, **23,** 1262 (1929); Canadian Pat. 266,538, Dec. 7, 1926; through *C.A.*, **21,** 917 (1927).
418. Krause, E., *Ber.*, **54,** 2060 (1921).
419. Krause, E., *Ber.*, **62,** 1877 (1929).
420. Krause, E., and A. von Grosse, *Die Chemie der metall-organischen. Verbindungen,* Borntraeger, Berlin, (1937) pp. 372-429.
421. Krause, E., and O. Schlöttig, *Ber.*, **58,** 427 (1925).
422. Krause, E., and O. Schlöttig, *Ber.*, **63,** 1381 (1930).
423. Krause, E., and N. Schmitz, *Ber.*, **52,** 2150 (1919).
424. Krause, E., and N. Schmitz, *Ber.*, **52,** 2165 (1919).
425. Krohn, I. T., U.S. Pat. 2,727,053 (to Ethyl Corp.), Dec. 13, 1955; *C.A.*, **50,** 10761 (1956).
426. Krohn, I. T., and H. Shapiro, U.S. Pat. 2,594,183 (to Ethyl Corp.), April 22, 1952; *C.A.*, **47,** 145 (1953).
427. Krohn, I. T., and H. Shapiro, U.S. Pat. 2,594,225 (to Ethyl Corp.), April 22, 1952; *C.A.*, **47,** 146 (1953).
428. Kropp, J. L., and M. Burton, *J. Chem. Phys.*, **37,** 1752 (1962).
429. La Paglia, S. R., *J. Mol. Spectr.*, **7,** 427 (1961).
430. La Paglia, S. R., *Spectrochim. Acta*, **18,** 1295 (1962).
431. Latham, K. G., Brit. Pat. 800,609, Aug. 27, 1958; through *C.A.*, **53,** 19880 (1959).
432. Lautsch, W. F., *Chem. Tech.* (*Berlin*), **10,** 419 (1958).
433. Law, H. D., *J. Chem. Soc.*, **101,** 1016 (1912).
434. Law, H. D., *J. Chem. Soc.*, **101,** 1544 (1912).
435. Law, H. D., *Proc. Chem. Soc.*, **28,** 98 (1912).
436. Lederer, K., *Ber.*, **49,** 349 (1916).
437. Leeper, R. W., *Iowa St. Coll. J. Sci.*, **18,** 57 (1943); through *C.A.*, **38,** 726 (1944).
438. Leeper, R. W., L. Summers, and H. Gilman, *Chem. Rev.*, **54,** 101 (1954).
439. Leersmakers, J. A., *J. Am. Chem. Soc.*, **55,** 4509 (1933).
440. Lehmkuhl, H., *Annals N.Y. Acad. Sci.*, **125,** 124 (1965).
441. Lehmkuhl, H., R. Schaefer, and K. Ziegler, *Chem. Ingr. Tech.*, **36,** 612 (1964); *C.A.*, **61,** 6637 (1964).
441a. Leigh-Smith, A., and H. O. W. Richardson, *Nature*, **135,** 828 (1935).
442. Leighton, P. A., and R. A. Mortensen, *J. Am. Chem. Soc.*, **58,** 448 (1936).

443. Lesbre, M., *Compt. rend.*, **210,** 535 (1940).

444. Letort, M., and A. Combe, *Rev. inst. franc. petrole*, **8,** 491 (1953); through *C.A.*, **48,** 3671 (1954); Proc. Third World Petroleum Congr., 1951, Sec. VII, 192–205; *C.A.*, **48,** 14171 (1954).

445. Lichtenwalter, M., *Iowa St. Coll. J. Sci.*, **14,** 57 (1939); through *C.A.*, **34,** 6241 (1940).

446. Linsk, J., U.S. Pat. 3,116,308 (to Standard Oil Co. of Indiana), Dec. 31, 1963; *C.A.*, **60,** 6867 (1964).

447. Linsk, J., and E. A. Mayerle, U.S. Pat. 3,155,602 (to Standard Oil Co. of Indiana), Nov. 3, 1964; *C.A.*, **62,** 2795 (1965).

448. Lippincott, E. R., and M. C. Tobin, *J. Am. Chem. Soc.*, **75,** 4141 (1953).

448a. Lodochnsikova, V. I., E. M. Panov, and K. A. Kocheshkov, *Zh. Obshch. Khim.*, **37,** 547. (1967)

449. Long, L. H., *J. Chem. Soc.*, **1956,** 3410.

450. Long, L. H., *Pure Appl. Chem.*, **2,** 61 (1961).

451. Lowig, C., *Ann. Chem.*, **88,** 318 (1853).

452. Lutz, H. D., *Z. Naturforsch*, **20b,** 1011 (1965).

453. Lyttle, W. B., and H. N. Rexroad, *Proc. West Va. Acad. Sci.*, **31–32,** 190 (1959–60); *C.A.*, **55,** 18322 (1961).

454. Madden, H. J., U.S. Pat. 2,727,052 (to Ethyl Corp.), Dec. 13, 1955; *C.A.*, **50,** 8709 (1956).

455. Maddox, M. L., S. L. Stafford, and H. D. Kaesz, *Applications of Nuclear Magnetic Resonance to the Study of Organometallic Compounds*, Vol. 3, F. G. A. Stone and R. West, Eds., Academic Press, N.Y., 1965, pp. 1–171.

456. Mador, I. L., *J. Chem. Phys.* **22,** 1617 (1954).

457. Mahler, E. R. W. de, Belg. Pat. 418,670, Dec. 31, 1936; through *C.A.*, **31,** 5816 (1937).

458. Maier, L., *Agnew. Chem.*, **71,** 161 (1959).

459. Makarova, L. G., and A. N. Nesmeyanov, *Zh. Obshch. Khim.*, **9,** 771 (1939); through *C.A.*, **34,** 391 (1940).

460. Makin, F. B., and W. A. Waters, *J. Chem. Soc.*, **1938,** 843.

461. Malatesta, A., *Can. J. Chem.*, **37,** 1176 (1959).

462. Malatesta, A., U.S. Pat. 3,045,035 (to "Montecatini" Societa Generale per l'Industria Mineraria Chimica), July 17, 1962; through *C.A.*, **57,** 16656 (1962).

463. Malatesta, L., and R. Pizzotti, *Gazz. chim. ital.*, **73,** 349 (1943); through *C.A.*, **41,** 1137 (1947).

464. Malinovskii, M. S., *Trudy Gor'kov. Gosudarst. Pedagog. Inst.*, **19405,** 39; *Khim. Referat. Zhur.*, **42,** 45 (1941); through *C.A.*, **37,** 3070 (1943).

465. Malý, J., and J. Teplý, *Chem. Zvesti*, **7,** 553 (1953); *C.A.*, **48,** 10401 (1954).

466. Manulkin, Z. M., *Zh. Obshch. Khim.*, **18,** 299 (1948); through *C.A.*, **42,** 6742 (1948).

467. Manulkin, Z. M., *Zh. Obshch. Khim.*, **20,** 2004 (1950); through *C.A.*, **45,** 5611 (1951).

468. Manulkin, Z. M., *Uzbek. Khim. Zhur., Akad. Nauk Uzbek. SSR*, **1958**(2), 41; through *C.A.*, **53,** 9112 (1959).

469. Manulkin, Z. M., *Uzbek. Khim. Zhur.*, **1960**(2), 66; through *C.A.*, **55,** 12330 (1961).

470. Marlett, E. M., *Annals N.Y. Acad. Sci.*, **125,** 12 (1965).

471. Marlett, E. M., P. Kobetz, and R. C. Pinkerton, 142nd Meeting of the American Chemical Society, Atlantic City, N.J., 1962, Abstract of Papers, p. 25N.

472. Mattson, G. W., U.S. Pat. 2,744,126 (to Ethyl Corp.), May 1, 1956; *C.A.*, **50,** 12801 (1956).
473. Matwiyoff, N. A., and R. S. Drago, *Inorg. Chem.*, **3,** 337 (1964).
474. Maxted, E. B., and K. L. Moon, *J. Chem. Soc.*, **1949,** 2171.
475. McCleary, R. F., and E. F. Degering, *Ind. Eng. Chem.*, **30,** 3634 (1938).
476. McDonald Blair, J., D. Bryce-Smith, and B. W. Pengilly, *J. Chem. Soc.*, **1959,** 3174.
477. McKay, T. W. (to Ethyl Corp.), Ger. Pat. 1,120,448, Dec. 28, 1961; through *C.A.*, **57,** 11234 (1962).
478. McKay, T. W., U.S. Pat. 3,088,885 (to Ethyl Corp.), May 7, 1963; *C.A.*, **59,** 10119 (1963).
479. Meals, R. N., *J. Org. Chem.*, **9,** 211 (1944).
480. Meerwin, H., G. Hinz, H. Majert, and H. Sönke, *J. prakt. Chem.*, **147,** 226 (1936).
481. Meinert, R. N., *J. Am. Chem. Soc.*, **55,** 979 (1933).
482. Melvin, W. S., and A. F. Nichols, U.S. Pat. 2,907,780 (to E. I. du Pont de Nemours & Co.), Oct. 6, 1959; *C.A.*, **54,** 3204 (1960).
483. Metal, J. E., U.S. Pat. 2,373,531 (to Shell Dev. Co.), April 10, 1945; through *C.A.*, **40,** 1024 (1946).
484. Meyer, M., *Chem. News*, **131,** 1 (1925); through *C.A.*, **19,** 2636 (1925).
485. Michel, J. M., U.S. Pat. 2,067,331 (to I. G. Farbenind. A.-G.), Jan. 12, 1937; through *C.A.*, **31,** 1603 (1937).
486. Midgley, T. A., Jr., U.S. Pat. 1,592,954 (to Ethyl Gasoline Corp.), July 20, 1926; *C.A.*, **20,** 3228 (1926).
487. Midgley, T., Jr., U.S. Pat. 1,622,228, March 22, 1927; through *C.A.*, **21,** 1546 (1927).
488. Milde, R. L., and H. A. Beatty, *Advances in Chemistry Series*, No. 23, American Chemical Society, Washington, D.C., 1959, p. 306.
489. Milazzo, G., *Gazz. chim. ital.*, **71,** 73 (1941); *C.A.*, **37,** 1928 (1943).
490. Miller, D. M., and C. A. Winkler, *Can. J. Chem.*, **29,** 537 (1951).
491. Mironov, V. F., and A. L. Kravchenko, *Dokl. Akad. Nauk SSSR*, **158,** 656 (1964).
492. Mironov, V. F., and A. L. Kravchenko, *Zh. Obshch. Khim.*, **34,** 1356 (1964); *C.A.*, **61,** 677 (1964).
493. Mishima, S., *Nippon Kagaku Zasshi*, **87,** 162 (1966); through *C.A.*, **65,** 15420 (1966).
494. Moedritzer, K., *Organometal. Chem. Rev.*, **1,** 179 (1966).
495. Möller, S., and P. Pfeiffer, *Ber.*, **49,** 2441 (1916).
496. Monroe, K. P., U.S. Pat. 1,645,389–90, Oct. 11, 1927; through *C.A.*, **21,** 3907 (1927).
497. Monroe, K. P., U.S. Pat. 1,661,809–10, March 6, 1928; through *C.A.*, **22,** 1367 (1928).
498. Monsanto Chemical Co., French Pat. 1,347,066, Dec. 27, 1963; through *C.A.*, **60,** 12055 (1964).
499. "Montecatini" Societa generale per l'Industria mineraria e chimica (by D. Costabello and C. Longiave); Italian Pat. 556,538, Feb. 6, 1957; through *C.A.*, **53,** 2093 (1959).
500. "Montecatini" Societa per l'Industria mineraria e chimica (by A. Malatesta), Ital. Pat. 560,499, April 5, 1957; through *C.A.*, **53,** 4134 (1959).
501. Mortensen, R. A., and P. A. Leighton, *J. Am. Chem. Soc.*, **56,** 2397 (1934).
502. Mortensen, R. A., and P. A. Leighton, *J. Am. Chem. Soc.*, **58,** 448 (1936).

503. Moss, R., and E. V. Browett, *Analyst*, **91,** 428 (1966).
504. Moss, R., and K. Campbell, *J. Inst. Petroleum*, **53,** 89 (1967).
505. Murin, A. N., V. D. Nefedov, V. M. Zaitsev, and S. A. Grachev, *Dokl. Akad. Nauk SSSR*, **133,** 123 (1960); *C.A.*, **54,** 24481 (1960).
506. Nagai, Y., *Proc. Imp. Acad.* (*Tokyo*), **3,** 664 (1927); through *C.A.*, **22,** 1231 (1928).
507. Nalco Chemical Co., Belg. Pat. 590,453, from *Tech. Translations*, **5,** 662 (1961).
508. Nalco Chemical Co. (by D. G. Braithwaite and W. Hanzel), Belg. Pat. 611,212, June 6, 1962; through *C.A.*, **57,** 16330 (1962).
509. Nalco Chemical Co. (by D. G. Braithwaite), Belg. Pat. 613,892, Aug. 13, 1962; through *C.A.*, **59,** 7559 (1963).
510. Nalco Chemical Co. (by D. G. Braithwaite, L. L. Bott, and K. G. Phillips), Belg. Pat. 671,841, March 4, 1966; through *C.A.*, **66,** 76157q (1967).
511. Nalco Chemical Co. (by D. G. Braithwaite), Ger. Pat. 1,197,086, July 22, 1965; through *C.A.*, **63,** 11023 (1965).
512. Nalco Chemical Co. (by D. G. Braithwaite), Ger. Pat. 1,202,790, Oct. 14, 1965; through *C.A.*, **64,** 3064 (1966).
513. Nalco Chemical Co. (by D. G. Braithwaite), Ger. Pat. 1,226,100, Oct. 6, 1966; through *C.A.*, **65,** 19690 (1966).
514. Nalco Chemical Co. (by D. G. Braithwaite), Ger. Pat. 1,231,242, Dec. 29, 1966; through *C.A.*, **66,** 95196t (1967).
515. Nalco Chemical Co. (by D. G. Braithwaite), Ger. Pat. 1,231,700, Jan. 5, 1967; through *C.A.*, **66,** 76155n (1967).
516. Narasimhan, P. T., and M. T. Rogers, *J. Chem. Phys.*, **34,** 1049 (1961).
517. National Aluminate Corp., Brit. Pat. 839,172, June 29, 1960; through *C.A.*, **54,** 24036 (1960).
518. Nazarova, L, M., *Zh. Obshch. Khim.*, **29,** 2671 (1959); *C.A.*, **54,** 12980 (1960).
519. Nazarova, L. M., *Zh. Obshch. Khim.*, **31,** 1119 (1961); *C.A.*, **55,** 23404 (1961).
520. Nazarova, L. M., and G. E. Aleksandrova, *Vysokomolekul. Soedin.*, **3,** 1823 (1961); through *C.A.*, **56,** 14462 (1962).
521. Nazarova, L. M., E. N. Kharlamova, G. E. Aleksandrova, and E. B. El'tekova, *Zh. Obshch. Khim.*, **31,** 3308 (1961); *C.A.*, **57,** 2242 (1962).
522. Nefedov, V. D., and N. A. Varshav, *Zh. Fiz. Khim. SSSR*, **28,** 981 (1954); *C.A.*, **49,** 7354 (1955).
523. Neiman, M. B., and V. A. Shushunov, *Zh. Fiz. Khim. SSSR*, **22,** 161 (1948); through *C.A.*, **42,** 5316 (1948).
524. Neiman, M. B., and V. A. Shushunov, *Dokl. Akad. Nauk SSSR*, **60,** 1347 (1948); through *C.A.*, **45,** 425 (1951).
525. Nesmeyanov, A. N., and R. Kh. Freidlina, *Ber.*, **67,** 735 (1934).
526. Nesmeyanov, A. N., and K. A. Kocheshkov, *Ber.*, **67,** 317 (1934).
527. Nesmeyanov, A. N., and K. A. Kocheshkov, *Bull. acad. Sci. URSS, Classe sci. chim.*, **1945,** 522; through *C.A.*, **42,** 5870 (1948).
528. Nesmeyanov, A. N., and L. G. Makarova, *Zh. Obshch. Khim.*, **7,** 2647 (1937); through *C.A.*, **32,** 2095 (1938).
529. Nesmeyanov, A. N., and L. G. Makarova, *Izvest. Akad. Nauk SSSR Otdel. Khim. Nauk*, **1954,** 380; through *C.A.*, **49,** 5346 (1955).
530. Nesmeyanov, A. N., T. P. Tolstaya, and L. S. Isaeva, *Dokl. Akad. Nauk SSSR*, **125,** 330 (1959); *C.A.*, **53,** 19927 (1959).
531. Neumann, W. P., and K. Kühlein, *Angew. Chem.*, **77,** 808 (1965).
532. Noltes, J. G., M. C. Henry, and M. J. Janssen, *Chem. Ind.* (*London*), **1959,** 298.

533. Norman, V., and T. P. Whaley, U.S. Pat. 3,018,194 (to Ethyl Corp.), Jan. 23, 1962; *C.A.*, **56,** 8402 (1962).
534. Nosek, J., *Collection Czech. Chem. Commun.*, **29,** 597 (1964); through *C.A.*, **60,** 8053 (1964).
535. Ogawara, R., and M. Sakiyama, *Kagaku no Ryoiki Zokan*, **No. 45,** 127 (1961); through *C.A.*, **55,** 17211 (1961).
536. Oikawa, H., *Nippon Kagaku Zasshi*, **84,** 1 (1963); through *C.A.*, **58,** 10873 (1963); **84,** 11 (1963); through *C.A.*, **58,** 10872 (1963).
537. Osberghaus, O., and R. Taubert, *Z. Physik. Chem.* (*Frankfurt N.F.*), **4,** 264 (1955).
538. Ol'dekop, Yu. A., and B. N. Moryganov, *Zh. Obshch. Khim.*, **23,** 2020 (1953); through *C.A.*, **49,** 3097 (1955).
539. Ol'dekop, Yu. A., and R. F. Sokolova, *Zh. Obshch. Khim.*, **23,** 1159 (1953); through *C.A.*, **47,** 12226 (1953).
540. Oosterhout, J. C. D., U.S. Pat. 2,155,678 (to Texas Co.), April 25, 1939; through *C.A.*, **33,** 6039 (1939).
541. Ormandy, W. R., *J. Inst. Petroleum Tech.*, **10,** 335 (1924); *C.A.*, **19,** 1194 (1925).
542. Overend, J., and J. R. Scherer, *J. Opt. Soc. America*, **50,** 1203 (1960).
543. Padberg, F. J., *Dissertation, Univ. of Aachen*, 1965.
544. Pagliarini, P., U.S. Pat. 2,848,471 (to Compagnia Italiana Petrolio S.p.A.), Aug. 19, 1958; *C.A.*, **52,** 19948 (1958); Brit. Pat. 918,519; through *C.A.*, **59,** 5196 (1963).
545. Pai, N. G., *Proc. Royal Soc.*, (*London*), **A149,** 29 (1935).
546. Paneth, F., and K. Herzfeld, *Z. Electrochem.*, **37,** 577 (1931).
547. Paneth, F., and W. Hofeditz, *Ber.*, **62,** 1335 (1929).
548. Paneth, F., W. Hofeditz, and A. Wunsch, *J. Chem. Soc.*, **1935,** 372.
549. Paneth, F., and W. Lautsch, *Naturwiss.*, **18,** 307 (1930).
550. Paneth, F., and W. Lautsch, *Ber.*, **64,** 2702 (1931).
551. Paneth, F., and W. Lautsch, *Ber.*, **64,** 2708 (1931).
552. Paneth, F., and W. Lautsch, *J. Chem. Soc.*, **1935,** 380.
553. Panov, E. M., and K. A. Kocheshkov, *Dokl. Akad. Nauk SSSR*, **85,** 1037 (1952); through *C.A.*, **47,** 6365 (1953).
554. Pascal, P., *Compt. rend.*, **218,** 57 (1944).
555. Pascal, P., *Bull. Soc. Chim., France*, **11,** 321 (1944).
556. Pascal, P., *Proprietes magnetique et constitution chimique*, *Grignard, Traite Chimie organique*, **2,** 571; through *C.A.*, **42,** 5731 (1948).
557. Pascal, P., F. Gallais, and J. F. Labarre, *Compt. rend.*, **256,** 335 (1963).
558. Pasynkiewicz, S., *Przemysl Chem.*, **43,** 534 (1964).
559. Pasynkiewicz, S., S. Malinowski, and J. Bitter, *Przemysl Chem.*, **44,** 500 (1963); through *C.A.*, **64,** 273 (1966).
560. Pearce, F. G., L. T. Wright, H. A. Birkness, and J. Linsk, U.S. Pat. 3,180,810 (to Standard Oil Co. of Indiana), April 6, 1965; through *C.A.*, **63,** 1816 (1965).
561. Pearsall, H. W., U.S. Pat. 2,414,058 (to Ethyl Corp.), Jan. 7, 1947; *C.A.*, **41,** 2430 (1947).
562. Pearson, T. G., P. L. Robinson, and E. M. Stoddart, *Proc. Royal Soc.* (*London*), **A142,** 275 (1933).
563. Pearson, T. H., S. M. Blitzer, D. R. Carley, T. W. McKay, R. L. Ray, L. L. Sims, and J. R. Zietz, *Advances in Chemistry Series*, No. 23, American Chemical Society, Washington, D.C., 1959, p. 299.
564. Pedinelli, M., R. Magri, and M. Randi, *Chim. Ind.* (*Milan*), **48,** 144 (1966); through *C.A.*, **64,** 12719 (1966).

565. Pedinelli, M., and M. Randi, *Chem. Ind.* (*Milan*), **46,** 172 (1964); through *C.A.*, **61,** 83 (1964).
566. Peterson, D. B., T. Arakawa, D. A. G. Wolmsley, and M. Burton, *J. Phys. Chem.*, **69,** 2880 (1965).
567. Pines, H., U.S. Pat. 2,850,513 (to Universal Oil Products), Sept. 2, 1958; *C.A.*, **53,** 4211 (1959).
568. Pinkerton, R. C., U.S. Pat. 3,028,325 (to Ethyl Corp.), April 3, 1962; *C.A.*, **57,** 4471 (1962).
569. Plunkett, R. J., U.S. Pat. 2,477,465 (to E. I. du Pont de Nemours & Co.), July 26, 1949; *C.A.*, **43,** 8398 (1949).
570. Polis, A., *Ber.*, **20,** 716 (1887).
571. Polis, A., *Ber.*, **20,** 3331 (1887).
572. Polis, A., *Ber.*, **21,** 3424 (1888).
573. Pollard, F. H., G. Nickless, and P. C. Uden, *J. Chromatog.*, **19,** 28 (1965).
574. Pratt, G. L., and J. H. Purnell, *Trans. Faraday Soc.*, **60,** 371 (1964).
575. Pratt, G. L., and J. H. Purnell, *Trans. Faraday Soc.*, **60,** 519 (1964).
576. Prilezhaeva, N., *Compt. rend. acad. sci. USSR*, **3,** 252 (1934); *C.A.*, **28,** 7166 (1934).
577. Pritchard, H. O., and H. A. Skinner, *Chem. Rev.*, **55,** 745 (1955).
578. Proffitt, D. K., and J. G. Sharron, U.S. Pat. 1,863,451 (to Ethyl Gasoline Corp.), June 14, 1932; through *C.A.*, **26,** 4066 (1932).
579. Ptashinskii, I. A., and M. K. Frolova, *Trudy, Vsesoyuz. Nauch.-Issledovatel. Inst. po Pererabotke Nefti i Gaza i Poluchen. Zhidkogo Topliva*, **1957,** No. 6, 181; through *C.A.*, **53,** 15538 (1959).
580. Puchinyan, E. A., *Tr. Tashkentsk. Farmatsevt. Inst.*, **1,** 310 (1957); through *C.A.*, **58,** 2464 (1963).
581. Puchinyan, E. A., *Tr. Tashkentsk. Farmatsevt. Inst.*, **2,** 311 (1960); through *C.A.*, **57,** 11228 (1962).
582. Puchinyan, E. A., and Z. M. Manulkin, *Dokl. Akad. Nauk Uz. SSR*, **1961,** No. 12, 51; *C.A.*, **58,** 543 (1963).
583. Pyk, S. D., U.S. Pat. 2,561,636 (to Associated Ethyl), July 24, 1951; *C.A.*, **47,** 8629 (1953).
584. Quinn, E. I., V. H. Dibeler, and F. L. Mohler, *J. Res. Natl. Bur. Stand.*, **57,** 41 (1956).
585. Ramsden, H. E., Brit. Pat. 824,944 (to Metal and Thermit Corp.); through *C.A.*, **54,** 17238 (1960).
586. Ramsden, H. E., and H. F. Shannon, U.S. Pat. 3, 156,716 (to Esso Research and Engineering Co.), Nov. 10, 1964; *C.A.*, **62,** 2794 (1965).
587. Randaccio, C., Ital. Pat. 500,102, Nov. 17, 1954; through *C.A.*, **51,** 10560 (1957).
588. Rao, C. N. R., J. Ramachandran, and A. Balasubramanian, *Can. J. Chem.*, **39,** 171 (1961).
589. Rao, C. N. R., J. Ramachandran, M. S. C. Iah, S. Somasekhara, and T. V. Rajakumar, *Nature*, **183,** 1475 (1959).
590. Razuvaev, G. A., *Zh. Obshch. Khim.*, **4,** 629 (1934); through *C.A.*, **29,** 2167 (1935).
591. Razuvaev, G. A., Yu. I. Dergunov, and N. S. Vyazankin, *Zh. Obshch. Khim.*, **31,** 998 (1961); *C.A.*, **55,** 23321 (1961).
592. Razuvaev, G. A., O. S. D'Yachkovskaya, N. S. Vyazankin, and O. A. Shchepetkova, *Dokl. Akad. Nauk SSSR*, **137,** 618 (1961); *C.A.*, **55,** 20925 (1961).

593. Razuvaev, G. A., M. S. Fedotov, T. N. Zaichenko, and N. A. Kul'vinskaya, *Sbornik Statei Obshch. Khim.*, **2,** 1514 (1953); through *C.A.*, **49,** 5346 (1955).
594. Razuvaev, G. A., V. V. Fetyukova, and M. A. Shubenko, *Sci. Reports Gorki State Univ.*, **15,** 91 (1941); Reference 4 in *Zh. Obshch. Khim.*, **26,** 1107 (1956), by G. A. Razuvaev and O. S. D'Yachkovskaya; *C.A.*, **50,** 16663 (1956).
595. Razuvaev, G. A., and M. M. Koton, *Ber.*, **66,** 854 (1933).
596. Razuvaev, G. A., and M. M. Koton, *Zh. Obshch. Khim.*, **5,** 361 (1935); through *C.A.*, **29,** 6217 (1935).
597. Razuvaev, G. A., G. G. Petukhov, R. F. Galiulina, and N. N. Shabanova, *Zh. Obshch. Khim.*, **34,** 3812 (1964); *C.A.*, **62,** 9163 (1965).
598. Razuvaev, G. A., G. G. Petukhov, and Yu. A. Kaplin, *Zh. Obshch. Khim.*, **33,** 2394 (1963); *C.A.*, **59,** 14014 (1963).
599. Razuvaev, G. A., G. G. Petukhov, Yu. A. Kaplin, and O. N. Druzhkov, *Dokl. Akad. Nauk SSSR*, **152,** 1122 (1963); *C.A.*, **60,** 1556 (1964).
600. Razuvaev, G. A., N. S. Vyazankin, Yu. I. Dergunov, and O. S. D'Yachkovskaya, *Dokl. Akad. Nauk SSSR*, **132,** 364 (1960); *C.A.*, **54,** 20937 (1960).
601. Razuvaev, G. A., N. S. Vyazankin, and O. S. D'Yachovskaya, *Zh. Obshch. Khim.*, **32,** 2161 (1962); *C.A.*, **58,** 7963 (1963).
602. Razuvaev, G. A., N. S. Vyazankin, and N. N. Vyshinskii, *Zh. Obshch. Khim.*, **29,** 3662 (1959); *C.A.*, **54,** 17015 (1960).
603. Razuvaev, G. A., N. S. Vyazankin, and N. N. Vyshinskii, *Zh. Obshch. Khim.*, **30,** 967 (1960); *C.A.*, **54,** 23646 (1960).
604. Razuvaev, G. A., N. S. Vyazankin, and N. N. Vyshinskii, *Zh. Obshch. Khim.* **30,** 4099 (1960); *C.A.*, **55,** 24546 (1961).
605. Reeves, L. W., and E. J. Wells, *Can. J. Chem.*, **41,** 2698 (1963).
606. Renger, G., *Ber.*, **44,** 337 (1911).
607. Renger, G., *Ber.*, **45,** 3321 (1912).
608. Riccoboni, L., *Gazz. chim. ital.*, **71,** 696 (1941); through *C.A.*, **36,** 6902 (1942).
609. Riccoboni, L., U. Belluco, and G. Tagliavini, *Ric. sci., Rend. Sez*, **A2,** 269 (1962); through *C.A.*, **58,** 13976 (1963).
610. Riccoboni, L., U. Belluco, and G. Tagliavini, *Ric. sci., Rend. Sez.* **A2,** 323 (1962); through *C.A.*, **58,** 13976 (1963).
611. Riccoboni, L., and L. Oleari, *Ric. sci., Rend. Sez.* **A3,** 1031 (1963); through *C.A.*, **60,** 4846 (1964).
612. Riccoboni, L., G. Pilloni, G. Plazzogna, and C. Bernardin, *Ric. sci., Rend. Sez.* **A3,** 1231 (1963); through *C.A.*, **60,** 15192 (1964).
613. Riccoboni, L., G. Pilloni, G. Plazzogna, and G. Tagliavini, *J. Electroanal. Chem.*, **11,** 340 (1966).
614. Rice, F. O., U.S. Pat. 2,087,656 (to Standard Oil Co. of Indiana), July 20, 1937; through *C.A.*, **31,** 6259 (1937).
615. Rice, F. O., Brit. Pat. 407,036, March 5, 1934; through *C.A.*, **28,** 4847 (1934).
616. Rice, F. O., W. R. Johnston, and B. L. Evering, *J. Am. Chem. Soc.*, **54,** 3529 (1932).
617. Riddle, J. M., U.S. Pat. 2,950,301 (to Ethyl Corp.), Aug. 23, 1960; *C.A.*, **55,** 3434 (1961).
618. Riddle, J. M., U.S. Pat. 2,950,302 (to Ethyl Corp.), Aug. 23, 1960; *C.A.*, **55,** 3433 (1961).
619. Rifkin, E. B., Twenty-Third Midyear Meeting American Petroleum Industries, Los Angeles, Calif., May, 1958.

620. Rifkin, E. B., and D. H. Ewen, U.S. Pat. 2,580,243 (to Ethyl Corp.), Dec. 25, 1951; *C.A.*, **46,** 2790 (1952).
621. Rifkin, E. B., and C. Walcutt, *Ind. Eng. Chem.*, **48,** 1532 (1956).
622. Robinson, G. C., *J. Org. Chem.*, **28,** 843 (1963).
623. Robinson, G. C., U.S. Pat. 3,057,897 (to Ethyl Corp.), Oct. 9, 1962; *C.A.*, **58,** 4598 (1963).
624. Robinson, W. T., U.S. Pat. 3,005,780 (to E. I. du Pont de Nemours & Co.), June 11, 1958; *C.A.*, **56,** 6247 (1962).
625. Roetheli, B. E., and I. B. Simpson, Brit. Pat. 797,093 (to Esso Research and Engineering Co.), June 25, 1958; through *C.A.*, **53,** 930 (1959).
626. Romm, F. S., *Zh. Obshch. Khim.*, **10,** 1784 (1940); through *C.A.*, **35,** 3880 (1941).
627. Ruddies, G. F., U.S. Pat. 2,415,444, Feb. 11, 1947; through *C.A.*, **41,** 2886 (1947).
628. Ruddies, G. F., U.S. Pat. 2,473,972, June 21, 1949; *C.A.*, **44,** 6233 (1950).
629. Rudel, H. W., U.S. Pat. 2,378,793 (to Standard Oil Dev. Co.), June 19, 1945; through *C.A.*, **39,** 4473 (1943).
630. Russell, G. A., *J. Am. Chem. Soc.*, **81,** 4815 (1959).
631. Russell, G. A., *J. Am. Chem. Soc.*, **81,** 4825 (1959).
632. Ryason, P. R., *Combust. Flame*, **7,** 235 (1963); through *C.A.*, **62,** 16293 (1965).
633. Saikina, M. K., *Uchenze Zapiski Kazan. Gosudarst. Univ. im. V. I. Ul'yanova–Lenina Khim.*, **116,** 129 (1956); through *C.A.*, **51,** 7191 (1957).
634. Sandler, S. R., and K. C. Tsou, *J. Phys. Chem.*, **68,** 300 (1964).
635. Sandler, S. R., and K. C. Tsou, *Intern. J. Appl. Radiation Isotopes*, **15,** 419 (1964); through *C.A.*, **61,** 9139 (1964).
636. Sandy, C. A., U.S. Pat. 3,113,955 (to E. I. du Pont de Nemours & Co.), Dec. 10, 1963; *C.A.*, **60,** 5550 (1964).
637. Saunders, B. C., and G. J. Stacey, *J. Chem. Soc.*, **1949,** 919.
638. Saunders, B. C., and T. S. Worthy, *J. Chem. Soc.*, **1953,** 2115.
639. Schaefer, J. H., *Chem. Eng.*, **57**(8), 102 (1950); **57**(8), 164 (1950).
640. Schepps, W., *Ber.*, **46,** 2564 (1913).
641. Schlenk, W., and J. Holtz, *Ber.*, **50,** 262 (1917).
642. Schmidt, M., and H. Schumann, *Z. anorg. allgem. Chem.*, **325,** 130 (1963).
643. Schmitz-DuMont, O., and B. Ross, *Z. anorg. allgem. Chem.*, **349,** 328 (1967).
644. Schneider, W. G., and A. D. Buckingham, *Discussions Faraday Soc.*, **34,** 147 (1962).
645. Schuler, M. J., U.S. Pat. 3,197,491 (to E. I. du Pont de Nemours & Co.), July 27, 1965; *C.A.*, **63,** 9985 (1965).
646. Scott, D. W., W. D. Good, and G. Waddington, *J. Phys. Chem.*, **60,** 1090 (1956).
647. Semerano, G., and L. Riccoboni, *Ber.*, **74,** 1089 (1941).
648. Semerano, G., and L. Riccoboni, *Z. Physik. Chem.*, **A189,** 203 (1941).
649. Semerano, G., L. Riccoboni, and F. Callegari, *Ber.*, **74,** 1297 (1941).
650. Semerano, G., L. Riccoboni, and L. Götz, *Z. Elektrochem.*, **47,** 484 (1941).
651. Setzer, W. C., R. W. Leeper, and H. Gilman, *J. Am. Chem. Soc.*, **61,** 1609 (1939).
652. Seyferth, D., and M. A. Weiner, *J. Am. Chem. Soc.*, **84,** 361 (1962).
653. Shapiro, A., and D. A. Olson, U.S. Pat. 2,969,329 (to Socony Mobil Oil Co., Inc.), Jan. 24, 1961; *C.A.*, **55,** 14898 (1961).
654. Shapiro, H., U.S. Pat. 2,535,235–7 (to Ethyl Corp.), Dec. 26, 1950; *C.A.*, **45,** 3865 (1951).

655. Shapiro, H., U.S. Pat. 2,597,754 (to Ethyl Corp), May 20, 1952; *C.A.*, **47,** 1183 (1953).
656. Shapiro, H., *Advances in Chemistry Series*, No. 23, American Chemical Society, Washington, D.C., 1959, pp. 290–8.
657. Shapiro, H., and E. G. DeWitt, U.S. Pat. 2,575,323 (to Ethyl Corp.), Nov. 20, 1951; *C.A.*, **46,** 5073 (1952).
658. Shapiro, H., and E. G. DeWitt, U.S. Pat. 2,635,106 (to Ethyl Corp.), April 14, 1953; *C.A.*, **48,** 2762 (1954); Ger. Pat. 937,350; *C.A.*, **53,** 222 (1959).
659. Shapiro, H., and I. T. Krohn, U.S. Pat. 2,688,628 (to Ethyl Corp.), Sept. 7, 1954; *C.A.*, **49,** 14797 (1955).
660. Shapiro, H., and H. R. Neal, U.S. Pat. 2,992,250–61 (to Ethyl Corp.), July 11, 1961; *C.A.*, **55,** 22799 (1961).
661. Shapiro, H., and H. R. Neal, U.S. Pat. 3,004,997–9 (to Ethyl Corp.), Oct. 17, 1961; *C.A.*, **56,** 3730 (1962).
662. Shappirio, S., U.S. Pat. 2,012,356 (Aug. 27, 1935); through *C.A.*, **29,** 6752 (1935).
663. Sheline, R. K., *J. Chem. Phys.*, **18,** 602 (1950).
664. Sheline, R. K., and K. S. Pitzer, *J. Chem. Phys.*, **18,** 595 (1950).
665. Shelomov, I. K., O. A. Osipov, and O. E. Kashireninov, *Zh. Obshch. Khim.*, **33,** 1056 (1963); *C.A.*, **59,** 11548 (1963).
666. Sheppard, H., *Trans. Faraday Soc.*, **51,** 1465 (1955).
667. Shimizu, K., *J. Chem. Soc. Japan, Pure Chem. Sect.*, **77,** 1284 (1956); through *C.A.*, **51,** 2386 (1957).
668. Shushunov, V. A., and Yu. N. Baryshnikov, *Dokl. Akad. Nauk SSSR*, **91,** 331 (1953); *C.A.*, **48,** 8658 (1954).
669. Shushunov, V. A., and Yu. N. Baryshnikov, *Zh. Fiz. Khim. SSSR*, **27,** 830 (1953); *C.A.*, **49,** 2838 (1955).
670. Shushunov, V. A., and Yu. A. Shlyapnikov, *Zh. Fiz. Khim. SSSR*, **22,** 157 (1948); through *C.A.*, **42,** 5316 (1948).
671. Shushunov, V. A., and N. A. Sokolov, *Tr. po Khim. i Khim. Tekhnol.*, **1,** 265 (1958); through *C.A.*, **54,** 6277 (1960).
672. Sieberth, H., *Z. anorg. allgem. Chem.*, **263,** 82 (1950).
673. Sieberth, H., *Z. anorg. allgem. Chem.*, **268,** 177 (1952).
674. Sieberth, H., *Z. anorg. allgem. Chem.*, **271,** 75 (1952).
675. Simons, J. H., R. W. McNamee, and C. D. Hurd, *J. Phys. Chem.*, **36,** 939 (1932).
676. Sims, L. L., S. M. Blitzer, D. R. Carley, S. E. Cook, E. C. Juenge, B. R. Lowrance, and T. H. Pearson, 133rd Meeting of the American Chemical Society, San Francisco, 1958, Abstract of Papers, p. 46L.
677. Singh, G., *J. Org. Chem.*, **31,** 949 (1966).
678. Skinner, H. A., "The Strengths of Metal-to-Carbon Bonds," in *Advances in Organometallic Chem.*, Vol. 2, F. G. A. Stone and R. West, Eds., Academic Press, N.Y., 1964, pp. 49–114.
679. Skinner, J. F., and R. M. Fuoss, *J. Phys. Chem.*, **68,** 2998 (1964).
680. Smaller, B., and M. S. Matheson, *J. Chem. Phys.*, **28,** 1169 (1958).
681. Smith, G. W., *J. Chem. Phys.*, **39,** 2031 (1963).
682. Smith, G. W., *J. Chem. Phys.*, **40,** 2037 (1964).
683. Smith, G. W., *J. Chem. Phys.*, **42,** 4220 (1966).
684. Solvay and Cie, Brit. Pat. 761,844, (Nov. 21, 1956); *C.A.* **51,** 10561 (1957).
685. Spialter, L., G. R. Buell, and C. W. Harris, *J. Org. Chem.*, **30,** 375 (1965).
686. Spice, J. E., and W. Twist, *J. Chem. Soc.*, **1956,** 3319.
687. Spiesecke, H., and W. G. Schneider, *J. Chem. Phys.*, **35,** 722 (1961).

688. Staehling, C., *Compt. rend.*, **157,** 1430 (1914).
689. Standard Oil Co. of Indiana (by J. M. Coopersmith, J. Linsk, E. Field, R. W. Carl and E. A. Mayerle), Ger. Pat. 1,157,616, Nov. 21, 1963; through *C.A.*, **61,** 1892 (1964).
690. Stasinevich, D. S., and A. L. Gol'dshtein, *Tr. po Khim. i Khim. Tekhnol.*, **3,** 209 (1960); through *C.A.*, **56,** 1469 (1962).
691. Staveley, L. A. K., H. P. Paget, B. B. Goalby, and J. B. Warren, *Nature*, **164,** 787 (1949).
692. Staveley, L. A. K., H. P. Paget, B. B. Goalby, and J. B. Warren, *J. Chem. Soc.*, **1950,** 2290.
693. Staveley, L. A. K., J. B. Warren, H. P. Paget, and D. J. Dowrick, *J. Chem. Soc.*, **1954,** 1992.
694. Stearn, A. E., *J. Am. Chem. Soc.*, **62,** 1630 (1940).
695. Stecher, J. L., U.S. Pat. 2,134,091 (to E. I. du Pont de Nemours & Co.), Oct. 25, 1938; through *C.A.*, **33,** 530 (1939).
696. Stecher, J. L., M. G. Amick, and C. E. Daniels, U.S. Pat. 2,047,391 (to E. I. du Pont de Nemours & Co.), July 14, 1936; through *C.A.*, **30,** 5929 (1936).
697. Stone, L. S., U.S. Pat. 2,923,740 (to Callery Chemical Co.), Feb. 2, 1960; *C.A.*, **54,** 15247 (1960).
698. Strohmeier, W., and K. Miltenberger, *Z. Physik. Chem. (Frankfurt)*, **17,** 274 (1958).
699. Strohmeier, W., and K. Miltenberger, *Chem. Ber.*, **91,** 1357 (1958).
700. Sudgen, S., *J. Chem. Soc.*, **1929,** 316.
701. Sugino, K., *Yuki Gøsei. Kyokai Shi*, **24,** 1170 (1966); through *C.A.*, **66,** 51572h (1967).
702. Sullivan, F. W., U.S. Pat. 2,087,660 (to Standard Oil Co. of Indiana), July 20, 1937; through *C.A.*, **31,** 6454 (1937).
703. Sullivan, F. W., U.S. Pat. 2,148,138 (to Standard Oil Co. of Indiana), (Feb. 2, 1939); through *C.A.*, **33,** 4012 (1939).
704. Sullivan, F. W., Jr., and L. Chalkey, Jr., U.S. Pat. 1,611,695 (to Standard Oil Co. of Indiana), Dec. 21, 1926 *C.A.*, **21,** 415 (1927); Ger. Pat. 505,688, Aug. 27, 1926; through *C.A.*, **25,** 1841 (1931).
705. Sullivan, F. W., and F. F. Duvoky, U.S. Pat. 1,938,547 (to Standard Oil Co. of Indiana), Dec. 5, 1933; through *C.A.*, **28,** 1182 (1934).
706. Sullivan, F. W., and V. Voorhees, U.S. Pat. 1,938,546 (to Standard Oil Co. of Indiana), Dec. 5, 1933; through *C.A.*, **28,** 1182 (1934).
707. Summers, L., *Iowa St. Coll. J. Sci.*, **26,** 292 (1952); through *C.A.*, **47,** 8673 (1953).
708. Syavtsillo, S. V., and A. F. Danilina, *Gigiena i Sanit.*, **1952,** 24; through *C.A.*, **47,** 2399 (1953).
709. Tafel, J., *Ber.*, **39,** 3626 (1906).
710. Tafel, J., *Ber.*, **42,** 3146 (1909).
711. Tafel, J., *Ber.*, **44,** 323 (1911).
712. Tagliavini, G., and U. Belluco, *Ric. sci. Rend. Sez.* **A32,** 76 (1962); through *C.A.*, **57,** 13785 (1962).
713. Tagliavini, G., U. Belluco, and L. Cattalini, *Ric. sci., Rend. Sez.* **A2,** 350 (1962); through *C.A.*, **59,** 2846 (1963).
714. Tagliavini, G., U. Belluco, G. Schiavon, and L. Riccoboni, *Ric. sci.*, **28,** 2349 (1958); through *C.A.*, **53,** 19851 (1959).
715. Tagliavini, G., G. Schiavon, and U. Belluco, *Gazz. chim. ital.*, **88,** 746 (1958); through *C.A.*, **53,** 19850 (1959).
716. Taimsalu, J., and J. L. Wood, *Trans. Faraday Soc.*, **59,** 1754 (1963).

717. Takenaka, T., and R. Goto, *Nippon Kagaku Zasshi*, **83,** 997 (1962); through *C.A.*, **58,** 10883 (1963).
718. Talalaeva, T. V., and K. A. Kocheshkov, *Zh. Obshch. Khim.*, **8,** 1831 (1938); through *C.A.*, **33,** 5819 (1939).
719. Talalaeva, T. V., and K. A. Kocheshkov, *Zh. Obshch. Khim.*, **12,** 403 (1942); through *C.A.*, **37,** 3068 (1943).
720. Talalaeva, T. V., and K. A. Kocheshkov, *Dokl. Akad. Nauk SSSR*, **77,** 621 (1951); through *C.A.*, **45,** 10191 (1951).
721. Tamborski, C., F. E. Ford, W. L. Lehn, G. J. Moore, and E. J. Soloski, *J. Org. Chem.*, **27,** 619 (1962).
722. Tamborski, C., E. J. Soloski, and S. M. Dec, *J. Organometal. Chem.*, **4,** 446 (1965).
723. Tanner, H., U.S. Pat. 2,635,105 (to Ethyl Corp.), April 14, 1953; *C.A.*, **48,** 2762 (1954).
724. Tanner, H., U.S. Pat. 2,635,107 (to Ethyl Corp.), April 14, 1953; *C.A.*, **48,** 2762 (1954).
725. Taylor, H. S., and W. H. Jones, *J. Am. Chem. Soc.*, **52,** 1111 (1930).
726. Tel'noi, V. I., and I. B. Rabinovich, *Zh. Fiz. Khim., SSSR*, **39,** 2076 (1965); through *C.A.*, **63,** 14146 (1965).
727. Teplý, J., and J. Malý, *Chem. Zvesti*, **7,** 463 (1953); through *C.A.*, **48,** 10401 (1954).
728. Terenin, A., *J. Chem. Phys.*, **2,** 441 (1934).
729. Terenin, A., and N. Prilzhaeva, *Acta Physicochem. URSS*, **1,** 759(1934); through *C.A.*, **29,** 7798 (1935).
730. Thomas, W. H., and S. E. Cook, U.S. Pat. 3,097,223 (to Ethyl Corp.), July 9, 1963; *C.A.*, **59,** 7289 (1963).
731. Thomas, W. H., and S. E. Cook, U.S. Pat. 3,098,090 (to Ethyl Corp.), July 16, 1963; *C.A.*, **59,** 7289 (1963).
732. Thomas, W. H., and S. E. Cook, U.S. Pat. 3,133,095–6 (to Ethyl Corp.), May 12, 1964; *C.A.*, **61,** 3147 (1964).
733. Thomas, W. H., and S. E. Cook, U.S. Pat. 3,133,100–2 (to Ethyl Corp.), May 12, 1964; *C.A.*, **61,** 4393 (1964).
734. Thomas, W. H., and S. E. Cook, U.S. Pat. 3,197,492 (to Ethyl Corp.), July 27, 1965; *C.A.*, **63,** 9730 (1965).
735. Thompson, H. W., *Proc. Royal Soc. (London)*, **A150,** 603 (1935); *J. Chem. Soc.*, **1934,** 790.
736. Thornton, D. P., Jr., *Petroleum Processing*, **7,** 846 (1952).
737. Tomilov, A. P., Yu. D. Smirnov, and S. L. Varshavskii, *Zh. Obshch. Khim.*, **35,** 391 (1965); *C.A.*, **63,** 5238 (1965).
738. Toms, F. H., and C. P. Mony, *Analyst*, **53,** 328 (1928); through *C.A.*, **22,** 3040 (1928).
739. Tsou, K. C., and S. R. Sandler, U.S. Pat. 3,244,637 (to Borden Co.), April 5, 1966; *C.A.*, **65,** 1732 (1966).
740. Tullio, V., U.S. Pat. 3,072,695 (to E. I. du Pont de Nemours & Co.), Jan. 8, 1963; *C.A.*, **58,** 13993 (1963).
741. Tullio, V., U.S. Pat. 3,072,694 (to E. I. du Pont de Nemours & Co.), Jan. 8, 1963; *C.A.*, **58,** 13992 (1963).
742. Ulrych, T. J., and R. D. Russell, *Geochim. Cosmochim. Acta*, **28,** 455 (1964); through *C.A.*, **61,** 2863 (1964).
743. Vertyulina, L. N., and I. A. Korshunov, *Khim. Nauka i Prom.*, **4,** 136 (1959); through *C.A.*, **53,** 12096 (1959).

744. Vignes, S., and A. Cottin, *Compt. rend. Cong. Ind. Gaz.*, **80,** 246 (1963); through *C.A.*, **62,** 14400 (1965).
745. Vogel, A. I., W. T. Cresswell, and J. Leicester, *J. Phys. Chem.*, **58,** 174 (1954).
746. Vogel, C. C., *J. Chem. Ed.*, **25,** 55 (1948).
747. Voorhees, V., U.S. Pat. 1,974,167 (to Standard Oil Co. of Indiana), Sept. 18, 1934; through *C.A.*, **28,** 6997 (1934).
748. Vorlander, D., *Ber.*, **58,** 1893 (1925).
749. Vyazankin, N. S., G. A. Razuvaev, and Yu. I. Dergunov, *Tr. po. Khim. i Khim. Tekhnol.*, **4,** 652 (1961); through *C.A.*, **58,** 543 (1963).
750. Vyshinskii, N. N., and N. K. Rudnevskii, *Tr. po Khim. i Khim. Tekhnol.*, **3,** 538 (1960); *C.A.*, **56,** 4262 (1962).
751. Vyshinskii, N. N., and N. K. Rudnevskii, *Optika i Spektroskopiya*, **10,** 797 (1961); through *C.A.*, **58,** 4059 (1963).
752. Wall, H. H., U.S. Pat. 3,158,636 (to Ethyl Corp.), Nov. 24, 1964; *C.A.*, **62,** 3870 (1965).
753. Wallen, L. L., *Iowa St. Coll. J. Sci.*, **29,** 526 (1955); through *C.A.*, **49,** 10418 (1955).
754. Walters, E. L., U.S. Pat. 2,361,337–8 (to Shell Dev. Co.), Oct. 24, 1944; through *C.A.*, **39,** 2643 (1945).
755. Waters, D. N., and L. A. Woodward, *Proc. Royal Soc.* (*London*), **A246,** 119 (1958).
756. Watt, G. W., *Chem. Rev.*, **46,** 317 (1950).
757. Wells, E. J., and L. W. Reeves, *J. Chem. Phys.*, **40,** 2036 (1964).
758. Weyer, K., *Dissertation*, *Aix-la-Chapelle*, (*1956*).
759. Whelen, M. S., *Advances in Chemistry Series*, No. 23, American Chemical Society, Washington, D.C., (1959), pp. 82–86.
760. Whitman, N., U.S. Pat. 2,657,225 (to E. I. du Pont de Nemours & Co.), Oct. 27, 1953; through *C.A.*, **48,** 2358 (1954).
761. Whittingham, G., *Nature*, **160,** 671 (1947).
762. Whittles, A. B. L., and W. F. Slawson, *Geochim. Cosmochim. Acta*, **29,** 142 (1965); through *C.A.*, **62,** 9918 (1965).
763. Wiczer, S. B., U.S. Pat. 2,960,515, Nov. 15, 1960; *C.A.*, **55,** 9282 (1961).
764. Widmaier, V. O., *Brennstoff-Chem.*, **34,** 83 (1953).
765. Willemsens, L. C., and G. J. M. van der Kerk, *Investigations in the Field of Organolead Chemistry*, Intern. Lead Zinc Res. Organ., N.Y., 1965.
766. Willemsens, L. C., J. C. Overeem, and G. J. M. van der Kerk, *Prog. Report XIV* (April 1–June 30 1963), to Intern. Lead Zinc Res. Organ., Project No. LC-18.
767. Williams, K., and H. Daudt, U.S. Pat. 1,550,940, Aug. 25, 1925; through *C.A.*, **20,** 209 (1926).
768. Wong, C. H., and V. Schomaker, *J. Chem. Phys.*, **28,** 1007 (1958).
769. Wunderlich, D. K., and L. N. Fussell, U.S. Pat. 3,159,557 (to Sinclair Research, Inc.), Dec. 1, 1964; through *C.A.*, **62,** 11432 (1965).
770. Wyler, J. A., U.S. Pat. 2,486,773 (to Trojan Powder Co.), Nov. 1, 1949; through *C.A.*, **44,** 1709 (1950).
771. Young, C. W., J. S. Kohler, and D. S. McKinney, *J. Am. Chem. Soc.*, **69,** 1410 (1947).
772. Youtz, M. A., U.S. Pat. 1,658,544, Feb. 7, 1928; through *C.A.*, **22,** 1164 (1928).
773. Yu'rev, Yu. K., M. A. Gal'bershtam, and I. I. Kandror, *Zh Obshch. Khim.*, **34,** 4116 (1964); through *C.A.*, **62,** 9163 (1965).

774. Zakharkin, L. I., and O. Yu. Okhlobystin, *Izves. Akad. Nauk SSSR, Otdel. Khim. Nauk*, **1959,** (1942); through *C.A.*, **54,** 9738 (1960).
775. Zartman, W. H., and H. Adkins, *J. Am. Chem. Soc.*, **54,** 3398 (1932).
776. Zasosov, V. A., and K. A. Kocheshkov. *Sb. Statei po Obshch. Khim., Akad. Nauk SSSR*, **1,** 278 (1953); through *C.A.*, **49,** 913 (1955).
777. Zasosov, V. A., and K. A. Kocheshkov, *Sb. Statei po Obshch. Khim., Akad. Nauk SSSR*, **1,** 285 (1953); through *C.A.*, **49,** 913 (1955).
778. Zechmeister, L., and J. Csabay, *Ber.*, **60,** 1617 (1927).
779. Zhdanov, G. S., and I. G. Ismailzade, *Zhur. Fiz. Khim. SSSR*, **24,** 1495 (1950); *C.A.*, **45,** 4112 (1951).
780. Zhdanov, G. S., and I. G. Ismailzade, *Dokl. Akad. Nauk SSSR*, **68,** 95 (1949); *C.A.*, **43,** 8764 (1949).
781. Ziegler, K., *Angew. Chem.*, **71,** 628 (1959).
782. Ziegler, K., *Angew. Chem.*, **72,** 565 (1960).
783. Ziegler, K., *Brennstoff-Chem.*, **40,** 209 (1959).
784. Ziegler, K., Belg. Pat. 617,628, Nov. 14, 1962; through *C.A.*, **60,** 3008 (1964).
785. Ziegler, K., Brit. Pat. 814,609, June 10, 1959; through *C.A.*, **53,** 17733 (1959); U.S. Pat. 2,985,568; *C.A.*, **53,** 17733 (1959).
786. Ziegler, K., Brit. Pat. 848,364, Sept. 14, 1960; through *C.A.*, **55,** 5199 (1961); French Pat. 1,139,719, July 4, 1957.
787. Ziegler, K., Brit. Pat. 864,393, April 6, 1961; through *C.A.*, **55,** 18398 (1961).
788. Ziegler, K., Brit. Pat. 864,394, April 6, 1961; through *C.A.*, **55,** 20962 (1961).
789. Ziegler, K., Brit. Pat. 923,652, April 18, 1963; through *C.A.*, **59,** 10117 (1963).
790. Ziegler, K. (by W. R. Kroll), Ger. Pat. 1,165,031, March 12, 1964; through *C.A.*, **61,** 10706 (1964).
791. Ziegler, K. (and H. Lehmkuhl), Ger. Pat. 1,127,900, April 19, 1962; through *C.A.*, **57,** 11235 (1962).
792. Ziegler, K. (and H. Lehmkuhl), Ger. Pat. 1,134,672, August 16, 1962; through *C.A.*, **58,** 550 (1963).
793. Ziegler, K. (and H. Lehmkuhl), Ger. Pat. 1,153,754, Sept. 5, 1963; through *C.A.*, **60,** 1794 (1964).
794. Ziegler, K. (and H. Lehmkuhl), Ger. Pat. 1,161,562, Jan. 23, 1964; through *C.A.*, **60,** 11623 (1964).
795. Ziegler, K. (and H. Lehmkuhl), Ger. Pat. 1,181,220, Nov. 12, 1964; through *C.A.*, **62,** 6156 (1965).
796. Ziegler, K. (and H. Lehmkuhl), Ger. Pat. 1,212,085, March 10, 1966; through *C.A.*, **64,** 19675 (1966).
797. Ziegler, K., (and H. Lehmkuhl), Ger. Pat. 1,666,196, March 26, 1964; through *C.A.*, **60,** 15910 (1964).
798. Ziegler, K., (and H. Lehmkuhl), *Z. anorg. allgem. Chem.*, **283,** 414 (1956).
799. Ziegler, K., (and H. Lehmkuhl), *Chem. Ingr. Tech.*, **35,** 325 (1963).
800. Ziegler, K., H. Lehmkuhl, and W. Grimme, Ger. Pat. 1,114,330; Belg. Pat. 575,595, Feb. 11, 1959; *C.A.*, **56,** 4524 (1962).
801. Ziegler, K., and O.- W. Steudel, *Ann. Chem.*, **652,** 1 (1962).
802. Zimmer, H., I. Heckenbleikner, O. A. Homberg, and M. Canzik, *J. Org. Chem.*, **29,** 2632 (1964).
803. Zimmer, H., and O. A. Homberg, *J. Org. Chem.*, **31,** 947 (1966).
804. Zinov'ev, Yu. M., and L. K. Soborovskii, *Zh. Obshch. Khim.*, **34,** 929 (1964); *C.A.*, **60,** 15904 (1964).
805. Zuliani, G., *Metano(Padua)*, **14,** 207 (1960); through *C.A.*, **62,** 5112 (1965).

VII Hexaalkyl- and hexaaryldilead compounds, R_3PbPbR_3

The first successful isolation and characterization of a pure compound having the composition R_3Pb is attributed to Krause and Schmitz (61) who reported the preparation of the *p*-tolyl and *p*-xylyl derivatives in 1919; preparation of the cyclohexyl derivative was reported in 1921 (57), followed by a report of the preparation of the pure ethyl derivative by Midgley and co-workers (74) in 1923. The aryl derivatives are relatively well characterized; being crystalline solids, they are easily purified. Most of the known alkyl derivatives are liquids, although hexamethyldilead melts at 38° and the newly synthesized neopentyl and neophyl compounds are crystalline solids (98,114). In general, the alkyl derivatives are not as well characterized as the aryl derivatives.

Because of their faint to strong yellow color both in the pure state and in solution, compounds having the overall composition R_3Pb were originally considered to be capable of existing as free radicals of the type $R_3Pb\cdot$. However, the preponderance of physical property measurements does not support the existence of $R_3Pb\cdot$ free radicals, so that the compounds are best represented as hexaorganodilead compounds containing a lead-to-lead bond.

Because of the presence of the lead–lead bond, the R_6Pb_2 compounds are highly reactive, relatively unstable compounds. Although stable to hydrolysis, they are readily oxidized by air, especially in the presence of light, to form organolead oxides and hydroxides. They also decompose thermally to form the tetraorganolead compound and lead metal, and therefore are precursors in many reactions used to synthesize R_4Pb compounds.

Only a small number of hexaorganodilead compounds is known; with one exception, all of these are of the symmetrical type in which all the organic groups are identical, R_6Pb_2. No concerted effort has been made to prepare unsymmetrical derivatives of the type R_3PbPbR_3' or $R_2R'PbPbR'R_2$. However, a recent report from Willemsens and van der Kerk (111) claims the successful preparation of (triphenylplumbyl)tri-*p*-tolyllead. These unsymmetrical derivatives will probably prove to be less stable than their symmetrical analogs.

Analogs of hexaorganodilead compounds are known in which one of the lead atoms is replaced by tin or germanium. (Triphenylstannyl)-triphenyllead is a white, crystalline solid which is stable at room temperature in the solid state (110). However, it is unstable to hydrolysis and decomposes to form tetraphenyltin and lead hydroxide. A series of (triphenylgermyl)triorganolead compounds of the type Ph_3GePbR_3 has also been prepared (78). These compounds are reasonably stable. However, attempts to prepare the trialkylgermyl analogs by a similar procedure were unsuccessful.

A second class of compounds containing a lead–lead bond is that of the type $(R_3Pb)_4Pb$. Only two such compounds are now known, in which R = phenyl or *p*-tolyl, and both were reported by Willemsens and van der Kerk in 1964 (109). However, a number of mixed metal derivatives of the type $(Ph_3M)_4M'$, where M and M′ are germanium, tin, or lead, has been reported by these same investigators (108). These polymetal compounds are discussed in Chapter XV.

The successful synthesis of compounds containing four lead-to-metal bonds suggests that the chemistry of R_6Pb_2 compounds and higher homologs has not been fully exploited. The research efforts of Willemsens, van der Kerk, and colleagues at the Institute for Organic Chemistry TNO, Utrecht, The Netherlands, have been especially successful in elucidating the mechanisms for formation of hexaorganodilead compounds in Grignard reactions, and their reactions with various reagents. Further efforts along these lines should be equally rewarding. Hexaorganodilead compounds are discussed in a recent review article on the synthesis and properties of the R_6M_2 derivatives of the Group IVb metals (31).

1. SYNTHESIS

Hexaorganodilead compounds are usually obtained as trace to major by-products in syntheses of tetraorganolead compounds. Thus, under certain conditions, hexaethyldilead is formed as a major by-product in the reaction of sodium–lead alloy or magnesium–lead alloy and ethyl chloride; it is also formed in the reaction of lead metal with magnesium metal and ethyl chloride. However, R_6Pb_2 compounds are best synthesized by one of the two general methods discussed below.

A. Reactions of lead(II) halides with Grignard or organolithium reagents

The reaction of lead halides with Grignard or organolithium reagents was discussed in Chapter III. As was mentioned there, this reaction proceeds with the intermediate formation of R_6Pb_2, which subsequently is decomposed to the R_4Pb compound and lead metal. In fact, with increasing chain length in the alkyl derivatives or with increasing ring substitution in the phenyl derivatives, disproportionation of the R_6Pb_2 intermediate becomes difficult, so that it becomes the main product of the reaction.

The PbX_2–RMgX reaction has been used to synthesize a variety of hexaorganodilead derivatives:

$$3PbX_2 + 6RMgX \rightarrow R_6Pb_2 + 6MgX_2 + Pb$$

Hexamethyldilead has been prepared in 61% yield by the reaction of methylmagnesium bromide and lead chloride in diethyl ether at −5 to −8° (21); the low reaction temperature is required to prevent disproportionation of the hexamethyldilead to tetramethyllead. With propyl and higher alkyl derivatives, the R_6Pb_2 compounds become sufficiently stable so that their isolation without gross contamination by R_4Pb becomes rather easy. Properties of the known R_6Pb_2 compounds are summarized in Table VII-1.

The hexaneopentyl and hexaneophyl derivatives were prepared recently via the Grignard route (98,114). Hexacyclohexyldilead and hexaphenethyldilead were obtained free of their R_4Pb analogs by Krause and Schlöttig (60) from the reaction of $PbCl_2$ and the respective Grignard reagent. On the other hand, preparation of hexabenzyldilead and of its *o*-fluoro-substituted derivative was not successful; they were apparently unstable and disproportionated to the respective R_4Pb compounds (10). However, hexa-*o*-bromobenzyl- and hexa-*o*-chlorobenzyldilead are relatively stable and have been described as deep red and light red crystalline solids, respectively.

In the aryl series, hexaphenyldilead and various substituted phenyl derivatives have been prepared via the Grignard procedure. Willemsens and van der Kerk (110) have shown that the co-formation of tetraphenyllead in the synthesis of hexaphenyldilead can be completely eliminated by avoiding the presence of an excess of phenylmagnesium bromide throughout the reaction, i.e., by adding the Grignard reagent to the slurry of lead(II) chloride in ether. Hexa-*m*-tolyldilead (75), and hexa-*p*-tolyldilead, hexa-*o*-tolyldilead, and hexa-*p*-xylyldilead (58) have also been prepared using the Grignard–$PbCl_2$ reaction; the *o*-tolyl and *p*-xylyl derivatives

Table VII-1 Hexaorganodilead Compounds, R_6Pb_2

COMPOUND (R IN R_6Pb_2)	COLOR	M.P., °C	REFERENCES[a]
n-Butyl	Yellow	Oil	27,35,110
o-Bromobenzyl	Deep red	d. >170	10
o-Chlorobenzyl	Orange red	d. >170	10
Cyclohexyl	Yellow	190 (d.)	27,32,57,60,80
o-Ethoxyphenyl		178–9 (d.)	11,32
p-Ethoxyphenyl		170–1 (d.)	11,32
Ethyl	Yellow	Oil, b. 100^2	17,20,32,48,74,91
Isopropyl		Oil	41
Mesityl		>325	11,32,66,77
o-Methoxyphenyl		198–201 (d.)	11,32
p-Methoxyphenyl		198–200 (d.)	11,32,38
Methyl	Yellow	38	21,22,99,107
1-Naphthyl	Yellow	d. 255	11,32
Neopentyl	Pale yellow	205–6 (d.)	98,114
Neophyl		132–3	114
n-Pentyl			53
β-Phenethyl	Yellow		60
Phenyl	Light yellow	d. 155–60	27,28,32,33,35,36,58,59, 62,69,102
n-Propyl		Oil	43
m-Tolyl		109 (d. 116–7)	11,27,32,75
o-Tolyl		d. 240	7,8,27,58
p-Tolyl		d. 193	8,11,32,35,58
m-Xylyl[b]	Pale yellow	233.5 (d.)	40
p-Xylyl[c]	Greenish yellow	220 (d.)	58,61,95

[a] Other references are given in text.
[b] *m*-Xylyl = 2,4-dimethylphenyl
[c] *p*-Xylyl = 2,5-dimethylphenyl

disproportionate only with great difficulty, so that they are easily obtained free of R_4Pb. The mesityl and 1-naphthyl derivatives have also been prepared, but in low yield (32); steric factors are postulated to be responsible for the poor yields.

Organolithium reagents can be used in place of Grignard reagents in the synthesis of R_6Pb_2 with satisfactory results. Thus, hexaphenyldilead has been prepared in 79% yield by the reaction of $PbCl_2$ and phenyllithium in diethyl ether at −10°, followed by refluxing the reaction mixture. Hexa-1-naphthyldilead was obtained in 10% yield using 1-naphthyllithium, as compared with a 6% yield when 1-naphthylmagnesium bromide was used (32). The use of organolithium as a reactant offers no apparent advantage over the Grignard reagent. In fact, the facile reaction of RLi compounds with R_6Pb_2 requires that an excess of RLi be avoided to minimize R_6Pb_2 loss via the reaction:

$$R_6Pb_2 + RLi \rightarrow R_4Pb +_2 R_3PbLi$$

B. Reactions of triorganolead halides with sodium or lithium metal

A second general method of synthesis of R_6Pb_2 compounds is the reaction of triorganolead halides with sodium metal:

$$2R_3PbCl + 2Na \xrightarrow{NH_3} R_6Pb_2 + 2NaCl$$

These reactions are usually conducted in liquid ammonia. An excess of sodium must be avoided to prevent conversion of the R_6Pb_2 to R_3PbNa.

Hexamethyldilead was prepared in this manner by Calingaert and Soroos (21), but the yield was only 7%. However, satisfactory yields of hexaethyldilead (21,32) and various hexaaryldilead derivatives have been obtained using this reaction (28,32). Triphenyllead iodide is reported to react more smoothly with sodium metal than does triphenyllead chloride, although reaction of triphenyllead chloride with sodium–lead alloy, Na_4Pb_9, gave hexaphenyldilead in nearly quantitative yield (28). The higher yield obtained with Na_4Pb_9 was attributed to the reduced activity of the sodium when used as the alloy, with consequent reduction in the vigor of the reaction.

The use of lithium metal in the above reaction has been little investigated. However, the reduction of triphenyllead halides with lithium metal in tetrahydrofuran has been used to prepare triphenylplumbyllithium and is postulated to proceed via intermediate formation of hexaphenyldilead (34,102), as indicated by the formation of a white cloudiness at an intermediate stage:

$$2R_3PbCl + 2Li \xrightarrow{THF} R_6Pb_2 \xrightarrow{2Li} 2R_3PbLi$$

Gilman's group (6,33) has shown that reaction of diphenyllead dichloride and diphenyllead diiodide with lithium metal in liquid ammonia produces hexaphenyldilead; sodium and calcium metals gave similar results.

The synthesis of unsymmetrical hexaorganodilead derivatives via the sodium metal reaction has been attempted unsuccessfully (33). Reduction of phenyldicyclohexyllead chloride or ethyldiphenyllead chloride with sodium metal in liquid ammonia did not yield the expected $R_2R'PbPbR'R_2$. With the former compound, a red coloration was produced; evaporation of the ammonia and extraction of the residue with chloroform yielded a red solution which was very sensitive to light. Reduction of ethyldiphenyllead chloride was not accompanied by any particular color change, but a yellow solution was obtained upon extraction of the reaction products with chloroform. However, the desired unsymmetrical hexaorganodilead compounds could not be isolated from either reaction mixture.

C. Other methods of synthesis

The preparation of hexaphenyldilead has been attempted by the reaction of triphenylplumbyllithium and triphenyllead chloride in diethyl ether. Instead, tetraphenyllead was formed; this was attributed to dissociation of the triphenylplumbyllithium into phenyllithium and diphenyllead.

$$Ph_3PbLi \rightleftharpoons Ph_2Pb + PhLi$$

$$PhLi + Ph_3PbCl \rightarrow Ph_4Pb + LiCl$$

Subsequent data argue against a gross dissociation of the Ph_3PbLi and there is seemingly no reason why hexaphenyldilead cannot be prepared via the triphenylplumbyllithium–triphenyllead chloride reaction. Furthermore, Willemsens and van der Kerk (111) have recently prepared (triphenylplumbyl)tri-*p*-tolyllead by the reaction of triphenylplumbyllithium with tri-*p*-tolyllead chloride; this is the only known unsymmetrical R_6Pb_2 compound. No details were given except that the product was stable and was isolated in good yield. Similar attempts to prepare (triphenylplumbyl)trimethyllead, $Ph_3PbPbMe_3$, and $Ph_3PbPb(Ph)Me_2$ (using Me_3PbCl and $Me_2PhPbCl$) gave phenyltrimethyllead and diphenyldimethyllead, respectively. Decomposition of the desired R_6Pb_2 products via phenyl group migration was postulated.

$$Ph_3PbPbMe_3 \rightarrow (Ph_2Pb) + PhPbMe_3$$

As was mentioned earlier, (triphenylstannyl)triphenyllead was successfully prepared by Willemsens and van der Kerk (110) as a white crystalline solid by the reaction of triphenylplumbyllithium and triphenyltin chloride in tetrahydrofuran at $-10°$. It decomposes at elevated temperature to tetraphenyltin and lead metal and is unstable to hydrolysis. A series of (triphenylgermyl)triorganolead derivatives has been prepared by Neumann and Kühlein (78) by the reaction of the appropriate triorganolead diethylamide compound with triphenylgermane. The triphenyllead, triisobutyllead, tri-*n*-butyllead, and tricyclohexyllead derivatives were prepared in this fashion; the tri-*n*-butyl derivative is an oil, the others are solids. (Table VII-2.)

Hexaethyldilead has been synthesized by electrolysis of a solution of triethyllead hydroxide (or triethyllead chloride and alkali) in ethanol using lead electrodes (74). Excellent results were obtained from this method using a compartmented cell to segregate the electrode processes, and using aqueous sodium hydroxide as solvent so that the hexaethyldilead separates as a second liquid phase (17,48). However, a recent report claims that this method yields a mixture of diethyllead and hexaethyldilead (92). Chemical reduction of an aqueous solution of triethyllead hydroxide with aluminum or zinc metal has also been used to prepare

Table VII-2 Mixed Metal Analogs of R_6Pb_2, $R_3Pb{-}MR'_3$

COMPOUND	COLOR	M.P., °C	REFERENCE
$Ph_3SnPbPh_3$	White	[a]	110
$Ph_3GePbPh_3$		227 (d.)	78
$Ph_3GePb(i\text{-}Bu)_3$		86	78
$Ph_3GePb(n\text{-}Bu)_3$		Oil	78
$Ph_3GePb(C_6H_{11})_3$		174 (d.)	78

[a] Becomes untransparent at 110°, then melts at 165–229°.

hexaethyldilead (48,91); some diethyllead is reported to be formed along with the hexaethyldilead (91).

Reduction of acetone with sodium–lead alloy is reported to yield hexaisopropyldilead (41). Tafel (101) and Ghira (29) have concluded that the product from the reaction of ethyl iodide and sodium–lead alloy, as reported by Löwig (70,71) and Klippel (53,54), was primarily hexaethyldilead and not tetraethyllead.

2. PHYSICAL PROPERTIES

Because of their relative instability, few physical properties measurements have been made on any of the R_6Pb_2 compounds and most of these measurements have been directed at an unequivocal answer to the question of dissociation of R_6Pb_2 to triorganolead radicals. This subject is discussed later in this chapter.

As was discussed earlier, electron diffraction measurements (99) gave a value of 2.88 Å for the lead–lead bond distance in hexamethyldilead and a value of 2.25 Å for the lead–carbon bond distance. From X-ray diffraction measurements, two values were calculated for the lead–lead bond distances in hexaphenyldilead, 2.75 ± 0.09 Å and 2.94 ± 0.09 Å (55). The existence of an ionic lattice containing Ph_2Pb^{+2} ions and dimeric

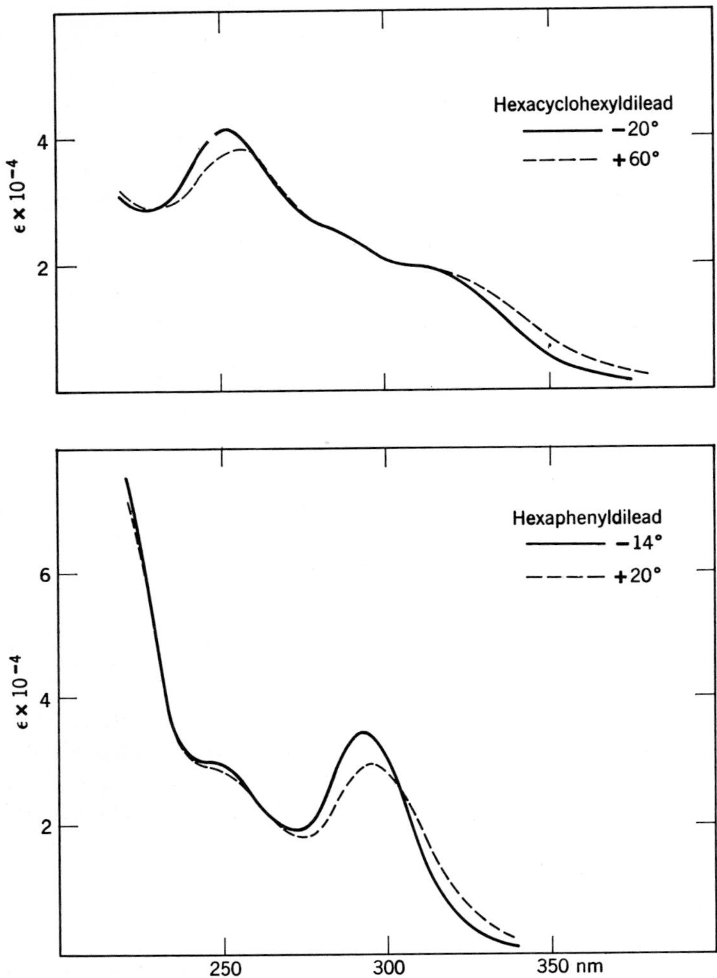

FIG. VII-1 Ultraviolet absorption spectra of hexacyclohexyldilead and hexaphenyldilead at different temperatures. [Taken from Drenth, Willemsens and van der Kerk, *J. Organometal. Chem.*, **2**, 279 (1964).]

Ph_4Pb^{-2} ions was proposed as a possible explanation of the two values for the lead–lead distance.

The ultraviolet absorption spectrum of hexaphenyldilead shows an intense band at 294 μ which originally was attributed to a strong interaction between phenyl groups on different lead atoms across the lead–lead bond through overlap of vacant *d*-orbitals of the lead atoms; hexaphenylditin shows a similar band at 247 μ which was ascribed to a similar interaction (44,45). A similar band, however, is also found in the ultraviolet spectrum of hexacyclohexyldilead (25), but at a somewhat shorter wavelength ($\sim$254 μ) than in hexaphenyldilead (Fig. VII-1). Since the cyclohexyl group cannot effectively interact with vacant *d*-orbitals of the lead atom, the presence of this characteristic band in hexacyclohexyldilead has been attributed to the valence electrons in the lead–lead bond. Similar bands are observed in the ultraviolet spectra of tetrakis-(triphenylplumbyl)-lead and other derivatives of the type $(Ph_3M)_4M'$ (25).

Further support for the assignment of these ultraviolet absorption characteristics to a lead–lead interaction in dilead and polylead derivatives has been offered based on the effect of temperature on the absorption spectra of hexaphenyldilead and hexacyclohexyldilead (26). At elevated temperatures both compounds show a broadening of their absorption bands at 294 and 254 μ, respectively. Since broadening of the ultraviolet absorption bands of molecules containing heavy metal atoms is a well-known phenomenon which is associated with the temperature-dependent distribution of molecules over vibrational levels, broadening of these bands in the spectra of hexaphenyldilead and hexacyclohexyldilead represents further evidence that they are associated with the metal–metal bond.

A detailed ultraviolet absorption spectrum of hexaethyldilead has not been published. However, it would be expected to be similar to that of hexaphenyldilead and hexacyclohexyldilead, with an absorption band resulting from the lead–lead bond. Absorption in the region of 320 μ has been reported to be useful for determining the concentration of hexaethyl-dilead in solution (94); this may be the absorption band attributable to the lead–lead bond.

The infrared spectrum of hexaphenyldilead is almost identical to that of tetraphenyllead and hexaphenylditin (Fig. VII-2) (46). Hexaphenyldilead shows the same characteristic bands at 1428 and 1052 cm^{-1} found in tetraphenyllead, which are attributed to the phenyl group perturbed by the heavy metal atom.

The infrared and Raman spectra of hexaethyldilead are reported to be very similar to those of tetraethyllead; this is believed to indicate that there is little interaction between the normal coordinates of the various ethyl groups (106).

The NMR spectra of hexaneopentyldilead and hexaneophyldilead (98,114) and hexa-*p*-tolyldilead (52a) have been reported. These were discussed in Chapter VI.2, dealing with the physical properties of tetraorganolead compounds. Paramagnetic resonance measurements have been made on hexaphenyldilead both in the solid state and in solution (18).

The dipole moment of hexaphenyldilead has been shown to be zero, as expected (67).

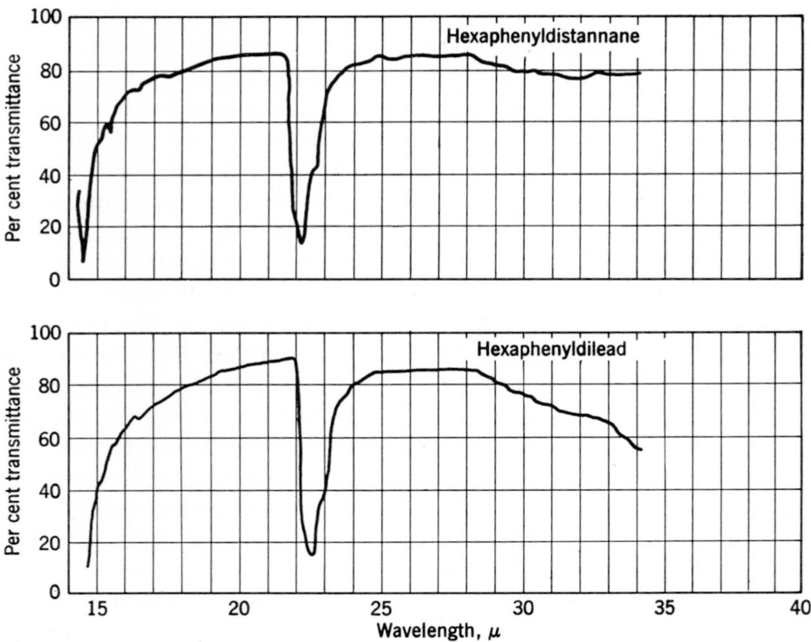

FIG. VII-2 Infrared absorption spectra of hexaphenyldilead and hexaphenylditin from 15–35 μ. [Taken from Harrah, Ryan and Tamborski, *Spectrochim. Acta*, **18,** 21 (1962).]

3. DISSOCIATION OF R_6Pb_2 COMPOUNDS

One of the most puzzling aspects of organolead chemistry has been the question of the dissociation of hexaorganodilead compounds into triorganolead radicals. In his original synthesis of the first such compounds, Krause noted their characteristic yellow color and proposed that these compounds would readily dissociate into $R_3Pb\cdot$ radicals. Cryoscopic measurements showed that the observed molecular weights of hexaphenyldilead (58), hexacyclohexyldilead (57), and hexa-*p*-xylyldilead and hexa-*p*-tolyldilead (61), decreased with increasing dilution in benzene; at 0.1 wt % in benzene, the observed molecular weight was very close to that of an $R_3Pb\cdot$ radical. The hypothesis of the existence of free $R_3Pb\cdot$ radicals was strengthened by the observation that the characteristic yellow color of the solid R_6Pb_2 compounds usually became intensified in solution. Other investigators also found evidence for the dissociation of hexaethyldilead (74), hexamesityldilead (66), and hexaphenyldilead (69), based on cryoscopic measurements. On the other hand, still others found that the observed molecular weight by cryoscopy appears to be dependent on the solvent used. Thus, in naphthalene and biphenyl solution the molecular weights of hexaphenyldilead and hexacyclohexyldilead corresponded to those of the "dimeric" R_6Pb_2 (72), while cryoscopic measurements on hexamesityldilead in biphenyl (77) showed that the degree of dissociation in this solvent is not nearly that indicated in chloroform or benzene solutions (66).

However, the existence of triorganolead radicals was not indicated by magnetic susceptibility measurements on dilute solutions of hexaphenyldilead (82), hexacyclohexyldilead (51,76), hexa-*o*-chlorobenzyldilead (10), or hexa-*o*-bromobenzyldilead (10). Also, electron paramagnetic resonance measurements on hexamesityldilead in dilute benzene solution did not give any evidence for the presence of organolead radicals, while magnetic susceptibility measurements on solid hexamesityldilead gave inconclusive results (77).

Because of the instability of organolead compounds, especially hexaorganodilead compounds, and because of the great effect of small amounts of impurities on cryoscopic measurements in dilute solutions, the concensus of opinion has been that no dissociation of R_6Pb_2 to $R_3Pb\cdot$ radicals occurs in solution. It should be pointed out that a similar situation exists with hexaorganoditin compounds, i.e., they give evidence both for and against dissociation, despite the fact that they are more stable than their organolead analogs.

Additional evidence against dissociation of R_6Pb_2 has been presented in recent years. No evidence was obtained by Belluco and co-workers (18) for an exchange reaction between hexaphenyldilead and triphenyllead chloride in benzene using radioactive lead to label the hexaphenyldilead. Cryoscopic and osmometric measurements on hexaphenyldilead in benzene solution have been carefully conducted by Willemsens and van der Kerk (110), who found that at the lowest concentrations ($0.001M$) used by Krause, the cryoscopic measurements did not give reproducible results; most of the values were near the calculated value of 877 for the molecular weight of hexaphenyldilead. Osmometric methods gave more reproducible results and indicated that hexaphenyldilead was not dissociated even at $0.001M$ concentrations in benzene. It has also been shown that solutions of hexaphenyldilead and hexacyclohexyldilead obey Beer's Law; therefore, the thermochromic behavior exhibited by these compounds is not a result of an equilibrium between R_6Pb_2 molecules and $R_3Pb\cdot$ radicals (26). Rather, the thermochromic effect, as was discussed earlier, is due to broadening of the absorption peak associated with the lead–lead bond at higher temperatures, thus causing increased absorption in the visible range of the spectrum.

A final argument against dissociation into $R_3Pb\cdot$ radicals is that hexaphenyldilead does not undergo free radical-type reactions. The addition of hexaphenyldilead to maleic anhydride does not occur, as was originally reported by Gilman and Leeper (33,65). No reaction occurred between hexaphenyldilead and isoprene, pyrrole, styrene, furan, and 1,4-diphenyl-1,3-butadiene during 30 days at room temperature (33); also hexaphenyldilead did not react with cyclohexene under ultraviolet irradiation nor with carbene (from catalytic decomposition of diazomethane) (110).

An alternate mode of dissociation of R_6Pb_2 has been postulated by several investigators which involves its dissociation into R_4Pb and R_2Pb:

$$R_6Pb_2 \rightleftharpoons R_4Pb + R_2Pb$$

Thus, dissociation of hexaphenyldilead into tetraphenyllead and diphenyllead has been proposed as a possible explanation for the formation of lead(II) acetate in nearly 50% yield in the reaction of hexaphenyldilead and acetic acid (62), and of lead(II) chloride in comparable yield in the reaction of hexaethyldilead and hydrogen chloride (16). The thermal decomposition of hexaethyldilead, based on spectrophotometric data, is postulated to proceed via intermediate formation of diethyllead (91,92) and the observation that lead exchange between hexaphenyldilead, tagged with Radium D, and tetraphenyllead occurs in benzene solution under very mild conditions has been interpreted to indicate that the hexaphenyldilead dissociates into tetraphenyllead and diphenyllead (15). Finally, the dissociation of hexaphenyldilead according to the above scheme has been proposed to account for the products from its reactions with metal halides and metals (85).

On the other hand, Willemsens and van der Kerk (110) point out that the hydrolytic stability of hexaphenyldilead argues against its facile dissociation into tetraphenyllead and diphenyllead; diphenyllead is readily hydrolyzed and would be decomposed to lead hydroxide. These same investigators did not observe a reaction between tetraphenyllead and diphenyllead to form hexaphenyldilead.

On the basis of all the foregoing evidence, it must be concluded that the dissociation of R_6Pb_2 (into $2R_3Pb\cdot$ or into $R_4Pb + R_2Pb$) does not occur under low temperature conditions at which the R_6Pb_2 is stable. Under more rigorous conditions, an irreversible dissociation takes place, leading to the formation of R_4Pb and lead metal, and possibly other products dependent upon the presence of some other reagent.

4. CHEMICAL PROPERTIES

Hexaorganodilead compounds exhibit a higher order of chemical reactivity and thermal instability than the corresponding tetraorganolead compounds because of the presence of the lead–lead bond. No attempt has been made to measure the strength of this bond, although a number of reactions is known which should be easily adaptable to thermodynamic measurements. However, the lead–lead bond energy has been estimated to be about 20 kcal from the heat of vaporization of lead metal (52). The various chemical reactions of R_6Pb_2 seem to follow either of two different initial paths, on the basis of the products formed.

$$R_6Pb_2 \longrightarrow 2R_3Pb\cdot$$

$$R_6Pb_2 \longrightarrow R_4Pb + R_2Pb$$

From the relative reactivities of the various aryl derivatives, steric factors are indicated to play a determining role in the stability and reactivity of the different R_6Pb_2 compounds.

A. Pyrolysis

R_6Pb_2 compounds decompose thermally in the well known manner, according to the equation:

$$2R_6Pb_2 \longrightarrow 3R_4Pb + Pb$$

This reaction is involved in the synthesis of R_4Pb compounds via the lead halide reaction. Thermal stability increases with increasing alkyl chain length and with increasing ring substitution in aryl derivatives. The relative stability of a number of hexaorganodilead derivatives has been shown to increase in the qualitative order (32,58): methyl, ethyl < *m*-tolyl, phenyl < *p*-tolyl, *p*-ethoxyphenyl, *p*-methoxyphenyl < *o*-tolyl, *o*-ethoxyphenyl, *o*-methoxyphenyl < cyclohexyl, mesityl, 1-naphthyl. Aryl derivatives are generally more stable than alkyl compounds. Thus, hexaethyldilead decomposes within a few days at room temperature in a closed container protected from light (32), whereas hexaphenyldilead has been stored without decomposition for five years under similar conditions (28) and hexamesityldilead is reported to be stable to 325° (32). The lower hexaalkyl derivatives are less stable but may be stored indefinitely at Dry Ice temperature (97).

Hexaorganodilead compounds are less stable in solution; hexacyclohexyldilead decomposes in benzene solution, even in the absence of light (2), and pyridine is reported to accelerate the disproportionation of hexa-*p*-tolyldilead (58). Various reagents also accelerate the disproportionation of hexaethyldilead; these include ethyl iodide (63), activated carbon (39), and various metal oxides, such as silica gel (73). Aluminum chloride also accelerates the disproportionation of hexaethyldilead (89); complete disproportionation occurred in 2–3 hr at room temperature in the presence of 1–1.5% aluminum chloride. Hexaethyldilead also undergoes disproportionation at room temperature in the presence of 2% 1,2-dibromoethane to form tetraethyllead and lead metal in near quantitative yield (88).

The kinetics of the thermal decomposition of hexaethyldilead have been studied in detail by Razuvaev and his colleagues (91,93). By spectrophotometric methods, decomposition was shown to proceed with initial formation of diethyllead which subsequently decomposed to lead metal and ethyl radicals according to the sequence:

$$(C_2H_5)_6Pb_2 \rightarrow (C_2H_5)_4Pb + (C_2H_5)_2Pb$$
$$(C_2H_5)_2Pb \rightarrow 2C_2H_5\cdot + Pb$$

The tetraethyllead formed in the initial step also slowly decomposes under the conditions used (105–170°) to form hexaethyldilead, according to:

$$2(C_2H_5)_4Pb \rightarrow (C_2H_5)_6Pb_2 + 2C_2H_5\cdot$$

This mode of disproportionation is vastly different from the conventional mode of disproportionation of hexaethyldilead in which all ethyl groups end up bonded to lead as tetraethyllead. The decomposition is catalyzed by the finely divided lead metal produced, so that the overall reaction is autocatalytic. Hexaethyldilead was completely decomposed in 5 hr at 105° to tetraethyllead, lead metal and a mixture of hydrocarbons formed from the ethyl radicals.

The disproportionation of hexaethyldilead in the presence of catalytic amounts of 1,2-dibromoethane proceeds at room temperature to yield tetraethyllead and lead metal in near quantitative amounts, without formation of gaseous by-products (88,104). With stoichiometric amounts of 1,2-dibromoethane, a more complex product mixture was obtained. At 70°, tetraethyllead was obtained as the major product (66%) along with triethyllead bromide, diethyllead dibromide, lead bromide, and lead metal; gaseous products were also formed, consisting of 83% ethylene, 7–8% ethane, and 8–9% butane (88). Reaction with 1,2-dibromopropane and ethyl bromide gave similar products, except that in the ethyl bromide reaction ethane was formed in slightly greater amounts than ethylene. At a higher reaction temperature (135–150°), triethyllead bromide was formed as the major product (83).

The reaction of hexaethyldilead and dibromoalkanes is speculated to proceed via a cyclic intermediate of the type

$$\begin{array}{ccc} (C_2H_5)_3Pb & \text{———} & Pb(C_2H_5)_3 \\ \vdots & & \vdots \\ Br & & Br \\ \backslash & & / \\ CH_2 & \text{- - - -} & H_2C \end{array}$$

from which triethyllead bromide and ethylene are subsequently formed. The triethyllead bromide could then decompose to the other products according to the following reactions:

$$2(C_2H_5)_3PbBr \longrightarrow (C_2H_5)_4Pb + PbBr_2 + 2C_2H_5\cdot$$
$$3(C_2H_5)_3PbBr \longrightarrow 2(C_2H_5)_4Pb + PbBr_2 + C_2H_5Br$$
$$2(C_2H_5)_3PbBr \longrightarrow (C_2H_5)_4Pb + (C_2H_5)_2PbBr_2$$

Decomposition of triethyllead bromide in 1,2-dibromoethane at 70° gave 8.9% lead bromide along with gaseous hydrocarbons (74% ethylene, 20% ethane, and 5.7% butane); in nonane, some lead bromide was also formed, but the ratio of ethylene to ethane was much lower (41.8% ethylene and 51% ethane).

It would appear that the actual reaction mechanism is not so simple as that proposed above. The formation of tetraethyllead in such large amounts (64–70%) indicates that a simple disproportionation reaction is the major process at 70° (91). Also, the formation of triethyllead bromide as the major reaction product at 135–150° argues against its being an intermediate to the other products at the lower reaction temperature, i.e., the higher reaction temperature should favor the decomposition of the triethyllead bromide to other products. Furthermore, triethyllead bromide reacts readily with hexaethyldilead at 70° (84), according to the equation:

$$(C_2H_5)_6Pb_2 + 2(C_2H_5)_3PbBr \longrightarrow 3(C_2H_5)_4Pb + PbBr_2$$

At a lower ratio of $(C_2H_5)_3PbBr/(C_2H_5)_6Pb_2$ than 2:1, some lead metal is also formed from simple disproportionation of the hexaethyldilead. Hexaethyldilead reacts with triethyltin chloride at room temperature to form lead chloride, tetraethyllead, and tetraethyltin. From these considerations, the reaction of hexaethyldilead and dibromoalkanes must be very complex; at least three or four competing reactions can occur so that an unequivocal interpretation of the mechanism is not possible from the available data.

The decomposition of hexaphenyldilead in the presence of 1,2-dibromoethane was also investigated by Krebs and Henry (62). Lead bromide was formed in 98% yield and tetraphenyllead in 81% yield after 15 min reflux of a solution of hexaphenyldilead in the dibromoethane, based on the stoichiometry:

$$2Ph_6Pb_2 + C_2H_4Br_2 \longrightarrow 3Ph_4Pb + PbBr_2 + C_2H_4$$

No lead metal was formed. These products suggested that the reaction proceeded via initial dissociation of hexaphenyldilead to tetraphenyllead and unstable diphenyllead.

$$2Ph_6Pb_2 \longrightarrow 2Ph_4Pb + 2(Ph_2Pb)$$
$$2(Ph_2Pb) \longrightarrow Ph_4pb + Pb$$

or $$2Ph_6Pb_2 \rightarrow 3Ph_4Pb + Pb$$

then $$Pb + BrC_2H_4Br \rightarrow PbBr_2 + C_2H_4$$

The key reaction then would be the reaction of lead metal and 1,2-dibromoethane. However, this sequence does not provide for the formation of triethyllead bromide as observed by Razuvaev's group with hexaethyldilead (88,104).

The pyrolysis and photolysis of hexaphenyldilead have been investigated in C^{14}-labeled benzene. Little or no exchange of phenyl groups was detected between hexaphenyldilead and benzene in 10 hr at 170°; complete decomposition of the hexaphenyldilead to tetraphenyllead, lead metal and biphenyl occurred, but the activity of the tetraphenyllead contained only 0.6–0.7% of the initial activity of the benzene. Photolytic decomposition of the same mixture gave tetraphenyllead having an activity equivalent to 0.9% of the initial activity of the benzene (86). The biphenyl formed as by-product in these reactions was derived equally from the benzene solvent and the hexaphenyldilead; 42–48% of the phenyl groups in the biphenyl contained C^{14} labeling. Very similar results were obtained using deuterium-labeled benzene or hexaphenyldilead; over 90% of the biphenyl was formed as $C_6H_5{\cdot}C_6D_5$ (87). Hexaphenyldilead crystallizes from benzene with one molecule of benzene of crystallization (58).

The pyrolysis of hexaphenyldilead has been investigated under hydrogen pressure in the presence of a Raney nickel or copper chromite catalyst (81). Pyrolysis occurred under mild conditions to form tetraphenyllead, lead metal and biphenyl. The formation of biphenyl, and not benzene, suggests that a simple thermal disproportionation occurred, catalyzed by the nickel or copper chromite, and that the hydrogen did not participate in the reaction. Powdered copper metal also catalyzed the thermal decomposition of hexaphenyldilead (85).

Calingaert, Soroos, and Shapiro (22) found that pyrolysis of a mixture of hexamethyldilead and hexaethyldilead at 100° produced all five possible R_4Pb compounds, but not in a statistical distribution. On the other hand, pyrolysis of a mixture of hexa-*p*-tolyldilead and hexa-*p*-methoxyphenyldilead gave only the two symmetrical R_4Pb compounds, tetra-*p*-tolyllead and tetra-*p*-methoxylphenyllead (32). Furthermore, Willemsens and van der Kerk (111) obtained a mixture of *p*-tolyltriphenyllead and phenyltri-*p*-tolyllead upon decomposition of (triphenylplumbyl)tri-*p*-tolyllead. No explanation has been offered for the vast difference in the behavior of these three systems.

B. Photolysis

The hexaorganodilead compounds decompose quickly in sunlight, and in the presence of diffuse daylight they become very sensitive to reaction with air. Thus, hexa-*p*-xylyldilead, both as a solid and in solution, is reported to be completely stable to air in the absence of light; exposure to air in the presence of light results in its complete decomposition (61). Hexacyclohexyldilead behaves similarly (57). The less stable derivatives, such as hexamethyldilead and hexaethyldilead, are unstable and decompose slowly, even in the dark; diffuse daylight, however, accelerates their decomposition.

The photolysis of hexaphenyldilead in labeled benzene was discussed in the preceding section. The oxidation of hexacyclohexyldilead by air under conditions of photolysis is discussed later. Except for these few papers, no investigations of the photolysis of R_6Pb_2 compounds have been reported.

C. Electrolytic reduction

The electrolytic reduction of hexaethyldilead has been investigated by DeVries and co-workers (24). In methanol–benzene, using lithium chloride as the supporting electrolyte, hexaethyldilead produced an anodic half wave potential of −0.24 V (versus the standard calomel electrode). Vertyulina and Korshunov (103) report that in absolute ethanol the half wave potential occurs at 1.8–2.0 V, at which the diffusion current is proportional to the concentration. A polarographic method for the determination of hexaethyldilead in the presence of tetraethyllead has been developed.

The electrolytic reduction of hexaphenyldilead has been investigated more recently in dimethoxyethane solution by Dessy and co-workers (23) who concluded that anion production occurs at a half wave potential of −2.0 V (versus a $Ag/AgNO_3$ electrode) according to:

$$Ph_3PbPbPh_3 \xrightarrow{2e^-} 2Ph_3Pb^-$$

Hexaphenylditin and hexaphenyldigermane, as expected, were reduced at higher potentials (−2.9 and −3.5 V, respectively), while hexaphenyldisilane did not undergo electrochemical reduction. The electrochemical reduction of hexaphenyldilead in the presence of triphenyltin chloride produced an unstable species, believed to be (triphenylstannyl)triphenyllead, which in turn underwent electrolytic reduction at a potential of

-2.4 V to yield hexaphenylditin (and Ph_3Pb^-). Electrolysis of hexaphenyldilead in the presence of silver perchlorate gave a green coloration attributed to the formation of triphenylplumbylsilver.

D. Reactions with various reagents

(*1*) *Reactions with oxygen and oxygen-containing compounds*

Most of the hexaorganodilead compounds react readily with air with cleavage of the lead–lead bond to form triorganolead oxides, which can react further to yield a complex mixture of products. As mentioned above, some R_6Pb_2 derivatives do not react with air in the absence of light, but are readily decomposed by air in the presence of light.

The reaction of hexaethyldilead with oxygen has been investigated by Aleksandrov and his colleagues (2,3,5), who also investigated similar reactions of hexaethylditin (4) and tetraethyllead (3). The hexaethyldilead was oxidized at an appreciable rate by oxygen at 50° and the reaction was strongly autocatalytic. An increase in oxygen pressure decreased the rate of oxidation but the rate was increased greatly by an increase in temperature at constant concentration or by an increase in concentration at constant temperature. The products of oxidation at 50° were shown to be lead oxide, tetraethyllead (1 mole per mole of hexaethyldilead), ethanol, acetaldehyde, and water. No peroxides were found in the oxidation products (3).

Oxygenation is characterized by an induction period which is eliminated by the addition of bis(triethyllead) oxide to the hexaethyldilead (5). The reaction was shown by spectrophotometric methods to proceed with formation of bis(triethyllead) oxide, the concentration of which increased with time, and was postulated to involve initial formation of bis(triethyllead) peroxide, which then reacts with hexaethyldilead to form bis(triethyllead) oxide; the latter in turn decomposes to tetraethyllead and lead oxide.

$$(C_2H_5)_6Pb_2 + O_2 \rightarrow (C_2H_5)_3PbOOPb(C_2H_5)_3$$

$$(C_2H_5)_3PbOOPb(C_2H_5)_3 + (C_2H_5)_6Pb_2 \rightarrow 2[(C_2H_5)_3Pb]_2O$$

$$2[(C_2H_5)_3Pb]_2O \rightarrow 2(C_2H_5)_4Pb + 2PbO + 2C_2H_6 + 2C_2H_4$$

The other products (acetaldehyde, ethanol, and water) probably arise from oxidation of ethyl radicals. Some triethyllead ethoxide was detected; this could arise by reaction of the bis(triethyllead) oxide with ethanol (5). The mechanism of oxidation of hexaethylditin is believed to be similar to that of hexaethyldilead in the initial stages, but different products result because of the difference in stability and in the mode of decomposition of bis(triethyltin) peroxide compared to bis(triethyllead) peroxide.

The oxidation of hexacyclohexyldilead by air apparently occurs only in the presence of light; dicyclohexyllead oxide, lead oxide, lead dioxide and dicyclohexyl ether were identified as products of reaction (50). The reaction mixture was found to oxidize hydriodic acid solution to form iodine (presumably by the lead peroxide formed). However, when the oxidation was stopped short of completion, the resultant mixture did not oxidize hydrogen iodide and tricyclohexyllead iodide was obtained as the only by-product. No firm conclusions were drawn concerning the effect of the light on the oxidation reaction except that its effect was not to dissociate the hexacyclohexyldilead to dicyclohexyllead, since no tetracyclohexyllead was formed in the reaction mixture. However, irradiation of hexacyclohexyldilead in a nitrogen atmosphere produced a deep red coloration which was attributed to the formation of dicyclohexyldilead and tetracyclohexyllead via a disproportionation-type decomposition.

The oxidation of hexacyclohexyldilead was accelerated by carbon tetrachloride; the main products were tricyclohexyllead chloride, dicyclohexyllead dichloride, carbon dioxide, and cyclohexyl chloride (50). Hexacyclohexyldilead in carbon tetrabromide gave an explosive reaction with air; dicyclohexyllead dibromide was identified as one of the products. The oxidation of hexa-*p*-xylyldilead gave similar products in these solvents. The formation of organolead halides in these solvents is suggestive of a free radical reaction involving the solvent.

Hexaethyldilead is oxidized by organic peroxides under mild conditions (1). With cumene hydroperoxide, facile reaction occurred at 15° with formation of triethyllead hydroxide. Aleksandrov and co-workers (1) proposed that the reaction proceeds according to the sequence:

$$(C_2H_5)_6Pb_2 + C_6H_5C(CH_3)_2OOH \rightarrow (C_2H_5)_3PbOH + (C_2H_5)_3PbO(CH_3)_2CC_6H_5$$

$$(C_2H_5)_3PbOH + C_6H_5C(CH_3)_2OOH \rightarrow (C_2H_5)_3PbOO(CH_3)_2CC_6H_5 + H_2O$$

$$\text{or}\quad (C_2H_5)_3PbO(CH_3)_2CC_6H_5 + C_6H_5C(CH_3)_2OOH \rightarrow (C_2H_5)_3PbOO(CH_3)_2CC_6H_5 + C_6H_5C(CH_3)_2OH$$

The amount of triethyllead hydroxide formed in the reaction was found to decrease with increasing cumene hydroperoxide concentration while the amount of triethyllead hydroxide obtained upon hydrolysis of the reaction mixture was found to increase with increasing concentration of the hydroperoxide. The proposed sequence was tested by reaction of triethyllead hydroxide with cumene hydroperoxide to form a golden yellow, thermally unstable liquid, $(C_2H_5)_3PbOO(CH_3)_2CC_6H_5$; this product was decomposed by water to yield the original reactants, so that the reaction is reversible.

Reaction of hexaethyldilead and $(C_2H_5)_3PbOO(CH_3)_2CC_6H_5$ at 15° gave bis(triethyllead) oxide and $(C_2H_5)_3PbO(CH_3)_2CC_6H_5$.

Hexaethyldilead is reported to undergo facile reaction with acyl peroxides at room temperature; with dibenzoyl peroxide, triethyllead benzoate was formed within a few minutes (1); similar reactions occurred with acetylbenzoyl peroxide and cyclohexyl percarbonate (90). No reaction occurred with azo-bis-isobutyronitrile, so that reactions with peroxides are speculated to proceed via a cyclic intermediate (and not free radicals) which breaks down homolytically. Willemsens and co-workers (113) report that hexabutyldilead did not react with dibenzoyl peroxide in diethyl ether or ether–benzene solution although it reacted with peracetic acid to give tributyllead acetate in moderate yield. Tributyllead acetate was prepared in better yield (55%) by reaction of hexabutyldilead with a mixture of acetic acid and hydrogen peroxide in acetone at room temperature (110).

$$(C_4H_9)_6Pb_2 + H_2O_2 + 2HOAc \rightarrow 2(C_4H_9)_3PbOAc + 2H_2O$$

Hexaorganodilead compounds, like other organolead compounds, are readily oxidized by potassium permanganate. The reaction of hexa-*p*-tolyldilead and $KMnO_4$ in acetone was reported by Austin (8) to yield tri-*p*-tolyllead acetate; this product was speculated to arise through intermediate formation of tri-*p*-tolyllead hydroxide, followed by its reaction with acetic acid resulting from oxidation of acetone. The oxidation of hexa-*o*-tolyldilead under identical conditions gave an oil which was assumed to contain the hydroxide. On the other hand, the oxidation of hexaphenyldilead by potassium permanganate in dry acetone is reported to yield bis(triphenyllead) oxide (and $K_2O{\cdot}2MnO_2$) (9,110); addition of a small amount of water to the reaction mixture converted the oxide to triphenyllead hydroxide. If wet acetone was used, the hydroxide was formed directly. A method for the determination of hexaethyldilead in tetraethyllead has been developed which is based on titration with potassium permanganate in acetone (42,64).

(2) *Reactions with halogens and acids*

Hexaorganodilead compounds undergo facile reaction with halogens and halogen acids to yield organolead halides. The *in situ* halogenation of R_6Pb_2 in the PbX_2–RMgX reaction mixture to form R_3PbX, followed by reaction of the R_3PbX with more RMgX, is a common step in the synthesis of many tetraorganolead compounds. The halogenation of hexaalkyldilead compounds to trialkyllead halides proceeds cleanly. Hexacyclohexyldilead is reported to be cleaved by iodine at room temperature to form tricyclohexyllead iodide in quantitative yield (57). However, the reactions of hexaaryldilead compounds with halogens are much more complex. Bromination of hexa-*p*-xylyldilead produced a mixture of R_3PbX and R_2PbX_2 (61), whereas bromination of hexa-*p*-tolyldilead in pyridine gave lead(II) bromide and tetra-*p*-tolyllead (58). Similarly, iodination of hexaphenyldilead in benzene yielded a mixture of diphenyllead diiodide and lead(II) iodide; in aqueous alcohol or pyridine, Krause and Reissaus (58) obtained triphenyllead iodide in quantitative yield.

The iodination of hexaphenyldilead has been investigated most recently by Willemsens and van der Kerk (110). Krause and Reissaus' claim of a near quantitative yield of triphenyllead iodide by iodination in the presence of water could not be confirmed. Various other solvents were also tested, but in all cases lead(II) iodide was formed in amounts varying from 7–32.5%. In dry chloroform, tetraphenyllead (57%) and lead(II) iodide (32.5%) were the major products, along with triphenyllead iodide (4%). The high yield of tetraphenyllead when dry chloroform was used as solvent was interpreted to indicate that $Ph_3PbPbPh_2I$ may be formed as an intermediate and initiates the disproportionation of hexaphenyldilead (110). Best results were obtained with wet tetrahydrofuran containing added potassium iodide; this system gave triphenyllead iodide in 83% yield. This high yield was attributed to the formation of the nucleophilic I_3^- ion, so that the reaction involved a nucleophilic attack by I_3^- instead of an electrophilic substitution by iodine.

The reaction of hexaaryldilead compounds with halogen acids is also complex. Generally, reaction proceeds according to the stoichiometry (where R = aryl) (8,32,38):

$$R_6Pb_2 + 3HX \rightarrow R_3PbX + PbX_2 + 3RH$$
$$R_3PbX + HX \rightarrow R_2PbX_2 + RH$$

Emeléus and Evans (27) determined the number of moles of hydrogen chloride to react in 5 min with various hexaorganodilead compounds, using 95% ethanol and dry chloroform as solvents. The results are shown in Table VII-3.

On the basis of the above data, it was concluded that in 95% ethanol the stoichiometry of the reactions involving the hexaaryldilead compounds agrees with that required for the equation:

$$R_6Pb_2 + 3HCl \rightarrow R_3PbCl + PbCl_2 + 3RH$$

and the authors proposed that the reaction proceeds by initial cleavage of a lead–carbon bond instead of the lead–lead bond, according to:

$$R_6Pb_2 + 2HCl \rightarrow R_4Pb_2Cl_2 \rightarrow R_4Pb + PbCl_2 + 2RH$$

or $$R_6Pb_2 + 3HCl \rightarrow R_3Pb_2Cl_3 \rightarrow R_3PbCl + PbCl_2 + 3RH$$

Table VII-3 Moles (0.1N) HCl *Reacting per Mole* R_6Pb_2 *in Ethanol and Chloroform (in 5 minutes at 20 ± 2°)*

R IN R_6Pb_2	95% ETHANOL	$CHCl_3$
Methyl	0.60[a]	2.97[a]
Cyclohexyl	0.175[a]	1.76[a]
Phenyl	2.91[a]	4.00[a]
m-Tolyl	2.91	
p-Tolyl	3.03	3.82 (in 2 min)
o-Methoxyphenyl	2.90	4.00[a]

[a] *Average value from two or more times.*
Taken from Emeléus and Evans, *J. Chem. Soc.*, **1964,** 511.

Belluco and colleagues (13) have concluded that in the reactions of hexaethyldilead with benzoyl chloride or acetyl chloride (which provide a controlled source of HCl), the reaction process takes place according to the sequence below at a molar ratio of $2HCl/(C_2H_5)_6Pb_2$:

$$(C_2H_5)_6Pb_2 \rightleftharpoons (C_2H_5)_4Pb + (C_2H_5)_2Pb \xrightarrow{2HCl} (C_2H_5)_4Pb + PbCl_2 + 2C_2H_6$$

At a mole ratio of $3HCl/(C_2H_5)_6Pb_2$, the reaction proceeds further with formation of triethyllead chloride and lead chloride. By analogy to Emeléus and Evans' scheme, the first reaction then could involve:

$$(C_2H_5)_6Pb_2 \xrightarrow{2HCl} (C_2H_5)_4Pb_2Cl_2 \rightarrow (C_2H_5)_4Pb + PbCl_2$$

Willemsens and van der Kerk (110) have rejected the postulated formation of $R_4Pb_2Cl_2$ and its subsequent decomposition to R_4Pb and $PbCl_2$ in these reactions on the basis that they did not obtain any tetraphenyllead in the reaction of equimolar quantities of hexaphenyldilead and hydrogen chloride; also, tetraphenyllead was recovered unchanged when hydrogen chloride was permitted to react with a mixture of hexaphenyldilead and tetraphenyllead. As an alternate possibility they suggested:

$$R_6Pb_2 + HCl \rightarrow R_3PbCl + R_2Pb + RH$$
$$R_2Pb + 2HCl \rightarrow PbCl_2 + 2RH$$

The reaction of hexaphenyldilead and acetic acid has been investigated by Krebs and Henry (62). In refluxing heptane, reaction occurred to produce benzene (90%), lead(II) acetate (41%), tetraphenyllead (25%), triphenyllead acetate (18%), and unchanged hexaphenyldilead (3%); in refluxing glacial acetic acid, only lead(II) acetate and diphenyllead diacetate were formed. Similar results were obtained using thiolacetic acid. On the basis of these results, and results from the reaction of hexaphenyldilead with 1,2-dibromoethane, discussed earlier, Krebs and Henry proposed that these reactions involve an initial dissociation of hexaphenyldilead into tetraphenyllead and diphenyllead:

$$Ph_6Pb_2 \rightarrow Ph_4Pb + (Ph_2Pb)$$
$$Ph_4Pb + HOAc \rightarrow Ph_3PbOAc + PhH$$
$$Ph_3PbOAc + HOAc \rightarrow Ph_2Pb(OAc)_2 + PhH$$
$$(Ph_2Pb) + 2HOAc \rightarrow Pb(OAc)_2 + 2PhH$$

This sequence would explain why the combined yield of ph_3PbX and ph_2PbX_2 from these and many other reactions of Ph_6Pb_2 is never above 50% (8,32).

Reaction of hexa-*p*-tolyldilead and isobutyric acid at 120–140° is reported to give di-*p*-tolyllead diisobutyrate in poor yield (56). The use of aqueous acetic or formic acid has been patented as a method of removal of hexaethyldilead from tetraethyllead (47). Hexaaryldilead compounds react with hot concentrated nitric acid to yield diaryllead dinitrates (56,79,96).

The results from the above investigations do not permit any clear interpretations concerning the mechanism of reaction of R_6Pb_2 compounds with acids and other reagents. Whether reaction involves initial cleavage of a lead–carbon bond (27,110), or initial rupture of the lead–lead bond by dissociation of R_6Pb_2 to R_4Pb and R_2Pb (13,62), or both, will require further elucidation.

(3) *Reactions with various other reagents*

Hexaphenyldilead reacts with sulfur at room temperature; after ten days at 20°, equimolar amounts of these two reactants in benzene gave bis(triphenyllead) sulfide in 59% yield, along with small amounts of diphenyllead sulfide and tetraphenyllead (13). Hexaethyldilead also reacted with SO_2Cl_2, $SOCl_2$, SCl_2, and S_2Cl_2 in toluene under mild conditions to yield triethyllead chloride in 97, 86, 67, and 65% yields, respectively (14). The decreasing yields were attributed to an increasing nucleophilic character of the sulfur atom in this group of compounds.

Hexaorganodilead compounds also react with a variety of metal salts. Hexaphenyldilead and hexa-*p*-tolyldilead undergo reaction with alcoholic silver nitrate at −78° to form green colored solids which are postulated to be triarylplumbylsilver compounds. Hexa-*o*-tolyl- and hexa-*p*-xylyldilead do not react with cold (−78°) silver nitrate solution; upon warming, gradual darkening of the reaction mixture occurs and a silver mirror is deposited (61). Hexacyclohexyldilead (112), hexabutyldilead (112), and hexaethyldilead (12,17) are reported to react with silver nitrate to yield a silver mirror directly, even at low temperatures.

Hexaethyldilead reduces silver nitrate, mercury(II) chloride, mercury(I) chloride, copper(II) chloride, and copper(I) chloride in methanol at room temperature (12); the products obtained are dependent on stoichiometry. Typical reactions are shown below with only the stoichiometry of the reactants being shown:

$$1(C_2H_5)_6Pb_2 + 1AgNO_3 \rightarrow (C_2H_5)_3PbNO_3 + Ag$$
$$+ 2HgCl_2 \rightarrow (C_2H_5)_3PbCl + Hg_2Cl_2$$
$$+ 1HgCl_2 \rightarrow (C_2H_5)_3PbCl + Hg$$
$$+ 1Cu_2Cl_2 \rightarrow (C_2H_5)_4Pb + PbCl_2 + Cu$$
$$+ 2CuCl_2 \rightarrow (C_2H_5)_4Pb + PbCl_2 + (C_2H_5)_3PbCl + Cu_2Cl_2$$
$$2(C_2H_5)_6Pb_2 + 1CuCl_2 \rightarrow (C_2H_5)_4Pb + PbCl_2 + Cu$$
$$+ 1Hg_2Cl_2 \rightarrow (C_2H_5)_4Pb + PbCl_2 + Hg$$
$$+ 1AuCl_3 \rightarrow (C_2H_5)_4Pb + PbCl_2 + Au$$
$$+ 1FeCl_3 \rightarrow (C_2H_5)_4Pb + PbCl_2 + FeCl_2$$

It is interesting to note that in some reactions triethyllead chloride (or nitrate) was formed; in others the products were a mixture of tetraethyllead and lead(II) chloride. Most interestingly, a 1:2 mole ratio of $(C_2H_5)_6Pb_2$ to $HgCl_2$ gave triethyllead chloride and mercury(I) chloride; at a 1:1 ratio the product was triethyllead chloride along with mercury metal. Yet, the reaction of hexaethyldilead and mercury(I) chloride gave a mixture of tetraethyllead and lead(II) chloride.

Razuvaev and co-workers (85) have shown that hexaphenyldilead reacts with copper(II) chloride and cobalt(II) chloride at elevated temperatures. With excess copper chloride, lead(II) chloride and chlorobenzene were formed; with a deficiency of copper chloride, benzene and tetraphenyllead were the main products. Cobalt chloride gave lead(II) chloride and chlorobenzene. These reactions were postulated to proceed via a Ph_6Pb_2–MCl_2 complex in which the first step involved rupture of the lead–lead bond to form Ph_3PbCl. Hexa-*m*-xylyldilead reacted with thallium(III) chloride to yield di-*m*-xylyllead dichloride (40). Hexa-*p*-methoxyphenyldilead (68) and hexacyclohexyldilead (9) both reduced iron(III) chloride to iron(II) chloride; a near quantitative yield of tricyclohexyllead chloride was obtained. Reaction of hexa-*p*-methoxyphenyldilead and iron(II) iodide gave lead(II) iodide and tetra-*p*-methoxyphenyllead (68).

Gilman and his group (32,96) have shown that hexaethyldilead reacts with magnesium bromide and magnesium iodide in diethyl ether to form triethyllead halide. This reaction has been postulated to be a possible source of triorganolead halides in the reaction of lead(II) halides and Grignard reagents. Similarly, a number of hexaorganodilead compounds has been shown to undergo reaction with a mixture of magnesium metal and magnesium iodide, the qualitative order of reactivity being related to the thermal stability of the R_6Pb_2 compound (32). The products obtained were R_4Pb and RCOOH, the latter product being formed upon carboxylation and hydrolysis of the reaction mixture. The following sequence was suggested to account for the observed products:

$$R_6Pb_2 + Mg_2I_2(Mg + MgI_2) \rightarrow 2R_3PbMgI$$

$$2R_3PbMgI \rightarrow 2R_2Pb + 2RMgI$$

$$2R_2Pb \rightarrow R_4Pb + Pb$$

$$2RMgI \xrightarrow{CO_2} 2RCO_2MgI \xrightarrow{H_2O} 2RCOOH$$

However, the more thermally stable R_6Pb_2 derivatives, e.g., cyclohexyl, mesityl, or *o*-methoxyphenyl, gave R_3PbI, instead of R_4Pb, as the major product.

Hexaphenyldilead and hexa-*o*-tolyldilead react readily with aluminum chloride; lead(II) chloride and tetraaryllead were the main products, along with arylaluminum chlorides (30). Hexaethyldilead and lead(IV) acetate at room temperature gave triethyllead acetate and lead(II) acetate (90). The facile reaction of hexaethyldilead with triethyllead bromide and triethyltin chloride to form tetraethyllead and lead(II) halide was discussed earlier.

In general, the reactions of hexaorganodilead compounds with metal halides are very similar to those of tetraorganolead derivatives. Reduction of the metal halide to a lower valence state or to the free metal usually occurs. A notable difference in the reactions of R_6Pb_2 and R_4Pb compounds is found in their reaction with mercury(II) halides. With R_6Pb_2, reduction to mercury(I) halide or mercury metal occurs; with R_4Pb compounds, a simple metathesis occurs to form R_3PbX and RHgX. A second notable difference is the nature of the lead products. In many reactions involving R_6Pb_2 and metal salts, a mixture of R_4Pb and PbX_2 is usually formed, instead of R_3PbX and/or R_2PbX_2. Hexaethyldilead, in combination with ethylaluminum dichloride and titanium(III) chloride, is reported to be a more active catalyst for the polymerization of propylene than is tetraethyllead in this same system (100).

Hexaphenyldilead reacts readily with the alkali and alkaline earth metals in liquid ammonia to form triphenylplumbylmetal derivatives, as illustrated by the formation of benzyltriphenyllead when benzyl chloride is added to the resultant reaction mixture (33). The reaction of hexaphenyldilead and lithium metal in tetrahydrofuran represents a convenient synthesis of triphenylplumbyllithium (102).

$$Ph_6Pb_2 + 2Li \longrightarrow 2Ph_3PbLi$$

Gilman and Moore (35) have shown that hexaphenyldilead and hexa-*p*-tolyldilead undergo exchange with butyllithium to form phenyllithium and *p*-tolyllithium, respectively. The amount of exchange was determined by carboxylation of the reaction mixture (to react with the RLi product) and isolation of the resultant benzoic or toluic acid. The relative order of reactivity of various aryllead compounds with butyllithium was found to increase in the order: $Ph_4Pb < Ph_6Pb_2 < (p\text{-tolyl})_4Pb < (p\text{-tolyl})_6Pb_2$.

Hexaphenyldilead and phenyllithium undergo facile reaction to form tetraphenyllead and triphenylplumbyllithium (19). By treatment of the reaction mixture from phenyllithium and excess hexaphenyldilead with benzyl chloride, tetraphenyllead and benzyltriphenyllead were formed in excellent yield (92 and 83%, respectively) (110). Hexaphenyldilead is also cleaved by phenylmagnesium bromide, but more slowly than by phenyllithium; dissociation of hexaphenyldilead by phenylmagnesium bromide is reported to be faster in tetrahydrofuran than in diethyl ether (110).

Willemsens and van der Kerk (110) have reported recently that hexaphenyldilead is cleaved by sodium ethoxide in benzene–ethanol to form lead(II) ethoxide and triphenyllead ethoxide in equimolar amounts.

$$Ph_6Pb_2 + OEt^- \longrightarrow Ph_3PbOEt + Ph_3Pb^-$$

$$Ph_3Pb^- + 3EtOH \longrightarrow Pb(OEt)_2 + OEt^- + 3C_6H_6$$

Thus, sodium ethoxide is not consumed in the reaction; the lead ethoxide arises from solvolysis of the triphenylplumbyl anion by the ethanol solvent.

Hexaphenyldilead is cleaved by diaryl disulfides to yield triphenyllead aryl sulfides, Ph_3PbSAr (110). The reaction is accelerated by water; an ionic mechanism is postulated wherein the water renders the disulfide more nucleophilic.

As was discussed earlier, Gilman and Leeper's (33,65) reported reaction of hexaphenyldilead and maleic anhydride, involving lead–lead cleavage by addition of triphenyllead across the double bond of maleic anhydride to form bis(triphenylplumbyl)succinic anhydride, could not be confirmed. Willemsens and van der Kerk (110) have concluded that the original reaction probably involved reaction of hexaphenyldilead with maleic acid impurity in the maleic anhydride to form diphenyllead maleate.

From the foregoing discussion, it is apparent that reactions of hexaorganodilead compounds with nucleophilic reagents tend to proceed cleanly, with exclusive cleavage of the lead–lead bond. With electrophilic reagents, the reactions are much more complex and involve attack at lead–carbon bonds as well as at the lead–lead bond. Also, Willemsens and van der Kerk (110) have found that reactions of hexaphenyldilead with electrophilic reagents are much more complex than reactions of hexabutyldilead, presumably because of more facile cleavage by electrophiles of phenyllead (or aryllead) bonds compared to butyllead (or alkyllead) bonds. The mechanism of formation of tetraorganolead and lead(II) halides (or carboxylates) in reactions of R_6Pb_2 with electrophilic reagents is not fully understood and requires further elucidation. Current activity in this area suggests that new developments should be forthcoming shortly. Reactions of hexaethyldilead and hexaethylditin involving homolytic cleavage of the metal–metal bond are the subject of a recent review (105).

REFERENCES

1. Aleksandrov, Yu. A., T. G. Brilkina, A. A. Kvasov, G. A. Razuvaev, and V. A. Shushunov, *Dokl. Akad. Nauk SSSR*, **129,** 321 (1959); *C.A.*, **54,** 7608 (1960).
2. Aleksandrov, Yu. A., T. G. Brilkina, and V. A. Shushunov, *Dokl. Akad. Nauk SSSR*, **136,** 89 (1961); through *C.A.*, **55,** 27027 (1961).
3. Aleksandrov, Yu. A., T. G. Brilkina, and V. A. Shushunov, *Tr. po Khim. i Khim. Tekhnol.*, **4,** 3 (1961); through *C.A.*, **56,** 492 (1962).
4. Aleksandrov, Yu. A., B. A. Radbil, and V. A. Shushunov, *Tr. po Khim. i Khim. Tekhnol.*, **3,** 388 (1960); through *C.A.*, **55,** 27023 (1961).
5. Aleksandrov, Yu. A., and N. N. Vyshinskii, *Tr. po Khim. i Khim. Tekhnol.*, **4,** 656 (1962); through *C.A.*, **58,** 3453 (1963).
6. Apperson, L. D., *Iowa St. Coll. J. Sci.*, **16,** 7 (1941); through *C.A.*, **36,** 4476 (1942).
7. Austin, P. R., *J. Am. Chem. Soc.*, **53,** 1548 (1931).
8. Austin, P. R., *J. Am. Chem. Soc.*, **53,** 3514 (1931).
9. Bähr, G., *Z. anorg. allgem. Chem.*, **253,** 330 (1947).
10. Bähr, G., and G. Zoche, *Chem. Ber.*, **88,** 542 (1955).
11. Bailie, J. C., *Iowa St. Coll. J. Sci.*, **14,** 8 (1939); through *C.A.*, **34,** 6241 (1940).
12. Belluco, U., and G. Belluco, *Ric. sci., Rend., Sez.* **A32,** 102 (1962); through *C.A.*, **57,** 13786 (1962).
13. Belluco, U., L. Cattalini, A. Peloso, and G. Tagliavini, *Ric. sci., Rend., Sez.* **A2,** 269 (1962); through *C.A.*, **59,** 1667 (1963).
14. Belluco, U., L. Cattalini, A. Peloso, and G. Tagliavini, *Ric. sci., Rend., Sez.* **A3,** 1107 (1963); through *C.A.*, **61,** 677 (1964).
15. Belluco, U., L. Cattalini, and G. Tagliavini, *Ric. sci., Rend., Sez.* **A2,** 110 (1962); *C.A.*, **57,** 13786 (1962).
16. Belluco, U., A. Peloso, L. Cattalini, and G. Tagliavini, *Ric. sci., Rend., Sez.* **A2,** 269 (1962); through *C.A.*, **59,** 1667 (1963).
17. Belluco, U., G. Tagliavini, and R. Barbieri, *Ric. sci.*, **30,** 1675 (1960); through *C.A.*, **55,** 14175 (1961).
18. Belluco, U., G. Tagliavini, and P. Favero, *Ric. sci., Rend., Sez.* **A2,** 98 (1962); through *C.A.*, **57,** 13786 (1962).

19. Bindschadler, E., and H. Gilman, *Proc. Iowa Acad. Sci.*, **48,** 273 (1941), through *C.A.*, **36,** 1595 (1942).
20. Calingaert, G., *Chem. Rev.*, **2,** 43 (1925).
21. Calingaert, G., and H. Soroos, *J. Org. Chem.*, **2,** 535 (1938).
22. Calingaert, G., H. Soroos, and H. Shapiro, *J. Am. Chem. Soc.*, **64,** 462 (1942).
23. Dessy, R. E., W. Kitching, and T. Chivers, *J. Am. Chem. Soc.*, **88,** 453 (1966).
24. DeVries, J. E., A. Lauw-Zecha, and A. Pellecer, *Anal. Chem.*, **31,** 1995 (1959).
25. Drenth, W., M. J. Janssen, G. J. M. van der Kerk, and J. A. Vliegenthart, *J. Organometal. Chem.*, **2,** 265 (1964).
26. Drenth, W., L. C. Willemsens, and G. J. M. van der Kerk, *J. Organometal. Chem.*, **2,** 279 (1964).
27. Emeléus, H. J., and P. R. Evans, *J. Chem. Soc.*, **1964,** 511.
28. Foster, L. S., W. M. Dix, and I. J. Gruntfest, *J. Am. Chem. Soc.*, **61,** 1685 (1939).
29. Ghira, A., *Gazz. chim. ital.*, **24,** 42 (1894).
30. Gilman, H., and L. D. Apperson, *J. Org. Chem.*, **4,** 162 (1939).
31. Gilman, H., W. H. Axtell, and F. K. Cartledge, "Catenated Organic Compounds of Silicon, Germanium, Tin and Lead" in *Advances in Organometallic Chemistry*, Vol. 4, F. G. A. Stone and R. West, Eds., Academic Press, New York (1966), pp. 63–77.
32. Gilman, H., and J. C. Bailie, *J. Am. Chem. Soc.*, **61,** 731 (1939).
33. Gilman, H., and R. W. Leeper, *J. Org. Chem.*, **16,** 466 (1951).
34. Gilman, H., O. L. Marrs and S. Y. Sim, *J. Org. Chem.*, **27,** 4232 (1962).
35. Gilman, H., and F. W. Moore, *J. Am. Chem. Soc.*, **62,** 3206 (1940).
36. Gilman, H., L. Summers, and R. W. Leeper, *J. Org. Chem.*, **17,** 630 (1952).
37. Gilman, H., L. Summers, and R. W. Leeper, *Chem. Rev.*, **54,** 101 (1954).
38. Gilman, H., and E. B. Towne, *J. Am. Chem. Soc.*, **61,** 739 (1939).
39. Gittens, T. W., and E. L. Mattison, U.S. 2,763,673 (to E. I. du Pont de Nemours & Co.), September 18, 1956, *C.A.*, **51,** 4414 (1957).
40. Goddard, A. E., *J. Chem. Soc.*, **123,** 1161 (1923).
41. Goldach, A., *Helv. Chim. Acta*, **14,** 1436 (1931).
42. Gol'dshtein, A. L., N. P. Lapisova, and L. M. Schtifman, *Zhur. Analit. Khim.*, **17,** 143 (1962); through *C.A.*, **57,** 5709 (1963).
43. Grüttner, G., and E. Krause, *Ber.*, **49,** 1415 (1916).
44. Hague, D. N., and R. H. Prince, *Proc. Chem. Soc.*, **1962,** 300.
45. Hague, D. N., and R. H. Prince, *J. Chem. Soc.*, **1965,** 4690.
46. Harrah, L. A., M. T. Ryan, and C. Tamborski, *Spectrochim. Acta*, **18,** 21 (1962).
47. Hedden, G. D., and B. A. Rausch, U.S. 3,072,696 (to E. I. du Pont de Nemours & Co.) Jan. 8, 1963; *C.A.*, **59,** 11560 (1963).
48. Hein, F., and A. Klein, *Ber.*, **71,** 2381 (1938).
49. Hein, F., and E. Nebe, *Ber.*, **75,** 1744 (1942).
50. Hein, F., E. Nebe, and W. Reimann, *Z. anorg. allgem. Chem.*, **251,** 125 (1943).
51. Jensen, K. A., and N. Clauson-Kaas, *Z. anorg. allgem. Chem.*, **250,** 277 (1943).
52. Kerk, G. J. M. van der, *Ind. Eng. Chem.*, **58,** 29 (1966).
52a. Kitching, W., V. G. Kumar Das, and P. R. Wells, *Chem. Commun.*, **1967,** 356.
53. Klippel, J., *Jahresber.*, **1860,** 380.
54. Klippel, J., *J. prakt. Chem.*, **81,** 287 (1860).
55. Koch, K., *Dissertation*, *Darmstadt*, 1961; Reference 189 in L. C. Willemsens, *Organolead Chemistry*, Intern. Lead Zinc Res. Organ., New York, 1964.
56. Kocheshkov, K. A., and E. M. Panov, *Izvest. Akad. Nauk SSSR, Otd. Khim. Nauk*, **1955,** 718; through *C.A.*, **50,** 7076 (1956).

57. Krause, E., *Ber.*, **54,** 2060 (1921).
58. Krause, E., and G. G. Reissaus, *Ber.*, **55,** 888 (1922).
59. Krause, E., and O. Schlöttig, *Ber.*, **58,** 427 (1925).
60. Krause, E., and O. Schlöttig, *Ber.*, **63,** 1381 (1930).
61. Krause, E., and N. Schmitz, *Ber.*, **52,** 2165 (1919).
62. Krebs, A. W., and M. C. Henry, *J. Org. Chem.*, **28,** 1911 (1963).
63. Krohn I. T., and H. Shapiro, U.S. 2,555,891 (to Ethyl Corp.) June 5, 1951; *C.A.*, **46,** 523 (1952).
64. Lapisova, N. P., and A. L. Gol'dshtein, *Zhur. Analit. Khim.*, **16,** 508 (1961); through *C.A.*, **56,** 8012 (1962).
65. Leeper, R. W., *Iowa St. Coll. J. Sci.*, **18,** 57 (1943); through *C.A.*, **38,** 726 (1944).
66. Lesbre, M., J. Satgé, and D. Voigt, *Compt. rend.*, **246,** 594 (1948).
67. Lewis, G. L., P. F. Oesper, and C. P. Smyth, *J. Am. Chem. Soc.*, **62,** 3242 (1940).
68. Lichtenwalter, M., *Iowa St. Coll. J. Sci.*, **14,** 57 (1939); through *C.A.*, **34,** 6241 (1940).
69. Lile, W. J., and R. C. Menzies, *J. Chem. Soc.*, **1950,** 617.
70. Löwig, C., *Ann. Chem.*, **88,** 318 (1853).
71. Löwig, C., *J. prakt. Chem.*, **60,** 304 (1853).
72. Malatesta, L., *Gazz. chim. ital.*, **73,** 176 (1943); through *C.A.*, **38,** 5128 (1944).
73. McDyer, T. W., and R. D. Closson, U.S. 2,571,987 (to Ethyl Corp.), Oct. 16, 1951; *C.A.*, **46,** 3556 (1952).
74. Midgley, T., Jr., C. A. Hochwalt, and G. Calingaert, *J. Am. Chem. Soc.*, **45,** 1821 (1923).
75. Mine, K., *J. Chem. Soc. Japan*, **55,** 1168 (1934); through *C.A.*, **29,** 7940 (1935).
76. Morris, H., and P. W. Selwood, *J. Am. Chem. Soc.*, **63,** 2509 (1941).
77. Müller, E., F. Günter, K. Scheffler, and H. Fettel, *Chem, Ber.*, **91,** 2888 (1958).
78. Neumann, W. P., and K. Kühlein, *Tetrahedron Letters*, **1966,** 3419.
79. Panov, E. M., and K. A. Kocheshkov, *Dokl. Akad. Nauk SSSR*, **85,** 1037 (1952); through *C.A.*, **47,** 6365 (1953).
80. Peters, R. A., U.S. 2,967,105 (to Stupa Corp.) Jan. 3, 1961; through *C.A.*, **55,** 9129 (1961).
81. Podall, H. E., H. E. Petree, and J. R. Zietz, *J. Org. Chem.*, **24,** 1222 (1959).
82. Preckel, R., and P. W. Selwood, *J. Am. Chem. Soc.*, **62,** 2765 (1940).
83. Razuveav, G. A., Yu. I. Dergunov, and N. S. Vyazankin, *Zh. Obshch. Khim.*, **31,** 998 (1961); *C.A.*, **55,** 23321 (1961).
84. Razuvaev, G. A., Yu. I. Dergunov, and N. S. Vyazankin, *Zh. Obshch. Khim.*, **32,** 2515 (1962); *C.A.*, **58,** 9111 (1963).
85. Razuvaev, G. A., M. S. Fedotov, T. B. Zavarova, and N. N. Bazhenova, *Tr. po Khim. i Khim. Tekhnol.*, **4,** 622 (1961), *C.A.*, **58,** 543 (1963).
86. Razuvaev, G. A., G. G. Petukhov, and Yu. A. Kaplin, *Zh. Obshch. Khim.* **33,** 2394 (1963); *C.A.*, **59,** 14014 (1963).
87. Razuvaev, G. A., G. G. Petukhov, Yu. A. Kaplin, and O. N. Druzhov, *Dokl. Akad. Nauk SSSR*, **152,** 1122 (1963); *C.A.*, **60,** 1556 (1964).
88. Razuvaev, G. A., N. S. Vyazankin, and Yu. I. Dergunov, *Zh. Obshch. Khim.*, **30** 1310 (1960); *C.A.*, **55,** 362 (1961).
89. Razuvaev, G. A., N. S. Vyazankin, Yu. I. Dergunov, and O. S. D'yachkovskaya, *Dokl. Akad. Nauk SSSR*, **132,** 364 (1960); *C.A.*, **54,** 20937 (1960).
90. Razuvaev, G. A., N. S. Vyazankin, and O. A. Shchepetkova, *Zh. Obshch. Khim.*, **30,** 2498 (1960); *C.A.*, **55,** 14290 (1961).

91. Razuvaev, G. A., N. S. Vyazankin, and N. N. Vyshinskii, *Zh. Obshch. Khim.*, **29,** 3662 (1959); *C.A.*, **54,** 17015 (1960).
92. Razuvaev, G. A., N. S. Vyazankin, and N. N. Vyshinskii, *Zh. Obshch. Khim.*, **30,** 967 (1960); *C.A.*, **54,** 23646 (1960).
93. Razuvaev, G. A., N. S. Vyazankin, and N. N. Vyshinskii, *Zh. Obshch. Khim.*, **30,** 4099 (1960); *C.A.*, **55,** 24546 (1961).
94. Rudnevskii, N. K., and N. N. Vyshinskii, *Izvest. Akad. Nauk SSSR Ser. Fiz.*, **23,** 1228 (1959); through *C.A.*, **54,** 6392 (1960).
95. Schmidt, H., *Medicine in its Chemical Aspects*, Vol. 3, Bayer, Leverkusen (1938), pp. 394–404.
96. Setzer, W. C., R. W. Leeper, and H. Gilman, *J. Am. Chem. Soc.*, **61,** 1609 (1939).
97. Shapiro, H., Unpublished Data.
98. Singh, G., *J. Org. Chem.*, **31,** 949 (1966).
99. Skinner, H. A., and L. E. Sutton, *Trans. Faraday Soc.*, **36,** 1209 (1940).
100. Stamicarbon, N. V., Netherland Appl. 288,760 March 10, 1965; through *C.A.*, **63,** 4418 (1965).
101. Tafel, J., *Ber.*, **44,** 323 (1911).
102. Tamborski, C., F. E. Ford, W. L. Lehn, G. J. Moore, and E. J. Soloski, *J. Org. Chem.*, **27,** 619 (1962).
103. Vertyulina, L. N., and I. A. Korshunov, *Khim. Nauka i Prom.*, **4,** 136 (1959); through *C.A.*, **53,** 12096 (1959).
104. Vyazankin, N. S., G. A. Razuvaev, and Yu. I. Dergunov, *Tr. po Khim. i Khim. Tekhnol.*, **4,** 652 (1961); through *C.A.*, **58,** 543 (1963).
105. Vyazankin, N. S., G. A. Razuvaev, Yu. I. Dergunov, and O. A. Shchepetkova, *Tr. po Khim. i Khim. Tekhnol.*, **4,** 58 (1961), through *C.A.*, **56,** 1468 (1962).
106. Vyshinskii, N. N., and N. K. Rudnevskii, *Tr. po Khim. i Khim. Tekhnol.*, **3,** 538 (1960); through *C.A.*, **56,** 4261 (1926).
107. Wiczer, S. B., U.S. 2,447,926, August 24, 1948; *C.A.*, **42,** 7975 (1948).
108. Willemsens, L. C., and G. J. M. van der Kerk, *J. Organometal. Chem.*, **2,** 260 (1964).
109. Willemsens, L. C., and G. J. M. van der Kerk, *J. Organometal. Chem.*, **2,** 271 (1964).
110. Willemsens, L. C., and G. J. M. van der Kerk, *Investigations in the Field of Organolead Chemistry*, Intern. Lead Zinc Res. Organ., New York (1965).
111. Willemsens, L. C., and G. J. M. van der Kerk, Progress Report No. 25, to Internat. Lead Zinc Res. Organ., Proj. LC-18 (January–March, 1966).
112. Willemsens, L. C., J. C. Overeem, and G. J. M. van der Kerk, Progress Report No. 14, to Intern. Lead Zinc Res. Organ. (April–June, 1963).
113. Willemsens, L. C., J. C. Overeem, and G. J. M. van der Kerk, Progress Report No. 16, to Intern. Lead Zinc Res. Organ. (October 1–December 30, 1963).
114. Zimmer, H., and O. A. Homberg, *J. Org. Chem.*, **31,** 947 (1966).

VIII Dialkyl- and diaryl- (two-covalent) lead compounds, R_2Pb

1. NATURE OF BONDING

The two-covalent state of the lead atom has been previously discussed in Chapter II. This state has been presented as arising from the $s^2p^2\ ^3P$ ground state of the atom, as shown in the structure **1**. The bond angle which

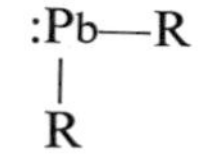

$$:\mathrm{Pb}(-\mathrm{R})_2$$

(1)

would be expected for this state is 90°, for p^2 bonding. Indeed, electron diffraction measurements by Lister and Sutton (18) on both lead and tin dihalides suggest that such molecules are angular.

A particular configuration which is more probable than **1** is shown in structure **2**, in which the molecule is assumed to have sp^2 hybridization and a vacant p orbital (7,33). In these structures, the nonbonding electrons

$$\mathrm{PbR_2}$$

(2)

exist as a lone pair, and there are only six electrons in the valence shell of the metal. This produces a tendency toward polymerization, which would result in a more stable octet of electrons.

Actually, however, insufficient experimental evidence is available on R_2Pb compounds to throw much light on the type of structure postulated.

Table VIII-1 R_2Pb *Compounds*

COMPOUND	PHYSICAL PROPERTIES	REFERENCES
$(CH_3)_2Pb$		19
$(C_2H_5)_2Pb$	oil, d_{20} 1.9653	21,22,24
$(i\text{-}C_3H_7)_2Pb$	oil	6,11,25,26,27
$(s\text{-}C_4H_9)_2Pb$	oil	23
$(3\text{-}C_5H_{11})_2Pb$	oil	23
$(cyclo\text{-}C_6H_{11})_2Pb$		11
$(C_6H_5)_2Pb$		1,4,5,8,9,13,16,24,29,30,32
$(o\text{-}CH_3C_6H_4)_2Pb$		16
$[(CH_3)_3C_6H_2]_2Pb$		9

Other Group IV metal compounds, such as diphenyltin (12) and diphenylgermanium (14), are much better known than the analogous lead compounds, and indeed, there is some question still that R_2Pb compounds are capable of independent stable existence except for the cyclopentadienyl derivatives discussed in Chapter IX. Ordinarily, the diphenyltin and diphenylgermanium compounds are polymeric, colored substances. Willemsens and van der Kerk have recently re-investigated the preparation of diphenyllead very carefully, with instructive results, as described below.

In 1943, Jensen and Clauson-Kaas (13) carefully investigated diphenyltin, with instructive results. It is monomeric when freshly prepared, but polymerizes readily to reach the pentamer stage or larger. Magnetic susceptibility measurements showed no paramagnetic effect at any stage of the polymerization. Likewise, its dipole moment was about 1.0 μ throughout the polymerization. Jensen and Clauson-Kaas suggested an explanation based on a biradical to explain these observations. A more likely interpretation has been advanced by Leeper et al. (17). They have pointed out that the polymeric materials would be expected to be diamagnetic. With the formal charge separation represented a large dipole moment should result. However, a decreased dipole moment such as was observed could be attributed to different configurations around the two tin atoms, one planar, the other pyramidal, resulting in some hybridization.

2. SYNTHESIS

It is generally assumed that R_2Pb compounds are the first intermediate products in the preparation of R_4Pb compounds by the reaction of lead halides with organometallic compounds:

$$PbX_2 + 2RMgX \rightarrow R_2Pb + 2MgX_2$$

This assumption has been fortified by the transient yellow, red, or brown colors that appear during the reactions; see, for example, Möller and Pfeiffer (20), and Krause and von Grosse (15). Despite this general assumption, however, no dialkyllead compounds of the R_2Pb type were claimed to be isolated until recently, in sharp contrast to the isolation of the dialkyltin compounds. The isolation of diaryllead compounds (16) has been claimed in only one investigation, but this "isolation" has been substantially disproved (29). (Table VIII-I.)

The history of the attempts to synthesize R_2Pb compounds is interesting. In 1907, Tafel (26,27) electrolytically reduced acetone in sulfuric acid, using a lead cathode. He obtained a dark red oil, assumed to contain diisopropyllead. The oil absorbed carbon dioxide and oxygen readily, giving decolorized products, and it added bromine directly to give diisopropyllead dibromide:

$$(C_3H_7)_2Pb + Br_2 \rightarrow (C_3H_7)_2PbBr_2$$

However, diisopropyllead could not be isolated because of its instability. Similar experiments were made with methyl ethyl ketone and diethyl ketone by Renger (23) in 1911, apparently producing di-*sec*-butyllead and di-3-pentyllead, respectively.

Fichter and Stein (6) in 1931 carried out reductions of acetone in dilute sulfuric acid, somewhat similar to Tafel's, but using NaPb, Na_2Pb, and Na_4Pb alloys as the cathodes instead of lead metal. The chief product was pinacol, but a small amount of volatile, unstable isopropyllead compound was apparently obtained. In the same year, Goldach (10) repeated the experiments using Na_4Pb cathodes at about $-10°$. A red-brown solution of diisopropyllead was apparently obtained, but was not isolated and characterized as such.

In 1965, Sekine and his co-workers (25) followed up carefully on the work of Tafel and electrolytically reduced acetone in the presence of various metal electrodes. It was indicated that the isopropyl radical was formed, and that with lead or mercury cathodes the colored diisopropylmetal compound could be isolated in a relatively stable form. The mercury compound was easily isolated and the structure was confirmed. For the lead compound, however, the measurement of physical properties was difficult due to its instability in air. The composition of diisopropyllead was clearly indicated from a series of decomposition studies in nitrogen, oxygen, benzene, ethyl ether, and 20% sulfuric acid, but the properties and characteristics were not described. The results also indicated that the organometallic compound was the source of the propane formed during the electrolysis.

In 1966, Tomilov and Klyuev (28) reduced acetone in 20% sulfuric acid on lead–tin cathodes. Organometallic compounds apparently resulted when the cathodes contained over 50% lead; with pure lead cathodes, good yields of isopropanol and pinacone were obtained instead.

In the 1920's, Calingaert (3) attempted without success to isolate dialkyllead compounds by the reaction of dialkyllead dihalides with sodium in liquid ammonia, and also by the reaction of Grignard reagents with lead chloride. It should be noted in this connection that the former reaction is an elegant method of preparing the dialkyltin compounds, but it appears on the basis of Calingaert's work to be useless for dialkyllead compounds. At a much later date, Apperson (1) attempted similarly, without success, to reduce diphenyllead dihalides with sodium in liquid ammonia, and also with lithium, aluminum, hydrazine, and by catalytic hydrogenation.

In 1942, Hein and Nebe (11) reacted tricyclohexylplumbylsodium with iodine:

$$(C_6H_{11})_3PbNa + I_2 \rightarrow (C_6H_{11})_3PbI + NaI$$

The iodine, dissolved in dry benzene, was added to the ether solution of the sodium salt. A white precipitate of sodium iodide was obtained, and the solution turned yellow with tricyclohexyllead iodide. After a short interval, a dark red color appeared, which they assumed to be dicyclohexyllead. Further titration with iodine resulted in slow clearing of the red solution. Dicyclohexyllead could have resulted from the postulated disproportionation reaction of hexacyclohexyldilead:

$$2(C_6H_{11})_3PbNa + I_2 \rightarrow [(C_6H_{11})_3Pb]_2 + 2NaI$$

$$[C_6H_{11}]_3Pb]_2 \rightarrow (C_6H_{11})_4Pb + (C_6H_{11})_2Pb$$

However, except for the red color, no evidence for the formation of dicyclohexyllead was observed.

In the aromatic series, the preparation of both diphenyllead and di-*o*-tolyllead was claimed in 1922 by Krause and Reissaus (16). They were supposedly prepared through the reaction of anhydrous lead chloride with the ethereal Grignard reagent at below 2°. At higher temperatures, decomposition of the diaryllead compound was rapid, and even at 2° the yield was only a few per cent. Most of the product consisted of R_4Pb and R_6Pb_2 compounds. An amorphous red solid was separated by extraction with benzene, fractional crystallization, and precipitation with alcohol. The product was obtained as amorphous, blood-red solids, soluble in ether and benzene, and reactive to oxygen and iodine. Although the "diphenyllead" was analyzed and was "proved" to be monomeric by cryoscopic molecular weight determinations in benzene, subsequent attempts to duplicate the preparation, both by Krause and Reissaus themselves, Willemsens and van der Kerk (29,32), Jensen and Clauson-Kaas, cited above, Glockling et al. (9), and Apperson (1) were not successful. Instead, hydrolysis occurred with formation of basic lead halide.

Hypothesizing that organolithium compounds would react more readily than the Grignard reagents with lead chloride at temperatures sufficiently low so that decomposition of R_2Pb product could be avoided, Gilman et al. (8), carried out the reaction of phenyllithium in ether with lead chloride at $-10°$. A set of equilibrium products was suggested:

$$(C_6H_5)_3PbLi \rightleftharpoons (C_6H_5)_2Pb + C_6H_5Li$$

(although this "equilibrium" has been shown to be very far to the left (29)). Some intermediate diphenyllead, apparently stable at low temperature, was indicated to be formed in solution, and the solution was bright yellow rather than red.

Hydrolysis gave a heavy yellow precipitate of hexaphenyldilead and a deposit of lead metal, apparently resulting from disproportionation. The intermediate product from *o*-tolyllithium was blood-red. The differences in color are probably not of real significance, since it is well recognized in the analogous tin series that the coloris de pendent on the degree of polymerization and may range from yellow to red to brown. Also, Krause and Reissaus did observe yellow intermediate solutions in their aryllead work.

In 1961, Robinson (24) attempted to prepare both diethyllead and diphenyllead by forcing several organometallic reagents rapidly through a column containing a large excess of divalent lead salt. Thus, any soluble intermediate lead products formed would be in minimal contact with unreacted organometallic reagent. No success was achieved in producing R_2Pb compounds; however, the work is interesting from the standpoint of the mechanism of reaction to form R_4Pb compounds. This reaction was discussed in Chapter II.

Glockling and his co-workers (9) have also attempted to prepare diphenyllead and dimesityllead by the reaction of lead halides with both Grignard and aryllithium reagents. Repetition of the experiments of Krause and Reissaus was unsuccessful; hydrolysis resulted in loss of color and the formation of inorganic lead compounds. In various modifications of Gilman's experiments, the phenyllithium or mesityllithium was reacted in tetrahydrofuran with lead chloride or lead bromide at $-20°$ or lower. These reactions are rapid, as are also those of the corresponding Grignard reagents in tetrahydrofuran. Yellow to red solutions resulted, dependent on the stoichiometry employed. The mesityl products were more stable than the phenyl compounds. Even so, heating produced tetramesityllead and lead metal. Invariably, attempts at isolation by hydrolysis led to inorganic products.

Willemsens and van der Kerk (30,32) have also carefully reinvestigated the Krause and Reissaus preparation of diphenyllead, with enlightening results. They too found that hydrolysis of the reaction mixture (from lead chloride and phenylmagnesium bromide) resulted in decomposition of any phenyllead compounds to lead oxide without appearance of a red color if the atmosphere and the water were oxygen-free. However, if the hydrolysis and oxidation were allowed to proceed simultaneously, a red color persisted, and a red compound of the composition and structure $[(C_6H_5)_3Pb]_4Pb$ was produced. The same compound was also produced by reaction of triphenylplumbyllithium and lead(II) chloride. Other Group IV metal compounds of similar structure were subsequently prepared (31). The ultraviolet absorption spectra of these compounds are very similar (5). On the basis of these findings, Willemsens and van der Kerk have concluded that the red compound isolated by Krause and Reissaus was probably their newly identified $(C_6H_5)_{12}Pb_5$ derivative.

Regardless of the many failures at R_2Pb isolation and characterization, recent publications by Razuvaev et al. (21,22) claim the successful preparation of diethyllead. The reduction of diethyllead dichloride with aluminum metal in 2.5*N* potassium hydroxide resulted in a bright red liquid. This product, presumably diethyllead, was stable at the temperature of liquid nitrogen, but it decomposed at room temperature to hexaethyldilead, gases, and lead metal. It is worthy of note that Apperson (1) in somewhat similar work attempted unsuccessfully to reduce diphenyllead dihalides with aluminum (as well as with lithium, sodium in liquid ammonia, hydrazine, and by catalytic hydrogenation) thus casting some doubt on the validity of Razuvaev's product as diethyllead.

Although the lowest member of the R_2Pb series, dimethyllead, has never been prepared or isolated, Miller and Winkler (19) indicated some time ago that it exists also. Using radioactive lead or tellurium mirrors in the Paneth technique, and passing dimethyl ether or acetone over the mirrors, they found that the rate of removal was the same for both metals, 7×10^{-8} mole/min. This suggested to them that dimethyllead and dimethyltellurium were the initial products formed.

There is also polarographic evidence for the transitory existence of diphenyllead, as indicated by a 1966 publication by Dessy, Kitching, and Chivers (4).

3. CHEMICAL PROPERTIES AND REACTIONS

Patently, from the discussion above, very few reactions of the R_2Pb compounds have been explored or understood. Rapid decomposition and disproportionation to lead metal and R_6Pb_2, and reactions of the R_2Pb products with oxygen, carbon dioxide, water, and bromine and iodine, have been demonstrated, as cited above. Disproportionation appears to be a primary reaction (29).

All the available evidence cited above, both with dialkyl- and diaryllead compounds, has shown that the R_2Pb compounds decompose much more rapidly than the analogous R_2Sn compounds. For example, diethyllead decomposed at room temperature, and diphenyllead, even had it existed at all under the Krause and Reissaus conditions, decomposed at about 100° on heating in a capillary tube. The yellow products, presumably diphenyllead intermediate, obtained by Gilman et al. and Glockling et al. cited above, from phenyllithium and lead chloride in the preparation of triphenylplumbyllithium, disproportionated readily on refluxing in solution to give lead metal and hexaphenyldilead in good yield. Also, refluxing a solution of diphenyllead with phenylmagnesium bromide produced hexaphenyldilead.

One definite addition reaction of diphenyllead has been demonstrated. Gilman, and Glockling, and their associates, cited above, have shown that diphenyllead reacts with phenyllithium to give triphenylplumbyllithium, $(C_6H_5)_3PbLi$. From the electronic configuration proposed for R_2Pb compounds, such reactions should be expected. Indeed, it would be interesting to test whether the R_2Pb compounds are capable of entering into reactions as ligand electron donors, for example, to form by direct addition mixed metal carbonyls, $R_2PbM(CO)_x$, mixed metal cyclopentadienyls, or R_2PbL_2, where L is a chelating ligand.

Willemsens and van der Kerk (29) have carried out the preparation of diphenyllead in tetrahydrofuran and in the presence of pyridine, two solvents which display a greater electron-donating capacity than ethyl ether. Their experiments indicated that the disproportionation proceeded much more slowly in these solvents, probably because of their stronger complexing power. They also showed that the loose complex between diphenyllead and phenylmagnesium bromide in ether is stabilized when tetrahydrofuran or pyridine is present; stable triphenyllead magnesium bromide is formed under these conditions when one mole of lead chloride is reacted with three moles of phenylmagnesium bromide. Benzyltriphenyllead was obtained in 85% yield by the addition of benzyl chloride to such reaction mixtures; by contrast, only 41% yield of benzyltriphenyllead was obtained when ether was the solvent.

Addition reactions of diisopropyllead with bromine and carbon dioxide have already been noted above (26,27). However, there is no clear-cut evidence that diphenyllead will add such reagents (29). Iodine gives lead iodide and iodobenzene; oxygen also gives cleavage instead of addition, and water yields lead oxide and benzene by cleavage:

$$(C_6H_5)_2Pb + 2I_2 \rightarrow PbI_2 + 2C_6H_5I$$

$$(C_6H_5)_2Pb + H_2O \rightarrow PbO + 2C_6H_6$$

Undoubtedly, further work on the R_2Pb compounds will show them to be reactive to a wide variety of reagents. Insofar as synthesis methods for dialkyllead compounds are concerned, Razuvaev's work on the new low-temperature method using aluminum to reduce R_2PbX_2 compounds should certainly be verified and extended, in order to make certain that the product obtained was not of the type $(Et_3Pb)_4Pb$, rather than Et_2Pb. Further, it would appear that additional careful investigations might yet lead eventually to the low-temperature isolation of diaryllead compounds from the reaction of aryllithium compounds with lead halides. Possibly, attempts to prepare di-9-phenanthryllead might be especially successful, since it has been established by Bähr and Gelius (2) that the di-9-phenanthryltin possesses unusual thermal stability; no decomposition occurred when this tin compound was heated to 360°. Moreover, the work of Willemsens and van der Kerk suggests that stabilization of diaryllead compounds might be obtained by coordination with bases.

REFERENCES

1. Apperson, L. D., *Iowa State Coll. J. Sci.*, **16,** 7 (1941); *C.A.*, **36,** 4476 (1941).
2. Bähr, G., and R. Gelius, *Ber.*, **91,** 829 (1958).
3. Calingaert, G., *Chem. Rev.*, **2,** 43 (1925).
4. Dessy, R. E., W. Kitching, and T. Chivers, *J. Am. Chem. Soc.* **88,** 453 (1966).
5. Drenth, W., M. J. Janssen, and G. J. M. van der Kerk, *J. Organometal. Chem.*, **2,** 265 (1964).
6. Fichter, F., and I. Stein, *Helv. Chim. Acta*, **14,** 1205 (1931); *C.A.*, **26,** 360 (1932).
7. Gillespie, R. J., and R. S. Nyholm, *Quart. Rev.*, **11,** 339 (1957).
8. Gilman, H., L. Summers, and R. W. Leeper, *J. Org. Chem.*, **17,** 630 (1952).
9. Glockling, F., K. Hooton, and D. Kingston, *J. Chem. Soc.*, **1961,** 4405.
10. Goldach, A., *Helv. Chim. Acta*, **14,** 1436 (1931).
11. Hein, F., and E. Nebe, *Ber.*, **75,** 1744 (1942).
12. Ingham, R. K., S. D. Rosenberg, and H. Gilman, *Chem. Rev.*, **60,** 459 (1960).
13. Jensen, K. A., and N. Clauson-Kaas, *Z. anorg. allgem. Chem.*, **250,** 277 (1943).
14. Kraus, C. A., and C. L. Brown, *J. Am. Chem. Soc.*, **52,** 4031 (1930).
15. Krause, E., and A. von Grosse, *Die Chemie der metall-organischen Verbindungen*, Borntraeger, Berlin, 1937, p. 409.
16. Krause, E., and G. G. Reissaus, *Ber.*, **55,** 888 (1922).
17. Leeper, R. W., L. Summers, and H. Gilman, *Chem. Rev.*, **54,** 101 (1954).
18. Lister, M. W., and L. E. Sutton, *Trans. Faraday Soc.*, **37,** 406 (1941).
19. Miller, D. M., and C. A. Winkler, *Can. J. Chem.*, **29,** 537 (1951); *C.A.*, **45,** 8403 (1951).

20. Möller, S., and P. Pfeiffer, *Ber.*, **49,** 2441 (1916).
21. Razuvaev, G. A., N. S. Vyazankin, and N. N. Vyshinskii, *Zh. Obshch. Khim.*, **29,** 3662 (1959); through *C.A.*, **54,** 17015 (1960).
22. Razuvaev, G. A., N. S. Vyazankin, and N. N. Vyshinskii, *Zh. Obshchei Khim.*, **30,** 967 (1960); through *C.A.*, **54,** 23646 (1960).
23. Renger, G., *Ber.*, **44,** 337 (1911).
24. Robinson, G. C., Ethyl Corp., Unpublished Data.
25. Sekine, T., A. Yamura, and K. Sugino, *J. Electrochem. Soc.*, **112,** 439 (1965).
26. Tafel, J., *Ber.*, **39,** 3626 (1907).
27. Tafel, J., *Ber.*, **44,** 323 (1911).
28. Tomilov, A. P., and B. L. Klyuev, *Elektrokhimia*, **2,** 1405 (1966); through *C.A.*, **66,** 71819x (1967).
29. Willemsens, L. C., and G. J. M. van der Kerk, *Investigations in the Field of Organolead Chemistry*, Intern. Lead Zinc Res. Organ., Inc., New York (1965).
30. Willemsens, L. C., and G. J. M. van der Kerk, *19th Intern. Congr. Pure Appl. Chem., London, Div. A.*, **1963,** 194.
31. Willemsens, L. C., and G. J. M. van der Kerk, *J. Organometal. Chem.*, **2,** 260 (1964).
32. Willemsens, L. C., and G. J. M. van der Kerk, *J. Organometal. Chem.*, **2,** 271 (1964).
33. Woodworth, R. C., and P. S. Skell, *J. Am. Chem. Soc.*, **81,** 3383 (1959).

IX Cyclopentadienyl-type lead compounds

1. NATURE OF BONDING

Only two examples of the class of simple bis(cyclopentadienyl)lead compounds have been synthesized, the parent compound, bis(cyclopentadienyl)lead, and the corresponding bis-methylcyclopentadienyl derivative. According to Fischer and his co-workers in their early papers (5–8,20), bis(cyclopentadienyl)lead is diamagnetic, with a dipole moment of $1.63 \pm 0.06\ \mu$. This compares with a value of $2.16\ \mu$ obtained subsequently by Strohmeier and Hobe (18), and $1.29 \pm 0.04\ \mu$ in cyclohexane by Fischer and Schreiner (6). This evidence alone indicated the compounds to be of the "normal" covalent organolead compound type with a localized sigma bond between lead and carbon atoms; i.e., like diphenyllead. This suggested either normal angular lead–carbon bonding or, less likely, perhaps a system with two unparallel, anti-podal, centrically-bonded rings. At any rate, the compound did not appear to possess the

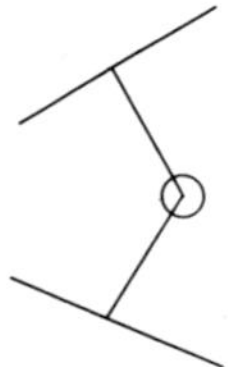

parallel "sandwich" configuration of ferrocene. On the other hand, Lindstrom and Barusch (14,15) showed that the infrared spectrum of this compound is remarkably similar to that of ferrocene, thus suggesting a ferrocene-type structure. Subsequently, Fritz and others in Fischer's laboratory (9) made infrared measurements on bis(cyclopentadienyl)lead [as well as bis(cyclopentadienyl)tin] in potassium bromide discs. On the basis of these spectra, they agreed with Lindstrom and Barusch regarding its ferrocene-type bonding.

Since the dipole moment data and the infrared evidence could be reconciled only by postulating an angular ring-bonded structure, Dave, Evans, and Wilkinson (4) undertook a further investigation of the infrared spectra of both the cyclopentadienyl and the methylcyclopentadienyl compounds, using tetrachloroethylene and carbon disulfide as solvents, and also measured the dipole moment and proton nuclear magnetic resonance spectra. These data confirmed that the compounds possess angular sandwich structures, the nuclear magnetic resonance spectrum consisting of a single sharp line at room temperature. This type of structure is the first of its kind to be established. Wilkinson regards the structure as somewhat analogous to that of the plumbous halides investigated by Lister and Sutton (16). In 1961 Fritz and Fischer (10) reinvestigated the infrared spectra of both bis(cyclopentadienyl)lead and its analogous tin compound, after Fritz and Schneider (11) had discussed the infrared spectra one year earlier. The measurements, made on Nujol and "Hostaflon" mulls in the range 2–40 μ, led these investigators to reconfirm the conclusion that the compounds do indeed possess an angular, sandwich-type configuration, but that there is predominantly sigma bonding, with some additional π-bonding between the metal atom and the rings.

Almenningen et al. (1) investigated the structure of gaseous bis(cyclopentadienyl)lead by means of electron diffraction. It was found that the angle between the pyramidal axes is $135 \pm 15°$, i.e., the angle between the C_5 planes being $45 \pm 15°$. The data did not provide significant evidence for a deviation from C_{5_v} symmetry of the $(C_5H_5)_2Pb$, although such an asymmetry could not be ruled out.

The electrical conductivities of several metal cyclopentadienyl compounds, including bis(cyclopentadienyl)lead, were measured by Strohmeier et al. (19) in three different ethers. They found that all the compounds investigated behaved as 1:1 electrolytes and that the dissociation constants, K_1, increase in the order of diethyl ether < tetrahydrofuran < dimethoxyethane. This is in general agreement with preparative experience, since chemical reactions involving cyclopentadienyl metal compounds proceed with better yields and rates in tetrahydrofuran than in diethyl ether.

Bombieri and Panattoni (2) obtained two crystalline modifications of bis(cyclopentadienyl)lead by sublimation in vacuum, from a sample prepared according to the original Fischer and Grubert method. Form 1, yellow in color, proved to be orthorhombic. The structure determination by X-ray crystallography has been completed (17):

$$a = 16.37 \pm 0.04\ \text{Å} \qquad \text{space group, } P_{Na}\ 2,$$
$$b = 16.02 \pm 0.03\ \text{Å} \qquad z = 4$$
$$c = 9.61 \pm 0.03\ \text{Å}$$

Within the polymer chain one cyclopentadienyl ring appears to be situated between two lead atoms with a mean Pb—C distance of 3.06 Å; the other ring is bonded to only one lead atom with a Pb—C distance of 2.76 Å. The distance between lead atoms in the chain is 5.636 Å. The rings are perpendicular to the axis between the lead atom and the center of the ring. Form 2, which was yellow-brown in color, was shown to be prismatic. The cell dimensions were calculated to be:

$$a = 8.42 \pm 0.02\ \text{Å} \qquad \alpha = 94°27' \pm 30'$$
$$b = 10.61 \pm 0.03\ \text{Å} \qquad \beta = 96°58' \pm 30'$$
$$c = 15.92 \pm 0.02\ \text{Å} \qquad \gamma = 90°7' \pm 30'$$

The authors report that the expected reflections were observed and that the space group is triclinic Pl or $P\bar{1}$.

More recently, Fritz and Schwarzhans (12) have synthesized a series of tetraorganolead compounds containing either one or two cyclopentadienyl groups. These compounds, the trimethyl- and triethylcyclopentadienyllead compounds and dimethylbis(cyclopentadienyl)lead, were prepared by reaction of the corresponding alkyllead chloride and cyclopentadienylsodium in benzene under nitrogen. The NMR spectra were obtained directly using the filtered solutions of the compounds. Trimethylcyclopentadienyllead was crystallized from absolute ether at 0°. Fritz and Schwarzhans (13) have also prepared diethylbis(cyclopentadienyl)lead, and a series of dimethyl-, diethyl-, and trimethyl-, and triethyllead compounds in which a methyl group was substituted in the cyclopentadienyl ring. These cyclopentadienyl derivatives of tetravalent lead are apparently quite unstable. They were isolated as yellow oils or solids which could be handled successfully at low temperatures, but dimethylbis(cyclopentadienyl)lead was shown to decompose at 0°, even in solution, into the somewhat more stable bis(cyclopentadienyl)lead. The trimethyl- and triethylcyclopentadienyllead compounds, however, were stable in solution at room temperature. Infrared and proton magnetic resonance spectra indicated that different chemical shifts were observed for the olefinic protons in the methylcyclopentadienyl compounds. It was concluded by the authors that in these compounds the organolead moiety is attached to the carbon atom bearing the methyl group. Moreover, it was concluded

$$\text{CH}_3 \text{ and } \text{R}_3\text{Pb} \text{ on the same ring carbon of cyclopentadiene}$$

that the cyclopentadienyl groups are sigma-bonded in such compounds.

Obviously, from the foregoing discussion it seems apparent that the last word may not yet have been said on the bonding and structure of cyclopentadienyllead compounds. Thus, the class of cyclopentadienyllead compounds still provides an area for further research. The compounds prepared to date are listed in Table IX-I.

2. SYNTHESIS

The bis(cyclopentadienyl)lead compounds have been prepared only from anhydrous lead(II) salts and cyclopentadienylsodium. Fischer and Grubert (7) prepared the unsubstituted compound from lead nitrate and the sodium compound dissolved in dimethylformamide:

$$2C_5H_5Na + Pb(NO_3)_2 \rightarrow (C_5H_5)_2Pb + 2NaNO_3$$

Lindstrom and Barusch (14,15) employed the reaction of cyclopentadienylsodium dissolved in liquid ammonia with lead chloride, and produced the bis(cyclopentadienyl)lead in 55% yield. Wilkinson and co-workers (4) preferred the use of lead acetate, claiming increased safety over the lead nitrate method, and higher yields.

The cyclopentadienyl derivatives of tetravalent lead have been synthesized only from cyclopentadienylsodium and the corresponding alkyllead chloride (12).

3. REACTIONS

Little is known concerning the reactions of the bis(cyclopentadienyl)-lead or the tetravalent lead derivatives; more work is obviously required. The bis compounds are relatively unstable organolead types, extraordinarily sensitive to oxygen in the manner of diorganolead compounds generally, according to all investigators. However, they are stable under nitrogen. They are yellow or yellow-brown solids, thus also being reminiscent of other diorganolead compounds. According to Fischer and Grubert, bis(cyclopentadienyl)lead is insoluble in and unhydrolyzed by water; on the other hand, Dave, Evans, and Wilkinson (4) report it to be very sensitive to traces of moisture. Surprisingly, the solid sinters (but does not melt) at 138–139°. It dissolves readily in organic solvents, such as benzene, ether, and isooctane, reacting slowly with the dissolved oxygen or water.

The cyclopentadienyl derivatives of tetravalent lead are unstable thermally, decomposing in at least one instance to the somewhat more stable bis compound (13). Otherwise, little is known of their reactions.

4. INDENYL AND FLUORENYL LEAD COMPOUNDS

In addition to the simple cyclopentadienyl-type lead compounds, indenyl and fluorenyl derivatives of tetravalent lead containing either two or three phenyl groups have been reported. These have been prepared by

Table IX-1 Cyclopentadienyl Lead Compounds

COMPOUND	REFERENCES
$(C_5H_5)_2Pb$	1,2,5–11,14,15,17–20
$(CH_3C_5H_4)_2Pb$	4
$(CH_3)_3PbC_5H_5$	12
$(C_2H_5)_3PbC_5H_5$	12
$(CH_3)_2Pb(C_5H_5)_2$	12
$(C_2H_5)_2Pb(C_5H_5)_2$	13
$(CH_3)_3Pb(C_5H_4CH_3)$	13
$(C_2H_5)_3Pb(C_5H_4CH_3)$	13
$(CH_3)_2Pb(C_5H_4CH_3)_2$	13
$(C_2H_5)_2Pb(C_5H_4CH_3)_2$	13
$(C_6H_5)_3Pb(C_9H_7)$	3
$(C_6H_5)_2Pb(C_9H_7)_2$	3
$(C_6H_5)_3Pb(C_{13}H_9)$	3
$(C_6H_5)_2Pb(C_{13}H_9)_2$	3

d'Ans and co-workers (3) from the corresponding phenyllead chlorides and the respective lithium compounds. Except for the qualitative demonstration that the compounds are sensitive to water and are easily cleaved by hydrogen chloride, little has been done to elucidate the nature of the bonding in these compounds. From the available evidence, the bonding appears to be of the sigma type. Attempts to prepare bis(indenyl)lead and bis(fluorenyl)lead, analogous to bis(cyclopentadienyl)lead, gave only lead metal and indene or fluorene, respectively.

REFERENCES

1. Almennigen, A., A. Haaland, and T. Motzfeldt, *J. Organometal. Chem.*, **7,** 97 (1967).
2. Bombieri, G., and C. Panattoni, *Acta Crystallogr.*, **20,** 595 (1966).
3. d'Ans, J., H. Zimmer, and M. V. Brauchitsch, *Chem. Ber.*, **88,** 1507 (1955).

4. Dave, L. D., D. F. Evans, and G. Wilkinson, *J. Chem. Soc.*, **1959,** 3684.
5. Fischer, E. O., *Angew. Chem.*, **69,** 715 (1957).
6. Fischer, E. O., and S. Schreiner, *Chem. Ber.*, **92,** 938 (1959).
7. Fischer, E. O., and H. Grubert, *Z. anorg. allgem. Chem.*, **286,** 237 (1956).
8. Fischer, E. O., and U. Piesbergen, *Z. Naturforsch.*, **11b,** 758 (1956).
9. Fritz, H. P., *Chem. Ber.*, **92,** 780 (1959).
10. Fritz, H. P., and E. O. Fischer, *J. Chem. Soc.*, **1961,** 547.
11. Fritz, H. P., and R. Schneider, *Ber.*, **93,** 1171 (1960).
12. Fritz, H. P., and K. E. Schwarzhans, *J. Organometal. Chem.*, **1,** 297 (1964).
13. Fritz, H. P., and K. E. Schwarzhans, *Chem. Ber.*, **97,** 1390 (1964).
14. Lindstrom, E. G., and M. R. Barusch, Brit. Pat. 774,529 (to Calif. Research Corp.) May 8, 1957.
15. Lindstrom, E. G., and M. R. Barusch, Abstract 77, pp. 46–50, 131st ACS Meeting, Div. of Organic Chem., Miami, Fla. April 7–12, 1957.
16. Lister, M. W., and L. E. Sutton, *Trans. Faraday Soc.*, **37,** 406 (1941).
17. Panattoni, C., G. Bombieri, and U. Croatto, *Acta Crystallogr.*, **21,** 823 (1966); through *C.A.*, **66,** 23057 (1967).
18. Strohmeier, W., and D. von Hobe, *Z. Elektrochem.*, **64,** 945 (1960).
19. Strohmeier, W., H. Landsfeld, and F. Gernert, *Z. Elektrochem.*, **66,** 823 (1962).
20. Weiss, E., *Z. anorg. allgem. Chem.*, **286,** 236 (1956).

X Unsaturated organolead compounds

By present definition, the category of unsaturated organolead compounds embraces those compounds in which the unsaturation is either adjacent to the lead atom or non-adjacent. Furthermore, the unsaturation may consist of either alkenyl or alkynyl groups, but the hexaalkyldilead- or trialkyllead-type of structures sometimes referred to as "unsaturated" in the older organometallic literature are excluded (see Chapter VII). Many interesting lead compounds containing alkenyl and alkynyl groups have been prepared (Table X-1). The polymerization of many of these compounds has been attempted, with frequent success; however, the simple vinyl and allyl lead compounds have not been homopolymerized successfully.

The history of unsaturated lead compounds dates back to 1918, when Grüttner and Krause (28) first prepared allyltriethyllead. However, most of the research activity involving the polymerization of unsaturated organolead compounds has taken place within the past several years. Copolymers derived from lead-containing monomers appear to have both theoretical and practical interest. They have some special properties attributable to the lead atom, such as ability to stop X-ray radiation, and rather high densities. In addition, they are frequently mechanically strong and easily formed. They also have several deficiencies: thus, they are generally inferior to the corresponding tin- and silicon-containing polymers in thermal stability, just as monomeric organolead compounds are generally inferior to organotin and organosilicon compounds in this respect; also they are frequently hazy rather than clear; this characteristic may also be due to the lack of thermal stability. In addition, polymers obtained from the organolead salts of unsaturated acids are susceptible to hydrolytic cleavage. However, further development work on organolead

Table X-1 Unsaturated Lead Compounds

COMPOUND	PROPERTIES	REFERENCES
	Acetylenic Compounds	
$[C_2Pb]_n$		16
$(CH_3)_3PbC{\equiv}CCH_3$	b.p.$_{16}$ 66°; d_{23} 1.82	3
$(CH_3)_3PbC{\equiv}CC_2H_5$	b.p.$_{4}$ 54°	2,3
$(C_2H_5)_3PbC{\equiv}CH$	b.p.$_{2.5}$ 110°	77
$(C_2H_5)_3PbC{\equiv}CCH_3$	b.p.$_{1.5}$ 81°	77
$(C_2H_5)_3PbC{\equiv}CC_2H_5$	b.p.$_{3}$ 105°	77
$(C_2H_5)_3PbC{\equiv}CC_6H_5$	b.p.$_{1}$ 66°	77
$(C_6H_5)_3PbC{\equiv}CH$	Oil	106
$(C_6H_5)_3PbC{\equiv}CCH_3$	m.p. 65°	77
$(C_6H_5)_3PbC{\equiv}CC_2H_5$	m.p. 64°	77
$(C_6H_5)_3PbC{\equiv}CC_6H_5$	m.p. 56°	77,113
$(C_6H_5)_3PbC{\equiv}C(cyclo\text{-}C_6H_{11})$	m.p. 84°	77
$(CH_3)_3PbC{\equiv}CCl$	b.p.$_{50}$ 102°, with dec.	109
$(C_6H_5)_3PbC{\equiv}CBr$	m.p. 95°	76
$(C_2H_5)_3PbC{\equiv}CPb(C_2H_5)_3$	b.p.$_{0.05}$ 135–40; $n_D^{20.5}$ 1.567; $d_{20.5}$ 1.904	8,30
$(cyclo\text{-}C_6H_{11})_3PbC{\equiv}CPb(cyclo\text{-}C_6H_{11})_3$	m.p. 136°	29,33
$(C_6H_5)_3PbC{\equiv}CPb(C_6H_5)_3$	m.p. 138.5°	8,30,76
$(o\text{-}CH_3C_6H_4)_3PbC{\equiv}CPb(o\text{-}CH_3C_6H_4)_3$	m.p. 121°	29,33
$(p\text{-}CH_3C_6H_4)_3PbC{\equiv}CPb(p\text{-}CH_3C_6H_4)_3$	m.p. 130	29,33
$(C_6H_5C_2H_4)_3PbC{\equiv}CPb(C_2H_4C_6H_5)_3$	m.p. 62°	29,33
$(cyclo\text{-}C_6H_{11}C{\equiv}C)_4Pb$	dec. 94°	35
$(p\text{-}BrC_6H_4C{\equiv}C)_4Pb$	dec. 155°	32
$(C_2H_5)_3PbC{\equiv}CCH{=}CH_2$	b.p.$_{6}$ 114°, n^{D20} 1.5564; d_{20} 1.5602	115

$(C_2H_5)_3PbC{\equiv}CC(CH_3){=}CH_2$	b.p.$_{3}$ 93–94°, n_D^{20} 1.5462; d_{20} 1.5849	115
$(C_2H_5)_3PbC{\equiv}CC{=}CHCH_2CH_2CH_2$ (ring)	b.p.$_{1.5}$ 95°	77
$(C_6H_5)_3PbC{\equiv}CC{=}CHCH_2CH_2CH_2$ (ring)	m.p. 50°	77
$(C_6H_5)_3PbC{\equiv}CC{=}CHCH_2CH_2CH_2CH_2$ (ring)		77
$(cyclo\text{-}C_6H_{11})_3Pb(C{\equiv}C)_2H$	b.p.$_{4}$ 115°, with dec.	34,77
$(cyclo\text{-}C_6H_{11})_3Pb(C{\equiv}C)_2C_2H_5$	m.p. 68°, b.p.$_{4}$ 175°, with dec.	34,77
$(cyclo\text{-}C_6H_{11})_3Pb(C{\equiv}C)_2C_3H_7$	b.p.$_{4}$ 140°	34,77
$(cyclo\text{-}C_6H_{11})_3Pb(C{\equiv}C)_2C_4H_9$	b.p.$_{4}$ 150°, with dec.	34,77
$(C_6H_5)_3Pb(C{\equiv}C)_2H$	m.p. 93°, with dec.	77,106
$(C_6H_5)_3Pb(C{\equiv}C)_2CH_3$	m.p. 55°	77
$(C_6H_5)_3Pb(C{\equiv}C)_2C_6H_5$	m.p. 81°	77
$(C_6H_5)_3Pb(C{\equiv}C)_2Pb(C_6H_5)_3$	m.p. 187°	77
$(C_2H_5)_3Pb(C{\equiv}C)_2Pb(C_2H_5)_3$	m.p. 80°	34
$(cyclo\text{-}C_6H_{11})_3Pb(C{\equiv}C)_2Pb(cyclo\text{-}C_6H_{11})_3$	m.p. 138	34
$(C_2H_5)_3Pb(C{\equiv}C)_2Pb(C_6H_5)_3$	m.p. 187°, with dec.	31
$(C_2H_5)_3Pb(C{\equiv}C)_3Pb(C_2H_5)_3$	m.p. 110°, with dec.	34
$(cyclo\text{-}C_6H_{11})_3Pb(C{\equiv}C)_3Pb(cyclo\text{-}C_6H_{11})_3$	dec. 148°	34
$(C_6H_5)_3Pb(C{\equiv}C)_3Pb(C_6H_5)_3$	dec. 150°	34
$(C_6H_5)_2Pb(C{\equiv}CC_6H_5)_2$		40
$(C_6H_5C{\equiv}C)_4Pb$		40
$[(CH_3)_3SiC{\equiv}C]_4Pb$	m.p. 108°C	56
	Vinyl Compounds	
$(CH_2{=}CH)_4Pb$	b.p.$_{0.6}$ 34°, 50°; $_{11}$ 69–70°; d_4^{20} 1.7882; n_D^{20} 1.5462, 1.5470, n_D^{26} 1.5430	7,10,12,17,41,42, 47,48,49,74,97, 98,101

(continued)

Table X-1 (continued)

COMPOUND	PROPERTIES	REFERENCES
$(CH_2{=}CCH_3)_4Pb$		96
$(CF_2{=}CF)_4Pb$	b.p.$_{.8}$ 51–2°, n_D^{18} 1.4193	110
$(C_2H_5)_3(CH_2{=}CH)Pb$		57,97
$(C_2H_5)_3[(CH_3)_2C{=}CH]Pb$	b.p.$_{.12}$ 99–101°	22,25,26
$(C_2H_5)_2(CH_2{=}CH)_2Pb$	b.p.$_{.1}$ 30°	7,57,97
$(CH_2{=}CH)_3(C_2H_5)Pb$		96
$(CH_2{=}CH)_3PbCl$	m.p., 119–21°	42,47
$(CH_2{=}CH)_3PbOOCCH_3$	m.p., 168–70°	47
$(CH_2{=}CH)_3PbOOCCH_2CH_3$	m.p., 132°	42
$(CH_2{=}CH)_3PbOOCCCl_3$	m.p., >300°	47
$(CH_2{=}CH)_3PbNO_3$	dec. at 80°	42
$(CH_2{=}CH)_2PbCl_2$	m.p., >300°	42,47
$(CH_2{=}CH)_2(C_2H_5)PbCl$		96
$(CH_2{=}CH)_2(C_6H_5)PbCl$		96
$(CH_2{=}CH)(C_2H_5)_2PbCl$		96
$(CH_2{=}CH)(C_6H_5)_2PbCl$		96
$(CH_2{=}CH)(C_2H_5)PbCl_2$		96
$(CH_2{=}CH)(C_6H_5)PbCl_2$		96
$(ClCH{=}CH)_2Pb(OOCCH_3)_2$		81,82
$(ClCH{=}CH)_2PbCl_2$		81,82
$(C_2H_5)_3PbCH{=}CHC_6H_5$	b.p.$_{.0.01}$ 94°	27
$(C_6H_5)_3(CH_2{=}CH)Pb$	m.p., 35–6°; b.p.$_{.0.002}$ 111–4°; b.p.$_{.0.05}$ 160–70°	37,47,102,103
$(C_6H_5)_3Pb(CH_2{=}CC_6H_5)$	m.p. 107–9°	24

Compound	Properties	References
$(C_6H_5)_2(CH_2{=}CH)_2Pb$	b.p.$_{0.002}$ 111–4°; b.p.$_{0.05}$ 120–55°	37,47,97
$(C_6H_5)(CH_2{=}CH)_3Pb$		96
Alkenyl Compounds Other Than Vinyl		
$(C_2H_5)_3(CH_2{=}CHCH_2)Pb$	b.p.$_{10}$ 86°, n_D^{20} 1.5423	21,24,27,28
$(C_2H_5)[CH_2{=}C(CH_3)CH_2]Pb$	b.p.$_{0.1}$ 45–50°	27
$(C_2H_5)_3(CH_2CH_2CH{=}CH_2)Pb$	b.p.$_{3}$ 78°, n_D^{20} 1.5230	24
$(C_6H_5)_3(CH_2{=}CHCH_2)Pb$	m.p., 76–7	1,67
$(C_6H_5)_3(CH_2CH_2CH{=}CH_2)Pb$	m.p., 84–6°	24
$(C_6H_5)_3(p\text{-}C_3H_5C_6H_4)Pb$	m.p., 206°	94
$(C_6H_5)_3[p\text{-}CH_2{=}C(CH_3)CONHC_6H_4CO_2]Pb$		51
$(C_6H_5)_2(CH_2{=}CHCH_2)_2Pb$		22,37
$(C_6H_5)_2(p\text{-}C_3H_5C_6H_4)_2Pb$	m.p., 194–6°	94
$(CH_2{=}CHCH_2)_4Pb$		21,23,67,96,105
$[(p\text{-}C_3H_5C_6H_4)_4Pb]_n$		95
$(C_6H_5)_2(CH_2CH_2CH{=}CH_2)PbCl$		24
$(C_6H_5)_2(p\text{-}C_3H_5C_6H_4)(p\text{-}CH_2BrCHBrCH_2C_6H_4)Pb$	m.p., 134–5°	94
$(C_6H_5)_2(p\text{-}C_3H_5C_6H_4)(p\text{-}CH_2OHCHOHCH_2C_6H_4)Pb$		94
$(C_2H_5)(CH_2CH_2CH{=}CH_2)PbCl_2$		24
$(CH_3)_3(CH{=}CHCN)Pb$		71
cis-$(C_6H_5)_3(CH{=}CHCOOC_2H_5)Pb$	m.p., 66–7°	112
cis-$(C_6H_5)_3(CH{=}CHCN)Pb$	m.p., 69–70°	112
Styrenyl Compounds		
$(CH_3)_3(p\text{-}CH_2{=}CHC_6H_4)Pb$	b.p.$_{0.0015}$ 60–1°; d_4^{20} 1.7278; n_D^{20} 1.6070	73,84
$(CH_3)_3[p\text{-}CH_2{=}C(CH_3)C_6H_4]$	b.p.$_{0.001}$ 69–71°; d_4^{20} 1.6816; n_D^{20} 1.5963	73,84
$(C_2H_5)_3(p\text{-}CH_2{=}CHC_6H_4)Pb$		91
$(C_2H_5)_3[p\text{-}CH_2{=}C(CH_3)C_6H_4]Pb$		58,93

(continued)

Table X-1 (continued)

COMPOUND	PROPERTIES	REFERENCES
$(C_6H_5)_3(p\text{-}CH_2{=}CHC_6H_4)Pb$	m.p., 105–7° (59); 107–9° (83); 110–1° (85)	59,62,65,72,83,85, 91,99
$(C_6H_5)_2(p\text{-}CH_2{=}CHC_6H_4)_2Pb$	m.p., 96–7°	72,83
$(p\text{-}CH_2{=}CHC_6H_4)_4Pb$	m.p., 165–70°	14,97
$[(p\text{-}CH_2{=}CHC_6H_4)_4Pb]_n$	See text	15
$(C_6H_5)_3(p\text{-}CH_2{=}CHC_6H_4CO_2)Pb$	m.p., 136–8°	61
	Acrylates and Methacrylates	
$(CH_3)_3PbO_2CC(CH_3){=}CH_2$	See text	69
$(C_3H_7)_3PbO_2CCH{=}CH_2$	m.p., 123°	100
$(C_6H_5)_3PbO_2CCH{=}CH_2$	m.p., 201–3°	20,60,63,66
$(C_6H_5)_3PbO_2CC(CH_3){=}CH_2$	m.p., 126–8°; 124°	20,52–55,60,63,64 66,108
$(C_6H_5)_2Pb[O_2CC(CH_3){=}CH_2]_2$	m.p., 204°; 224°; dec. 205–12°	52–54,68,89,90
$p\text{-}CH_3C_6H_4Pb[O_2CC(CH)_3{=}CH_2]_3$		89

polymers may yet give rise to practical uses, especially since the copolymerization of some unsaturated lead compounds with many other polymerizable monomers has become easy. Moreover, terpolymerization has not as yet been reported with lead compounds, and there are no reports of attempts to use the lead atom for crosslinking or coordination.

1. LEAD ACETYLIDES

The simple parent acetylides, or "carbides," of lead are comparatively unknown. As far back as 1923, Durand (16), in a study of the double decomposition of metallic acetylides and salts in aqueous media, added calcium carbide to a solution of lead acetate in water. The resulting reaction precipitated some gray-white solid, analysis of which indicated the empirical composition C_2Pb. This "lead acetylide" was apparently polymeric and stable to air and water, but was not further characterized. It may be worthy of further study.

2. ORGANOLEAD ACETYLIDES

During the period since 1954, several workers, principally Hartmann and his associates, have investigated the syntheses of several organolead and other organometal derivatives of acetylene. In 1954, Beermann and Hartmann (8,30) prepared bis(triethyllead) acetylide, $Et_3PbC{\equiv}CPbEt_3$, and bis(triphenyllead) acetylide, by means of the reaction of triethyllead chloride or triphenyllead chloride with monosodium acetylide in liquid ammonia. The analogous Grignard reactions were unsuccessful.

The bis(triethyllead) compound was found to be sensitive to water, whereas the phenyl derivative was not, except to alkaline solutions. In general, both compounds were quite reactive. For example, both compounds were cleaved by iodine to form the R_3PbI compound, silver ions reacted to give silver acetylide, and Grignard reagents resulted in a metathetical reaction.

$$R_3PbC{\equiv}CPbR_3 + 2I_2 \rightarrow 2R_3PbI + IC{\equiv}CI$$

$$R_3PbC{\equiv}CPbR_3 + 2AgNO_3 \rightarrow AgC{\equiv}CAg + 2R_3PbNO_3$$

$$R_3PbC{\equiv}CPbR_3 + 2R'MgX \rightarrow XMgC{\equiv}CMgX + 2R_3PbR'$$

In 1959, Hartmann and Eschenbach (29,33) described the preparation of a number of additional compounds of the type $R_3PbC{\equiv}CPbR_3$, in which the R groups were the *o*-tolyl, cyclohexyl, and phenethyl. Further, in 1961, Hartmann et al. (31) synthesized a number of metal compounds of butadiyne (diacetylene), containing germanium, tin, lead, or antimony, including bis(triphenyllead) diacetylide,

$$(C_6H_5)_3PbC{\equiv}C{-}C{\equiv}CPb(C_6H_5)_3$$

prepared from triphenyllead bromide and disodium diacetylide in liquid ammonia. This compound was stable to air and water, and the infrared spectrum was measured. The corresponding symmetrical ethyl and cyclohexyl compounds were prepared in the same manner in 83 and 13% yields, respectively (34). In a similar reaction, the analogous tri-acetylene lead compounds also were synthesized. The phenyl compound and the cyclohexyl compounds were both obtained in good yield, 65 and 52%, respectively. Aqueous silver nitrate cleaved the lead–carbon bonds, and the compounds were photosensitive, but they were not affected by water or warm alkali. Hartmann and Komorniczyk (34) also synthesized nondistillable compounds of the type $(cyclo\text{-}C_6H_{11})_3Pb(C{\equiv}C)_2R$ (R = H, C_2H_5, C_3H_7, and C_4H_9) by adding tricyclohexyllead iodide dissolved in ether to the sodium alkylbutadiyne, $Na(C{\equiv}C)_2R$, in liquid ammonia; the yields were in the range 54–80%. The tricyclohexylleadbutadiyne was found to disproportionate easily at room temperature:

$$2(cyclo\text{-}C_6H_{11})_3PbC{\equiv}CC{\equiv}CH \rightarrow [(cyclo\text{-}C_6H_{11})_3PbC{\equiv}C{-}]_2 + HC{\equiv}CC{\equiv}CH$$

Monosubstituted derivatives, $R_3PbC{\equiv}CH$, were not isolated by Beermann and Hartmann. At the time, this was tentatively attributed to rapid disproportionation to the disubstituted compounds and acetylene. However, in 1963, Masson and co-workers (77) successfully prepared a series of monosubstituted acetylides, using essentially Hartmann's method. In general, the products were stable, non-disproportionating solids, easily recrystallizable from pentane. Both sodium and magnesium acetylides, suspended in refluxing toluene or pentane, were used for the preparative work:

$$(C_6H_5)_3PbCl + NaC{\equiv}CCH_3 \rightarrow (C_6H_5)_3PbC{\equiv}CCH_3 + NaCl$$

The compounds synthesized were of the general formula $Ph_3PbC{\equiv}CR$, where R = methyl, ethyl, phenyl (see also Chapter XII), cyclopent-1-enyl, and cyclohex-1-enyl. Similarly, triethyllead acetylide, $Et_3PbC{\equiv}CH$, was also prepared together with the analogous methyl, ethyl, phenyl, and cyclopent-1-enyl compounds; yields were in the range 30–53%. Some of the alkyne compounds were converted by treatment in boiling liquid ammonia to polyacetylenes, for example, the butadiynyl compound, $Ph_3PbC{\equiv}CC{\equiv}CH$, and the corresponding pentadiynyl and phenyl-butadiynyl compounds. Ultraviolet and infrared data were obtained on many of the compounds. A complete description of Masson's work, including the preparation of bis(triphenyllead) acetylide and triphenyllead bromoacetylide, $Ph_3PbC{\equiv}CBr$, was published in 1965 (76).

Trimethyllead acetylides were not described until 1965, when two patents appeared by Ballinger (2,3). Ballinger prepared trimethyllead methylacetylide, $(CH_3)_3PbC{\equiv}CCH_3$, and the corresponding ethylacetylide via lithium acetylide:

$$(CH_3)_3PbCl + LiC{\equiv}CC_2H_5 \rightarrow (CH_3)_3PbC{\equiv}CC_2H_5 + LiCl$$

These trimethyllead alkylacetylides are claimed to be useful antiknock agents. Similarly, Steingross and Zeil (109) prepared trimethyllead chloroacetylide, adding lithium chloroacetylide to a suspension of trimethyllead chloride in ether:

$$(CH_3)_3PbCl + LiC{\equiv}CCl \rightarrow (CH_3)_3PbC{\equiv}CCl + LiCl$$

The compound was sufficiently stable to allow measurement of its infrared spectrum.

Zavgorodnii and Petrov (115) synthesized two lead-containing alkenynyl compounds, $Et_3PbC{\equiv}CCH{=}CH_2$ and $Et_3PbC{\equiv}CC(CH_3){=}CH_2$, by the reaction of triethylplumbyl sodium (prepared in liquid ammonia) and the corresponding alkenylbromoacetylene. They were reported to be stable only in the absence of air. Attempted additions at the enyne system were unsuccessful, as might be expected in lead-containing systems.

Hartmann and his associates prepared two tetraacetylides of lead, as well as of other Group IVb metals, by a new method. Tetrakiscyclohexylethynyllead, $(C_6H_{11}C{\equiv}C)_4Pb$ (35), was prepared in 22.5% yield by reaction between lithium cyclohexylacetylide, $LiC{\equiv}CC_6H_{11}$, and rubidium hexachloroplumbate, Rb_2PbCl_6, in tetrahydrofuran at room temperature. The *p*-bromophenylacetylide compound, $(p\text{-}BrC_6H_4C{\equiv}C)_4Pb$ (32), was prepared in 22% yield by a similar type of reaction in refluxing tetrahydrofuran.

In 1966, Henry and Pant (40) prepared what appeared to be diphenylbis(phenylethynyl)lead, from diphenyllead dichloride and lithium phenylacetylide, and tetrakis(phenylethynyl)lead, from potassium hexachloroplumbate and sodium phenylacetylide, in tetrahydrofuran, but the compounds were not purified.

One tetra-acetylide containing a second metal is known: $[(CH_3)_3SiC{\equiv}C]_4Pb$, prepared recently by Komarov and Yarosh (56), apparently from $(CH_3)_3SiC{\equiv}CMgBr$ and $PbCl_4$.

Simonnin (106) has studied the nuclear magnetic resonance spectra of a number of acetylenic derivatives of elements of Groups IV, V, and VI, including $(C_6H_5)_3PbC{\equiv}CH$ and $(C_6H_5)_3PbC{\equiv}CC{\equiv}CH$. Based on the observed chemical shifts of the acetylenic protons at infinite dilution in carbon tetrachloride or deuterochloroform, heteroatom–proton coupling through the acetylenic system was found to occur in the lead compounds, as well as in analogous phosphorus, silicon, and tin derivatives. Also, Petrov and Lebedev (92) examined the nuclear magnetic resonance spectra of $Et_3PbC{\equiv}CCH{=}CH_2$ and $Et_3PbC{\equiv}CC(CH_3){=}CH_2$, together with the spectra of the parent hydrocarbons and the silicon and tin analogs; Zavgorodnii and Petrov (115) also obtained the infrared spectra of these compounds.

Acetylide compounds were reviewed briefly in 1967 by Davidson and Henry (13).

3. ALKENYLLEAD COMPOUNDS

A considerable number of alkenyllead derivatives has been reported in the past decade. In the immediately following discussion, however, vinyllead compounds are omitted for the most part, and are treated in the following sub-sections, separately from other alkenyllead compounds, simply for convenience. It should be recognized that the term vinyl is frequently used broadly, both to cover true vinyllead compounds in which the vinyl group is contiguous to the lead atom, and also vinyl aromatic compounds of the styrenyl- or vinylphenyllead type.

The allyllead compounds are not as stable thermally as the vinyl or the usual alkyllead compounds, nor have the simple allyl compounds been homopolymerized successfully. Since Grüttner and Krause's (28) original synthesis of allyltriethyllead, the preparation of this compound has been repeated several times; however, compounds with more than one allyl group are clearly much less stable. The preparation of diallyldiphenyllead from allyl Grignard reagent and diphenyllead dibromide has been attempted without success by both Gilman and Eisch (22) and Henry and Noltes (37). Furthermore, several attempts to synthesize tetraallyllead by several investigators by means of Grignard reactions and disproportionations have been uniformly unsuccessful (21,23,67,105,111), although this compound is mentioned, together with many other unsaturated lead compounds, in a recent British patent by Ramsden (96). This synthesis problem appears to be deserving of further effort.

Allyltriethyllead is commonly prepared by the reaction of a triethyllead halide with an allyl Grignard reagent:

$$Et_3PbCl + H_2C{=}CHCH_2MgCl \rightarrow H_2C{=}CHCH_2PbEt_3 + MgCl_2$$

It has also been prepared from triethylplumbylsodium and allyl chloride by Glockling and Kingston (27). The compound is a fairly stable, colorless liquid, similar to tetraethyllead in its common properties. Its phenyl analog, allyltriphenyllead, was prepared in an analogous manner in good yield from triphenyllead chloride and allyl Grignard reagent by Austin (1), and later by others (24,67). As might have been expected, the phenyl derivative is a stable solid under ordinary conditions.

The general method of preparation used by Glockling and Kingston is illustrated by the method used for allyltriethyllead. Triethylplumbyl sodium was prepared in a solution of liquid ammonia and ether at $-80°$. This was then reacted with a solution of allyl chloride in ether.

$$NaPbEt_3 + H_2C{=}CHCH_2Cl \rightarrow Et_3PbCH_2CH{=}CH_2 + NaCl$$

In 1962, Puchinyan and Manulkin (94,95) prepared *p*-allylphenyllead compounds as well as similar *p*-allylphenyltin compounds. *p*-Allylphenyltriphenyllead, $p\text{-}CH_2{=}CHCH_2C_6H_4Pb(C_6H_5)_3$, was prepared in approximately 62% yield from triphenyllead chloride and *p*-allylphenylmagnesium bromide; bis(*p*-allylphenyl)diphenyllead was prepared similarly in 52% yield from diphenyllead dichloride. Tetra-*p*-allylphenyllead was also successfully prepared as a partially polymerized product by the reaction of lead(II) chloride and the corresponding Grignard reagent (95). The product was obtained in 41% yield, as an amorphous, non-fusible material having a molecular weight of about 1270.

Certain types of organolead alkenyl compounds containing additional functionality were prepared in 1965 by Willemsens (112), utilizing the addition of triphenylplumbyllithium to compounds containing reactive triple bonds, followed by hydrolysis. For example, the following reactions were carried out:

$$LiPbPh_3 + HC{\equiv}CCOOEt \rightarrow cis\text{-}Ph_3PbCH{=}CHCOOEt$$

$$LiPbPh_3 + HC{\equiv}CCN \rightarrow cis\text{-}Ph_3PbCH{=}CHCN$$

Leusink and van der Kerk (71) achieved similar results by the addition of trimethyllead hydride to cyanoethyne; different (α and β) adducts were formed depending on whether ether or butyronitrile was used as a solvent. Similar products were obtained directly by reaction of tributyllead hydride with unsaturated molecules (see Chapter XI).

Seyferth and Weiner (102) showed that allyllithium can be prepared by the reaction of allyltriphenyllead with phenyllithium in diethyl ether.

Aside from the allyl compounds, several additional trialkyl- and triarylalkenyllead compounds are known. Gilman's group (24) prepared triethyl-3-butenyllead, $Et_3PbCH_2CH_2CH{=}CH_2$, triphenyl-3-butenyllead and triphenyl-β-styrenyllead. These were prepared from 3-butenylmagnesium bromide, or β-styrenylmagnesium bromide, respectively, and the corresponding R_3PbCl compound, in 77–99% yields. As had been expected from analogy to the alkyl analogs, the first compound is a liquid and the latter two are solids. Hydrogen chloride cleaved two ethyl groups from the triethylbutenyllead, and gave the salt, ethyl-3-butenyllead dichloride. However, in the aryl series, only one phenyl group was split off from triphenyl-3-butenyllead in this type of reaction, giving diphenyl-3-butenyllead chloride. The triphenyl-β-styrenyllead, a vinylic compound, lost mainly the β-styrenyl group, leaving triphenyllead chloride.

Glockling also investigated the cleavage of unsymmetrical alkenyl lead compounds. Allyltriethyllead and triethyl-α-styrenyllead (27), as well as *i*-but-1-enyltriethyllead (25,26), were prepared. This latter compound reacted with alcoholic silver nitrate with preferential cleavage of the alkenyl group to form *i*-but-1-enylsilver.

$$Et_3PbCH{=}CMe_2 + Ag^+ \rightarrow Et_3Pb^+ + Me_2C{=}CHAg$$

This organosilver compound is worthy of special comment because of its remarkable stability. Alkylsilver compounds such as methylsilver decompose at very low temperatures ($-78°$), but isobutenylsilver was shown to deposit silver only very slowly at $-30°$. Moreover, Glockling and Kingston (27) showed that α-styrylsilver, a dark red solid, decomposed completely to silver and polymeric organic products only after it had been heated for several hours in refluxing ethanol, and it was stable in air for several hours at room temperature. The unusual stability of these two compounds was ascribed to the presence of a double bond adjacent to the silver atom. In general, Glockling and Kingston concluded that the silver compound appeared to be less stable as the stability of the R radical became greater, so that in the two extreme cases of the methallyl and benzyl radicals, no silver compounds at all could be obtained under the conditions of the experiments.

Gilman, Towne, and Jones, (24) showed that in both the triethyl- and the triphenylallyllead compounds, the allyl group is readily cleaved by halogens and halogen acids, to form the R_3PbX compound under mild conditions. On halogenation, addition to the double bond, if any occurs, is undetectable. On the other hand, Austin (1) showed that oxidation of allyltriphenyllead with potassium permanganate gave the novel functional compound 1-(triphenylplumbyl)-2,3-propanediol, $(C_6H_5)_3$-$PbCH_2CHOHCH_2OH$. Remarkably, Puchinyan and Manulkin (94,95) found that bromination of bis(*p*-allylphenyl)diphenyllead at room temperature gave diphenyl-*p*-allylphenyl-*p*-2,3-dibromopropylphenyllead, $(C_6H_5)_2Pb(p\text{-}C_3H_5C_6H_4)(C_6H_4CH_2CHBrCH_2Br)$, rather than the expected cleavage of a lead–carbon bond. Cleavage may have been prevented by steric factors in this instance.

No systematic study has been made of the reactivity of alkenyllead compounds with hydrogen chloride. The available data indicate that, except for the allyl group, the reactivity decreases with increasing displacement of the double bond from the lead–carbon bond. The mechanism of cleavage of organolead compounds by halogen acids was discussed in detail in Chapter VI; as was indicated there, the mechanism appears to involve an electrophilic attack at the most electron-rich carbon bound to lead and is not strictly dependent on or indicative of the relative electronegativity of the organic group itself, as originally suggested by Kharasch and his students (50). The allyl group is unique among alkenyl groups in that is it exceedingly reactive with both electrophilic and nucleophilic reagents. Explanations for this reactivity were discussed in Chapter VI.

Koton and his associates (65,67) showed that allyltriphenyllead is thermally stable up to only 160°, whereas allyltriphenyltin is stable at higher temperatures. At temperatures above 160° decomposition begins with the formation of tetraphenyllead, biallyl and metallic lead, according to the following overall scheme:

$$4(C_3H_5)Pb(C_6H_5)_3 \rightarrow 3(C_6H_5)_4Pb + [(C_3H_5)_4Pb] \rightarrow 2C_6H_{10} + Pb$$

Tetraallyllead was speculated to be an intermediate disproportionation product.

The Koton group also found that allyltriphenyllead (like allyltriphenyltin) is not polymerizable thermally below 160°, either in the absence of initiators, or in the presence of either dibenzoyl peroxide, *tert*-butyl peroxide, various Zeigler catalysts, or 2,2′-azo-bis-isobutyronitrile. Instead, these initiators reduce the decomposition temperature, to as low as room temperature in the case of Ziegler catalysts. Moreover, in experiments on copolymerization, it was found that allyltriphenyllead inhibits the polymerization of methyl methacrylate and styrene. The compound yielded triphenyllead iodide on treatment with iodine, as was expected from previous work.

A. Vinyllead compounds

Vinyl Grignard reagents were described as early as 1902, but the yields were very poor and the reagents were not useful for the preparation of vinylmetal compounds. However, after Normant's discovery (88) in 1957 that vinyl Grignard reagents can be prepared easily in tetrahydrofuran, the first lead compound containing a simple unsubstituted vinyl group was reported independently by several investigators. Many other vinylmetal compounds were reported at about the same time. In 1959, several papers were published within the space of a few months on the synthesis and properties of tetravinyllead. Such vinylmetal compounds promised to be of both theoretical and practical importance, since the vinyl group may allow both p_π-d_π bonding and interesting polymerizations of lead-containing molecules. Vinylmetal compounds have been reviewed by Seyferth (101) in 1962, and also by Kaesz and Stone (49) in 1960.

Originally, Maier (74) published on methods involving the reaction of either lead tetrachloride or ammonium hexachloroplumbate with vinylmagnesium bromide at $-20°$, to produce tetravinyllead in 4–17% yields.

$$PbCl_4 + 4H_2C{=}CHMgBr \rightarrow (H_2C{=}CH)_4Pb + 4MgClBr$$

In a similar type of reaction, Bartocha and co-workers (5–7) prepared tetravinyllead from potassium hexachloroplumbate and the Grignard reagent in yields of about 70%; diethyldivinyllead was prepared from diethyllead dichloride and vinyl Grignard reagent.

In a subsequent publication Maier (75) obtained yields as high as 85% using lead(II) chloride instead of a tetravalent lead salt, according to the following equation:

$$2PbCl_2 + 4H_2C{=}CHMgBr \rightarrow (H_2C{=}CH)_4Pb + 4MgClBr + Pb$$

In a more detailed paper, Juenge and Cook (47) also described the reaction of lead chloride (or lead acetate) with vinylmagnesium halide in tetrahydrofuran. In addition to tetravinyllead, they prepared triphenylvinyllead and diphenyldivinyllead by the reaction of vinylmagnesium bromide with triphenyllead chloride or diphenyllead dichloride in tetrahydrofuran. Henry and Noltes (37) also prepared these latter compounds in a similar manner.

Also, in 1960, Korshak and his co-workers (57) used the reaction of vinyl Grignard with ethyllead halides to extend the work of Bartocha and Gray (7) on diethyldivinyllead. They prepared both diethyldivinyllead and triethylvinyllead.

The preparation of tetravinyllead from vinylmagnesium halide and lead tetrachloride in tetrahydrofuran and hydrocarbon solvents has been patented by Ramsden and Shannon (98). A patent application has been filed in the Netherlands on the preparation of mixed tetravinylalkyl lead compounds by a redistribution reaction between tetravinyllead and tetraalkyllead (18). A second patent application in the Netherlands (19) claims the preparation of mixed tetravinylalkyl lead compounds by electrolysis of the systems:

$$Pb + RMgX + CH_2{=}CHCl \rightarrow R_x(CH_2{=}CH)_yPb$$

$$Pb + CH_2{=}CHMgX + RCl \rightarrow R_x(CH_2{=}CH)_yPb$$

Tetravinyllead and the mixed alkyl- and arylvinyllead compounds have been patented by Juenge (46).

Only cleavage reactions of true vinyllead compounds are known. No addition reactions or homopolymerizations have been reported to date. Copolymerization, however, has been reported, as described below.

In general, tetravinyllead appears to be somewhat similar to tetraethyllead in its common physical properties, but less stable on the whole. It is reasonably stable on storage, but sensitive to water. If kept below room temperature, it is stable for several months. However, at higher temperatures it decomposes to give a yellow solid. Like most vinyl-metal compounds, tetravinyllead is more susceptible to electrophilic cleavage than the corresponding ethyl compounds. Thus, tetravinyllead is cleaved more easily than tetraethyllead. For example, although tetraethyllead reacts slowly with acetic acid in the absence of a catalyst at room temperature (105), tetravinyllead reacts more rapidly under these conditions to produce trivinyllead acetate. Juenge and Cook (47) studied the cleavage reactions of tetravinyllead with chlorine, hydrogen chloride and carboxylic acids. Reaction of tetravinyllead with hydrogen chloride and chlorine gas gave trivinyllead chloride and divinyllead dichloride, respectively. Trivinyllead trichloroacetate was also prepared by the reaction between tetravinyllead and trichloroacetic acid. This reaction occurs with explosive violence, but it is easily controlled under the proper conditions. Juenge and Cook also found that vinyllead compounds can react explosively during combustion analysis. Furthermore, they compared the melting points of several solid vinyl compounds with the corresponding ethyllead compounds. No particular pattern was apparent. However, surprisingly, divinyllead dichloride and trivinyllead trichloroacetate did not melt below 300°.

In addition to vinyl Grignard reagents, vinylmercury and vinyltin compounds have been used for the synthesis of vinyl derivatives of metals. In the particular case of lead, this type of synthesis has been applied to bis-(chlorovinyl)lead diacetate, by Nesmeyanov and Borisov (81,82).

$$(ClCH{=}CH)_2Hg + Pb(OAc)_4 \rightarrow (ClCH{=}CH)_2Pb(OAc)_2 + Hg(OAc)_2$$

Bis(β-chlorovinyl)lead dichloride, which reacts with mercury to yield chlorovinylmercuric chloride, as shown below, also was prepared. The available data indicate that these reactions proceed exclusively with strict retention of configuration, so that a *trans*-chlorovinyllead compound leads to a *trans*-chlorovinylmercury compound.

$$(ClCH{=}CH)_2PbCl_2 + 2Hg \rightarrow 2ClCH{=}CHHgCl + Pb$$

Sterlin and his associates (110) have prepared tetrakis(perfluorovinyl)-lead, $(CF_2{=}CF)_4Pb$ (see Chapter XII, Organofunctional Lead Compounds).

Henry and Noltes (38,39) studied the uncatalyzed addition of triphenyltin hydride and diphenyltin dihydride to a series of mono- and divinyl and allyl compounds of the Group IV elements. It was found that both normal and abnormal additions occurred, depending on the particular reaction involved. Normal addition may be represented by

$$(C_6H_5)_3SnH + (C_6H_5)_3MCH{=}CH_2 \rightarrow (C_6H_5)_3SnCH_2CH_2M(C_6H_5)_3$$

where M = Si, Ge, or Sn. In the case of lead, no addition products were formed; both triphenylvinyllead and triphenylallyllead were reductively cleaved to metallic lead.

Seyferth and his associates have shown that tetravinyllead (17,48) and triphenylvinyllead (102–104) react with phenyllithium in ether solution to form vinyllithium:

$$(C_6H_5)_3PbCH{=}CH_2 + C_6H_5Li \rightarrow (C_6H_5)_4Pb + CH_2{=}CHLi$$

The reactions were found to be suitable for the preparation of vinyllithium, but tetravinyltin is preferable to tetravinyllead because of its greater stability and more ready availability. Triphenylvinyltin and phenyllithium undergo a similar exchange reaction; however, with the corresponding silane and germane, phenyllithium adds across the double bond instead. The differences in the reactions of the vinyl derivatives of the Group IVb metals have been reviewed by Seyferth (101). Vinyllithium has also been prepared from metallic lithium (43,48) by a similar type of reaction:

$$(CH_2{=}CH)_4Pb + 4Li \rightarrow 4CH_2{=}CHLi + Pb$$

However, this type of reaction does not go smoothly, probably because it is heterogeneous.

Just as lithium cleaves vinyl groups from tetravinyllead, sodium in liquid ammonia has recently been demonstrated to be capable of the same type of action by Holliday and Pendlebury (41). However, the general reaction in the case of sodium in liquid ammonia, both with tetravinyllead and the tri- and divinyllead chlorides, tends to take place according to:

$$R_xPbCl_{4-x} + e_v^- + NH_3 \rightarrow R_{x-1}Pb + RH + NH_2^-$$

Unless long reaction times are used, all vinyl groups are not removed, and if insufficient sodium is added to achieve complete cleavage, some tetravinyllead is found in the products. The data suggest that cleavage becomes more difficult as the vinyl groups decrease to only one, and this may be due to the formation of polyanionic species $(PbR_x)_n^{n-}$, according to the authors' speculation.

In 1967, Holliday and Pendlebury (42) studied the cleavage of tetravinyllead further, using bromine, hydrogen chloride, acetic acid, propionic acid, and silver nitrate as reactants. Trivinyllead salts resulted in most cases when the cleavage was controlled, except with hydrogen chloride, from which either the trivinyl- or the divinyllead salt could be produced readily. More drastic cleavage with bromine or HBr gave lead(II) bromide, as expected.

Tetravinyllead has been used by Maier (75) to synthesize monovinylhalophosphines, -arsines, -stibines, and -bismuthines in good yields by direct vinylation reactions, according to the equation:

$$(CH_2{=}CH)_4Pb + 2MX_3 \xrightarrow{CCl_4} (CH_2{=}CH)_2PbX_2 + 2CH_2{=}CHMX_2$$

where M = P, As, Sb, or Bi, and X = Cl or Br. These reactions are similar to those of tetraethyllead with these metal halides.

In a physical-chemical investigation of vinyl lead compounds, Henry and Noltes (37) examined the infrared spectra of triphenylvinyllead and triphenylallyllead, in a series of Group IVb metal compounds. A characteristic peak was observed for the vinyl group, with the peak shifting, as in other organometal compounds, to higher wave lengths as the mass of the metal atom increased. The spectra of vinyl compounds of some Group IVb (and Group Vb) metals have also been studied by Chumaevskii (11), but lead was not included in the study.

The nuclear magnetic resonance spectra of tetravinyllead and many other compounds have been considered by Brügel et al. (10). Also, the long-range metal–proton coupling constants for Pb^{207} in this compound (and other vinyl–metal compounds) have been measured by Clawley and Danyluk (12). A linear correlation of coupling constants with atomic number was observed for the Group IVb metals. The *gem* metal proton spin-spin coupling constants were relatively large.

4. POLYMERIZATION OF VINYL AND STYRENYL LEAD COMPOUNDS

As suggested by the preceding discussion of unsaturated lead compounds, a number of groups in several countries have been studying unsaturated lead compounds in the past few years primarily in order to investigate their usefulness in polymerization. Perhaps the most intensive work has been carried out in the USSR, the Netherlands, and the U.S.; much of this investigation has been concentrated on *p*-vinylphenyl (or styrenyl) compounds, whose polymerization gives chains with pendant lead atoms, and these studies are made possible by the recent availability of *p*-styrenyl Grignard reagents through the work of Leebrick and Ramsden (70). A further discussion of organolead polymer uses is given in Chapter XVII.

The organolead polymers have been very briefly reviewed by Wojnarowski (114) in 1962, together with similar tin and germanium products. The organopolymers of the Group IVb elements have also been reviewed briefly by Ingham and Gilman (44) in 1962, and very briefly by Braun (9), who covered macromolecular metal–organic compounds more broadly. Henry and Davidson (36) reviewed Group IVb-type polymers more fully in 1965.

As indicated above, true vinyllead compounds (C═C—Pb bonding) have never been homopolymerized. However, copolymerizations of these compounds have been carried out successfully. Korshak et al. (57) showed that both diethyldivinyllead and triethylvinyllead decompose to lead metal under the action of peroxide polymerization catalysts at 120–130°. However, with either styrene or α-methylstyrene present, copolymerization products containing 4.5–6% lead were obtained at high pressures.

It is much easier to homopolymerize the styrenyllead compounds than to copolymerize the simple vinyl compounds. In 1959, Koton and his colleagues (62) reported, without giving details, on the synthesis of triphenyl-*p*-styrenyllead, a compound which was subsequently discussed at the 1960 IUPAC Symposium on Macromolecular Chemistry in Moscow (65). The styrenyl (or *p*-vinylphenyl) compound was prepared from triphenyllead chloride and *p*-styrenylmagnesium chloride.

$$Ph_3PbCl + p\text{-}CH_2{=}CHC_6H_4MgCl \longrightarrow Ph_3PbC_6H_4CH{=}CH_2$$

The compound melted at 105–107° (59) (initially reported as melting at 86–88° and 88–90°), and it homopolymerized readily; in fact, the triphenyl-*p*-styrenyllead (and its tin analog) polymerized more readily than styrene itself. It also copolymerized at 60° with styrene or methyl methacrylate in conversions of $<25\%$ to yield transparent and plastic films, using 2,2′-azo-bis-isobutyronitrile initiator.

In 1960, the synthesis and polymerization of this same compound, as well as triethyl-*p*-styrenyllead, was described by Pars and co-workers (91). In what has become perhaps the most generally used method of synthesis of styrenyl-lead compounds, they prepared these compounds in a modification of the Koton method by reacting the appropriate R_3PbCl compound with styrenylmagnesium chloride in tetrahydrofuran. $Ph_3PbC_6H_4CH{=}CH_2$, a white crystalline solid, shows an infrared spectrum peak at 1060 cm^{-1}, the characteristic phenyl vibrational absorption shift due to perturbation by lead, according to Noltes and Henry (85). Its ultraviolet spectrum showed λ_{max}^{EtOH} 248 mμ (ε 30,800). The monomer was shown to be quite soluble in organic solvents, in marked contrast to tetraphenyllead. Polymerization of the monomer was accomplished with *tert*-butyl hydroperoxide initiator.

In contrast to triphenyl-*p*-styrenyllead, isolation of triethyl-*p*-styrenyllead was accomplished only with difficulty, owing to a tendency toward spontaneous polymerization during distillation. The liquid monomer was found to be stable at room temperature in the presence of an inhibitor, but polymerizable in the absence of an inhibitor (or by exposure to ultraviolet light in the presence of inhibitor) to a hard, transparent polymer.

Also, in 1960, Noltes, Budding, and van der Kerk (83) studied triphenyl-*p*-styrenyllead, and diphenyldi-*p*-styrenyllead, as well as other members of a series of Group IVb metal compounds. Moreover, they investigated the synthesis and polymerization of the *p*-trimethyllead derivatives of styrene and α-methylstyrene (84). These compounds polymerized easily in the presence of 2,2′-azo-bis-isobutyronitrile at 70–90° to give approximately 60% yields of clear, colorless solids. The structures were confirmed to be organolead substituted polystyrenes by their infrared spectra. Surprisingly, the rates of polymerization decreased in the order Pb > Si > C > Ge > Sn. The unusually fast rate of polymerization of the lead compounds was attributed to the occurrence of homolytic lead–carbon bond cleavage during polymerization, thus providing "extra" chain initiators. This assumption would also account for the observed lower solubility of the lead polymers because of crosslinking of the chains. Good films could not be prepared from the lead-containing polymers.

Further, in 1960, Korshak and his group (57) studied several of the compounds mentioned above, and also the previously unknown triethyl-*p*-(α-methylstyrenyl)lead (58,93). This compound was prepared by the following reaction scheme:

$$BrC_6H_4COCH_3 \xrightarrow[Et_2O]{CH_3MgI} BrC_6H_4C(CH_3)_2OH \xrightarrow[Al_2O_3,\ 325^\circ]{-H_2O}$$

$$BrC_6H_4C(CH_3){=}CH_2 \xrightarrow[THF]{Mg,\ Et_3PbCl} Et_3PbC_6H_4C(CH_3){=}CH_2$$

The α-methylstyrenyl compound was polymerized at 80° for 6 hr under 6000 atm pressure with 2,2′-azo-bis-isobutyronitrile as catalyst. The polymer was obtained in 50% yield, with partial thermal destruction during polymerization. It was a colorless, hard material, soluble in benzene and toluene, with an intrinsic viscosity in toluene of $[\eta] = 0.46$. If the monomer was not purified before polymerization, a rubbery material was obtained instead. This demonstration of rubbery qualities may possibly be significant for future investigations in the organometallic polymer field. The corresponding germanium-containing polymer was more stable, but the tin polymer was less stable.

The Utrecht group (73) reported in 1965 on a study of both the trimethyl-*p*-styrenyl- and trimethyl-*p*-α-methylstyrenyllead compounds (as well as of the corresponding derivatives of the other Group IVb elements). The homopolymerization products at 70–90°, using 2,2′-azo-bis-isobutyronitrile initiator, were clear, colorless solids which swelled rather than dissolved in chloroform and aromatic solvents, provided the molecular weight was sufficiently high. The rate of polymerization was dependent on the nature of the metal atom, the sequence being Pb > Si > C > Ge > Sn, but these may be partly the same data reported by Noltes, Budding, and van der Kerk (84) in 1960.

In 1965, Sandler et al. (99) restudied the copolymerization of triphenyl-*p*-styrenyllead, as well as of the corresponding tin compound, with both styrene and vinylstyrene. Copolymers containing mole fractions of organometallic as low as 10–30% were found to be soluble in benzene. A higher content of organometallic monomer gave crosslinked copolymers even at low conversions. These were not soluble in any organic solvents tested, but swelled to a great extent in benzene, suggesting a loose network. Small amounts of the organometallics in the copolymerization with styrene indicated that these compounds are better accelerators than either tetraphenyllead or tetraphenyltin. The pure triphenyl-*p*-styrenyllead and tin compounds, when homopolymerized with the usual azo-bis-isobutyronitrile initiator, gave almost identical time versus dilatometer height curves, indicating that the polymerization rates were similar.

Van der Kerk's group also prepared styrenyl polymers of lead in which tin was present as a second metal (72,86,87). These were prepared by the addition of organotin hydride to a styrenyllead moiety:

$$Ph_2SnH_2 + (p\text{-}CH_2{=}CHC_6H_4)_2PbPh_2 \longrightarrow (\text{—}Ph_2SnCH_2CH_2C_6H_4PbPh_2)C_6H_4CH_2CH_2\text{—})_n$$

$$p\text{-}HSn(R_2)C_6H_4\text{—}Sn(R_2)H + (p\text{-}CH_2{=}CHC_6H_4)_2PbPh_2 \longrightarrow (\text{—}(R_2)SnC_6H_4Sn(R_2)CH_2CH_2C_6H_4Pb(Ph)_2C_6H_4CH_2CH_2\text{—})_n$$

The properties of these bimetallic polymers and related lower molecular weight derivatives are discussed in Chapter XV.

Some investigation has been made of tetra-*p*-styrenyllead (14,15,97). In 1960, Drefahl et al. (15) described in a brief note the compounds formed by reaction of *p*-bromostyrenyl Grignard reagent and lead tetraacetate or ammonium hexachloroplumbate. The products from the hexachloroplumbate took the form of mixtures of Pb(*p*-styrenyl)$_n$, where $n = 2$, 3, or 4. The product of the lead tetraacetate reaction was easily polymerizable to yellow polymers; these were soluble in organic solvents and exhibited fluorescence in solution.

Monomeric tetrakis(*p*-vinylphenyl)lead (or tetra-*p*-styrenyllead) was isolated by Drefahl and Lorenz (14) in 1964. It was prepared, together with the corresponding tin and mercury compounds, by means of a Wittig reaction between the metal-substituted *p*-benzaldehyde compound and triphenylphosphinemethylene. The lead compound was colorless and showed characteristic vinyl absorption bands in the infrared region. It polymerized readily, both as a homopolymer and as a copolymer with styrene. The starting material, tetra-*p*-benzaldehydelead, $(p\text{-OHCC}_6\text{H}_4)_4\text{Pb}$, was synthesized in 75% yield by the reaction of the tetra-*p*-benzaldehyde-ethyleneacetallead in tetrahydrofuran and water with toluenesulfonic acid. Tetra-*p*-benzaldehydeethyleneacetallead,

$$\left(\begin{array}{l} \text{H}_2\text{C}{-}\text{O} \\ \;\;|\qquad\quad \text{C}\overset{\text{H}}{}{-}\text{C}_6\text{H}_4 \\ \text{H}_2\text{C}{-}\text{O} \end{array}\right)_4\text{Pb}$$

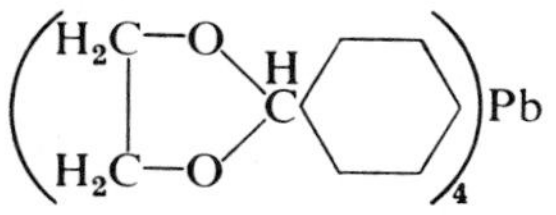

was synthesized by reacting magnesium and *p*-bromobenzaldehyde-ethyleneacetal in tetrahydrofuran to form a Grignard reagent and then reacting this with ammonium hexachloroplumbate; the lead compound was obtained in 28% yield.

To conclude this discussion of the polymerization of styrenyllead compounds, it should be mentioned that styrene has been shown to incorporate a considerable amount of tetraphenyllead in a certain type of copolymerization. Presumably, this involves a free radical dissociation of some kind. Baroni and his associates (4) reported in 1964 that the solubility of tetraphenyllead in styrene at 20° is only 0.5%, but at 140° is as high as 46% by weight. When a solution of 30% of tetraphenyllead in styrene was polymerized at 145° to 160° for 10 hr, a transparent, light- and air-stable polymer resulted. This polymer displayed satisfactory mechanical properties. However, similar polymerization at 200° yielded a dark, non-uniform mass.

5. POLYMERIZATION OF ACRYLATE-TYPE AND OTHER UNSATURATED LEAD COMPOUNDS

Unsaturated esters containing no metal atom have a long and important history in polymerization chemistry, as for example, in the history of methyl methacrylate and its polymers. In organolead chemistry also, some unusual polyester compounds have been reported as a result of the interest in organolead polymerizations.

Langkammerer (69) patented some Group IVb metal derivatives of α,β-unsaturated acids as early as 1941, and included the reaction of trimethyllead bromide with potassium methacrylate, which gave trimethyllead methacrylate. The claim was made that these compounds can be homopolymerized, or polymerized in admixture with other monomers, such as styrene, to give products that are useful for films, coatings, impregnating materials, and as molding compounds.

Following this, Saunders and Stacey (100) prepared tripropyllead acrylate in 1949, and Panov and Kocheshkov (89,90) synthesized diphenyllead dimethacrylate and *p*-tolyllead trimethacrylate in 1952, but no serious early effort was made to polymerize these compounds. However, in the period 1958–1961, Koton and his co-workers (60,63,66), and also Kochkin et al. (52–54), published several papers on the preparation and polymerization of organolead acrylates and methacrylates. The compounds prepared by Koton included triphenyllead acrylate and methacrylate, obtained in a generally useful method by treating the appropriate acid with the triphenyllead hydroxide:

$$R_3PbOH + R'CH{=}CR''COOH \rightarrow R'CH{=}CR''COOPbR_3$$

In a similar manner, Kochkin prepared the diphenyllead dimethacrylate in 60% yield, as well as the triphenyllead methacrylate in 54% yield. The compounds polymerized easily to solid polymers in the presence of benzoyl peroxide or azo-bis-isobutyronitrile, and also copolymerized easily with methyl methacrylate. The triphenyllead methacrylate was shown by the Koton group (64) to copolymerize more readily than the corresponding tin compound, and to be more stable thermally than the corresponding mercury compound. The organolead compounds were copolymerized with styrene and methyl methacrylate to yield mechanically strong polymers (108). It was also shown by the Kochkin group (55,107) that the introduction of organolead methacrylates into vinyl chloride compositions significantly increased the resistance of the polyvinyl chloride to heat and light.

Some additional patent literature has also appeared on the organolead acrylates and methacrylates. A 1960 Russian patent (54) states that dialkyllead oxides will react with acrylic-type acids to yield the corresponding diesters:

$$R_2PbO + 2R'CH{=}CR''COOH \rightarrow (R'CH{=}CR''COO)_2PbR_2 + H_2O$$

Further, Lane (68) has claimed in a 1962 patent that diesters may be prepared from the tetrasubstituted lead compounds by acid cleavage:

$$R_4Pb + 2R'CH{=}CR''COOH \rightarrow R_2Pb(OOCR''C{=}CHR')_2$$

In 1960, Fuji Chemical Co. (20) obtained coverage in a Japanese patent on lead polymethacrylate and acrylate types of compounds as radioactive shielding materials.

Non-acrylate types of organolead esters have also been prepared and polymerized. Koton and Kiseleva (61) synthesized the triphenyllead ester of *p*-vinylbenzoic acid in 52% yield by means of the following reactions:

$$p\text{-}ClMgC_6H_4CH{=}CH_2 \xrightarrow{CO_2,\ H_2SO_4} p\text{-}CH_2{=}CHC_6H_4COOH$$

$$p\text{-}CH_2{=}CHC_6H_4COOH + Ph_3PbOH \rightarrow p\text{-}CH_2{=}CHC_6H_4COOPbPh_3$$

The ester polymerized at 100° to at least 75% of a yellowish insoluble polymer, when reaction was initiated with 2,2′-azo-bis-isobutyronitrile. However, the rate of polymerization and thermal stability of the polymer in the temperature range 150–300° were inferior to the analogous products containing mercury, tin, or antimony. The lead-containing ester monomer decomposed on being heated to 150–250°, disproportionating to tetraphenyllead. On cleavage with alcoholic HCl, the ester yielded triphenyllead chloride.

Also, in 1962, the compound $p\text{-}CH_2{=}C(CH_3)CONHC_6H_4COOPbPh_3$ was prepared in 58% yield by Kiseleva et al. (51) from triphenyllead hydroxide. The compound was polymerized to a slightly colored, infusible polyester in the presence of 2,2′-azo-bis-isobutyronitrile, but both the ease of polymerization and the thermal stability of the polymer were inferior in comparisons with the corresponding tin and mercury compounds.

Jones and Mital (45) patented copolymers prepared from an unsaturated aliphatic acid or anhydride with a polyalkenyl compound containing silicon, tin, germanium, or lead. These copolymers are claimed to be useful in printing inks, household polishes, and as thickening agents for secondary crude oil recovery, but no examples are given for lead.

Italian, German, and French patents have been obtained by Montecatini (Natta et al.) on polymers derived from alkenyl silicon, germanium, tin, and lead compounds (78–80). The patents claim polymeric compositions from monomers of the type

$$\begin{array}{c} R' \\ | \\ R{-}Pb{-}(CH_2)_nCH{=}CH_2 \\ | \\ R'' \end{array}$$

in which n is 0, 1, 2, 3, or 4, and R, R′, and R″ are aliphatic, cycloaliphatic, or aromatic groups, prepared with such catalysts as $TiCl_3$, $TiCl_4$, or VCl_4 with R_3Al. However, no examples are given for lead in the specifications. For the other metals, for which examples are given, linear, highly crystalline, isotactic polymers were obtained. Various copolymers were also prepared.

REFERENCES

1. Austin, P. R., *J. Am. Chem. Soc.*, **53,** 3514 (1931).
2. Ballinger, P., U.S. Pat. 3,206,490 (to California Research Corp.), Sept. 14, 1965.
3. Ballinger, P., U.S. Pat. 3,185,553 (to California Research Corp.), May 25, 1965.
4. Baroni, E. E., S. F. Kilin, T. N. Lebsadze, I. M. Rozman, and V. M. Shonuja, *At. Energ. USSR*, **17**(6), 497 (1964); through *C.A.*, **62,** 10605 (1965).
5. Bartocha, B., U.S. Pat. 3,100,217 (Aug. 6, 1963).
6. Bartocha, B., C. M. Douglas, and M. Y. Gray, *Z. Naturforsch.*, **14b,** 809 (1959).
7. Bartocha, B., and M. Y. Gray, *Z. Naturforsch.*, **14b,** 350 (1959).
8. Beermann, C., and H. Hartmann, *Z. anorg. allgem. Chem.*, **276,** 20 (1954).
9. Braun, D., *Angew. Chem.*, **73,** 197 (1961).
10. Brügel, W., T. Ankel, and F. Krückeberg, *Z. Elektrochem.*, **64,** 1121 (1960); through *C.A.*, **55,** 10075 (1961).
11. Chumaevskii, N. A., *Tr. Komis, po Spektroskopii, Akad. Nauk SSSR*, **3**(1), 84 (1964); through *C.A.*, **64,** 10592 (1966).
12. Clawley, S., and S. S. Danyluk, *J. Phys. Chem.*, **68,** 1240 (1964).
13. Davidson, W. E., and M. C. Henry, *Chem. Rev.*, **67,** 73 (1967).
14. Drefahl, G., and D. Lorenz, *J. prakt. Chem.*, **24,** 106 (1964).
15. Drefahl, G., G. Plötner, and D. Lorenz, *Angew. Chem.*, **72,** 454 (1960).
16. Durand, J. F., *Comp. rend.*, **177,** 693 (1923).
17. Ethyl Corp., Brit. Pat. 876,008, Aug. 30, 1961.
18. Ethyl Corp., Netherlands Pat. Application 6,505,907, Nov. 12, 1965; *C.A.*, **64,** 14005 (1966).
19. Ethyl Corp., Netherlands Pat. Application 6,507,727, Dec. 23, 1965; *C.A.*, **64,** 17640 (1966).
20. Fuji Chemical Co., Japan. Pat. 2360(60) (by T. Nishimura and H. Sato); *C.A.*, **55,** 5186 (1961).
21. Gilman, H., and E. Bindschadler, *J. Org. Chem.*, **18,** 1675 (1953).
22. Gilman, H., and J. Eisch, *J. Org. Chem.*, **20,** 763 (1955).
23. Gilman, H., J. Robinson, and C. G. Stuckwisch, Unpublished Data.
24. Gilman, H., E. B. Towne, and H. C. Jones, *J. Am. Chem. Soc.*, **55,** 4689 (1933).
25. Glockling, F., *J. Chem. Soc.*, **1955,** 716.
26. Glockling, F., *J. Chem. Soc.*, **1956,** 3640.
27. Glockling, F., and D. Kingston, *J. Chem. Soc.*, **1959,** 3001.
28. Grüttner, G., and E. Krause, *Ann.*, **415,** 338 (1918).
29. Hartmann, H., Ger. Pat. 1,061,322, July 16, 1959.
30. Hartmann, H., Ger. Pat. 1,062,244, July 30, 1959.
31. Hartmann, H., E. Dietz, K. Komorniczyk, and W. Reiss, *Naturwiss.*, **48,** 570 (1961).
32. Hartmann, H., and M. K. El A'ssar, *Naturwiss.*, **52,** 304 (1965).
33. Hartmann, H., and W. Eschenbach, *Naturwiss.*, **46,** 321 (1959).
34. Hartmann, H., and K. Komorniczyk, *Naturwiss.*, **51,** 214 (1964).
35. Hartmann, H., and K. Meyer, *Naturwiss.*, **52,** 303 (1965).
36. Henry, M. C., and W. E. Davidson, *Annals N.Y. Acad. Sci.*, **125,** 172 (1965).
37. Henry, M. C., and J. G. Noltes, *J. Am. Chem. Soc.*, **82,** 555 (1960).
38. Henry, M. C., and J. G. Noltes, *J. Am. Chem. Soc.*, **82,** 558 (1960).
39. Henry, M. C., and J. G. Noltes, *J. Am. Chem. Soc.*, **82,** 561 (1960).

40. Henry, M. C., and B. C. Pant, Quarterly Report No. 22, Project LC-28, Intern. Lead Zinc Res. Organ., New York (Dec. 31, 1966).
41. Holliday, A. K., and R. E. Pendlebury, *J. Chem. Soc.*, **1965,** 6659.
42. Holliday, A. K., and R. E. Pendlebury, *J. Organometal. Chem.*, **7,** 281 (1967).
43. Honeycutt, J. B., Jr., U.S. Pat. 3,059,036 (to Ethyl Corp.), Oct. 16, 1962.
44. Ingham, R. K., and H. Gilman, "Organopolymers of Group IV Elements," in *Inorganic Polymers*, F. G. A. Stone and W. A. G. Graham, Eds., Academic Press, New York, 1962, Chapt. 6, p. 321.
45. Jones, J. F., and A. J. Mital, Brit. Pat. 836,755 (to B. F. Goodrich Co.), June 9, 1960.
46. Juenge, E. C., U.S. Pat. 3,071,607 (to Ethyl Corp.), Jan. 1, 1963.
47. Juenge, E. C., and S. E. Cook, *J. Am. Chem. Soc.*, **81,** 3578 (1959).
48. Juenge, E. C., and D. Seyferth, *J. Org. Chem.*, **26,** 563 (1961).
49. Kaesz, H. D., and F. G. A. Stone, "Vinylmetallics," in *Organometallic Chemistry*, H. Zeiss, Ed., Reinhold, New York (1960), Chapt. 3, p. 88.
50. Kharasch, M. S., and A. L. Flenner, *J. Am. Chem. Soc.*, **54,** 674 (1932).
51. Kiseleva, T. M., M. M. Koton, and G. M. Chetyrkina, *Izv. Akad. Nauk SSSR, Otd. Khim. Nauk*, **1962,** 1798; through *C.A.*, **58,** 7962 (1963).
52. Kochkin, D. A., *Dokl. Akad. Nauk SSSR*, **135,** 857 (1960); through *C.A.*, **55,** 11341 (1961).
53. Kochkin, D. A., L. V. Luk'yanova, and E. B. Reznikova, *Zh. Obshch. Khim.*, **33,** 1945 (1963); through *C.A.*, **59,** 11551 (1963).
54. Kochkin, D. A., Y. P. Novichenko, G. I. Kuznetsova, and L. V. Laine, USSR Pat. 133,224 (Aug. 26, 1960); through *C.A.*, **55,** 11923 (1961).
55. Kochkin, D. A., S. G. Veryenyikyina, and I. P. Chyekmaryeva, *Dokl. Akad. Nauk SSSR*, **139,** 1375 (1961); through *C.A.*, **56,** 7343 (1962).
56. Komarov, N. V., and O. G. Yarosh, *Zh. Obshch. Khim.*, **37,** 264 (1967); through *C.A.*, **66,** 95126v (1967).
57. Korshak, V. V., A. M. Polyakova, and M. D. Suchkova, *Vysokomol. Soedin.*, **2,** 13 (1960); through *C.A.*, **55,** 361 (1960).
58. Korshak, V. V., A. M. Polyakova, and E. S. Tambovtseva, *Vysokomol. Soedin.*, **1,** 1021 (1959); through *J. Polymer Sci.*, **42,** 284 (1960); *C.A.*, **54,** 22438 (1960).
59. Koton, M. M., and L. F. Dokukina, *Vysokomol. Soedin.*, **6,** 1791 (1964); through *C.A.*, **62,** 282 (1965).
60. Koton, M. M., and F. S. Florinskii, *Zh. Obshch. Khim.*, **32,** 3057 (1962); through *C.A.*, **55,** 11341 (1961).
61. Koton, M. M., and T. M. Kiseleva, *Izvest. Akad. Nauk SSSR, Otdel. Khim. Nauk*, **1961,** 1783; through *C.A.*, **56,** 8740 (1962).
62. Koton, M. M., T. M. Kiseleva, and F. S. Florinskii, *Izvest. Akad. Nauk. SSSR, Otdel. Khim. Nauk*, **1959,** 948; through *C.A.*, **54,** 1378 (1960).
63. Koton, M. M., T. M. Kiseleva, and F. S. Florinskii, *Vysokomolekul. Soedin.*, **2,** 1639 (1960); through *C.A.*, **55,** 26522 (1961).
64. Koton, M. M., T. M. Kiseleva, and F. S. Florinskii, *Mezhdunarod. Simpozium po Makromol. Khim., Dokl.*, Moscow, 1960, Sektsiya, **1,** 167 (1960); through *C.A.*, **55,** 7272 (1961).
65. Koton, M. M., T. M. Kiseleva, and F. S. Florinskii, *Izvest. Akad. Nauk SSSR, Otdel. Khim. Nauk*, **1959,** 948; through *C.A.*, **54,** 1378 (1960).
66. Koton, M. M., T. M. Kiseleva, and F. S. Florinskii, *J. Polymer Sci.*, **52,** 237 (1961); through *C.A.*, **54,** 22436 (1960).

67. Koton, M. M., T. M. Kiseleva, and N. P. Zapevalova, *Zh. Obshch. Khim.*, **30,** 186 (1960); through *C.A.*, **54,** 22436 (1960).
68. Lane, E. S., Brit. Pat. 907,775 (Oct. 10, 1962).
69. Langkammerer, C. M., U.S. Pat. 2,253,128 (to E. I. du Pont de Nemours & Co.) (Aug. 19, 1941).
70. Leebrick, J. R., and H. E. Ramsden, *J. Org. Chem.*, **23,** 935 (1958).
71. Leusink, A. J., and G. J. M. van der Kerk, *Rec. trav. chim.*, **84,** 1617 (1965).
72. Leusink, A. J., J. G. Noltes, H. A. Budding, and G. J. M. van der Kerk, *Rec, trav. chim.*, **83,** 609 (1964).
73. Leusink, A. J., J. G. Noltes, H. A. Budding, and G. J. M. van der Kerk, AFML Report TR-65-192, R & TD, Air Force Systems Command, WPAFB, Ohio (Dec., 1965).
74. Maier, L., *Angew. Chem.*, **71,** 161 (1959).
75. Maier, L., *Tetrahedron Letters*, **1959,** No. 6, 1.
76. Masson, J.-C., and P. Cadiot, *Bull. Soc. Chim. France*, **1965,** 3518.
77. Masson, J.-C., M. L. Quan, W. Chodkiewicz, and P. Cadiot, *Compt. rend.*, **257,** 1111 (1963).
78. "Montecatini" Societa generale per l'Industria mineraria e chimica, Ital. Pat. 589,299, March 4, 1959.
79. "Montecatini" Societa generale per l'Industria mineraria e chimica, Ger. Pat. 1,091,760, Oct. 27, 1960.
80. "Montecatini" Societa generale per l'Industria mineraria e chimica, French Pat. 1,217,343, 1961.
81. Nesmeyanov, A. N., *Bull. Soc. Chim. France*, **1946,** 569.
82. Nesmeyanov, A. N., and A. E. Borisov, *Tetrahedron*, **1,** 158 (1957).
83. Noltes, J. G., H. A. Budding, and G. J. M. van der Kerk, *Rec. trav. chim.*, **79,** 408 (1960).
84. Noltes, J. G., H. A. Budding, and G. J. M. van der Kerk, *Rec. trav. chim.*, **79,** 1076 (1960).
85. Noltes, J. G., and M. C. Henry, *Chem. Ind.* (*London*), **1959,** 298.
86. Noltes, J. G., and G. J. M. van der Kerk, *Rec. trav. chim.*, **80,** 623 (1961).
87. Noltes, J. G., and G. J. M. van der Kerk, *Chimia*, **16,** 122 (1962).
88. Normant, H., *Bull. Soc. Chim. France*, **1957,** 728.
89. Panov, E. M., and K. A. Kocheshkov, *Dokl. Akad. Nauk SSSR*, **85,** 1037 (1952); through *C.A.*, **47,** 6365 (1953).
90. Panov, E. M., and K. A. Kocheshkov, *Zh. Obshch. Khim.*, **25,** 489 (1955); through *C.A.*, **50,** 3271 (1956).
91. Pars, H. G., W. A. G. Graham, E. R. Atkinson, and C. R. Morgan, *Chem. Ind.* (London), **1960,** 693.
92. Petrov, A. A., and V. B. Lebedev, *Zh. Obshch. Khim.*, **32,** 657 (1962); through *C.A.*, **57,** 14612 (1962).
93. Polyakova, A. M., V. V. Korshak, and E. S. Tambovtseva, *Vysokomol. Soedin.*, **3,** 662 (1961); through *C.A.*, **55,** 27953 (1961).
94. Puchinyan, E. A., and Z. M. Manulkin, *Dokl. Akad. Nauk Uz. SSR*, **19**(3), 47 (1962); through *C.A.*, **57,** 13788 (1962).
95. Puchinyan, E. A., and Z. M. Manulkin, *Tr. Tashkentsk. Formatsevt. Inst.*, **3,** 427 (1962); through *C.A.*, **61,** 677 (1964).
96. Ramsden, H. E., Brit. Pat. 824,944 (to Metal & Thermit Corp.), Dec. 9, 1959.
97. Ramsden, H. E., U.S. Pat. 3,109,851 (to M & T Chemicals), Nov. 5, 1963.

98. Ramsden, H. E., and H. F. Shannon, U.S. Pat. 3,156,716 (to Esso Research & Engineering Co.), Nov. 10, 1964.
99. Sandler, S. R., J. Dannin, and K. C. Tsou, *J. Polymer Sci., A.*, **3,** 3199 (1956).
100. Saunders, B. C., and G. J. Stacey, *J. Chem. Soc.*, **1949,** 919.
101. Seyferth, D., "Vinyl Compounds of Metals," in *Progress in Inorganic Chemistry*, F. A. Cotton, Ed., Interscience, New York, 1961, Vol. III.
102. Seyferth, D., and M. A. Weiner, *J. Am. Chem. Soc.*, **83,** 3583 (1961).
103. Seyferth, D., and M. A. Weiner, *J. Am. Chem. Soc.*, **84,** 361 (1962).
104. Seyferth, D., and M. A. Weiner, U.S. Pat. 3,085,120 (to Ethyl Corp.), April 9, 1963.
105. Shapiro, H., Unpublished Data.
106. Simonnin, M.-P., *J. Organometal. Chem.*, **5,** 155 (1966).
107. Shostakovskii, M. F., V. N. Kotrelev, D. A. Kochkin, G. I. Kuznetsova, S. P. Kalinina, and V. V. Borisenko, *Zhur. Prikladi Khim.*, **31,** 1434 (1958), through *C.A.*, **53,** 3040 (1959).
108. Shostakovskii, M. F., V. N. Kotrelev, D. A. Kochkin, G. I. Kuznetsova, I. P. Kalinina, L. V. Laine, A. I. Borisova, and V. V. Borisenko, *Dokl. Na Mezdunarod. Simpoziumye Makromol. Khim., Moskve, 1960*, **52,** 223 (1961).
109. Steingross, W., and W. Zeil, *J. Organometal. Chem.*, **6,** 109 (1966).
110. Sterlin, R. N., S. S. Dubov, W. Li, L. P. Vakhomchik, and I. L. Knunyants, *Obshchestva im D.I. Mendeleeva*, **6,** 110 (1961); through *C.A.*, **55,** 15336 (1961).
111. Vijayarahavan, K. V., *J. Indian. Chem. Soc.*, **22,** 227 (1945).
112. Willemsens, L. C., Report No. 23, July 1–Sept. 30, 1965, Project No. LC-18, Intern. Lead Zinc Res. Organ., New York, N.Y.
113. Willemsens, L. C., and G. J. M. van der Kerk, *Investigations in the Field of Organolead Chemistry*, Intern. Lead Zinc Res. Organ., New York 1965.
114. Wojnarowski, T., *Polimery*, **7**(5), 165 (1962); through *C.A.*, **58,** 3509 (1963).
115. Zavgorodnii, V. S., and A. A. Petrov, *Dokl. Akad. Nauk SSSR*, **143,** 855 (1962); through *C.A.*, **57,** 3466 (1962).

XI Organolead salts and related compounds

1. TYPES OF ORGANOLEAD IONS

The organolead compounds discussed thus far contain exclusively lead–carbon or lead–lead bonds. However, the largest class of organolead compounds is that in which one or more of the lead bonds is to an anionic moiety other than carbon and is represented by the generic formulas R_3PbX, R_2PbX_2, and $RPbX_3$, in which X is a group containing a bonding element from Groups V, VI, or VII of the periodic table. Stable derivatives of the type XR_2PbPbR_3 derived from hexaorganodilead are unknown, although they have been postulated as intermediates in reactions of R_6Pb_2 with halogen acids and metal halides. Also, divalent lead compounds of the type RPbX have not been isolated.

The first compound of the type R_3PbX was prepared by Löwig (246) in 1853. Bis(triethyllead) carbonate was obtained when an alcoholic solution of the product from the sodium-lead alloy–ethyl iodide reaction was allowed to evaporate in air; by treatment of the carbonate (or hydroxide) with the appropriate acid, Löwig also prepared triethyllead sulfate, nitrate chloride, bromide, and iodide. In 1860, Klippel (188) prepared triethyllead formate, acetate, butyrate, benzoate, tartrate, oxalate, cyanide, and cyanate. A large number of such compounds is now known. Leeper, Summers, and Gilman (227) estimated that some 300 or 400 R_3PbX and R_2PbX_2 compounds were known in 1954, about half of which were halide derivatives. Of greater importance than the mere number of compounds synthesized, however, is the fact that several new types of compounds have been synthesized in recent years. These newer derivatives include organolead hydrides (R_3PbH), oxides $(R_3Pb)_2O$, alkoxides, peroxides, azides, amides (R_3PbNR_2), phosphides (R_3PbPR_2), selenides, and tellurides, to mention a few.

These compounds range from thermally stable, high-melting crystalline solids, e.g., the halides and salts of oxyacids, to unstable, air- and water-sensitive derivatives, e.g., the hydrides and phosphides. Many of these compounds have not been fully characterized. Most of the physical property measurements have been made using the halide and carboxylate salts, but even here the data are inconclusive. Based on dipole moment measurements, the ionic character of the lead–halogen bond in such compounds as trimethyllead chloride, triethyllead bromide, and triphenyllead chloride is considered to approximate that of the lead-to-halogen bond in the divalent lead halides (351). The high degree of ionic character of the organolead halides is indisputable in view of the large number of metathesis reactions they undergo, analogous to those of the divalent lead salts, and which are very useful in the synthesis of other R_3PbX and R_2PbX_2 derivatives (where X is a nonhalide). Yet, triphenyllead chloride does not exhibit any conductivity in dimethylformamide or pyridine solution (345,346), nor do solutions of triethyllead chloride and trimethyllead chloride in tetramethylsulfoxide and *N,N*-dimethylacetamide, although they do form 1:1 complexes with these solvents (254). The infrared spectra of trimethyllead acetate, formate, and monochloroacetate have been interpreted to indicate that, even in the solid state, their structure consists of a planar trimethyllead cation associated with a carboxylate anion (276). However, this latter conclusion has been challenged on the basis of the infrared spectra of solutions of these compounds (181).

No firm conclusions can be drawn concerning the exact nature (i.e., ionic versus covalent) of the Pb–X bond in these R_3PbX compounds. The nature of the bond will be strongly dependent on the nature of the X group. When X is an anion of a strong acid, the R_3PbX and R_2PbX_2 compounds should possess a high degree of ionic character although they may not exist as discrete R_3Pb^+ and X^- ions in the solid state or in nonpolar solvents. Certainly in water solutions, and in other highly polar solvents, the halide and hydroxide derivatives exist as solvated cations. On the other hand, the properties of R_3PbX derivatives of very weak acids, e.g., hydrides, phosphides, and amides, indicate that they are largely covalent compounds having little, if any, ionic character.

A second class of organolead "salts" of the type R_3PbM (and R_2PbM_2) is known in which lead is bonded to an alkali or alkaline earth metal. These compounds are even less well characterized than the R_3PbX and R_2PbX_2 derivatives discussed above. They are best represented as being complexes of diorganolead with another organometal moiety, such as an organosodium or organolithium compound. They are postulated as being highly ionic in character, the charge distribution being $R_3Pb^{\delta-}\ M^{\delta+}$, so that the organolead moiety is anionic in character. Such compounds have proved useful for the synthesis of various unsymmetrical tetraorganolead compounds.

The preparation and properties of both classes of compounds are discussed separately below.

2. ORGANOLEAD HYDRIDES, R_3PbH, R_2PbH_2

It is clear from our knowledge of the inorganic Group IVb metal hydrides that the lead–hydrogen bond is not as stable as the metal–hydrogen bonds of the lighter metals in this group; for example, a comparison of silane and germane with plumbane, which cannot be isolated, makes this obvious. In addition, organosilicon and organotin hydrides have also been well known for some time, but organolead hydrides were isolated for the first time as recently as 1959 by Duffy and Holliday (86–88). Numerous prior attempts had been made to prepare such compounds without success (93,102,179,294,359,375,396). Since 1959, despite an expansion of this pioneer work on trialkyllead hydrides, no aryllead hydrides or monoalkyllead trihydrides have been isolated.

Holliday and his co-workers systematically studied reactions that might lead to alkyllead borohydrides [see also Holliday and Jeffers (170)], and they investigated the reaction of trimethyllead chloride and potassium borohydride in liquid ammonia. After reaction at low temperature, a volatile compound was distilled off at $-5°$. This material proved to be trimethyllead hydride, a compound melting at $-104°$ and believed at first to decompose slowly above $-100°$ to a red solid, possibly pentamethyldilead hydride.

$$Me_3PbCl + KBH_4 \xrightarrow[-33°]{NH_3} [Me_3PbBH_4 \cdot XNH_3]$$
$$\xrightarrow{-5°} Me_3PbH + H_3B \cdot NH_3 + (X - 1)NH_3$$

Since this discovery of trimethyllead hydride, an easier method of synthesis has been developed by Becker and Cook (41), who reduced the corresponding chloride with lithium aluminum hydride in dimethyl ether at $-78°$. Neumann and Kühlein (270) used the same method with diethylene glycol dimethyl ether as the solvent. This method is similar to the preparation of alkyltin hydrides (93), except that it is carried out at lower temperature, which is critical. In addition to trimethyllead hydride, triethyllead hydride and the dimethyl- and diethyllead dihydrides were prepared in an analogous manner. These compounds were shown to decompose readily at 0°, the monohydrides in accordance with the overall equation:

$$4R_3PbH \rightarrow 3R_4Pb + 2H_2 + Pb$$

Duffy, Feeny, and Holliday (86) found that at temperatures above −37° trimethyllead hydride gave hexamethyldilead as an intermediate decomposition product which subsequently decomposed to tetramethyllead and lead metal (64). This type of decomposition was not observed for triethyllead hydride. It was shown further that the hydrides react with diazoalkanes and add to ethylene in much the same way as do the other Group IVb organometal hydrides, and that they react with hydrogen chloride to yield hydrogen quantitatively. Henry (157) has discussed briefly the synthesis of organolead hydrides and their addition to olefinic and acetylenic linkages, reactions which furnish a general method of preparing organolead compounds containing a variety of reactive functional groups.

The two methyllead hydride compounds have also been described in a brief note by Amberger (11), who also employed lithium aluminum hydride reduction of the halides.

The lead–hydrogen bond appears to be polar in nature. This is supported by the nuclear magnetic resonance spectrum (79), which suggests the existence of the equilibrium:

$$Me_3PbH \rightleftarrows Me_3Pb^- + Me_3PbH_2^+$$

The proton magnetic resonance of trimethyllead hydride has also been examined by Flitcroft and Kaesz (94), who concluded that the proton bonded to lead is less shielded in this compound than the protons bonded to carbon, and this was also true for the R_3SnH and R_2SnH_2 derivatives which have been recently re-investigated by Van der Kelen et al. (369).

In 1965, Neumann and Kühlein (13,270,271) extended the syntheses of R_3PbH compounds with larger alkyl groups to the *n*-propyl, *n*-butyl, *i*-butyl, and cyclohexyl compounds, and also prepared di-*n*-butyllead dihydride. These compounds were prepared in 70–90% yields by reacting the corresponding alkyllead chlorides with lithium aluminum hydride in diethylene glycol dimethyl ether at −60°. The hydrides were shown to be good reducing agents. For example, they readily reduced alkyl and alkenyl bromides and iodides, such as benzyl chloride, ω-bromoacetophenone, allyl bromide, bromobenzene, and iodobenzene, to the respective hydrocarbon at 0°.

$$R_3PbH + R'X \rightarrow R'H + R_3PbX$$

Carbon tetrachloride similarly was reduced to chloroform or dichloromethane, and 1,2-dibromoethane was converted to ethylene. In such reactions the lead compounds appear to be more reactive than their tin analogs; and in fact, tributyllead hydride was demonstrated to be capable of reducing triethyltin chloride:

$$Bu_3PbH + Et_3SnCl \rightarrow Bu_3PbCl + Et_3SnH$$

The alkyllead hydrides are also capable of reducing other types of organic compounds, e.g., benzaldehyde to benzyl alcohol, acetyl chloride to ethyl acetate, and nitrobenzene and nitrosobenzene to azoxybenzene.

Neumann and Kühlein also investigated the scope of the hydroplumbation reaction as a source of organofunctional lead compounds by reacting tributyllead hydride with acrylonitrile, methyl acrylate, stryrene, phenylacetylene, phenyl isothiocyanate, and phenyl isocyanate at 0 to $-20°$. This reaction provides an excellent method for introducing functionality into organolead compounds. Tributyllead hydride is more stable than the lower trialkyllead hydrides and is therefore more convenient to use.

$$n\text{-}Bu_3PbH + CH_2{=}CHCN \rightarrow n\text{-}Bu_3PbCH_2CH_2CN$$

$$n\text{-}Bu_3PbH + CH_3OOCCH{=}CH_2 \rightarrow n\text{-}Bu_3PbCH_2CH_2COOCH_3$$

$$n\text{-}Bu_3PbH + C_6H_5CH{=}CH_2 \rightarrow n\text{-}Bu_3PbCH_2CH_2C_6H_5$$

$$n\text{-}Bu_3PbH + C_6H_5C{\equiv}CH \rightarrow trans\text{-}Bu_3PbCH{=}CHC_6H_5$$

$$n\text{-}Bu_3PbH + C_6H_5N{=}C{=}O \rightarrow n\text{-}Bu_3PbN(CHO)C_6H_5$$

The phenyl isocyanate product tends to react further, while *n*-octene did not react with tributyllead hydride under the usual conditions.

In the 1965–66 period, Creemers and co-workers (74) stated that reactions of the type

$$R_3PbX + R'_3SnH \rightleftharpoons R_3PbH + R'_3SnX$$

are fast equilibrium reactions, and it appeared from low temperature studies that the equilibrium is to the left if X is chlorine, bromine, alkoxide, thioalkyl, or thioaryl. In the case of trialkyllead acetates, however, the equilibrium lies to the right, thus affording still another means of preparing trialkyllead hydrides. Moreover, it was shown that adducts of triethyllead and tributyllead hydrides could be prepared from acetylenic compounds either (*1*) by mixing the trialkyllead acetate and an organotin hydride in tetrahydrofuran at $-80°$, followed by adding an unsaturated compound; or (*2*) by mixing the trialkyllead acetate and an unsaturated compound in tetrahydrofuran, followed by adding the organotin hydride at −20°, for example:

(*1*) $(C_4H_9)_3PbOOCCH_3 + (C_6H_5)_3SnH \xrightarrow[-80°]{HC\equiv CCN,\ THF}$

$$\underset{(C_4H_9)_3Pb}{\overset{H}{}}\!\!>C{=}C<\!\!\underset{CN}{\overset{H}{}} + \underset{H}{\overset{H}{}}\!\!>C{=}C<\!\!\underset{CN}{\overset{Pb(C_4H_9)_3}{}} + (C_6H_5)_3SnOOCCH_3$$

(*2*) $(C_4H_9)_3PbOOCCH_3 + HC{\equiv}CCOOCH_3 \xrightarrow[-20°]{(C_2H_5)_3SnH,\ THF}$

$$\underset{(C_4H_9)_3Pb}{\overset{H}{}}\!\!>C{=}C<\!\!\underset{COOCH_3}{\overset{H}{}} + (C_2H_5)_3SnOOCCH_3$$

In reactions with cyanoacetylene it was found that both the cis-β and the α adducts were formed, and at −20° the alkyltin hydrides were sufficiently less reactive toward the acetylenic compounds than were the lead hydrides, so that competitive reactions did not occur. At room temperature, however, the corresponding organotin hydride adducts were formed in 10–20% yield (232). Preliminary experiments have indicated that adducts may be obtained from trimethyllead hydride (233) and dibutyllead dihydride by this general method. The triethyllead hydride adduct was prepared from cyanoacetylene, while the tributyllead hydride adduct was prepared from both cyanoacetylene and methyl propiolate. The organolead hydrides prepared to date are listed in Table XI-1.

3. ORGANOLEAD HYDROXIDES, OXIDES, ALKOXIDES, AND SILANOLATES, R_3PbOH, $(R_3Pb)_2O$, $R_2Pb(OH)_2$, R_2PbO, R_3PbOR', AND $R_3PbOSiR_3'$

Although some members of the alkyl and aryllead hydroxide and oxide classes have been known for many years, some important representative compounds have been synthesized for the first time only recently, while others are still unknown. For example, triethyllead oxide and triphenyllead oxide were first prepared after 1960. The monoalkyl and aryl compounds, $RPb(OH)_3$ and $(RPb)_2O_3$, are probably unknown although Lesbre (228,231) has formulated alkylplumbonic acids in the form $RPb(OH)_3$. The pure diaryllead dihydroxides also are unknown, and the dialkyllead oxides have probably never been obtained although they are mentioned in the older literature.

Trialkyllead hydroxides have been known since 1853, when Löwig first prepared triethyllead hydroxide (246). This compound and the trimethyl-, tripropyl-, tributyl-, triisobutyl-, triisoamyl-, and tricyclohexyllead hydroxides have received much study (21,25,51,53,54,61,62, 90,116,138,143,150–153,156,182,206,214,216,218,250,255,257,260,291, 319,321).

The trialkyllead hydroxides usually are hygroscopic solids which melt with decomposition, and are quite unstable to air; consequently they must be stored under an inert atmosphere. The compounds are easily prepared, usually by reaction of the analogous halides with silver oxide or alcoholic sodium or potassium hydroxide, or, if preferred, by electrolysis of the halides according to the method of Hein and Klein (153). The alcoholic alkali method does not remove all halide unless benzene is used as a solvent (61). The metathetical reaction is exemplified by the equation:

$$R_3PbX + MOH \rightarrow R_3PbOH + MX$$

The trialkyllead hydroxides are clearly ionic in nature, being weakly basic in aqueous solution. They are thus capable of absorbing carbon dioxide from the air to form carbonates, releasing ammonia from ammonium salts, and precipitating heavy metal hydroxides from solutions of the salts. Since they are basic in character, the hydroxides have found use in the preparation of numerous alkyllead salts by metathetical reactions with acids. This use has been discussed in detail by Saunders and his associates (150,151,257,319,321) and by Calingaert, Dykstra, and Shapiro (61). Triethyllead hydroxide has also been used to synthesize a mixed Pb–Fe compound, $Et_2PbFe(CO)_4$ (145), and other lead–metal derivatives.

The polarographic reduction of triethyllead hydroxide was examined by Korshunov and Malyugina (206) in buffered solutions in the pH range 1.86–11.58. The results suggested that the reduction proceeded through the triethyllead cation.

Bis(triethyllead) oxide was probably prepared for the first time in 1960, by Aleksandrov, Brilkina, and Shushunov (6–8). This compound was also referred to by Löwig (247) and by Klippel (187,188), but these investigators apparently had prepared triethyllead hydroxide instead. Aleksandrov et al. prepared the compound by the dehydration of triethyllead hydroxide with sodium dispersed in benzene:

$$2Et_3PbOH \xrightarrow{Na} (Et_3Pb)_2O$$

Bis(triethyllead) oxide was described as a yellow-green liquid which decomposed slowly even at room temperature to give ethane, ethylene, butane, tetraethyllead, and unidentified solid products. On being heated to 70–90°, the compound decomposed quickly to give the same mixture of products. The infrared spectrum of the compound was found to give an

Table XI-1 Organolead Hydrides, Hydroxides, Oxides, Alkoxides, Siloxanes, Peroxides, and Plumbonic Acids

COMPOUND	PROPERTIES	REFERENCES
	Organolead Hydrides	
$(CH_3)_3PbH$	m.p. ~−106°	11,41,86–88,94,232,233
$(C_2H_5)_3PbH$	m.p. −145°	41,74,157,232
$(C_3H_7)_3PbH$		13,270,271
$(C_4H_9)_3PbH$		13,74,232,270,271
$(i\text{-}C_4H_9)_3PbH$		13,270,271
$(cyclo\text{-}C_6H_{11})_3PbH$		13,270,271
$(CH_3)_2PbH_2$		11,41
$(C_2H_5)_2PbH_2$		41
$(C_4H_9)_2PbH_2$		13,270,271
	Triorganolead Hydroxides	
$(CH_3)_3PbOH$		153,216,227,255
$(C_2H_5)_3PbOH$		25,51,53,54,61,62,116,138,143,150–153,156, 182,206,214,216,218,227,250,257,260,291, 319,321
$(C_3H_7)_3PbOH$	dec. 140	152,216,227,257,291,319,321
$(C_4H_9)_3PbOH$		227,291,321
$(i\text{-}C_4H_9)_3PbOH$	dec. 115	216,227
$(i\text{-}C_5H_{11})_3PbOH$	liq.	143,216,227
$(cyclo\text{-}C_6H_{11})_3PbOH$		25,214,227
$(C_6H_5)_3PbOH$		25,45,138,165,216,218,227,250,255,305, 341,377
$(p\text{-}CH_3C_6H_4)_3PbOH$		216,227

$(C_6H_5)_2(o\text{-}HOOCC_6H_4)PbOH$		216,227
$(C_6F_5)_3PbOH$	dec. 185°	163
	Diorganolead Dihydroxides	
$(C_2H_5)_2Pb(OH)_2$		5,8,61,62,151,227
$(C_4H_9)_2Pb(OH)_2$		182,227
$(C_6H_5)_2Pb(OH)_2$		202
$(C_6H_5)_2(OH)PbOOCCH(CH_3)_2$	dec. 200°	398
	Triorganolead Oxides	
$[(C_2H_5)_3Pb]_2O$	liq., dec. r.t.	6,7,8,20,25
$[(C_6H_5)_3Pb]_2O$		48,138,227,247,255
	Diorganolead Oxides	
$(CH_3)_2PbO$		139,227,334
$(C_2H_5)_2PbO$		227,247
$(i\text{-}C_3H_7)_2PbO$		135,227,355
$(C_6H_5)_2PbO$		202,227,269,297,397,398
$(C_{10}H_{21})_2PbO$		398
$[(C_6H_5)_2PbOOCCH_3]_2O$		198,285
$[(C_6H_5)_2PbOOCCH(CH_3)_2]_2O$	d. 240°	198,285
$[(C_6H_5)_2PbOOCC_3H_7]_2O$	d. 225°	398
$[(p\text{-}CH_3OC_6H_4)_2PbOOCCH_3]_2O$		398

(continued)

Table XI-1 (continued)

COMPOUND	PROPERTIES	REFERENCES
	Organolead Alkoxides	
$(CH_3)_3PbOCH_3$	m.p. 90–2°, with dec.	288
$(C_2H_5)_3PbOCH_3$	dec. 60–70°	6–8,288
$(C_2H_5)_3PbOC_2H_5$	m.p. 18°	6–8,104,227,288,372
$(C_2H_5)_3PbOC_3H_7$		288
$(C_2H_5)_3PbOC_6H_5$	b.p.$_{0.5}$ 75°	66,104,227
$(C_2H_5)_3PbOCH_2C_6H_4$		6–8
$(C_2H_5)_3PbOC(CH_3)_2C_6H_5$	oil	6–8
$(CH_3)_3PbO(C_2H_4O)_2CH_3$		128
$(C_6H_5)_3PbOC_6H_4NO$		342
$(C_6H_5)_3PbOC_6H_5$	113°	49
	Plumbosiloxanes	
$(CH_3)_3PbOSi(CH_3)_3$	b.p.$_{720}$ 172°, m.p. −1°	327–333
$(CH_3)_3PbOSi(C_2H_5)_3$	b.p. 65°	328
$(C_2H_5)_3PbOSi(CH_3)_3$	b.p.$_{10}$ 111.5°	328
$(C_2H_5)_3PbOSi(C_2H_5)_3$	b.p. 100°	328
	Triorganolead Peroxides	
$(CH_3)_3PbOOC(CH_3)_3$	m.p. 81–3°	315,353
$(CH_3)_3PbOOC(CH_3)_2C_6H_5$	m.p. 55–7°	315,353

Compound		References
Triorganolead Peroxides (continued)		
$(C_2H_5)_3PbOOC(CH_3)_3$	m.p. 34–6°	4,6,353
$(C_2H_5)_3PbOOC(CH_3)_2C_6H_5$		4,6,353
p-$[(C_2H_5)_3PbOOC(CH_3)_2]C_6H_4C(CH_3)_2OOPb(C_2H_5)_3$		4,6
$(C_6H_5)_3PbOOC(CH_3)_3$	m.p. 97–9°	312,314,353
$(C_6H_5)_3PbOOC(C_6H_5)_3$	m.p. 110–2°	312,314,353
$(C_6H_5)_3PbOOC(CH_3)_2C_6H_4$	m.p. 113–5°	312,314,353
$(C_6H_5)_3PbOOPb(C_6H_5)_3$	m.p. 166–7°	315,353
Organoplumbonic Acids		
$CH_3Pb(OH)_3$		227,228,231
$C_2H_5Pb(OH)_3$		227,228,231
$C_3H_7Pb(OH)_3$		227,228,231
i-$C_3H_7Pb(OH)_3$		227,228,231
$C_4H_9Pb(OH)_3$		227,228,231
$C_6H_5CH_2Pb(OH)_3$		227,228,231
$C_3H_5Pb(OH)_3$		227,228,231
C_6H_5PbOOH		280,281
p-$CH_3C_6H_4PbOOH$		280,281
p-IC_6H_4PbOOH		245

intense absorption band at ~630 cm^{-1} (372a). The compound could be easily hydrolyzed back to the hydroxide, or alcoholyzed to alkoxides. The alcoholysis reaction can be represented by the equation:

$$(R_3Pb)_2O + R'OH \rightarrow R_3PbOH + R_3PbOR'$$

The methyl, ethyl, benzyl, and dimethylphenyl alkoxide derivatives were prepared in this manner; these also could be hydrolyzed to the corresponding hydroxides quite readily. [There are very few older references to trialkyllead alkoxides; triethyllead ethoxide and triethyllead phenoxide are reported in the Leeper, Summers, and Gilman review (227) as having been prepared in double decomposition reactions from the corresponding sodium alkoxide.] The oxide also reacted with hydroperoxides to form alkyllead peroxides (see below). The alkyllead alkoxides and peroxides were unstable and readily hydrolyzed or decomposed.

It has long been generally understood that alcohols, like water, do not affect the carbon–lead bond in tetraorganolead compounds. However, an interesting claim with respect to a new method of preparation of organolead alkoxides has been made by Pedinelli, Magri, and Randi (288), who stated that tetramethyllead and tetraethyllead will react with alcohols at 100° to yield alkoxides:

$$R_4Pb + R'OH \rightarrow R_3PbOR' + RH$$

Tetramethyllead was reported to be more reactive than tetraethyllead, and the order of alcohol reactivity was found to be MeOH > EtOH > iso-PrOH = *n*-BuOH = 0. The authors observed under ordinary conditions the expected hydrolysis of the alkoxides:

$$R_3PbOR' + H_2O \rightarrow R_3PbOH + R'OH$$

Using a more conventional approach, trimethyllead methoxide was prepared from trimethyllead chloride and sodium methoxide in tetrahydrofuran.

Other trialkyllead alkoxides derived from glycols have been mentioned in the patent literature by Giraitis, Kobetz, and Shapiro (128). For example, the trimethyllead derivative of the monomethyl ether of diethylene glycol, $Me_3PbO(C_2H_4O)_2Me$, was prepared from Me_3PbCl and $NaO(C_2H_4O)_2Me$ by refluxing these reactants in toluene.

Diorganolead dialkoxides have not been successfully isolated, but a few monophenyllead alkoxides have been recently reported. These are mixed oxides of the type, $PhPb(OR)(oxine)_2$ (175), where oxine is 8-hydroxyquinolinate.

The infrared absorption spectra of a number of trialkylmetal alkoxides of Group IVb metals, including triethyllead ethoxide, have been examined by Vyshinskii and co-workers (372). Semion and Kravtsov (342) have published a preliminary X-ray diffraction study of triphenyllead *p*-nitrosophenylate; the crystals were shown to be monoclinic.

Diethyllead dihydroxide is the only compound of the dialkyllead dihydroxide class which has been isolated. It was prepared by Calingaert, Dykstra, and Shapiro (61,62), and subsequently by Heap, Saunders, and Stacey (151), as well as by Aleksandrov, Brilkina, and Shushunov (5). The compound is readily prepared from diethyllead dichloride and silver oxide. In general, it is similar in its properties to triethyllead hydroxide, and is also weakly basic in nature.

The thermal decompositions of triethyllead hydroxide and diethyllead dihydroxide have been studied by Aleksandrov et al. (5). As might be expected, the compounds decomposed to give each other as intermediates, and, finally, to yield lead oxide, tetraethyllead, and ethylene, ethane, and butane. From the products obtained, decomposition probably involves both free radical and disproportionation-type reactions.

The alkyllead hydroxides can apparently be formed also by oxidation of the hexaalkyldilead compounds, although this is not a suitable method of synthesis. Aleksandrov and his colleagues (8) obtained both triethyllead hydroxide and diethyllead dihydroxide upon liquid phase oxidation of hexaethyldilead at 50° with oxygen. The reaction was strongly autocatalytic and gave a variety of products including one-half mole of tetraethyllead per mole of hexaethyldilead, lead oxide, ethyl alcohol, acetaldehyde, and water. No peroxides were found; this was attributed to their rapid decomposition under the reaction conditions employed.

Dialkyllead oxides were referred to in the older literature but they were not isolated and charrized. Clearly, actemore definitive work is needed on this class of compounds.

Only a few triaryllead hydroxides are known. The triphenyl-, tri-*p*-tolyl-, and *o*-carboxyphenyldiphenyllead hydroxides are stable crystalline solids (25,45,138,165,216,218,250,255,304,341,377). The triphenyl- and tritolyllead hydroxides are weak bases and insoluble in water, in contrast to the trialkyllead hydroxides; they also do not have sharp melting points. The methods of preparation are analogous to those used for the trialkyllead hydroxides; for example, triphenyllead chloride may be reacted with alcoholic potassium hydroxide to yield the triphenyllead hydroxide. In contrast, however, the diaryllead dihydroxides are as yet unknown, possibly because they are unstable at room temperature. The diaryllead oxides instead form very readily even under hydrolytic conditions (see below).

An interesting reaction of triphenyllead hydroxide was discovered by Reichle (304) in 1962, who found that the hydroxyl group could be replaced at room temperature in 48 hr by reaction with carbon disulfide to give stable, solid bis(triphenyllead) sulfide in 79% yield. Similarly, the phenyl tin, mercury, and arsenic sulfides were prepared, but the germanium, silicon, and boron analogs could not be made.

Another interesting reaction of triphenyllead hydroxide has been employed by Henry and Krebs (165) to prepare triphenyllead mercaptides:

$$(C_6H_5)_3PbOH + RSH \rightarrow (C_6H_5)PbSR + H_2O$$

The thiobutyl compound was thus prepared in 65% yield by shaking the mixture for 24 hr in ether.

Matwiyoff and Drago (255) investigated the donor and acceptor properties of triphenyllead hydroxide (and of the corresponding tin compound, as well as of trimethyl- and triethyllead phenoxide); the results could not be correlated with Rochow's new electronegativity scale (see Chapter VI). The observed trends in the donor-acceptor properties of the R_3MOH compounds were attributed to changes in the extent of π-bonding between the central metal and the oxygen as being the dominant effect. No hydrogen bonding interaction could be detected between the triphenylmetal hydroxide compounds and ether. In fact, a reversible dehydration reaction was observed with ether in every instance, as well as with dimethyl sulfoxide and dimethylformamide. An interesting observation made in the course of this work was that the trimethyl- and triethyllead hydroxides could be titrated with phenol in the presence of molecular sieves, suggesting a good analytical method for such hydroxides.

West and Baney (377) have measured the acidity and basicity of Ph_3MOH compounds as hydrogen bond donors and acceptors, where M is C, Si, Ge, Sn, or Pb. The trends indicated that dative π-bonding from oxygen to metal is strong in Ph_3SiOH, weaker in Ph_3GeOH, and negligible in the other compounds.

Bock and Deister (45) investigated the extraction of a number of anions such as Cl^-, Br^-, I^-, PO_4^{-3}, AsO_4^{-3}, SeO_3^{-2}, and others from neutral or slightly acid aqueous solution with a chloroform solution of triphenyllead hydroxide. In a similar manner, Schweitzer and McCarty (341) investigated the extraction of several anions into organic solvents containing triphenyllead hydroxide (or the corresponding tin, antimony, or arsenic compound). Chloroform and methyl isobutyl ketone were suitable solvents for the extraction of several common anions. Several species involved in the processes were identified, such as triphenyllead chloride, bromide, iodide, acetate, propionate, and benzoate; their stability constants were determined and equilibrium constants for the reactions were ascertained.

Bis(triphenyllead) oxide was prepared in relatively low purity by Brilkina et al. (48) by shaking a solution of triphenyllead hydroxide in benzene with a sodium dispersion, analogous to the method for triethyllead oxide. The isolated solid, obtained by evaporation of the filtrate, was about 70% pure, as judged from the molecular weight.

Bis(triphenyllead) oxide has been reported by Bähr (25), but it is likely that he prepared the triphenyllead hydroxide instead, since an aqueous method of preparation was employed: the oxidation of hexaphenyldilead with potassium permanganate in acetone solution followed by hydrolysis of the product. Willemsens and van der Kerk (386) obtained bis(triphenyllead) oxide when the same reaction was carried out in dry acetone.

In contrast to the relatively unknown diaryllead dihydroxides, diphenyllead oxide is synthesized easily, for example, by the hydrolysis of a diphenyllead salt (7,269,297,397,398). The compound is an insoluble white powder, which decomposes without melting, which suggests a polymeric structure. However, Kochkin et al. (202) have hydrolyzed diphenyllead dichloride with potassium hydroxide in mixed methanol–ethanol solution to give what was believed to be a mixture of $[(C_6H_5)_2PbO]_x$ and $(C_6H_5)_2Pb(OH)_2$.

Kocheshkov and co-workers (198,285,398) have prepared mixed diaryllead oxide esters by aqueous hydrolysis of diaryllead dicarboxylates dissolved in dry acetone, followed by the addition of diazomethane in ether, according to the overall equation:

$$2Ph_2Pb(OOCR)_2 + 2H_2O + 2CH_2N_2 \longrightarrow (Ph_2PbOOCR)_2O + 2MeOOCR + 2N_2$$

The acetate and isobutyrate compounds were prepared in 72% yield in 1 hr by this procedure. The same reaction was obtained using diazoethane or butane, but the yields were usually lower (285,398). 2-Naphthyl- and *p*-methoxyphenyllead derivatives were prepared in similar fashion using diazomethane (398). The products are apparently stable solids; most decomposed at temperatures ranging from 192 to 240°. The compounds were easily converted back to the original dicarboxylates by treatment with the appropriate acid.

Hydrolysis of diphenyllead diisobutyrate in aqueous acetone in the absence of diazomethane gave the simple basic isobutyrate, $Ph_2Pb(OH)(OOCC_3H_7\text{-}i)$ in 11% yield (398); similarly, diphenyllead diacetate was hydrolyzed to the basic acetate (285).

Some interesting "alkoxide" derivatives of the alkylplumbosiloxane type have been prepared by Schmidbaur and his associates as part of a series of such compounds of the Group IVb metals.

Schmidbaur and associates (327–333) obtained hexamethylplumbosiloxane by means of the reaction:

$$(CH_3)_3PbBr + (CH_3)SiONa \rightarrow (CH_3)_3PbOSi(CH_3)_3 + NaBr$$

The reaction, carried out in anhydrous ether at room temperature, gave high yields of product which had an offensive odor; it was stable in dry air even in light and survived distillation, but was very sensitive to water and reacted explosively with oxygen at above 140°, forming lead metal. Infrared data indicated a systematic shift of the SiOM band to higher wavelength in the sequence Si, Ge, Sn, Pb. Thermal stability and reactivity in heterolytic reactions were shown to decrease in the same order, presumably because of increasing mass, radii, and decreasing electronegativity of the heteroatoms.

Schmidbaur and Hussek (328) also prepared the hexaalkylplumbosiloxanes, $(CH_3)_3PbOSi(C_2H_5)_3$, $(C_2H_5)_3PbOSi(CH_3)_3$, and $(C_2H_5)_3PbOSi(C_2H_5)_3$, using essentially the same method as in the earlier work, and the properties of these compounds were found to be similar. Aluminum chloride reacted explosively with the hexamethylplumbosiloxane, yielding trimethyllead chloride quantitatively, along with a preponderance of trimethylsiloxyaluminum dichloride and a smaller amount of trimethylchlorosilane and aluminum oxychloride:

$$\begin{aligned} 2(CH_3)_3PbOSi(CH_3)_3 + 2AlCl_3 &\rightarrow 2(CH_3)_3PbCl + [(CH_3)_3SiOAlCl_2]_2 \\ &\rightarrow 2(CH_3)_3PbCl + 2(CH_3)_3SiCl \\ &\qquad + 2[AlOCl]_x \end{aligned}$$

The infrared and nuclear magnetic resonance spectra of the hexaalkylplumbosiloxanes were also discussed. The organolead hydroxides, oxides, alkoxides, and siloxanes prepared to date are listed in Table XI-1.

4. ORGANOLEAD PEROXIDES, R_3PbOOR

Only a few organolead peroxides are known. These are all R_3Pb— types; to date no diorganolead peroxides have been synthesized. Unlike the organolead hydroxides and oxides, the triorganolead peroxides are relatively new compounds, synthesized for the first time by Rieche and Dahlmann (312,314) in 1959. Subsequently, the preparation and reactions of organometallic peroxy compounds were reviewed in 1960 both by Brilkina and Shushunov (50), and by Rieche (313) (particularly the organometallic peroxides of germanium, tin, and lead), and more recently Sosnovsky and Brown (353) have reviewed this subject. The lead compounds are listed in Table XI-1.

Rieche and Dahlmann's methods of synthesis correspond to the equations:

$$(1) \quad Ph_3PbBr + NaNH_2 + ROOH \rightarrow Ph_3PbOOR + NH_3 + NaBr$$

$$(2) \quad ROOH + CH_3ONa \rightarrow ROONa + CH_3OH$$

$$Ph_3PbBr + ROONa \rightarrow Ph_3PbOOR + NaBr$$

By these methods, crystalline triphenyllead peroxide derivatives from *tert*-butyl, cumyl, and triphenylmethyl hydroperoxides were prepared easily in 74–97% yields at $-5°$ to room temperature, using equimolar proportions. The products were unstable, easily hydrolyzed, and reacted with HCl to form R_3PbCl compounds. Although they decomposed violently on heating, they could be stored without decomposition at low temperature in the absence of moisture.

In 1964 Rieche and Dahlmann (315) prepared several additional peroxy compounds of lead (and germanium and tin) using the method based on the sodium salt of the hydroperoxide as the preferred route to the new lead compounds, the trimethyllead peroxide derivatives of *tert*-butyl and cumyl hydroperoxides. In addition, $(C_6H_5)_3PbOOPb(C_6H_5)_3$, an unstable compound, preferably was prepared from hydrogen peroxide solution in ether and triphenylalkoxylead at 0°. The previous syntheses of organolead peroxides by Rieche and Dahlmann were also described in detail in this paper.

Aleksandrov, Brilkina, and Shushunov (4,6) prepared the triethyllead derivatives of *tert*-butyl and cumyl hydroperoxides, and also *p*-($Et_3PbO_2CMe_2)C_6H_4CMe_2O_2PbEt_3$, by the reaction of triethyllead hydroxide with the corresponding hydroperoxides:

$$Et_3PbOH + ROOH \rightarrow Et_3PbOOR + H_2O$$

Apparently, and surprisingly, hydrolysis caused no difficulty in this synthetic method, although the ethyl compounds were on the whole even less stable than the phenyl derivatives, and decomposed slowly at room temperature. Shushunov and Brilkina (348) also reacted triethyllead hydroxide and acetyl or benzoyl hydroperoxide in ether in the cold to give unidentified peroxidic lead compounds; on decomposition at room temperature, the peroxides gave principally the corresponding triethyllead esters. When triphenyllead hydroxide was treated in the same way, it yielded principally diphenyllead oxide. However, the analogous tin reactions went smoothly, yielding crystalline Ph_3SnO_2Ac from AcO_2H, and Ph_3SnOO_2CEt from $EtCO_2OH$.

Aleksandrov's group has shown that peroxidic lead compounds are formed in the liquid-phase oxidation of tetraethyllead with oxygen at 90°, along with ethane, ethylene, butane, acetaldehyde, triethyllead hydroxide, diethyllead dihydroxide, lead oxide, and water as coproducts.

5. ORGANOPLUMBONIC ACIDS, $[RPbOOH]_n$ or $[RPb(OH)_3]$

An interesting class of lead-containing analogs of benzoic acid, PhPbOOH, was discovered by Panov and Kocheshkov in 1952 (280,281). Benzene and toluene plumbonic acids were prepared in 60–80% yields by the hydrolysis with dilute alkali or alcoholic ammonium hydroxide of the corresponding monoaryllead triesters of acetic and butyric acids. The *p*-iodophenyl derivative was prepared similarly (245).

$$ArPb(OAc)_3 \xrightarrow{\text{hydrolysis}} ArPbOOH$$

Willemsens and van der Kerk (379) repeated the preparation of benzene-plumbonic acid, using a slightly modified procedure in which phenyllead triacetate was added slowly to one equivalent of dilute aqueous sodium hydroxide, and suggested that this procedure gives a purer product. Willemsens and van der Kerk found that their product had better solubility in dilute acids owing to the higher purity. Panov and Kocheshkov found that the arylplumbonic acids are very weak acids, infusible and amorphous, but soluble in strong alkali. They appeared to react with sulfuric acid, and with organic acids to give the corresponding esters. On drying *in vacuo*, the plumbonic acids were partly dehydrated to give unidentified polymeric materials.

Apparently benzeneplumbonic acid has the initial structure

$$C_6H_5Pb(OH)_3,$$

but it spontaneously loses water and becomes converted to a polymeric material approximating the formula $(C_6H_5PbOOH)_3$ and containing some —Pb—O—Pb—O— chains and bridges. As the poly-condensation progresses, the polymer becomes more insoluble and less acid-soluble, and the color changes from white to yellow.

The formation of alkylplumbonic acids had been described briefly by Lesbre at an earlier date (228,231), according to either of the following formulations:

$$RI + NaPbOOH \rightarrow RPbOOH + NaI$$

or

$$RI + NaPb(OH)_3 \rightarrow RPb(OH)_3 + NaI$$

The reactions were run at 5°, using an alkaline solution of lead oxide catalyzed with a trace of sodium iodide and with various alkyl halides. Alkylplumbonic acids with methyl, ethyl, propyl, *i*-propyl, butyl, benzyl, and allyl groups were prepared. The compounds were described as thermally unstable, decomposition beginning as low as 35°; on heating in a sealed tube, the compounds generally decomposed:

$$RPb(OH)_3 \rightarrow PbO + ROH + H_2O$$

However, the benzyl compound was reported to give some tetrabenzyllead on decomposition.

More investigation of the exact constitution and the reactions of the plumbonic acids would be of interest, due to the complexity of the forms described. The organoplumbonic acids prepared to date are given in Table XI-1.

6. LEAD–SULFUR, –SELENIUM, AND –TELLURIUM BONDED COMPOUNDS

A number of alkyl- and aryllead–sulfur bonded compounds have been known for many years, starting with Klippel's work in 1860. However, in recent years a large number of new lead–sulfur compounds have been prepared and new syntheses devised. Abel and Armitage (2) have reviewed the subject recently. In addition to the lead–sulfur bonded compounds, several lead–selenium and lead–tellurium bonded compounds have been prepared in recent years. These compounds are listed in Table XI-2.

The triorganolead and diorganolead sulfides are usually prepared by metathetical reactions of the type:

$$2R_3PbX + Na_2S \rightarrow (R_3Pb)_2S + 2NaX$$

or $$(C_6H_5)_2Pb(OOCCH_3)_2 + H_2S \rightarrow [(C_6H_5)_2PbS]_3 + 2CH_3COOH$$

Krebs and Henry (222) treated hexaphenyldilead with sulfur in benzene to prepare bis(triphenyllead)sulfide and diphenyllead sulfide, and Reichle (304) has employed the reaction of carbon disulfide with triphenyllead hydroxide to obtain bis(triphenyllead)sulfide.

The alkyl compounds are somewhat unstable, and tend to oxidize slowly in air. Indeed, this oxidation furnishes the most convenient preparation of the trialkyllead sulfates (61). The aryl compounds are more stable than their alkyl counterparts.

Much of the recent work in lead–sulfur bonded compounds has been done by Henry and his associates. In 1961, Krebs and Henry (221) started an investigation of the previously unknown sodium triphenyllead sulfide, $(C_6H_5)_3PbSNa$. The compound was prepared from triphenyllead chloride and excess sodium sulfide at below 30°:

$$(C_6H_5)_3PbCl + Na_2S \rightarrow (C_6H_5)_3PbSNa + NaCl$$

Table XI-2 Lead–Sulfur, -Selenium, and -Tellurium Bonded Compounds

LEAD–SULFUR BONDED COMPOUNDS	PROPERTIES	REFERENCES
$[(CH_3)_3Pb]_2S$		2,3,26,31
$[(C_2H_5)_3Pb]_2S$	liq., m.p. −45.1°; d. 2.05; n_D^{25} 1.6249	2,61,188,227
$[(C_6H_5)_3Pb]_2S$	m.p. 139–41°	2,138,164,221,227,304, 336,379
$(CH_3)_2PbS$		2,139,227
$CH_3(C_2H_5)PbS$		2,142,227
$(i\text{-}C_3H_7)_2PbS$		2,227,355
$(cyclo\text{-}C_6H_{11})_2PbS$	dec. >100°	137,227
$[(C_6H_5)_2PbS]_3$	dec. 80–90°	227,240,297
$(p\text{-}CH_3C_6H_4)_2PbS$	m.p. 98°, with dec.	2,227,298
$C_3H_7(i\text{-}C_4H_9)PbS$		2,142,227
$C_3H_7(i\text{-}C_5H_{11})PbS$		2,142,227
$(CH_3)_3PbSCH_3$	b.p.$_{0.75}$ 43°; b.p.$_{0.01}$ 32° n_D^{20} 1.6116; 1.6131	2,3,27,30
$(CH_3)_3PbSC_2H_5$	b.p.$_{0.05}$ 36°; n_D^{20} 1.5918	2,3
$(CH_3)_3PbSCH_2COOCH_3$	liq., dec. 100°	28,32
$(CH_3)_3PbSCH_2CONH_2$		29
$(CH_3)_2(C_2H_5)PbSCH_2CONH_2$		2,29
$(CH_3)_2C_7H_{15}PbSCH_2CONH_2$		2,29
$(C_2H_5)_3PbSC_2H_5$	b.p.$_{0.075}$ 76–8°	2,150,227,257
$(C_2H_5)_3PbSC_4H_9$	b.p.$_{0.5}$ 82–3°	104,227
$(C_2H_5)_3PbSC_7H_{15}$	b.p.$_{0.5}$ 114–5°	2,104,227

$(C_2H_5)_3PbSCOCH_3$	m.p. 45°	2,151
$(C_2H_5)_3PbSCH_2CONH_2$		2,29
$(C_2H_5)_2(i\text{-}C_3H_7)PbSCH_2CONH_2$		2,29
$(C_4H_9)_3PbSCH_2CONH_2$		2,29
$(CH_3)_3PbSCSN(CH_3)_2$	m.p. 92°	379
$(C_4H_9)_3PbSCH_2COOCH_2CH_2CH(CH_3)CH_2C(CH_3)_3$	liq.	379
$(C_6H_5)_3PbSCH_3$	m.p. 108–9°	2,164,165,221,289,379
$(C_6H_5)_3PbSC_2H_5$	m.p. 67–8°	2,164,165,289,379
$(C_6H_5)_3PbSC_3H_7$	m.p. 57–8°	2,164,165,289
$(C_6H_5)_3PbSC_4H_9$	dec. at b.p., n_D^{20} 1.6500	2,164,165,289
$(C_6H_5)_3PbSC_{10}H_7$	m.p. 73–5°	289
$(C_6H_5)_3PbSC_{10}H_{21}$		289
$(C_6H_5)_3PbSCH_2C_6H_5$	m.p. 82–3°	164,165,289
$(C_6H_5)_3PbSC_6H_5$	m.p. 106–7°	164,165,289,379
$(C_6H_5)_3PbSCOCH_3$	m.p. 92–3°; 90–1°	2,164,165,289,379
$(C_6H_5)_3PbSCOC_6H_5$	m.p. 93–4°	164,165,289
$(C_6H_5)_3PbSCH_2CH_2OH$	m.p. 78–9°	2,78
$(C_6H_5)_3PbSCH_2CH_2NH_2$	m.p. 40°	2,78
$(C_6H_5)_3PbSCH_2CH{=}CH_2$	m.p. 45°	2,78
$(C_6H_5)_3PbSCH_2COOCH_3$	m.p. 87–8°; 85°	5,78,289
$(C_6H_5)_3PbSC$ (benzothiazol-2-yl: N, S ring)	m.p. 58°	289
$(C_6H_5)_3PbSC$ (benzoxazol-2-yl: N, O ring)	m.p. 65°	289

(continued)

Table XI-2 (continued)

LEAD–SULPHUR BONDED COMPOUNDS	PROPERTIES	REFERENCES
$(C_6H_5)_3PbSCH_2CONH_2$	m.p. 124–5°	2,78
$(C_6H_5)_3PbSC_6F_5$	m.p. 91°	2,78
$(C_6H_5)_3PbSC_6H_4$-*p*-Cl	m.p. 102°	2,78,379
$(C_6H_5)_3PbSC_6H_4$-*p*-NO_2	m.p. 111°	2,78
$(C_6H_5)_3PbSC_6H_4$-*p*-NH_2	m.p. 135–6°	2,78
$(C_6H_5)_3$Pb-(17-β-mercaptotestosterone)	m.p. 125° or 152°	2,78
$(C_6H_5)_3PbSNa$	dec., r.t.	221
$(C_6H_5)_2Pb(SC_6H_5)_2$	m.p. 75°	78
$(C_6H_5)_2Pb(SC_6H_4$-*p*-$Cl)_2$	m.p. 71–2°	78
$(C_6H_5)_2Pb$(S–C(=N)(S) benzothiazolyl)$_2$	m.p. 148–50°	78
$(C_6H_5)_2Pb[SC(O)CH_3]_2$	m.p. 94–5°	165,289
$(C_6H_5)_3PbSP(S)(OCH_3)_2$	m.p. 89–91°	379
$(C_6H_5)_3PbSGe(C_6H_5)_3$	m.p. 128–9°	2,336
$(C_6H_5)_3PbSSn(C_6H_5)_3$	m.p. 92°	2,336
$(C_6H_5)_3PbSC(O)C_6H_4C(O)SPb(C_6H_5)_3$	dec. 174°	162
$(C_6H_5)_3PbSCH_2CH_2SPb(C_6H_5)_3$	m.p. 141–2°	78

$(C_6H_5)_3PbSC\overset{N-N}{\underset{S}{}}CSPb(C_6H_5)_3$ (1,3,4-thiadiazole-2,5-diyl)	dec. 205°	78
$(C_6H_5)_2PbSCH_2CH_2S$ (ring)		78
$[(C_6H_5)_2PbSCH_2C(O)O]_n$ (ring)		162
$(C_6H_5)_2Pb[SC(=N-C_6H_4-S)]_2$ (2-benzothiazolylthio)	m.p. 152–3°	289
Lead–Selenium Bonded Compounds		
$(C_6H_5)_3PbSeLi$		2,336
$(CH_3)_3PbSeCH_3$	b.p.$_{.3}$ 75° d_{20} 2.42	2,310,311
$(CH_3)_3PbSeC_6H_5$	d_{20} 2.60	2,310,311
$(C_6H_5)_3PbSePb(C_6H_5)_3$		2,336
$(C_6H_5)_3PbSeGe(C_6H_5)_3$		336
$(C_6H_5)_3PbSeSn(C_6H_5)_3$		336
Lead–Tellurium Bonded Compounds		
$(C_6H_5)_3PbTeLi$		2,336
$(C_6H_5)_3PbTePb(C_6H_5)_3$		2,336
$(C_6H_5)_3PbTeGe(C_6H_5)_3$		336
$(C_6H_5)_3PbTeSn(C_6H_5)_3$		336

It was shown to be hydrolytically and thermally unstable, slowly decomposing even at room temperature to bis(triphenyllead)sulfide and sodium sulfide. The sodium salt gave the new compound triphenyllead methylmercaptide in 68% yield upon treatment with methyl iodide:

$$(C_6H_5)_3PbSNa + CH_3I \rightarrow (C_6H_5)_3PbSCH_3 + NaI$$

The mercaptide was synthesized more directly in practically quantitative yield from lead(II) methylmercaptide and triphenyllead chloride in benzene at reflux:

$$2(C_6H_5)_3PbCl + Pb(SCH_3)_2 \rightarrow (C_6H_5)_3PbSCH_3 + PbCl_2$$

The same investigators (164,165) extended their work subsequently to include an investigation of alternate routes to various triphenyllead alkylmercaptide compounds, $(C_6H_5)_3PbSR$. Of four possible routes tested, the reaction of triphenyllead chloride with lead(II) alkylmercaptide described above proved best, although the reaction of triphenyllead hydroxide with butyl mercaptan was also quite effective for synthesizing the butylmercaptide compound:

$$(C_6H_5)_3PbOH + C_4H_9SH \rightarrow (C_6H_5)_3PbSC_4H_9 + H_2O$$

The $(C_6H_5)_3PbSR$ compounds prepared included the methyl, ethyl, propyl, butyl, benzyl, acetyl, and benzoyl derivatives. Infrared spectra characteristic of organolead compounds were obtained. Treatment of the thioalkyl compounds with hydrogen chloride apparently gave mixtures of all the possible chloride products. In addition, yields as high as 81% of the diphenyllead diacetylmercaptide, $(C_6H_5)_2Pb(SCOCH_3)_2$, were obtained by refluxing diphenyllead dichloride with the corresponding lead mercaptide in toluene:

$$(C_6H_5)_2PbCl_2 + Pb(SCOCH_3)_2 \rightarrow (C_6H_5)_2Pb(SCOCH_3)_2 + PbCl_2$$

Poorer yields of this same dimercaptide were obtained by treating tetraphenyllead with acetyl mercaptan. An attempt to synthesize the triphenyllead disulfide, $(C_6H_5)_3PbSSPb(C_6H_5)_3$, from triphenyllead chloride and sodium polysulfide gave bis(triphenyllead) sulfide, while reaction of triphenyllead chloride with lead allylmercaptide did not result in the expected triphenyllead allylmercaptide, $(C_6H_5)_3PbSCH_2CH{=}CH_2$.

Davidson, Hills, and Henry (78) have reported a number of additional functional triphenyllead alkylmercaptides, $(C_6H_5)_3PbSR$ (as well as similar germanes and stannanes). These compounds were commonly prepared in 70–90% yield at room temperature, using the route:

$$(C_6H_5)_3PbX + RSH + \text{base} \rightarrow (C_6H_5)_3PbSR + \text{base}\cdot HX$$

The usual base employed was pyridine or triethylamine. It was found that the previous method of choice, the reaction of R_3PbCl with $Pb(SR')_2$, was relatively ineffective for these functional compounds. These triphenyllead alkylmercaptides are of some special interest due to the functionality within the R groups, which included CH_2CH_2OH, $CH_2CH_2NH_2$, $CH_2CH{=}CH_2$, CH_2COOCH_3, CH_2CONH_2, C_6F_5, p-ClC_6H_4, p-$NO_2C_6H_4$, p-$NH_2C_6H_4$, and the 17-β-mercaptotestosterone.

Similar reactions of diphenyllead dichloride with thiols in the presence of base showed that generally the bisthiodiphenylplumbanes are thermally unstable (78). Only three such $(C_6H_5)_2Pb(SR)_2$ compounds were isolated successfully, where R = C_6H_5, p-ClC_6H_4, and

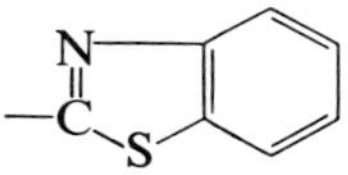

N
‖
—C
S

However, two other such compounds have been described (162):

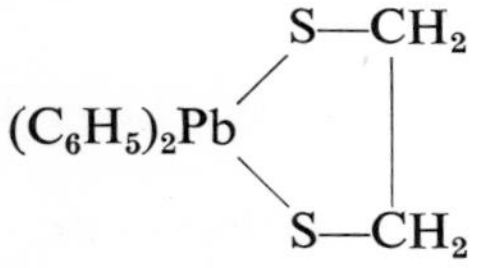

S—CH_2
$(C_6H_5)_2Pb$
S—CH_2
and
S—CH_2
$(C_6H_5)_2Pb$
O—C=O
n

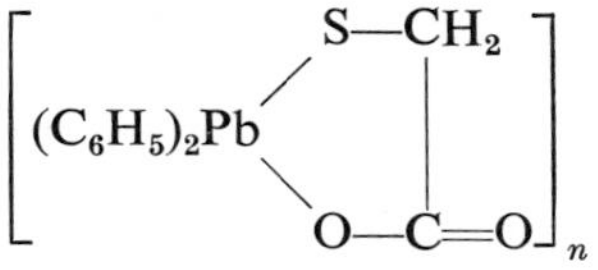

A quantitative study of the decomposition of the bis(phenylthio)-diphenylplumbane, one of the more stable compounds, showed that the decomposition occurred mainly according to the overall equation:

$$3(C_6H_5)_2Pb(SC_6H_5)_2 \rightarrow 2(C_6H_5)_3PbSC_6H_5 + Pb(SC_6H_5)_2 + C_6H_5SSC_6H_5$$

By using a mole ratio of triphenyllead halide to dithiol of 2:1, Davidson, Hills and Henry were able to prepare two dithiobistriphenylplumbanes, $(C_6H_5)_3PbSRSPb(C_6H_5)_3$, where R = CH_2CH_2 and

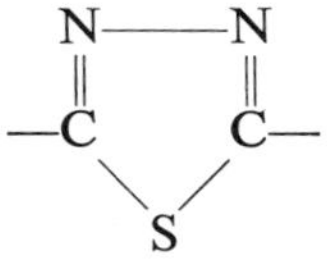

N——N
‖ ‖
—C C—
S

A number of lead–sulfur and lead–selenium bonded compounds have been patented by Ballinger and Richardson, for the most part because of interest in their antiknock properties. The kinds of compounds prepared and patented include $[(CH_3)_3Pb]_2S$, from trimethyllead chloride and sodium sulfide (26,31); $(CH_3)_3PbSCH_3$, from trimethyllead chloride, methyl mercaptide, and sodium hydroxide (27,30); $(CH_3)_3PbSCH_2$-$COOCH_3$, in the same way from the corresponding mercaptide (28);

$(CH_3)_3PbSCH_2CONH_2$, $(C_2H_5)_3PbSCH_2CONH_2$, $(CH_3)_2(C_2H_5)PbSCH_2$-$CONH_2$, i-$C_3H_7(C_2H_5)_2PbSCH_2CONH_2$, $(C_4H_9)_3PbSCH_2CONH_2$, and $[CH_3(CH_2)_6](CH_3)_2PbSCH_2CONH_2$, from the corresponding trialkyllead chloride and $HSCH_2CONH_2$ (29); $(CH_3)_3PbSCH_2COOCH_3$, from trimethyllead chloride and methyl thioglycolate upon treatment with ethanol and sodium carbonate solution in water (32); and $(CH_3)_3PbSeCH_3$, and $(CH_3)_3PbSeC_6H_5$, from trimethyllead chloride and methyl- or phenylselenium magnesium bromide in ether (310,311).

Using the lead alkylmercaptide method of Krebs and Henry, Perilstein (289) has recently prepared and patented a number of lead–sulfur bonded compounds which are reported to be useful as lubricating oil additives.

Schumann and Schmidt (336) have examined the infrared spectra of a number of $(C_6H_5)_3MXM'(C_6H_5)_3$ compounds, in which the M and M′ metals were germanium, tin, or lead, and the X groups were sulfur, selenium, or tellurium. The general method of synthesis used for the lead compounds was based on the reaction of triphenyllead chloride with a lithium triphenylmetal sulfide, selenide, or telluride:

$$(C_6H_5)_3PbCl + LiXM(C_6H_5)_3 \rightarrow (C_6H_5)_3PbXM(C_6H_5)_3 + LiCl$$

The lead compounds prepared in this manner included the Ph_3Pb— derivatives of —$SGePh_3$, —$SSnPh_3$, and —$SPbPh_3$; of —$SeGePh_3$, —$SeSnPh_3$, and —$SePbPh_3$; and of —$TeGePh_3$, —$TeSnPh_3$, and —$TePbPh_3$. The metal–sulfur vibrations in these compounds appeared at 300–400 cm^{-1}; the metal–selenium and metal–tellurium vibrations could not be detected down to 250 cm^{-1}.

Abel, Armitage, and Brady (1) repeated the preparation of methylthiotrimethyllead, ethylthiotrimethyllead, and bis(trimethyllead) sulfide and analogs of the other Group IVb elements for purposes of studying their relative basicities by measurement of the observed shift of the C–D stretching mode of deuterochloroform solutions of these compounds. The shift of the C–D stretching mode progressively increased in each homologous series in going from Si < Se < Sn < Pb. The shifts observed for the tin and lead compounds were interpreted as indirect evidence against the high electronegativity value assigned to lead based on the proton chemical shifts observed in the tetramethyl metal derivatives. Abel and Brady (3) examined the NMR spectra of methylthiotrimethyllead and bis(trimethyllead) sulfide, along with those of the analogs of the other Group IVb metals. The τ values of the metal–methyl protons in both series of compounds followed qualitatively the order Si > Sn > Ge > Pb > C; this is the same order as previously found for the tetramethyl derivatives of these elements.

7. ORGANOLEAD SALTS

The organolead compounds discussed in this section are derivatives of strong to weak acids and are commonly referred to as organolead salts. The large majority of these derivatives exhibit properties characteristic of divalent lead salts. They undergo metathesis-type reactions with various acids, bases, and metal salts, and are very useful for the preparation of other organolead salts. They also form coordination compounds in which the lead atom exhibits a coordination number of four to six, similar to the coordination compounds of inorganic lead salts.

Organolead salts of strong acids are stable to hydrolysis and exhibit a high degree of ionic character. On the other hand, organolead derivatives of very weak "acids", e.g., alkoxides, amides, and phosphides, exhibit a high degree of covalent character. They are readily hydrolyzed and tend to undergo insertion type reactions, instead of metatheses; they also tend to be less thermally stable than salts of strong acids. Hence, organolead salts of weak acids are more similar in their properties to tetravalent lead salts of carboxylic acids, such as lead(IV) acetate, than to divalent lead salts.

The largest number of organolead salts are derivatives of the types R_3PbX and R_2PbX_2, in which X is the anion of a strong to weak acid. Mono-organolead salts of the type $RPbX_3$ generally are unstable; the only $RPbX_3$ derivatives which are known are the aryllead tricarboxylates. These latter compounds are relatively stable and are fairly well characterized.

A number of organolead salts is known in which two or more different organic groups are present, i.e., unsymmetrical derivatives of the type $R_2R'PbX$ or $RR'PbX_2$. Such compounds are important intermediates in the synthesis of the unsymmetrical tetraorganolead compounds discussed earlier. A few organolead salts are also known in which two different anions are present. These are primarily the halides, prepared by halogenation of a triorganolead halide (95,143). However, they enjoy no special utility. A series of unsymmetrical diorganolead oxinate derivatives has also been prepared of the type $R_2Pb(Ox)(X)$, where X is halide or carboxylate (172,173).

Numerous methods of synthesis are known for R_3PbX and R_2PbX_2 compounds, so that the problem is one of choosing the best route to the desired compound. On the other hand, $RPbX_3$ compounds require special methods of synthesis; for this reason, their synthesis and reactions are discussed separately. The various known R_3PbX, R_2PbX_2 compounds are listed in Tables XI-3 to XI-10; the organolead derivatives of nitrogen and the other Group Vb elements are listed separately in Table XI-11.

Table XI-3 Symmetrical Triorganolead Halides

COMPOUND	MELTING POINT, °C	REFERENCES
Tribenzyllead bromide	dec. ~130	219,220,227
Tribenzyllead chloride		220,227
Tributyllead bromide	30–4	220,227,291,386
Tributyllead chloride	106.5–8.5; 109–11; 113	12,52,90,227,252, 257,291,321,386
Tri-*sec*-butyllead chloride	dec. 130	59,227,306
Tributyllead fluoride	dec. 145–8	388
Tricyclohexyllead bromide	dec. 210	214,227
Tricyclohexyllead chloride	206; dec. 236	59,214,227,264
Tricyclohexyllead fluoride	dec. 198	216,227
Tricyclohexyllead iodide	91.7; dec. 125	106,214,227
Tricyclopropyllead chloride	173–4	184
Trisdodecyllead chloride	63.5	227,259
Triethyllead bromide	103–4 (dec.); 105	17,106,227,303
Triethyllead chloride	dec. 120, 172; 165–7	12,51,55,61,121, 133,134
	subl.$_{0.01}$ 105–15	139,227,251, 256,263,290, 345,393
Triethyllead fluoride	dec. 240	216,227
Triethyllead iodide	19–20	61,213,227
Trihexadecyllead chloride	78–80°	227,259
Triisoamyllead bromide	132–3 (dec.)	70,143,215,227
Triisoamyllead chloride		187,188,227
Triisoamyllead fluoride	dec. 251	216,227
Triisoamyllead iodide	100	187,188,227
Triisobutylleadbromide	107–8 (dec.)	70,77,143,215,227
Triisobutyllead chloride	122 (dec.); dec. 145	143,150,227
Triisobutyllead fluoride	dec. 230	216,227
Triisobutyllead iodide		143,227
Triisopropyllead bromide		135,227
Triisopropyllead chloride		227,355
Triisopropyllead iodide		227,355
Trimesityllead bromide	145–6	130
Trimesityllead iodide	200–1	106,227
Tris(*p*-ethoxyphenyl)lead iodide	99–100	106,227
Tris(*o*-methoxyphenyl)lead chloride	167–71	227,264
Tris(*o*-methoxyphenyl)lead iodide	122–3	106,227
Tris(*p*-methoxyphenyl)lead bromide		125,227
Tris(*p*-methoxyphenyl)lead chloride	152–3	125,227
Trimethyllead bromide	131, 133 (dec.) 126–7	12,58,139,227, 349,386

(continued)

Table XI-3 (*continued*)

COMPOUND	MELTING POINT, °C	REFERENCES
Trimethyllead chloride	190 (dec); subl.$_{0.01}$ 187–95	12,52,58,61,139, 171,227
Trimethyllead fluoride	dec. 305	12,216,227,389
Trimethyllead iodide		58,63,227
Trineopentyllead bromide	166	349
Trineopentyllead chloride	207–8	349
Trineopentyllead iodide		400
Trineophyllead iodide		400
Trioctadecyllead chloride	82–3	227,259
Tripentafluorophenyllead bromide	119	163
Triphenethyllead bromide	91–2	219,227
Triphenyllead bromide	166	70,138,215,218, 227,386
Triphenyllead chloride	206; 210	42,119,138,194, 218,227,236, 258,386
Triphenyllead fluoride	305–8; dec. 318 subl. ~240	42,70,215,216, 227,286,364, 365,388
Triphenyllead iodide	138–9; 142	17,95,106,138, 227,386
Tripropyllead bromide	76–8; 81–2	59,227,291,321
Tripropyllead chloride	dec. 123–5; m.p. 135; 133–4; 137; subl. 100–10$_{0.01}$	12,52,58,139,227, 291,321
Tripropyllead fluoride	dec. 235	70,215,216,227
Tritetradecyllead chloride	74–5	227,259
Tri-*m*-tolyllead bromide	146–7	106,227
Tri-*o*-tolyllead bromide	126–7; 129–30	19,20,227
Tri-*o*-tolyllead chloride	141–2	20,105,227
Tri-*p*-tolyllead chloride	140–1	20,227
Tri-*p*-tolyllead fluoride	dec. 280	216,227
Tri-*p*-tolyllead iodide	115	106,217,227
Trivinyllead chloride	119–21; subl.$_1$ 90	183
Tri-2,5-xylyllead bromide	177	217,219a,227
Tri-2,5-xylyllead chloride	167.5; dec. 195	217,227

Table XI-4 Symmetrical Triorganolead Carboxylates

COMPOUND	MELTING POINT, °C	REFERENCES
Bis(tributyllead) adipate	152–3	386
Bis(tributyllead) fumarate		24,227
Bis(triethyllead) *d*-camphorate		120,227
Bis(triethyllead) oxalate dihydrate		187,188,227
Bis(triphenyllead) acetylene dicarboxylate	dec. ~215	160
Bis(triphenyllead) maleate	198–9; 300 (dec.)	113,227,320
Tris(triethyllead) citrate	>350	120,227
Triamyllead acetate	55–6.5	350,386
Tributyllead acetate	86	47,74,227,321,386
Tributyllead bromoacetate	54–5°	227,321
Tributyllead chloroacetate	60	227,321
Tributyllead β-chloropropionate	65–6; 72–3	227,321,386
Tributyllead crotonate	119	227,321
Tributyllead iodoacetate	83	227,321
Tributyllead 2-ethylhexanoate	72–6	47,386
Tributyllead laurate	6(n_D^{20} = 1.5019)	386
Tributyllead methacrylate	75–7	386
Tributyllead propionate	79–80	227,321
Tributyllead trichloroacetate	119	227,321
Tributyllead undecanoate		47
Tricyclohexyllead acetate	dec. ~120	386
Tricyclopentyllead acetate	dec. ~110	386
Tridodecyllead acetate	59	227,259
Triethyllead acetate	160; 159–63	9,51,120,187,188, 192,321,348,386
Triethyllead acid oxalate	>300	120,227
Triethyllead acid succinate	304 (dec.)	104,227
Triethyllead acid tartrate		187,188,227
Triethyllead acrylate	sinters 120	150,227
Triethyllead *p*-aminobenzoate monohydrate	84–6	111,120,227
Triethyllead *p*-anisate	97–8	120,227
Triethyllead anthranilate	96	227,321
Triethyllead benzoate	127	51,151,188,227, 348
Triethyl β-benzoylacrylate	139–41 (dec.)	120,227
Triethyllead bromoacetate	121	51,150,227
Triethyllead *m*-bromobenzoate	113–4	120,227
Triethyllead *o*-bromobenzoate	134–5	120,227
Triethyllead *p*-bromobenzoate	127–8	120,227
Triethyllead butyrate	107.9–8.5	11,51,187,188,227

(continued)

Table XI-4 (continued)

Triethyllead caproate	93; 94.7–5.8	10,51,93,227,386
Triethyllead caprylate	85–7.5	51,227
Triethyllead chloroacetate	147	51,150,227
Triethyllead *p*-chlorobenzoate	123–4	120,227
Triethyllead α-chlorocrotonate	153–5	120,227
Triethyllead β-chloropropionate	106	150,227
Triethyllead cinnamate	122–3 (dec.)	120,227
Triethyllead crotonate	135–6	150,227
Triethyllead dibromoacetate	98.6–101.8	51,227
Triethyllead dichloroacetate	113.5–4.5	51,227
Triethyllead diphenylacetate	164–5	120,227
Triethyllead *N*-ethylcarbazole-2-carboxylate	195 (dec.)	120,227
Triethyllead 2-ethylhexylmaleate		24,227
Triethyllead ethyl oxalate	55	150,227
Triethyllead ethyl thioglycolate	b.p.$_{0.5}$ 94–6	227
Triethyllead 9-fluorenecarboxylate	dec. 195–208	120,227
Triethyllead fluoroacetate	180.5 (dec.)	227,256,320
Triethyllead formate		187,188,227
Triethyllead furoate	156–7 (dec.)	111,120,227
Triethyllead furylacrylate	132–3 (dec.)	111,120,227
Triethyllead heptoate	90–90.8	51,227
Triethyllead hexyl maleate		24,227,309
Triethyllead *m*-iodobenzoate	135–6	120,227
Triethyllead *o*-iodobenzoate	138.5–9	120,227
Triethyllead *p*-iodobenzoate	129.5–30.5	120,227
Triethyllead isobutyrate	119.4–121.8	51,227
Triethyllead isovalerate	119.4–119.8	51,227
Triethyllead lepidine-2-carboxylate	153–5	120,227
Triethyllead *N*-methylanthranilate	133 (dec.)	120,227
Triethyllead methyl thiosalicylate	b.p.$_{0.008}$ 91–3	104,227
Triethyllead β-2-naphthoyl-propionate	134.5	120,227
Triethyllead *o*-nitrobenzoate	142–3 (dec.)	120,227
Triethyllead *m*-nitrobenzoate	172–3	120,227
Triethyllead *p*-nitrobenzoate	168–9 (dec.)	120,227
Triethyllead pelargonate	88–90	51,227
Triethyllead phenylacetate	101–6	111,117,227
Triethyllead *N*-phenylanthranilate	124.5–5.0	120,227
Triethyllead phenylpropiolate	149–50 (dec.)	120,227
Triethyllead propionate	141 (dec.)	51,150,227

(continued)

Table XI-4 (continued)

COMPOUND	MELTING POINT, °C	REFERENCES
Triethyllead salicylate	93	111,117,120,150, 227
Triethyllead stearate	64–5	386
Triethyllead styphnate		57
Triethyllead thioacetate	45	151,227
Triethyllead thiosalicylate	97–9	117,227
Triethyllead trichloroacetate	141	56,150,227
Triethyllead trifluoroacetate	143	175a
Triethyllead triphenylacetate	134–6 (dec.)	120,227
Triethyllead valerate	115.5–7.0	51,227
Triheptyllead acetate	25–30	350,386
Trihexyllead acetate	49–57	181,350,386
Triisoamyllead acetate	135–8	350,386
Triisobutyllead acetate	101–10; 117	150,227,386
Triisobutyllead butyrate	119	150,227
Triisobutyllead propionate	118	150,227
Trimethyllead acetate	183–4; 192–4; 194	61,150,181,227, 276,386
Trimethyllead chloroacetate	167; 169	150,227,276
Trimethyllead 2-ethylhexanoate	108–9	386
Trimethyllead formate	113	150,227,276
Trimethyllead isovalerate	160	150,227
Trimethyllead methacrylate		225,227
Trimethyllead trichloroacetate	>220	150,227
Trimethyllead trifluoroacetate	153	175a
Trioctyllead acetate	34; 38–40	350,386
Trioctyllead laurate	52–4	386
Triphenyllead acetate	204–6; 206–7	20,22,227,386
Triphenyllead acetoacetate	dec. 108–14	386
Triphenyllead acid maleate	207	113,227
Triphenyllead acrylate	201–3	209,211,212
Triphenyllead benzoate	117–20; 122–4	210,386
Triphenyllead caprate	116–8	49
Triphenyllead caproate	dec. 158–68	386
Triphenyllead cyanoacetate	dec. 147–55	386
Triphenyllead ethyl benzylmalonate	131–2	189,190,227,399
Triphenyllead 2-ethylhexanoate	104–6	386

(continued)

Table XI-4 (continued)

Triphenyllead ethyl malonate	dec. 159–60	189,190,227,386
Triphenyllead *p*-hydroxybenzoate	162–4	386
Triphenyllead isobutyrate	188–90	209,211,212
Triphenyllead laurate	91; 115–8	227,386
Triphenyllead *p*-methacrylamido-benzoate	227–30	186
Triphenyllead methacrylate	115–7; 123.5; 126–8	202,209,211,212
Triphenyllead myristate	102–3	227
Triphenyllead orotate	350 (dec.)	204,205
Triphenyllead palmitate	110	227
Triphenyllead phenylpropiolate	dec. 125–7	386
Triphenyllead picolinate		386
Triphenyllead stearate	101–9; 112	227,386
Triphenyllead thioacetate	90–1; 92–3	165,386
Triphenyllead thiobenzoate	93–4	165
Triphenyllead 6-uracilacetate	350 (dec.)	204,205
Triphenyllead *p*-vinylbenzoate	136–8	210
Tripropyllead acetate	116; 128	150,151,227,291, 321,386
Tripropyllead acrylate	123	227,321
Tripropyllead anthranilate	57–8	227,321
Tripropyllead bromoacetate	93–4	227,321
Tripropyllead butyrate	105–6	227,321
Tripropyllead chloroacetate	109–10	151,227,321
Tripropyllead β-chloropropionate	99–100	227,321
Tripropyllead crotonate	135	227,321
Tripropyllead iodoacetate	88–9	227,321
Tripropyllead isovalerate	110–1	227,321
Tripropyllead propionate	121–2	227,321
Tripropyllead trichloroacetate	139–40	227,321
Tripropyllead xanthate	57.5	227,321
Tri-*p*-tolyllead acetate	161–2	20,227
Trivinyllead acetate	169–70; subl.$_1$ 100	183
Trivinyllead trichloroacetate	>300; subl.$_1$ 180–95	183

Table XI-5 Other Symmetrical Triorganolead Salts

COMPOUND	MELTING POINT, °C	REFERENCES
Bis(triethyllead) carbonate		51,61,62,227
Bis(triethyllead) chromate	dec. 190	51,227
Bis(triethyllead) fluorophosphonate	>260	227,322
Bis(triethyllead) sulfate		54,61,62,176,187, 188,227
Bis(triisoamyllead) sulfate		188,227
Bis(trimethyllead) sulfate		100a
Bis(triphenyllead) carbonate	231	227
Bis(tripropyllead) sulfate		227,291
Tris(triphenyllead) diphosphate		297
Tributyllead-2-naphthalenesulfonate	68	227,321
Tributyllead-*p*-toluenesulfonate	81–2	227,321
Tridodecyllead nitrate	44–5	227,259
Triethyllead 1-amino-4-naphthalene-sulfonate	238–40 (dec.)	120,227
Triethyllead 2-amino-5-toluene-sulfonate	210 (dec.)	120,227
Triethyllead bicarbonate		61,227
Triethyllead *d*-camphor-10-sulfonate	dec. 172	120,227
Triethyllead cyanate		150,227
Triethyllead cyanide	189 (dec); 194	51,61,150,188, 227,345
Triethyllead cyclohexanesulfinate	132–4	120,227
Triethyllead dicresylphosphate		146,147
Triethyllead dicyclohexylphosphate		146
Triethyllead dihydrogenphosphate	>250	52,117,187,198, 227
Triethyllead diisoamylphosphate		146,147
Triethyllead ethanesulfinate	dec. >130	100
Triethyllead ethyloleylphosphate		146,147
Triethyllead ferricyanide		51,227
Triethyllead ferrocyanide		51,227
Triethyllead fluorophosphonate	260	227,322

(continued)

Table XI-5 (continued)

COMPOUND	MELTING POINT, °C	REFERENCES
Triethyllead 2-naphthalenesulfonate	152	150,227
Triethyllead nitrate		177,227,247
Triethyllead selenocyanate	33–4; 29.5–30.5	23,120,150,227
Triethyllead tetraphenylborate		178,293
Triethyllead thiocyanate	26.5–7.0; 35	120,150,188,227
Triethyllead *p*-toluenesulfinate	86–8	120,227
Triethyllead *o*-toluenesulfonate	189	150,227
Triethyllead *p*-toluenesulfonate	170	111,117,150,227
Triethyllead *p*-totylthiosulfonate	109	120,227
Triisobutyllead *o*-toluenesulfonate		150,227
Triisobutyllead *p*-toluenesulfonate		150,227
Trimethyllead methanesulfinate	122–4 (dec.)	100
Trimethyllead methanesulfonate	dec. 137–40	100a
Trimethyllead nitrate		127,227
Trimethyllead selenocyanate		23
Trimethyllead *p*-toluenesulfonate	>220	150,227
Tri(*m*-nitrophenyl)lead nitrate		113,227
Triphenyllead cyanide	dec. 250	23,89
Triphenyllead dimethylarsinate		158
Triphenyllead fulminate	174–5	40
Triphenyllead hyponitrite	222–4	227
Triphenyllead isocyanate		161
Triphenyllead nitrate	dec. 220–5	113,127,227
Triphenyllead selenocyanate		23
Triphenyllead thiocyanate	dec. 230	89
Tripropyllead cyanide	135 (dec.)	227,321
Tripropyllead fulminate	Explodes	40
Tripropyllead-2-naphthalene-sulfonate	126–7	227,321
Tripropyllead *o*-toluene-sulfonate	86–7	150,227,321
Tripropyllead *p*-toluene-sulfonate	73–4.5; 82–3	150,227,321

A. Triorganolead and diorganolead salts, R_3PbX, R_2PbX_2

(*1*) *Synthesis*

Most of the methods of synthesis of R_3PbX and R_2PbX_2 compounds involve the use of an organolead compound as starting material, since few direct syntheses from lead metal or inorganic lead salts are known. This is one area in which the organometallic chemistry of lead differs greatly from that of the other Group IVb metals. This difference results from: (*1*) the greater stability of divalent lead salts of inorganic acids as compared to tetravalent lead salts, (*2*) the greater stability of organometallic derivatives of tetravalent lead as compared to those of divalent lead, and (*3*) the gross instability of most $RPbX_3$ compounds which would be formed as the

Table XI-6 Unsymmetrical Triorganolead Salts

COMPOUND	MELTING POINT, °C	REFERENCES
$R_2R'PbX$ *Compounds*[a]		
Diamylmethyllead chloride	dec. 121–2	182,227
Dicyclohexylphenyllead chloride	dec. 205	113,227
Diethyl-ε-bromoamyllead bromide		141,227
Diethyllbutyllead chloroacetate	82–3	151,227
Diethylmethyllead chloride		65,227
Diethylphenyllead chloride		113,227,301
Diethylpropyllead chloroacetate	112	151,227
Diethylpropyllead propionate	86	151,227
Diethyl-2-thienyllead chloride		301
Diethylvinyllead chloride		301
Dimethylethyllead chloride		60,61,65,227
Diphenylamyllead chloride	123 (dec.)	18,227
Diphenylbenzyllead bromide	143 (dec.)	219,227
Diphenylbenzyllead chloride	Sinters 157	126,227
Diphenyl (*o*-carbomethoxyphenyl)lead chloride	170–1	116,227
Diphenyl (*o*-carboxyphenyl)lead chloride	210–20	116,227
Diphenylcyclohexyllead bromide	135	218,227
Diphenyl (3-diethylaminopropyl)lead chloride		122,354
Diphenyl (3-diethylaminopropyl)lead chloride hydrochloride		122,354

[a] For $R_2R'PbX$ compounds prepared but not isolated, see References 142 and 144.

initial intermediate in the alkylation or arylation of a tetravalent lead compound. For these reasons, most organolead salts are best synthesized from the R_4Pb or R_6Pb_2 compounds.

(*a*) Syntheses from R_4Pb and halogens or acids

The most common method of synthesis of organolead salts of halogen or carboxylic acids is the reaction of a tetraorganolead compound with halogen, halogen acid, or carboxylic acid in an inert solvent. Reaction can be effected in stepwise fashion to cleave either one or two lead–carbon bonds according to:

$$R_4Pb + X_2 \rightarrow R_3PbX + RX$$

$$R_3PbX + X_2 \rightarrow R_2PbX_2 + RX$$

or

$$R_4Pb + HX \rightarrow R_3PbX + RH$$

$$R_3PbX + HX \rightarrow R_2PbX_2 + RH$$

Table XI-6 (continued)

COMPOUND	MELTING POINT, °C	REFERENCES
$R_2R'PbX$ *Compounds*		
Diphenylethyllead bromide	119, dec. 130–5	218,227
Diphenylethyllead chloride	dec. 146–7	109,227,301
Diphenyl-9-fluorenyllead chloride	160	14
Diphenylmethyllead bromide	118, dec. 124	218,227
Diphenylphenethyllead bromide	119 (dec.)	219,227
Diphenylpropyllead chloride	141 (dec.)	18,227
Diphenyl-2-pyridyllead iodide	137–40	110,227
Diphenyl-2-thienyllead chloride		301
Diphenylvinyllead chloride		301
Diphenyl-2,5-xylyllead bromide	90	218,227
Dipropylethyllead chloride		151,227,257
Di-2-thienylethyllead chloride		301
Di-2-thienylphenyllead chloride		301
Di-*o*-tolylphenyllead bromide	117–8	19,227
Di-*o*-tolylphenyllead chloride	113–4	21,227
Divinylethyllead chloride		301
Divinylphenyllead chloride		301
$RR'R''PbX$ *Compounds*[b]		
Cyclohexylmethylphenyllead bromide	93–4	218,227
Phenylpropyl-*o*-tolyllead chloride	103–4	18,227

[b] For $RR'R''PbX$ compounds prepared but not isolated, see Reference 142.

The extent of the reaction of tetraalkyllead compounds with halogens can be restricted to the formation of the R_3PbX derivative by conducting it at Dry Ice temperature in an inert organic solvent. On the other hand, the reaction of tetraaryllead compounds with halogens is not as selective; some R_4Pb and R_2PbX_2 are always formed along with the R_3PbX product. However, these by-products are easily removed by selective crystallization and the triaryllead salt can be obtained as a pure product. When the halogenation reaction is conducted at higher temperatures, i.e., at about

Table XI-7 Symmetrical Diorganolead Dihalides

COMPOUND	MELTING POINT, °C	REFERENCES
Diamyllead dibromide	dec. 89	182,227
Diamyllead dichloride	dec. 123–5	182,227
Di-*dl*-amyllead dichloride	dec. 67	182,227
Dibutyllead dibromide		77,182,227
Di-*sec*-butyllead dibromide		227,306
Dibutyllead dichloride	dec. 180	90,182,227,386
Di(*p*-carbethoxyphenyl)lead dichloride	>270	227,264
Di(2-chlorovinyl)lead dichloride	dec. 163–7	227,268
Dicyclohexyllead dibromide	dec. 142; dec. 225	137,155,214,227
Dicyclohexyllead dichloride	dec. 180	137,155,227
Dicyclohexyllead diiodide	dec. ~98	137,154,214,227
Dicyclopropyllead dichloride	250° (dec.)	184
Diethyllead dibromide		62,139,227,262, 302,386
Diethyllead dichloride		62,118,133,139, 155,193,227, 262,393
Diethyllead diiodide		139,227
Diisoamyllead dibromide	Sinters 93	143,227
Diisoamyllead dichloride	dec. 108	143,227
Diisobutyllead dibromide	dec. 102–3	143,227
Diisobutyllead dichloride	dec. >100	143,227
Diisopropyllead dibromide		144,227,355
Diisopropyllead dichloride	Unstable	227,355
Diisopropyllead diiodide		227,355
Dimesityllead dibromide	198–9	130
Di(*o*-methoxyphenyl)lead dichloride	187–8.5	227,264
Di(*p*-methoxyphenyl)lead dichloride		125,227
Di(*p*-methoxyphenyl)lead diiodide	122–3	106,227
Dimethyllead dibromide	Unstable	112,139,227

(*continued*)

$-10°$, the reaction of both tetraalkyllead and tetraaryllead compounds proceeds to the formation of R_2PbX_2, along with some PbX_2; the complete conversion of R_4Pb to R_2PbX_2 is usually indicated by a distinct coloration of the solution by the presence of free halogen. At higher temperatures or with prolonged reaction times, halogenation will proceed beyond the R_2PbX_2 stage to yield RX and PbX_2.

The reaction of R_4Pb compounds with chlorine and bromine is exothermic and must be conducted carefully to prevent an uncontrolled decomposition. For this reason, this system is not recommended for the preparation of large quantities of R_3PbX or R_2PbX_2 compounds even though the reaction can be moderated by the use of selective solvents such

Table XI-7 (continued)

COMPOUND	MELTING POINT, °C	REFERENCES
Dimethyllead dichloride		139,193,227
Dimethyllead difluoride		398
Dimethyllead diiodide	Unstable	139,227
Dimethyllead bis(trifluoroacetate)	180	175a
Di(*m*-nitrophenyl)lead dibromide		191,227
Di(*m*-nitrophenyl)lead dichloride	Sinters 250; dec. 285–8	113,227
Di(*m*-nitrophenyl)lead diiodide	dec. 135	113,227
Di-3-pentyllead dibromide		227,306
Diphenyllead dibromide		70,220,227,262, 321,386,397
Diphenyllead dichloride	dec. 284–6	21,106,107,119, 193–195,220, 227,236,249, 267,268,297, 376,386,397
Diphenyllead difluoride		227,343
Diphenyllead diiodide	101–3	227,290,296,298, 343,386,397
Dipropyllead dichloride	dec. 228	107,227,321
Di-2-thienyllead dichloride	dec. 202	124, 227
Di-*o*-tolyllead dibromide	150–1	19,227
Di-*o*-tolyllead dichloride	178–9	20,21,227
Di-*p*-tolyllead dibromide		227,298
Di-*p*-tolyllead dichloride		20,21,227,298
Divinyllead dichloride		183
Di-2,4-xylyllead dichloride		131,227
Di-2,5-xylyllead dibromide	120 (dec.)	219a,227

Table XI-8 *Symmetrical Diorganolead Carboxylates*

COMPOUND	MELTING POINT, °C	REFERENCES
Diamyllead diacetate	81–4	386
Dibutyllead diacetate	54; 103	386,394
Dibutyllead bis(β-chloropropionate)	73–4	386
Dibutyllead di(2-ethylhexanoate)	127–8	386
Dibutyllead dilaurate	34–5	47,386
Dibutyllead dipropionate		394
Di(*p*-carbethoxyphenyl)lead diacetate	207–8	227,264
Di(2-chlorovinyl)lead diacetate	dec. 115–30	227,267,268
Di(2-chlorovinyl)lead dibenzoate	dec. 204–4.5	227,268
Dicyclopropyllead bis(chloroacetate)		184
Dicyclopropyllead bis(dichloroacetate)	179–80 (dec.)	184
Dicyclopropyllead diacetate		184
Diethyllead diacetate	130; 200–1	9,227,256,264,280, 386
Diethyllead dibenzoate	168	151,227
Diethyllead di(*m*-bromobenzoate)	178–9	120,227
Diethyllead di-*N*-butylanthranilate	169–9.5 (dec.)	120,227
Diethyllead bis(chloroacetate)	176–7; 180 (dec.)	151,227,257,280, 386
Diethyllead di-*p*-chlorobenzoate	185 (dec.)	120,227
Diethyllead dinicotinate	143 (dec.)	120,227
Diethyllead di-*m*-nitrobenzoate	179–80	120,227
Diethyllead dithioacetate	84.5–5	151,227
Diethyllead di-*p*-toluate	186 (dec.)	120,227
Diethyllead bis(trichloroacetate)	151 (dec.)	151,227
Diethyllead bis(trifluoroacetate)	185–6	175a
Diethyllead bis(trimethylacetate)	dec. 149	175a
Diheptyllead diacetate	<60 (grease)	386
Dihexyllead diacetate	<55 (grease)	386
Di(*o*-methoxyphenyl)lead diacetate	191–3	227,264
Di(*p*-methoxyphenyl)lead diacetate		125,227
Dimethyllead diacetate monohydrate	170–2	386
Di-2-naphthyllead diacetate	235.0–6.5	227,242,264
Diphenyllead diacetate	195; 200–1; 201–10	47,158,192,193, 207,220,227, 264,280,282, 297,386
Diphenyllead diacetate dihydrate		227,297
Diphenyllead diacetate hydroacetate	201–10	386
Diphenyllead dibenzoate	231–2; 240	196,208,227,280, 282,386
Diphenyllead dibutyrate	132–4	208,227
Diphenyllead dicaproate	Oil	208,227

(continued)

Table XI-8 (continued)

COMPOUND	MELTING POINT, °C	REFERENCES
Diphenyllead bis(chloroacetate)	dec. 185–205	227,280,282,386
Diphenyllead bis(ethyl maleate)	128–9	47,386
Diphenyllead di(2-ethylhexanoate)	137–42	386
Diphenyllead diformate	>230	207,227
Diphenyllead diformate monohydrate		227,297
Diphenyllead di(α-hydroxybutyrate)	198–201 (dec.)	208,227
Diphenyllead di(α-hydroxypropionate)		208
Diphenyllead diisobutyrate	201–3; 204–5; 210–7	198,209,211,212, 280,282,285, 386
Diphenyllead diisovalerate	166–8	208,227
Diphenyllead dilactate	212–5 (dec.)	208,227
Diphenyllead dilaurate	85	386
Diphenyllead maleate		47,386
Diphenyllead malonate monohydrate		386
Diphenyllead dimethacrylate	204–5; dec. 205–12; 225	196,199,202,209, 227,280,282
Diphenyllead dinicotinate	dec. 200; 200 (dec.)	200,201,204,205
Diphenyllead di-*p*-nitrobenzoate	256 (dec.)	132,227
Diphenyllead oxalate trihydrate	dec. 293–5	132,227,386
Diphenyllead phthalate monohydrate		386
Diphenyllead dipropionate	170–2; 181.5–2.5	52,132,208,227, 280,282,386
Diphenyllead pyridine-2,5-dicarboxylate		201
Diphenyllead distearate	98–9; 100–1	196,208,227,280, 282,394
Diphenyllead bis(trichloroacetate)	dec. 210, 213 (dec.)	132,175a,227
Diphenyllead bis(trifluoroacetate)	307	175a
Diphenyllead bis(trimethylacetate)	dec. 210–31	175a
Diphenyllead di-6-uracilacetate	dec. 330; 330 (dec.)	200,204,205
Diphenyllead divalerate	168–70	132,227
Di-2-thienyllead bis(chloroacetate)	dec. 174	283
Di-2-thienyllead diisobutyrate	dec. 192	283
Dipropyllead diacetate	122; 123–6	151,227,321
Di-*p*-tolyllead diacetate	183.5; 209–10	197,227,280,298
Di-*p*-tolyllead diformate	dec. 233	227,298
Di-*p*-tolyllead diisobutyrate	202–3	197,227,280

Table XI-9 Other Symmetrical Diorganolead Salts

COMPOUND	MELTING POINT, °C	REFERENCES
Diamyllead dinitrate		182,227
Diamyllead dinitrate dihydrate	Sinters 96–8	182,227
Diamyllead sulfate	Sinters 177	182,227
Dibutyllead dinitrate	123–5	182,227
Dibutyllead dinitrate dihydrate	Sinters 116–7	182,227
Di(2-chlorovinyl)lead sulfate	dec. 171–2	227,268
Diethyllead carbonate		61,62,227
Diethyllead dicyanide		344,345
Diethyllead dinitrate		177,227
Diethyllead bis(ethanesulfinate)	dec. 135	100,279
Diethyllead bis(ethanesulfonate)	dec. 172–4	100a
Diethyllead selenite	>286	51,227
Diethyllead di-*o*-toluenesulfonate	dec. 165	151,227
Diethyllead di-*p*-toluenesulfonate		151,227
Diethyllead sulfite		151,227,257
Diisopropyllead chromate		227,355
Diisopropyllead dinitrate		227,355
Dimethyllead chromate		139,220,227
Dimethyllead bis(methanesulfinate)		100,279
Dimethyllead bis(methanesulfonate)	dec. 165–8	100,100a
Di(*m*-nitrophenyl)lead dinitrate dihydrate		69,113,227,371
Diphenyllead bis(allylarsonate)		158
Diphenyllead bis(benzylarsonate)		158
Diphenyllead chromate		227,297
Diphenyllead dicyanide	dec. 245–55	227,397
Diphenyllead diiodate		132,227
Diphenyllead bis(methylarsonate)		158
Diphenyllead dinitrate		227,296,297,370
Diphenyllead dinitrate dihydrate		69,192,207,227, 296,297
Diphenyllead bis(*m*-nitrophenylarsonate)		158
Diphenyllead bis(phenylarsonate)		158
Diphenyllead diselenocyanate		23
Diphenyllead dithiocyanate		227,297
Dipropyllead dinitrate dihydrate	87–9	182,227
Di-*o*-tolyllead dinitrate		19,227
Di-*p*-tolyllead chromate		227,298
Di-*p*-tolyllead dinitrate trihydrate		196,227,280,298
Tris(diphenyllead) diphosphate		220,227,297

Table XI-10 Unsymmetrical Diorganolead Salts

COMPOUND	MELTING POINT, °C	REFERENCES
$RR'PbX_2$ *Compounds*		
(ε-Chloroamyl)ethyllead dichloride	dec. 120	141,227
(3-Chloro-2-hydroxypropyl)phenyllead dichloride	dec. ~135	386
Ethylisoamyllead dichloride	dec. 128	142,227
Ethylisobutyllead dibromide	Unstable	142,227
Ethylmethyllead chromate		142,227
Ethylmethyllead dichloride		142,227
Ethylmethyllead diiodide		142,227
Ethylphenyllead dichloride		301
Ethyl-2-thienyllead dichloride		301
Ethylvinyllead dichloride		301
Isoamylisobutyllead dibromide	dec. ~95	143,220,227
Isoamylpropyllead dichloride		142,227
Isobutylpropyllead dibromide		142,227
Isobutylpropyllead dichloride		143,220,227
Phenylpropyllead dichloride		18,227
Phenyl-2-thienyllead dichloride		301
Phenyl-*o*-tolyllead dibromide	116–7	19,227
Phenyl-*o*-tolyllead dinitrate dihydrate		19,227
Phenylvinyllead dichloride		301
R_2PbXX' *Compounds*		
Bis(diphenyllead) basic carbonate		227,297
Diethyllead bromide chloride		220,227
Dicyclohexyllead bromide chloride		155*a*
Diisobutyllead bromide chloride	dec. 110	143,220,227
Diphenyllead azide acetate	dec. 232–5	160
Diphenyllead basic acetate		198,285
Diphenyllead basic cyanide		227
Diphenyllead basic isobutyrate	dec. >200	398
Diphenyllead basic nitrate		227,296
Diphenyllead chloride acetate		158
Diphenyllead chloride dimethylarsinate		158
Diphenyllead chloride methylarsonate		158
Diphenyllead chloride propylarsonate		158
Di-*p*-tolyllead basic nitrate		227,298

Table XI-11 *Organolead Derivatives of Nitrogen and Other Group V Elements*

COMPOUND	MELTING POINT, °C	REFERENCE
A. R_3Pb—N< *compounds*		
Tributyllead-3-amino-1,2,4-triazole	116–8	386
Tributyllead benzimidazole	104–5	386
Tributyllead benzotriazole	103	386
Tributyl(carboxy-1-naphthylamino)lead methyl ester		79
Tributyllead cyanamide	Oil	386
Tributyl[[2-cyano-3-(diethylamino)-3-phenylpropenylidene]amino]lead		274
Tributyl[(2-cyano-3-phenylpropylidene)amino] lead		274
Tributyl(1-cyclohexyl-3,3-diethylureido)lead		274
Tributyl(1,2-dicyclohexyl-3,3-diethylguanidino)lead		274
Tributylplumbyldiethylamine		273
Tributyl[[α-(diethylamino)benzylidine]-amino]lead		274
Tributyllead imidazole	48–50	74,386
Tributyllead-2-mercaptobenzimidazole	67	386
Tributyl(1,3,3-triethyl-2-thioureido)lead		274
Tricyclohexylplumbyldiethylamine		273
Triethyllead benzenesulfonamide	132	227,319
Triethyllead benzimidazole	189–91	386
Triethylplumbyldiethylamine		273
Triethyllead 2,3-dihydro-1,4-phthalazinedione		238
Triethyllead *N*-ethylcarbazole-2-carboxylate	195 (dec.)	227,257
Triethyllead ethylenesulfonanilide	116	227,257,319
Triethyllead imidazole		74
Triethylplumbylisobutylamine	Oil	80
Triethyllead methylsulfonamide	97	227,319
Triethyllead methylsulfonanilide	115.5	227,319
Triethyllead phthalimide	131	150,227,257
Triethyllead sulfanilamide	171	120,227,319
Bis(triethylleadsulfonanilide) methane	180	227,257,319
Triethyllead tetrachlorophthalimide		239
Triethyllead *o*-toluenesulfonamide	133	227,319
Triethyllead *p*-toluenesulfonamide	127	227,257,319
Triethyllead *p*-toluenesulfonanilide	134	227,319
Triethyllead *p*-toluenesulfon-*p*-bromoanilide	117	227,319

(continued

Table XI-11 (continued)

COMPOUND	MELTING POINT, °C	REFERENCE
Triethyllead-*p*-toluenesulfon-*p*-chloroanilide	111.5	227,319
Triethyllead 1,2,4-triazole		386
Triisobutylplumbyldiethylamine		273
Triisobutyllead imidazole	167	386
Trimethyllead azide	dec. 165	361,363
Trimethylplumbylbis(trimethylsilyl)amine	b.p. 85–7	324
Trimethyllead cyanamide	dec. 126–7	386
Trimethyllead isocyanate	dec. 210–20	361
Trimethyllead isothiocyanate	dec. 145	361
Trimethyllead phthalimide		136
Trimethylplumbyl(trimethylsilyl)methylamine		325
Triphenyllead-3-amino-1,2-4-triazole	dec. 248–56	386
Triphenyllead azide	dec. 185, 186–7, 187.5–8.5	235–7,306,362
Triphenyllead benzimidazole	dec. ~270	386
Triphenyllead benzotriazole	dec. 275–6	386
(Carboxy-1-naphthylamino)triphenyllead methyl ester		79
Triphenyllead (4,5-dicarbomethoxy-1,2-3-triazole)	198–9.5	160
Triphenylplumbyldiethylamine		273
Triphenyllead imidazole	dec. ~285	386
[(1-Methoxy-*N*-1-naphthylformimidoyl)-1-naphthylamino]triphenyllead		79
Triphenyllead 1,2,4-triazole	dec. ~265	386
Triphenyl[(2,2,2-trichloro-1-methoxy-ethylidene)amino]lead		79
Triphenyl[[2,2,2-trichloro-*N*-(triphenyl-plumbyl)acetimidoyl]oxy]lead		79
Tripropyllead benzenesulfonamide	96	227,257,319
Tripropylplumbyldiethylamine		273
Tripropyllead methylsulfonamide	67	227,257,319
Tripropyllead phthalimide		150,227,319
Tripropyllead sulfanilamide	101	227,319
Tripropyllead *p*-toluenesulfonamide	100–1	227,257,319
Tripropyllead *p*-toluenesulfonanilide	104	227,319
Tripropyllead *p*-toluenesulfon-*p*-bromoanilide	117	227,319

(continued)

Table XI-11 (continued)

COMPOUND	M.P.	REFERENCE
Tripropyllead *p*-toluenesulfon-*p*-chloroanilide	123	227,319
B. $R_2PbN(\langle)_2$ *Derivatives*		
Diphenylplumbylbis(diethylamine)		273
Diphenyllead diazide	dec. 145	160,236,237
C. R_3Pb *Derivatives With Other Group V Elements*		
Bis(triphenylplumbyl)phenylphosphine	110 (dec.)	338
Tris(trimethylplumbyl)arsine	43–5	335
Tris(trimethylplumbyl)phosphine	46–7	335
Tris(triphenylplumbyl)arsine	158	335
Triphenylplumbyldiphenylarsine	d. 115	337
Triphenylplumbyldiphenylphosphine		338
Triphenylplumbyldiphenylphosphonate	dec. >300°	338
Triphenylplumbyldiphenylstibine	dec. 115	337
Tris(triphenylplumbyl)phosphine	110 (dec.)	335,338
Tris(triphenylplumbyl)stibine	150	335
Triphenylplumbyl bis(triphenylstannyl)-phosphine	171–2	33
Triphenylplumbyl(triphenylstannyl)-phenylphosphine	110	339

as pyridine. Thus, Grüttner (138) reported that he obtained triphenyllead bromide as the major product from the reaction of tetraphenyllead with bromine in pyridine at room temperature; the moderating effect of the pyridine was attributed to its ability to form a complex with bromine. However, Willemsens and van der Kerk (386) have reported that the reaction of bromine and tetraphenyllead in pyridine at $-15°$ gave substantial amounts of diphenyllead dibromide, along with large amounts of unreacted tetraphenyllead; the yield of triphenyllead bromide was always less than 50%. The role of solvent in these halogenation reactions merits further investigation; diethyl ether, carbon disulfide, and ethyl acetate are usually used as solvents.

The reactions of R_4Pb compounds with halogen acids and carboxylic acids are also convenient for the synthesis of R_3PbX and R_2PbX_2 derivatives. These reactions tend to be less vigorous and therefore can be conducted at higher temperatures than halogenation reactions. In reactions involving halogen acids, the selective formation of R_3PbX or

R_2PbX_2 is usually controlled by the choice of solvent and/or the reaction temperature. Thus, triphenyllead chloride can be prepared readily by addition of hydrogen chloride to tetraphenyllead in warm chloroform until the first appearance of insoluble diphenyllead dichloride (119); diphenyllead dichloride can be obtained by treating tetraphenyllead with a saturated solution of hydrogen chloride in benzene. Triphenyllead bromide has been prepared by a similar procedure ($Ph_4Pb + HBr$), except that ethanol is reported to be a better solvent than chloroform because of the higher solubility of triphenyllead bromide in ethanol (386). Triethyllead chloride has been prepared by bubbling hydrogen chloride into an ether solution of tetraethyllead at room temperature (118). However, hexane is a better solvent for this reaction; it affords high yields of a purer product because of the lower solubility of triethyllead chloride in hexane (61). With both solvents, the triethyllead chloride precipitates almost immediately upon addition of the hydrogen chloride. The reaction is monitored by periodic filtration of the triethyllead chloride formed, followed by further treatment of the filtrate with more hydrogen chloride. Diethyllead dichloride can be prepared by passing hydrogen chloride into a toluene solution of tetraethyllead at room temperature; the intermediate triethyllead chloride is soluble in toluene so that reaction proceeds easily to the R_2PbX_2 stage. At higher temperatures, the diethyllead dichloride becomes contaminated with lead chloride (61,393).

Organolead salts have also been prepared by reaction of R_4Pb compounds with aqueous solutions of halogen acids and other inorganic acids. Such reactions were used as early as 1859 by Buckton (55) and Cahours (58) to prepare triethyllead chloride from tetraethyllead and concentrated aqueous hydrochloric acid. With halogen acids the aqueous system offers no particular advantage over non-aqueous systems, and the reactions tend to be slower because of the lower solubility of the R_4Pb compound in the aqueous acid. However, Browne and Reid (51) concluded that the reaction of tetraethyllead with aqueous hydrochloric acid is an excellent method for the synthesis of triethyllead chloride.

The reaction of a tetraorganolead compound with aqueous nitric acid has been used to prepare tri- and diorganolead nitrates (51,177,296,298). Care must be taken in reactions of tetraalkyllead compounds with concentrated nitric acid because of the possibility of an explosive decomposition; violent decompositions have been encountered with the reaction of tetraethyllead with concentrated nitric acid. Organolead salts of inorganic acids other than halogen acids are generally best prepared by reaction of an organolead oxide or hydroxide with the aqueous acid, or by metathesis reactions with other metal salts.

$$R_3PbOH + HX \rightarrow R_3PbX + H_2O$$

$$R_3PbX' + NaX \rightarrow R_3PbX + NaX'$$

Metathesis reactions are discussed later.

Organolead carboxylates are prepared readily by the reaction of a tetraorganolead compound with a carboxylic acid. In the alkyl series, this reaction has been used primarily for the methyl and ethyl derivatives. Control of the reaction to produce the desired R_3PbX or R_2PbX_2 product is usually accomplished by control of temperature or stoichiometry. With the weaker carboxylic acids it becomes harder to stop the reaction at the R_3PbX stage, since the more stringent reaction conditions which are required tend to promote the formation of the R_2PbX_2 derivative. However, this problem can be minimized by conducting the reaction in the presence of silica gel which serves as a catalyst (51). The reaction of tetraalkyllead compounds with various carboxylic acids was investigated in some detail by Gilman and co-workers (120), who concluded that there was no relationship between the acidity of the acid and its reactivity with tetraalkyllead compounds. The nature of the solvent was found to be the most critical factor in determining whether R_3PbX or R_2PbX_2 was formed. Although strongly polar solvents were expected to favor R_2PbX_2 formation, and nonpolar or weakly polar solvents to favor R_3PbX formation, the opposite was found. Thus, reaction of tetraethyllead with *m*-nitrobenzoic acid in ethanol gave the triethyl derivative, even when excess acid was used. On the other hand, in benzene solvent a vigorous reaction occurred to yield diethyllead di(*m*-nitrobenzoate). The Gilman paper gives an excellent discussion of the problems encountered in the synthesis of various organolead carboxylates.

(*b*) Syntheses from R_4Pb with metal salts

Organolead halides are usually formed in reactions of tetraorganolead compounds with a wide variety of metal halides. The reaction of R_4Pb compounds with metal halides yields R_3PbX, R_2PbX_2, or PbX_2, depending on reaction conditions and stoichiometry. These reactions are discussed in Chapter VI. However, they have not enjoyed general utility for the synthesis of organolead halides.

Tetraorganolead compounds react readily with metal carboxylates, such as mercury(II) carboxylates or lead(IV) acetate. The reaction of tetraaryllead compounds with mercury(II) carboxylates can be controlled to produce $RPbX_3$, R_2PbX_2, or R_3PbX by proper adjustment of stoichiometry.

(*c*) Syntheses from hexaorganodilead compounds

Hexaorganodilead compounds, like the tetraorganolead derivatives, react readily with halogens, acids, and metal salts to yield organolead salts. However, these reactions are generally less satisfactory for the synthesis of R_3PbX and R_2PbX_2 compounds than reactions involving the R_4Pb analogs. A complex mixture of products is usually obtained, and in most cases PbX_2 is formed in appreciable amounts. Reactions of R_6Pb_2 compounds are discussed in Chapter VII.

A better synthesis of an R_3PbX derivative from an R_6Pb_2 compound is obtained by oxidation of the R_6Pb_2 prior to formation of the R_3PbX. Oxidation of hexaphenyldilead with potassium permanganate in dry acetone gives bis(triphenyllead) oxide; triphenyllead hydroxide is formed if wet acetone is used (25,386). The acetone solution of the oxide or hydroxide can then be reacted with hydrochloric acid to yield triphenyllead chloride (386). Tributyllead acetate has been prepared in a similar manner. Alternatively, the oxidation of hexabutyldilead with hydrogen peroxide in the presence of acetic acid has been used to prepare tributyllead acetate (386). Both methods should be generally applicable to the synthesis of triorganolead carboxylates.

(*d*) Syntheses from inorganic lead salts

As was discussed earlier, few direct syntheses of organolead salts via the alkylation of inorganic lead salts or lead metal are known. This situation results from the instability of monoorganolead salts which would be formed as the initial intermediate in the reaction of a tetravalent lead salt with a reactive organometallic compound. The $RPbX_3$ formed initially is unstable and decomposes into RX and PbX_2, so that subsequent alkylation leads to formation of R_6Pb_2 or R_4Pb via a conventional divalent lead salt reaction. Thus,

$$PbX_4 + RM \rightarrow RPbX_3 + MX$$

$$RPbX_3 \rightarrow RX + PbX_2$$

$$3PbX_2 + 6RM \rightarrow R_6Pb_2 \text{ (or } R_4Pb) + Pb + 6MX$$

The main exception to this sequence is the reaction of a lead(IV) carboxylate with diarylmercury to form a stable aryllead tricarboxylate. This system can be carried beyond the monoaryllead derivative to form the diaryl- or triaryllead carboxylates as well. Similarly, the arylation of diaryllead dihalides with diarylmercury yields triaryllead halides (21,57). Reactions of lead(IV) carboxylates with diarylmercury compounds are usually carried out in chloroform solvent at ambient temperature (264):

$$Pb(OOCR)_4 + Ar_2Hg \rightarrow ArPb(OOCR)_3 + ArHgOOCR$$

$$ArPb(OOCR)_3 + Ar_2Hg \rightarrow Ar_2Pb(OOCR)_2 + ArHgOOCR$$

There are isolated examples of the direct synthesis of triorganolead halides by the reaction of a lead(II) halide with a Grignard reagent. Triaklyllead halides have been obtained from the reaction of lead(II) halides with long chain alkyl Grignard reagents; Meals (259) prepared tri(dodecyl), tri(tetradecyl), tri(hexadecyl), and tri(octadecyl)lead chloride in this fashion. Similarly, trineopentyllead chloride was formed, along with hexaneopentyldilead, in the reaction of lead(II) chloride with neopentylmagnesium chloride(349), and tri- and diethyllead chloride and acetate were obtained as the major products when solutions of ethylmagnesium chloride and diethylzinc were passed rapidly through a bed of lead chloride or acetate in a glass column (316).

It has been suggested that the triorganolead halides formed in the above reactions may actually arise from acid hydrolysis during work-up of the crude reaction mixture in which an R_6Pb_2 or R_4Pb compound is present as the primary product (386). This is unlikely, however, since an acid hydrolysis step was not used to work-up the reaction mixture from which trineopentyllead chloride was isolated (349). The mechanism of R_3PbX formation in these particular reactions has not been defined; a likely possibility is the reaction of an R_6Pb_2 intermediate with the by-product magnesium halide (106).

Triphenyllead chloride was formed upon reduction of the phenyldiazonium chloride complex of lead(IV) chloride with powdered zinc or copper metal; analogous treatment of the lead(II) chloride complex gave diphenyllead dichloride. The phenyldiazonium bromide–lead(II) bromide complex did not react under similar conditions (195,269). This reaction is restricted to the synthesis of aryllead halides and even then the yields are very low.

The direct reaction of lead(IV) isobutyrate with thiophene has been effected; di(2-thienyl)lead diisobutyrate was isolated after the reaction mixture had been allowed to stand for a prolonged period at room temperature (283). Lead(IV) acetate also reacts directly with anisole and with 1,3-dimethoxybenzene to yield *p*-methoxyphenyllead triacetate and 1,3-dimethoxyphenyllead triacetate, respectively (148,299). These reactions are postulated to involve an electrophilic attack on the phenyl group by the lead tetraacetate.

(*e*) Metathesis reactions involving organolead compounds

A large number of organolead salts has been synthesized by metathesis reactions involving (*1*) the reaction of an organolead halide or carboxylate with an inorganic salt or (*2*) the reaction of an organolead oxide or hydroxide with the appropriate acid. These reactions are very similar to metathesis reactions of inorganic lead salts and are governed generally by the same principles which govern these latter reactions, i.e., similar techniques can be employed to shift the equilibrium and force the reaction in the desired direction.

Metathesis reactions have been employed extensively by Gilman (120) and Saunders (150,151,257,319–321) and their colleagues in studies of the sternutatory properties of alkyllead salts. Typical metathesis reactions are illustrated by the equations:

$$R_3PbOH + HX \rightarrow R_3PbX + H_2O$$

$$R_3PbCl + MX \rightarrow R_3PbX + MCl$$

$$R_3PbCl + Ag_2O \rightarrow R_3PbOH \xrightarrow{HX} R_3PbX$$

$$R_3PbCl + AgX \rightarrow R_3PbX + AgCl$$

These reactions are usually carried out in polar solvents, such as alcohol, ether or water, or mixed solvents, the choice of which depends on the inorganic salt or acid used. In many cases, proper choice of solvent is essential for getting good yields of a pure product.

Another metathesis-type reaction which has proved useful for the synthesis of organolead salts is the reaction of an organolead halide with a weak acid in the presence of a strong Lewis base as solvent, the role of the Lewis base being to react with the hydrogen halide by-product and force the reaction to completion.

$$R_3PbX + HX \xrightarrow[\text{pyridine}]{R_3N \text{ or}} R_3PbX + HX{\cdot}\text{Amine}$$

Compounds prepared in this manner include triphenylplumbyldiphenylphosphine from triphenyllead chloride and diphenylphosphine in triethylamine (338) and alkyl- and arylthiotriphenyllead from triphenyllead chloride and the corresponding thiol in an amine solvent (78,159,166). The synthesis of the thiol derivatives was discussed earlier in this chapter.

Many new R_3PbX derivatives have been prepared in the last few years. In addition to triphenylplumbyldiphenylphosphine, the diphenylarsine and stibine analogs have also been prepared (from Ph_3PbCl and $NaMPh_2$ in liquid ammonia), as well as tris(triphenylplumbyl)phosphine, $(Ph_3Pb)_3P$, and its arsine and stibine analogs (335,337). Mixed metal derivatives of the type $(Ph_3Sn)_2P(PbPh_3)$ and $(Ph_3Sn)P(Ph)(PbPh_3)$ are also known (339). Several triaryl- and trialkylplumbyldiethylamine derivatives have been prepared by Neumann and Kühlein (273,274) by reaction of triorganolead chloride or iodide with lithium diethylamide in diethyl ether or tetrahydrofuran; these compounds slowly decompose at room temperature and are unstable to hydrolysis. Amine derivatives were prepared earlier by DePree (80) by the reaction of a triorganolead halide with the sodium salt of a substituted amide in benzene solvent. Trimethylplumbylbis(trimethylsilyl)amine and trimethylplumbyltrimethylsilylmethylamine have been prepared in similar fashion by Scherer and Schmidt (324,325).

Triphenyllead azide, diphenyllead diazide, and trimethyllead azide have been prepared recently by reaction of an organolead halide with sodium azide (236,305,362) or by neutralization of the organolead hydroxide with hydrazoic acid (235–237,362). Willemsens (386) prepared a number of triorganolead derivatives of nitrogen heterocycles by reaction of the heterocycle with a triorganolead hydroxide. Organolead compounds containing a lead–nitrogen bond are included in a recent review of organometallic nitrogen compounds of some of the Group IVb metals (248).

Other recent syntheses of organolead salts include triphenyllead isocyanate (161) from triphenyllead chloride and silver isocyanate in dimethyl sulfoxide. This compound has also been prepared from triphenyllead chloride (but not iodide) and potassium thiocyanate in ethanol (89). Trimethyllead isocyanate and isothiocyanate have been prepared by reaction of trimethyllead chloride with the respective silver salt in benzene (361). Triphenyllead fulminate (40) has been prepared from triphenyllead hydroxide in chloroform and aqueous sodium cyanate; tripropyllead fulminate (40) was obtained by neutralization of an aqueous solution of tripropyllead hydroxide with a solution of sodium cyanate in sulfuric acid. Triethyllead thiocyanate and selenocyanate were obtained by reaction of triethyllead acetate with potassium thiocyanate or selenocyanate in water (120) and triphenyllead selenocyanate has been prepared by the reaction of triphenyllead chloride with a mixture of potassium selenocyanate and sodium acetate in boiling ethanol (23). Both triphenyllead selenocyanate and diphenyllead diselenocyanate were obtained by the reaction of tetraphenyllead or hexaphenyldilead with selenium selenocyanate in hot chloroform, depending on stoichiometry; triethyllead and trimethyllead selenocyanate were prepared in similar fashion from the tetraalkyllead compounds (23). From the infrared spectra, it was concluded that the selenocyanate moiety was bound to lead through the selenium atom. In contrast, the infrared spectrum of the product obtained from the reaction of hexaphenylditin with selenium selenocyanate indicated the selenocyanate group was bound to tin through the nitrogen atom, i.e., $Ph_3SnNCSe$ or triphenyltin isoselenocyanate (23).

Triphenyllead cyanide has been prepared by reaction of ethereal triphenyllead iodide with aqueous potassium cyanide; however, triphenyllead thiocyanate could not be prepared in similar fashion using potassium thiocyanate (89). The preparation and properties of the organolead pseudohalides are included in two recent review articles on organometallic pseudohalides (226,363). A number of phenyllead arsinates and arsonates has been prepared by Henry and co-workers (158,166) by the reaction of triphenyl- or diphenyllead halide with sodium arsinate or arsonate, or by the reaction of the halide with the free acid in pyridine.

Triethyllead tetraphenylborate was precipitated as an insoluble solid from the reaction of triethyllead chloride and sodium tetraphenylborate in water (178). This reaction has been proposed as a rapid gravimetric method for the analysis of trialkyllead salts.

(*f*) Other synthesis methods

Other less general methods of synthesis of organolead salts are worthy of mention. The reaction of dialkyllead sulfite with an aqueous acid is reported to be an excellent method of synthesis of dialkyllead salts (151). The dialkyllead sulfite is derived from the reaction of the tetraalkyllead compound with moist sulfur dioxide. The reaction of dry sulfur dioxide with tetraethyllead and tetramethyllead has been shown to yield the corresponding trialkyllead alkanesulfinate or dialkyllead bis(alkanesulfinate), dependent on temperature and solvent (100,279). The alkanesulfinates are oxidized by air to the respective sulfonates (100); alkanesulfonates have been prepared by direct reaction of sulfur trioxide with tetramethyl- and tetraethyllead (100a). Alkane- and arenesulfinates and -sulfonates have also been prepared by metathesis reactions involving triorganolead hydroxide (321) or acetate (120), and by the reaction of a tetraorganolead compound with an arenesulfonic acid (150).

Another class of compounds which is reported to provide a superior synthesis of dialkyllead salts is the novel compound having the formula $[R_4Pb(NO)_2]^{2+} \cdot 2NO_3^-$. Compounds of this type, in which R is ethyl or propyl, were obtained as crystalline solids from the reaction of dinitrogen tetroxide with tetraalkyllead at 0° in diethyl ether (167,168). Both derivatives are soluble in acetone, water, alcohol and acetic acid, and insoluble in ether and nonpolar solvents. Reaction with aqueous hydrochloric acid or sulfuric acid destroys the nitrosyl complex (the ir spectra indicated the presence of nitrosyl groups) to form the dialkyllead dichloride or dialkyllead sulfate derivative.

Several new organolead derivatives containing a lead–nitrogen bond have been prepared in recent years by the reaction of tributylplumbyldiethylamine with certain unsaturated organic compounds (274). Thus, reaction of tributylplumbyldiethylamine with isocyanates or isothiocyanates gave the tributyllead derivative of a substituted urea or thiourea.

Dicyclohexylcarbodiimide reacted with tributylplumbyldiethylamine according to the equation:

$$Bu_3PbNEt_2 + C_6H_{11}N{=}C{=}NC_6H_{11} \rightarrow \underset{R_3Pb}{C_6H_{11}\overset{|}{N}}{-}\underset{NEt_2}{\overset{|}{C}}{=}N{-}C_6H_{11}$$

Benzal-malodinitrile reacted via 1,4 addition to give $R_3PbN{=}C{=}C(CN){-}CH(NEt_2)Ph$. All of these exothermic reactions proceeded smoothly at 0–20°.

Tributyllead hydride and methoxide also react with unsaturated organic compounds to give organolead products, as mentioned earlier in this chapter. Thus, tributyllead hydride (273) is reported to react with alkenyl nitrile compounds to give keteneimines, and triphenyllead methoxide and tributyllead methoxide (79) are reported to react with dinaphthylcarbodiimide, naphthyl isocyanate, phenyl isothiocyanate, chloral, carbon disulfide, and trichloroacetonitrile. The latter reactions proceed rapidly at room temperature and are usually exothermic. They involve cleavage of the lead–oxygen bond of the alkoxide and addition of the fragments across a C=N, C=O, C=S, or C≡N moiety. The resultant triphenyllead products were stable at room temperature under nitrogen but the tributyllead compounds were unstable and blackened within a few days (79).

The direct oxidation of bis(triethyllead) sulfide in air gives bis(triethyllead) sulfate in good yield (61,176); triorganolead hydroxides absorb carbon dioxide to form the bicarbonate derivative, which can be converted to the carbonate by storage over activated alumina (61).

(2) *Physical properties*

Organolead salts of the type R_3PbX and R_2PbX_2 vary from highly crystalline, high melting, ionic-type compounds to low melting covalent-like solids or liquids. The melting points are strongly dependent on the nature of both the organic group R and the acid anion X. With halide derivatives, the melting point decreases in a homologous series in going from the fluoride to iodide derivatives. Thus, triphenyllead fluoride decomposes at 318° without melting, while triphenyllead chloride melts at 206°, the bromide at 166°, and the iodide at 138–139°. Similarly, triethyllead fluoride decomposes at 240° without melting, triethyllead bromide melts at 103–104°, and the iodide is an oil which freezes at 19–20°; triethyllead chloride decomposes below its melting point. Aryllead salts are generally higher melting than their alkyl analogs. Triethyllead cyanide melts at 189° and the thiocyanate derivative melts at 35°, while triphenyllead cyanide and thiocyanate decompose at 230–250° without melting.

With alkyllead salts, the melting points decrease with increasing chain length of the alkyl group; this effect is the reverse of that exhibited by *tetra*alkyllead compounds.

The salts of the lower alkyl derivatives of simple acids exhibit good solubility in water; as expected, solubility in water decreases with increasing alkyl chain length. Triorganolead salts of inorganic acids generally are more soluble in organic solvents than are the corresponding diorganolead salts. The solubility of organolead salts is as much dependent on the nature of the acid anion as on the nature of the organic group bound to lead. A large number of alkyllead salts is insoluble in water, making possible their facile preparation by metathesis reactions in aqueous solutions. Gilman's group (120) prepared a series of triethyllead salts by reaction of triethyllead acetate with the sodium salt of the respective acid in aqueous solution.

Aryllead salts tend to be more soluble in organic solvents, such as chloroform, than in water. Schweitzer and McCarty (341) determined the stability constants of a series of triphenyllead salts by measuring the extraction coefficient of the respective sodium salt of the anion under investigation between chloroform (or methyl isobutyl ketone) and a water solution of triphenyllead hydroxide. Similar studies were conducted by Bock and Deister (45).

Trialkyllead salts are especially strong sternutators (i.e., they induce sneezing). The sternutatory properties of organolead salts were investigated extensively by Gilman (120) and Saunders (150,151,257,319–321), and their co-workers in a series of papers. The sternutatory activity of R_3PbX compounds was found to increase with increasing chain length of the alkyl group in going from methyl to propyl, then to decrease with butyl; trimethyllead salts were found to exhibit negligible sternutatory activity. The activity of trialkyllead salts is very much stronger than that of their dialkyllead analogs; diethyllead salts possess negligible sternutatory activity. The sternutatory activity is also strongly dependent on the nature of the acid anion. Salts of organic acids are usually stronger sternutators than are those of inorganic acids; triethyllead thiocyanate and selenocyanate show no sternutatory activity. *N*-Tripropyllead methanesulfonamide is one of the most powerful lead sternutators; it is reported to be only slightly less potent than the best arsenic compounds.

Few systematic studies have been made of the physical properties of organolead salts. The *dipole moments* of several organolead halides have been reported in two papers (234,250) with generally poor agreement (Table XI-12).

The large values for the dipole moments of the organolead halides have been interpreted to indicate the presence of a high degree of ionic character in the lead–halogen bond. From the dipole moments, it has been concluded that the organolead halides are about 25–34% ionized, making them comparable to the lead(II) halides (351,352). However, *conductometric measurements* on solutions of triphenyllead chloride in dimethylformamide and pyridine show it to be a nonconductor, indicating that ionization to Ph_3Pb^+ and Cl^- does not occur to an appreciable extent (364,365). A low conductivity was obtained with the pyridine solution, but this was attributed to impurities in the pyridine. Matwiyoff and Drago (254) have shown

Table XI-12 Dipole Moments of Organolead Halides

COMPOUND	DIPOLE MOMENT, DEBYES[a]
$(C_2H_5)_3PbCl$	4.39 (4.66)
$(C_2H_5)_3PbBr$	4.46 (4.88)
$(CH_3)_3PbCl$	4.47
$(C_2H_5)_2PbCl_2$	4.70
$(C_6H_5)_3PbCl$	4.21 (2.32)
$(C_6H_5)_3PbBr$	4.21 (0.81)
$(C_6H_5)_3PbI$	3.73

[a] Taken from Lewis, Oesper, and Smyth, *J. Am. Chem. Soc.*, **62,** 3243 (1940). Values in parentheses taken from Malatesta and Pizzotti, *Gazz. chim. ital.*, **73,** 349 (1943); through *C.A.* **41,** 1137 (1947).

that solutions of triethyllead chloride, trimethyllead chloride, and the analogous tin compounds are nonconductors in tetramethylene sulfoxide and dimethylacetamide even though these compounds form one-to-one addition compounds with these solvents. The conductivities of triphenyllead chloride and analogs of the other Group IVb metals in liquid hydrogen chloride at $-95°$ have been measured by Peach and Waddington (286); triphenyllead chloride showed the highest specific conductance (2.2×10^{-6} ohm^{-1} cm^{-1}). However, its conductance is only about 2% that of triphenylmethyl chloride, so that ionization to Ph_3Pb^+ and Cl^- did not occur to any considerable extent. The organolead halides are excellent conductors in highly polar solvents, such as water, where strong solvation can occur. In aqueous hydrochloric acid solution, triethyllead chloride and diethyllead dichloride form anionic complexes of the type $R_3PbCl_2^-$ and $R_2PbCl_3^-$. These and other complexes of organolead salts are discussed later.

The *ultraviolet absorption spectrum* of triethyllead chloride exhibits continuous absorption, accompanied by appreciable decomposition (307).

A strong band is found at 493 cm^{-1} in the *infrared spectra* of solid trimethyllead chloride and dimethyllead dichloride; this band has been assigned to the lead–carbon asymmetric stretch by Shier and Drago (347). No band for the symmetric stretch was observed, so that the trimethyllead species is concluded to be nearly planar and the dimethyllead species nearly linear in the solid compounds. This is consistent with structures in which the chloride ions are bridging groups in both compounds, the trimethyllead species being trigonal bipyramidal and coordinated to two bridging chlorides and the dimethyllead species being octahedral and coordinated to four bridging chlorides. The infrared spectrum of triethyllead chloride has been determined for both the solid compound and for its solution in carbon tetrachloride (373). A band occurs at 940 cm^{-1} which increases in intensity in going from solution to the solid state. The infrared spectrum of tripropyllead chloride, as well as those of various methyl- and ethyllead halides, has been determined and compared to the spectra of the corresponding tetraalkyl derivatives (12).

The *infrared spectra* of a number of new organolead salts have been reported in recent years. Triphenyllead azide, trimethyllead azide and diphenyllead diazide show absorption bands in the regions 2934–2946, 1261–1280, and 650–655 cm^{-1} (237,360–362). The first two bands were assigned to stretching modes of the azide group while the third band was concluded to be a bending mode. Similar bands, slightly displaced, are found in the analogous organotin azides. The infrared spectra of trimethyllead azide, isocyanate, and isothiocyanate have been determined and compared to the spectra of the corresponding derivatives of the other Group IVb metals (361). The three vibrational modes internal to the anion and the metal–anion stretching modes all show a shift to progressively higher wave lengths with increasing atomic weight of the Group IVb metal. The infrared spectra of triphenylplumbylbis(trimethylsilyl)-amine (324), trimethylplumbyltrimethylsilylmethylamine (325), triphenyllead fulminate (40), and triphenyllead thiocyanate (89) have also been reported.

The infrared spectra of solid trimethyllead acetate, formate, and chloroacetate have been interpreted by Okawara and Sato (276) to indicate that these compounds, even in the solid state, consist of a planar trimethyllead cation in association with the carboxylate anion. However, Janssen and co-workers (181) concluded from the infrared spectra of trimethyl- and trihexyllead acetate that these compounds exist as coordination polymers in which the lead atom is penta-coordinated, with bonds to the three alkyl groups and to two oxygen atoms belonging to two different acetate groups. This conclusion was based on the observation that the absorption bands of the carboxylate group are shifted in going from the solid compounds to solutions in carbon tetrachloride; whereas the positions of the carboxylate bands in the solid compounds are very close to those found for salt-like carboxylates, e.g., in sodium acetate, the bands observed in carbon tetrachloride solutions were shifted more than half way toward the normal positions for organic esters. The absorption spectra of oganotin carboxylates show similar effects, so that a similar polymeric structure is proposed; on the other hand, germanium analogs do not exhibit this band shift.

The infrared spectra of the trimethyltin carboxylates were originally interpreted to be indicative of an ionic compound containing a planar trimethyltin cation associated with a carboxylate anion (277). However, the spectra of trimethyltin perchlorate, nitrate, and carbonate have been concluded to be more consistent with that of a compound containing covalently bonded groups, and the spectrum of the perchlorate derivative is consistent with the presence of a bidentate perchlorate group (275). Similar conclusions have been drawn from the infrared spectra of Me_3SnBF_4, Me_3SnAsF_6, and Me_3SnSbF_6 (69,149).

The preponderance of data on the infrared spectra of the trialkyltin salts seems to support a covalent structure involving five-coordinate tin atoms. However, X-ray diffraction analysis of trimethyltin cyanide indicated it to have an ionic structure containing Me_3Sn^+ and CN^-, even with the very weakly basic cyanide ions (326). Tobias (366) has reviewed the subject of σ-bonded organometallic cations and concludes that the available evidence supports an ionic structure for trialkyllead and trialkyltin compounds with strongly basic anions. This conclusion is based primarily on data for the organotin compounds, since little data are available for the organolead compounds. Certainly, the organolead salts would be expected to exhibit more ionic character than their tin analogs, but the meager data available suggest that these compounds are largely covalent in character, except in strongly solvating solvents. Further investigation is needed to permit an unequivocal interpretation of the nature of the bonding in organolead salts.

The *nuclear magnetic resonance spectra* of a series of trimethyllead halides were determined by Fritz and Schwarzhans (99); the proton chemical shifts were shown to increase with increasing polarity of the lead–halogen bond, while the $J_{^{207}Pb—C—H}$ coupling constants progressively decreased. If a planar Me_3Pb^+ cation were assumed to be present in trimethyllead fluoride, and a trigonal pyramid structure assumed for trimethyllead iodide, it was concluded that the variation in J values paralleled the variation of the C—Pb—C angle in the trimethyllead moiety. The proton shift and coupling constants for trimethyllead hydroxide lay between the fluoride and chloride (Table XI-13).

The NMR spectra of trineopentyllead chloride and bromide (349) show that the protons of the neopentyl moiety couple fairly strongly with the lead atom. The NMR spectrum of trimethylplumbylbis(trimethylsilyl)-amine has been compared to those of the trimethylstannyl and trimethylgermyl analogs, $(Me_3M)N(SiMe_3)_2$ (324). The NMR (and infrared) spectra indicate that the Si–N bond is strengthened by inductive effects exerted by the trimethylmetal moiety, the inductive effect being greatest with lead (Pb > Sn > Ge > Si) because of more facile interaction of the unshared electron pair of nitrogen with the *d*-orbitals of the larger Group IV metals.

Table XI-13 NMR Data for Trimethyllead Halides[a]

TRIMETHYLLEAD DERIVATIVE	δCH_3, CPS	$J_{^{207}Pb-C-H}$ IN CPS
Fluoride	−80	81
Hydroxide	−92	76
Chloride	−97.5	70
Bromide	−105	68
Iodide	−110.5	63

[a] For saturated $CHCl_3$ solutions, relative to tetramethylsilane. Taken from Fritz and Schwarzhans, *Chem. Ber.*, **97**, 1390 (1964).

Examination of triphenyllead azide by *differential thermal analysis* indicated the existence of three different polymorphs (71a). Polymorph I was obtained upon recrystallization from benzene; it was stable below 48° and was reversibly converted to II above this temperature. II was stable above 48° up to the decomposition temperature of 177°. III was formed upon recrystallization from chloroform or ethanol or by compressing I to 10,000 psi at room temperature; III was irreversibly converted to II above 110°.

Mixtures of various methyllead chlorides, ethyllead chlorides, ethyllead nitrates, and phenyllead chlorides have been separated successfully by *paper chromatography* using a butanol solution saturated with aqueous hydrochloric acid (33,34). Analyses of mixtures of different tetraorganolead compounds have been accomplished by conversion of the mixture to the R_3PbCl or R_3PbBr compounds and subjecting the resulting mixture of halide derivatives to paper chromatography (287,323). Paper chromatographic and ion exchange techniques have also been employed to demonstrate the presence of anionic organolead halide complexes in aqueous hydrochloric acid solutions.

(3) *Chemical properties*

Organolead salts generally are thermally unstable and decompose at elevated temperatures via a disproportionation reaction,

$$2R_3PbX \xrightarrow{\Delta} R_4Pb + R_2PbX_2$$
$$2R_2PbX_2 \xrightarrow{\Delta} R_3PbX + (RPbX_3)$$
$$(RPbX_3) \longrightarrow RX + PbX_2$$

so that the ultimate products of decomposition of both R_3PbX and R_2PbX_2 compounds are RX, PbX_2 and R_4Pb. This mode of decomposition is well documented and occurs with a variety of organolead salts (9,17,21,62,90, 257,302). The R_2PbX_2 compounds tend to be less stable than their R_3PbX analogs. Most organolead salts decompose below their melting point. Surprisingly, triphenyllead azide shows no explosive properties; however, an explosive decomposition occurred at 190–200° with diphenyllead diazide in the absence of an inert solid diluent (237). Tetraphenyllead, nitrogen, and biphenyl were identified as products from the thermal decomposition of triphenyllead azide (306). Tripropyllead fulminate also decomposed explosively without melting (40) and a product presumed to be dimethyllead diperchlorate exploded readily at room temperature when dried (98). However, a series of methyllead perchlorate compounds has been isolated as coordination complexes with pyridine and other Lewis bases (347); these perchlorate complexes melted without decomposition. The use of the triethyllead derivative of trinitroresorcinol in blasting caps is claimed in the patent literature (57) and the pyrolysis of diethyllead dicyanide is described in the patent literature as a method of preparing pure lead(II) cyanide (344).

The alkyllead halides, except for the fluorides, are unstable and tend to decompose slowly at room temperature even in the absence of light. Because of this instability, samples of organolead halides should be recrystallized after extended storage to remove lead(II) halide impurities. Furthermore, because of the relative instability of organolead salts in general, any interpretation of the mechanism of reactions involving organolead salts should consider the possibility of a simple thermal disproportionation.

The thermal disproportionation of triorganolead salts is a reversible process:

$$2R_3PbX \rightleftharpoons R_2PbX_2 + R_4Pb$$

Thus, Austin (21) obtained triphenyllead chloride in 86% yield by heating a mixture of tetraphenyllead and diphenyllead dichloride in an inert solvent; triethyllead chloride was obtained in 39% yield from tetraethyllead and diethyllead dichloride. On the other hand, tetraphenyllead and di-*o*-tolyllead dichloride did not react under similar conditions. In dimethyl sulfoxide, triphenyllead chloride was obtained in only 15% yield in 24 hr at 95° (223a); a shift in the equilibrium resulting from solvation effects of the dimethyl sulfoxide was postulated to explain this poor yield.

A second type of decomposition of organolead salts has been reported by Calingaert and co-workers (62) which involves the formation of a hydrocarbon as one of the products; this type of decomposition was observed during steam distillation of organolead halides. It does not involve a simple hydrolysis, since the hydrocarbon formed is that resulting from dimerization of the organic group, instead of simple protonation, i.e., R—R is formed instead of RH.

$$2R_3PbX \xrightarrow{\text{steam}} R_4Pb + PbX_2 + R{-}R$$

$$R_2PbX_2 \xrightarrow{\text{steam}} PbX_2 + R{-}R$$

The decomposition of triethyllead chloride and triethyllead bromide in the presence of steam gave butane in amounts equivalent to 33 and 20% decomposition, respectively, of the R_3PbX, based on the first reaction above.

A third type of thermal decomposition of organolead salts has been observed which involves the decarboxylation of certain triorganolead carboxylates according to:

$$R_3PbOOCCHR'COOR'' \rightarrow R_3PbCHR'COOR'' + CO_2$$

This reaction was discovered originally by Kocheshkov and Aleksandrov (189) in 1934. Decomposition of triphenyllead ethylmalonate at 165° gave $Ph_3PbCH_2COOC_2H_5$ in 40% yield; the triphenyllead derivative of the ethyl ester of benzyl malonic acid gave $Ph_3PbCH(CH_2C_6H_5)COOC_2H_5$ in 35% yield. Similar decompositions have been demonstrated by Willemsens and van der Kerk (386) for triphenyllead cyanoacetate and propiolate, but not for the acetoacetate and picolinate. This type of decomposition appears to be limited to a carboxylate salt having a strongly electronegative function in the β-position with respect to the carboxylate group and represents a convenient method of synthesizing organolead compounds containing functionally substituted organic groups.

Triorganolead halides undergo redistribution-type reactions with tetraorganolead compounds and other organolead halides (60,65). Thus, triethyllead chloride and trimethyllead chloride undergo redistribution of alkyl groups (and also chloride groups) to yield a mixture of ethylmethyllead halides (65):

$$(CH_3)_3PbCl + (C_2H_5)_3PbCl \longrightarrow (CH_3)_i(C_2H_5)_{3-i}PbCl$$

No catalyst is required to effect the redistribution of alkyl or aryl groups between organolead halides, since organolead halides are catalysts in the redistribution of alkyl groups between tetraalkyllead compounds (60). Redistribution reactions of organolead halides yield a statistical distribution of products, analogous to the R_4Pb compounds (60). However, redistribution reactions of organolead halides are accompanied by simple thermal disproportionation, so that a more complex reaction is involved than in the redistribution of tetraorganolead compounds (261).

The majority of organolead salts are stable to hydrolysis as evidenced by their facile preparation in aqueous media. R_3PbX compounds generally are more hydrolytically stable than their R_2PbX_2 analogs. The azides are surprisingly hydrolytically stable, as are many of the thiol derivatives. However, derivatives of very weak acids, e.g., Ph_3PbPPh_2, $Me_3PbN(SiMe_3)_2$, and Et_3PbOEt, are hydrolytically unstable and are readily hydrolyzed under mild conditions to yield the hydroxide. The hydrolysis of dimethyllead dichloride in aqueous sodium perchlorate has been investigated by measurement of solution pH as a function of concentration (98); the data indicated that no appreciable hydrolysis occurred below about pH 5. Three species were detected: $Me_2Pb(OH)_2$, $Me_2Pb(OH)_3^-$, and dimeric $[(CH_3)_2Pb(OH)_2Pb(CH_3)_2]^{2+}$. The NMR and infrared spectra of the nonhydrolyzed solutions of dimethyllead dinitrate indicated that the aquodimethyllead cation has a linear skeleton; weak interactions were detected between the dimethyllead cation with water molecules and nitrate ions in the first coordination sphere.

Diaryllead dicarboxylates undergo hydrolysis under relatively mild conditions. At room temperature in acetone solution containing about the stoichiometric amount of water, hydrolysis occurs within a few hours, accompanied by the precipitation of a basic salt of the composition, $Ar_2Pb(OH)OOCR$ (285,398). In the presence of stoichiometric amounts of a diazoalkane, hydrolysis gave compounds of the composition $[Ar_2Pb(OOCR)]_2O$ (198,285,398); the preparation of these compounds was discussed earlier in this chapter. Because of their facile hydrolysis, diorganolead dicarboxylates are best recrystallized from solutions containing a small amount of the free acid (386).

R_3PbX and R_2PbX_2 compounds undergo facile reactions with sodium and lithium metal which are useful for the preparation of R_6Pb_2 and R_3PbM compounds (see Chapters VII and XI.8). The stoichiometry of the reaction of trimethyllead chloride with potassium and lithium was discussed in Chapter VI.2, as were reactions of organolead salts with various electrophilic reagents.

Organolead salts react with certain inorganic metal salts to form organometallic derivatives of other metals. Typical of such reactions are:

$$R_2PbCl_2 + SnCl_2 \rightarrow R_2SnCl_2 + PbCl_2 \qquad (193);\; R = Me, Et, Ph$$

$$Et_3PbCl + TlCl_3 \rightarrow Et_2PbCl_2 + EtTlCl_2 \qquad (134)$$

$$Et_3PbOAc + Hg(OAc)_2 \rightarrow Et_2Pb(OAc)_2 + EtHgOAc \qquad (280)$$

A recent patent (71) claims that decomposition with ozone is an efficient way of removing triethyllead chloride from water; this reaction is reported to be applicable to the decontamination of aqueous streams used to scrub impurities from tetraethyllead produced via the conventional alloy reaction.

The reaction of diphenyllead dichloride with dilithioazobenzene, $C_6H_5N(Li)$—$N(Li)C_6H_5$, gave lead metal and azobenzene and not octaphenyl-1,2,4,5-tetraza-3,6-diplumbacyclohexane in which the lead would be bonded to nitrogen in a heterocyclic ring system (101). Contrariwise, diphenyldichlorosilane and diphenyldichlorogermane gave the heterocyclic product, while diphenyltin dichloride gave diphenyltin polymer and azobenzene.

Dessy and co-workers (81) have shown that triphenyllead chloride and diphenyllead dichloride (193) react rapidly with mercury metal at ambient temperature in dimethoxyethane solvent to yield diphenylmercury and lead metal; the acetate salts did not react under similar conditions. Reaction of diphenyllead dichloride and di-*o*-anisyllead dichloride with lead metal or sodium–lead alloy in refluxing xylene gave the corresponding triaryllead chloride (264); however, the products probably arose from a simple thermal decomposition of the organolead halide and not from a chemical reduction. Reaction of diphenyllead dichloride with hydrazine in ethanol gave tetraphenyllead in 54% yield (107); a higher yield (74%) was obtained using triphenyllead chloride. Diphenyllead diacetate also reacted with hydrazinium sulfate to form tetraphenyllead (264).

The reduction of di-*m*-nitrophenyllead dibromide with iron metal, titanium(III) chloride, or tin(II) chloride in hydrochloric acid is reported to give aniline in high yield, along with unspecified lead-containing products (191); on the other hand, careful addition of titanium(III) chloride to di-*m*-nitrophenyllead oxide in alcoholic hydrochloric acid is reported to give di-*m*-aminophenyllead dichloride (334). The reaction of triethyllead acetate with ferrocene, titanium(III) chloride, iron(II) acetylacetonate, and vanadium(III) acetylacetonate gave ethane in varying amounts, dependent on the solvent (374). Ethane formation was attributed to the formation of free ethyl radicals which extracted a proton from the solvent. Little or no ethylene or butane was formed in these systems.

The reaction of diphenyllead dinitrate with hot concentrated nitric acid or with a mixture of nitric acid and sulfuric acid gave di-*m*-nitrophenyllead dinitrate in low yield (68,371). Similarly, diphenyllead diacetate is reported to react with nitric acid at room temperature to yield di-*m*-nitrophenyllead dinitrate (334).

Triethyllead chloride and diethyllead dichloride undergo an insertion reaction with diazomethane in the presence of copper bronze to yield chloromethyltriethyllead and unstable di(chloromethyl)diethyllead (391, 392); with diazoethane, triethyllead chloride gave unstable (α-chloroethyl)triethyllead, which decomposed to tetraethyllead and lead(II) chloride during work-up. No reaction was obtained between lead(II) halides or lead(IV) halides with diazomethane, but lead(IV) acetate was reduced to lead(II) acetate.

A Russian patent (295) describes the reaction of trialkyllead chlorides with olefin oxides in liquid ammonia to yield hydroxyalkyltrialkyllead. Willemsens and van der Kerk (387) have shown that triphenyllead acetate reacts with ketene at room temperature in ethanol solvent to yield ethyl-(triphenylplumbyl)acetate, $Ph_3PbCH_2COOC_2H_5$; no reaction occurred in diethylether, and it was speculated that this was caused by the low solubility of triphenyllead acetate in ether. Ketene did not react with triethyllead acetate, triphenyllead chloride, or diphenyllead diacetate. Triphenyl- or triethyllead halides did not react with triethyl phosphite at elevated temperatures although trialkyltin halides did react to form trialkyltin dialkylphosphites (17).

Neumann and Kühlein (274) have described reactions of tributylplumbyl-diethylamine with various unsaturated organic compounds; these reactions were discussed under Other Synthesis Methods in Chapter VII.

(Triphenylplumbyl)diphenylphosphine and other phosphine derivatives have been prepared recently but their reactions have not been investigated (335,338,339). They react with alcoholic potassium hydroxide in the presence of air to form triphenyllead hydroxide and potassium orthophosphate. (Triphenylplumbyl)diphenylarsine and (triphenylplumbyl)-diphenylstibine have also been prepared (335,337). These latter compounds react readily with sulfur; they are also oxidized by air or alcoholic aqueous hydrogen peroxide to yield $Ph_3PbOAsPh_2$ and $Ph_3PbOSbPh_2$, respectively (337).

Trialkyllead halides undergo polarographic reduction involving a one-electron process which is ascribed to the reduction of the triorganolead cation to the R_3Pb radical (or R_6Pb_2) (72,73,308,367). A polarographic method for the determination of tetraethyllead has been developed which is based on its prior conversion to triethyllead halide.

Dessy and co-workers (81) have investigated the electrochemistry of the phenyl derivatives of the Group IVb metals of the type R_3MX and R_2MX_2. Diphenyllead diacetate was shown to undergo reduction in two one-electron steps at the dropping mercury electrode; these steps were postulated to involve the sequence:

$$Ph_2Pb(OAc)_2 \xrightarrow[-1.1\,V]{e^-} Ph_2Pb(OAc) + OAc^- \xrightarrow[-1.6\,V]{e^-} Ph_2Pb + OAc^-$$

Electrolysis of diphenyllead diacetate at a potential of −2.1 V gave diphenylmercury in quantitative yield. Similar results were obtained with triphenyllead acetate, i.e., diphenylmercury was formed in quantitative yield:

$$Ph_3PbOAc \xrightarrow[-1.4\,V]{e^-} Ph_3Pb^0 + OAc^-$$

It was concluded that arylation of mercury metal by the Ph_3Pb^0 species occurred faster than a coupling reaction to form hexaphenyldilead.

Electrochemical reduction of the triphenylmetal halides of the Group IVb metals becomes more facile with increasing atomic weight of the Group IVb metal. Electrochemical reduction of triphenyltin chloride gave hexaphenylditin, while the germanium and silicon analogs gave the Ph_3MH derivative. On the other hand, triphenyllead chloride reacted with mercury metal at room temperature in the absence of an applied potential, so that its electrochemical reduction could not be investigated.

The β decay of $Ph_3Pb^{210}I$ has been investigated and the decay products analyzed by paper chromatography (390). Diphenylbismuth iodide was detected as one of the major products along with varying amounts of "inorganic" bismuth, depending on the solvent used to develop the chromatogram. The results obtained by chromatography did not compare too well with results obtained by extraction of the decay products with a carrier (otherwise unidentified).

B. Monoorganolead salts, $RPbX_3$

The preparation of a monoorganolead salt of the type $RPbX_3$ was originally claimed by Lesbre (229,230) who reported the synthesis of an alkyllead trihalide according to the reaction:

$$RI + CsPbCl_3 \rightarrow RPbCl_3 + CsI$$

Subsequent attempts to confirm Lesbre's results were unsuccessful (62,85,280) and showed that alkyllead trihalides are too unstable to permit their isolation. Therefore, the first successful synthesis of an $RPbX_3$ compound is usually credited to Panov and Kocheshkov (280), who reported the synthesis of phenyllead triacetate in 1952. Several other

$RPbX_3$ compounds have been synthesized since then, but all are compounds in which R is an aryl group and X is a carboxylate group. Where R is aryl and X is halide, or where R is alkyl, the resultant $RPbX_3$ compound is unstable and disproportionates into RX and PbX_2 (75). $RPbX_3$ compounds are probably formed as unstable intermediates in many reactions involving the dealkylation of R_2PbX_2 compounds as evidenced by the formation of RX and PbX_2 as ultimate products in many of these reactions.

A number of aryllead tricarboxylates has now been prepared and characterized. They are usually obtained as low melting solids or oils and range in color from colorless to light yellow. The properties of these compounds are very different from those of R_2PbX_2 and $RPbX_3$ compounds. Thus, they require special methods of synthesis and they exhibit pronounced oxidative activity. Also, their mode of thermal decomposition appears to be different from that of R_3PbX or R_2PbX_2. In many respects, they are more similar in their properties to the lead tetracarboxylates than to other organolead salts. The known aryllead tricarboxylates are given in Table XI-14.

(*1*) *Synthesis*

(*a*) Reactions of $Ar_2PbX_2 + HgX_2$

Phenyllead tricarboxylates were originally prepared by Panov and Kocheshkov (280) by reaction of a diaryllead dicarboxylate with a mercury(II) carboxylate according to:

$$Ar_2Pb(OOCR)_2 + Hg(OOCR)_2 \rightarrow ArPb(OOCR)_3 + ArHgOOCR$$

A number of aryllead tricarboxylates has been synthesized via the $Ar_2Pb(OOCR)_2$–$Hg(OOCR)_2$ system. The reaction proceeds readily at room temperature and is usually conducted in an inert solvent, such as chloroform. The by-product arylmercury carboxylate is precipitated from the reaction mixture as the chloride by addition of HCl.

(*b*) Reactions of $Pb(OOCR)_4$ and Ar_2Hg

A second general method of synthesis of aryllead tricarboxylates is the reaction of a lead(IV) carboxylate with a diarylmercury compound (75,196,197,241–243,245,284,300).

$$Pb(OOCR)_4 + Ar_2Hg \rightarrow ArPb(OOCR)_3 + ArHgOOCR$$

The products are identical to those obtained in the $R_2Pb(OOCR)_2$–$Hg(OOCR)_2$ system above and the reaction is carried out under very

Table XI-14 Aryllead Tricarboxylates, $ArPb(OOCR)_3$

COMPOUND	MELTING POINT, °C	REFRACTIVE INDEX, n^{20}	REFERENCES
$PhPb(OOCCH_3)_3 \cdot H_2O$	77–86		386
$PhPb(OOCCH_3)_3$	101–2		148,196,244,280,284
	103–5		75
$PhPb[OOCCH(CH_3)_2]_3$	77–8		196,280,284
	77–9		386
$PhPb(OOCC_6H_5)_3$	149.5–51		196,280
$PhPb[OOCCH(C_2H_5)C_4H_9]_3$			278
$PhPb(OOCC_{11}H_{23})_3$	42		278,386
$PhPb(OOCCH_2Cl)_3$	Oil	1.5958	265,266
$PhPb[OOCCH(Br)CH_3]_2(OOCCH_3)$	Oil	1.5855	265,266
$PhPb[OOCCH(Br)CH_3]_3$	Oil	1.5815	265,266
$PhPb(OOCCH_2CH_2Br)_3$	Oil	1.5825	265,266
$PhPb(OOCCH_2CH_2CH_2Cl)_3$	Oil	1.5630	265,266
$PhPb[OOC(CH_2)_{10}Br]_3$	Oil	1.5350	265,266
$PhPb(OOCCH_3)_2[OOCCH(Br)CH_3]$	Oil	1.5878	265,266
p-$CH_3C_6H_4Pb(OOCCH_3)_3$	86–8		75,278,280,284
p-$CH_3C_6H_4Pb(OOCC(CH_3){=}CH_2)_3$	dec. 120		280
p-$CH_3C_6H_4Pb(OOCCH_2CH_2Cl)_3$	Oil	1.5734	197,265,266
p-$CH_3OC_6H_4Pb(OOCCH_3)_3$	135–40		75
	139–41		243
p-$CH_3OC_6H_4Pb(OOCC_2H_5)_3$	75–6		243
p-$CH_3OC_6H_4Pb(OOCC_6H_5)_3$	164		243
$[2,4(CH_3O)_2C_6H_3]Pb(OOCCH_3)_3$	146–9 (dec.)		299
$[2,5(CH_3O)_2C_6H_3]Pb(OOCCH_3)_3$	165–7		300
p-$IC_6H_4Pb(OOCCH_3)_3$	110–2		245
p-$IC_6H_4Pb(OOCC_2H_5)_3$	96		245
p-$IC_6H_4Pb[OOCCH(CH_3)_2]_3$	78		245
p-$IC_6H_4Pb(OOCC_6H_5)_3$	177–8		245
1-$C_{10}H_7Pb(OOCCH_3)_3$	165–6		386
	168–9		241
1-$C_{10}H_7Pb[OOCCH(CH_3)_2]_3$	99.5–101		241
1-$C_{10}H_7Pb(OOCC_6H_5)_3$	173–4		241
2-$C_{10}H_7Pb(OOCCH_3)_3$	130–5		386
	108–9		242
2-$C_{10}H_7Pb(OOCC_2H_5)_3$	84–5		242
2-$C_{10}H_7Pb(OOCCH_2CH_2Cl)_3$	Oil	1.6068	265,266
2-$C_{10}H_7Pb(OOCC_6H_5)_3$			278

similar conditions. This system is no more or less convenient than the other; the choice of method of synthesis of a specific compound will depend primarily on the availability of the required reagents.

Criegee and co-workers (75) investigated the reaction of lead(IV) acetate with a variety of dialkyl- and diarylmercury compounds, including benzylphenylmercury, and concluded that the reaction involves an electrophilic attack of the lead(IV) acetate on the mercury compound, as evidenced by the fact that the products from the phenylbenzylmercury reaction were phenyllead triacetate and benzylmercury acetate. Also the rate of reaction of lead(IV) acetate with various R_2Hg compounds was shown to decrease in the order *p*-anisyl > phenyl > *n*-butyl > benzyl > phenethyl > neophyl. This is the same order as has been observed for cleavage of diorganomercury compounds by acetic acid.

(*c*) Other synthesis methods

The synthesis of 2,4-dimethoxyphenyllead triacetate by reaction of lead(IV) acetate in excess 1,3-dimethoxybenzene has been reported by Preuss and Janshen (299); *p*-methoxyphenyllead triacetate has been prepared similarly (148). Presumably, this type of reaction can be extended to any benzene derivative which contains a relatively reactive hydrogen on the ring. The successful preparation of di-2-thienyllead diisobutyrate (283) from thiophene and lead(IV) isobutyrate suggests that 2-thienyllead triacetate is an intermediate and may be reasonably stable. Its successful preparation would represent the first stable $RPbX_3$ derivative in which the organic moiety is not an aryl group.

Aryllead tricarboxylates might also be prepared by the reaction of a tetraaryllead compound with lead(IV) carboxylate since diphenyllead diacetate has been prepared in this manner (193). Willemsens and van der Kerk (386) attempted the synthesis of phenyllead triacetate by this method. At room temperature the reaction was very slow even in the presence of mercury(II) acetate; at higher temperature, considerable decomposition occurred. However, *p*-tolyllead triacetate was successfully prepared in 80% yield from the reaction of di-*p*-tolyllead diacetate with lead(IV) acetate for 40 hr in acetic acid at 70° and in the presence of a small amount of mercury acetate (388). The arylation of lead(IV) acetate by phenyllithium proved to be unsatisfactory for the synthesis of phenyllead triacetate (386).

Several aryllead tricarboxylates have been prepared by transesterification-type reactions involving an aryllead tricarboxylate and a carboxylic acid, or by reaction of a carboxylic acid with an areneplumbonic acid (241,245,280). These latter compounds in turn are derived by neutralization of an aryllead tricarboxylate with aqueous ammonia in absolute alcohol (196,242,245).

$$ArPb(OOCR)_3 + 3R'COOH \longrightarrow ArPb(OOCR')_3 + 3RCOOH$$

$$ArPbOOH + 3R'COOH \longrightarrow ArPb(OOCR')_3 + 2H_2O$$

(2) *Physical and chemical properties*

Few physical properties measurements have been reported for the aryllead tricarboxylates. They exist as relatively low melting, crystalline solids, or as oils. Willemsens and van der Kerk (386) isolated the laurate, stearate, ethylhexanoate, α-chloroacetate, β-chloropropionate, and γ-chlorobutyrate derivatives as oils (379). Phenyllead triacetate, although reported to be a crystalline solid (75,280), could not be crystallized; the triacetate was eventually isolated as a crystalline hydrate by addition of a small amount of water (386).

The infrared spectra of a number of monoaryl- and diaryllead carboxylates have been investigated by Bogomolov and co-workers (46); the absorption band in the 430–460 cm^{-1} region was assigned to the lead–carbon bond.

The aryllead tricarboxylates exhibit surprisingly good thermal stability, compared to the trends in thermal stability indicated by the R_3PbX and R_2PbX_2 compounds, but the α-halocarboxylates are fairly unstable and decompose at room temperature within a few days (379). No systematic investigation has been made of the pyrolysis of aryllead tricarboxylates. The available data suggest that their mode of decomposition is different from that of R_2PbX_2 compounds to form RX and PbX_2. Thus, heating *p*-methoxyphenyllead triacetate in benzene at 80° gave 4-methoxybiphenyl as the major product; phenyllead triacetate in anisole gave all three methoxybiphenyl derivatives under similar conditions (148). These products are indicative of a free radical decomposition.

The decomposition of phenyllead triacetate in hot acetic acid gave lead(II) acetate, and not lead(IV) acetate; the latter would be formed if simple acetolysis of the phenyl–lead bond occurred. The decomposition of 2,5-dimethoxyphenyllead triacetate in hot acetic acid also gave lead(II) acetate, along with carbon dioxide, dimethoxybenzene, dimethoxybenzoic acid, dimethoxybenzaldehyde, and monoacetoxy derivatives of dimethoxybenzene (300); on the other hand, *p*-methoxyphenyllead triacetate in acetic acid gave only small amounts of anisole after 2 days at 80° (148). *p*-Tolyllead triacetate appears to be stable to acetolysis as evidenced by the fact that it could be isolated in 80% yield by heating di-*p*-tolyllead diacetate and lead(IV) acetate in acetic acid for 40 hr at 70° (388). Hydrogen sulfide decomposed 2,5-dimethoxyphenyllead triacetate in methanol–chloroform solution (148); dimethoxybenzene was identified as the major product.

Aryllead tricarboxylates are weak oxidizing agents. Thus, they are reported to oxidize potassium iodide to iodine (386); on the other hand, phenyllead triacetate is reported to react with sodium iodide in absolute ethanol to form lead(II) iodide and iodobenzene (244). They also give a green coloration with phenothiazine (386); a similar coloration is produced by lead(IV) acetate, but not by other organolead compounds. Triphenylstibine was oxidized by phenyllead triacetate in chloroform solution to give $Ph_3Sb(OAc)_2$ and $Ph_2Pb(OAc)_2$; with Ph_2SbOAc, the products were $Ph_2Sb(OAc)_3$, $Pb(OAc)_2$, and $Ph_2Pb(OAc)_2$ (244).

Aryllead tricarboxylates are hydrolytically unstable. Exposure to air for several days causes the melting point to become indefinite (196). Upon dissolution in acetone, phenyllead triacetate rapidly loses acetate groups to form a precipitate of a mixed phenyllead acetate oxide polymer (380). The aryllead tricarboxylates are less stable in polar solvents than in the solid state or in nonpolar solvents. Yet, phenyllead triacetate is sufficiently stable to hydrolysis to permit its isolation as a monohydrate (386). Two recent patents (265,266) describe the use of aryllead tricarboxylates as catalysts for the preparation of polyurethan foams in isocyanate–polyol and isocyanate–water reactions. This application has also been described by Overmars and van der Want (278).

C. Coordination compounds of organolead salts

The ability of organolead salts to form coordination compounds has been long known. Polis (296–298) prepared hydrates of diaryllead dinitrates and dicarboxylates as early as 1887. Pyridine complexes of diphenyllead dihalides and dinitrate, $Ph_2PbX_2 \cdot 4pyr$, were described by Pfeiffer and co-workers (291) in 1916. The latter complexes liberated pyridine upon exposure to air. Pfeiffer and co-workers also described a diammine complex of diphenyllead dibromide, $Ph_2PbBr_2 \cdot 2NH_3$, and in 1939 Foster and co-workers (97) detected by manometric methods the formation of a number of poorly defined complexes of triphenyllead chloride and ammonia of the type $Ph_3PbCl \cdot xNH_3$, where x was 1.3, 1.8, 2.7, and 9.65. Little further attention was devoted to the coordination chemistry of organolead compounds until the early 1960's. Since then, several new coordination compounds of organolead salts have been isolated and their structures elucidated. Complexes of organolead halides and carboxylates are now known with such ligands as pyridine (169,174,347), 1,10-phenanthroline (169,174), 2,2′-bipyridine (169), hexamethylphosphoramide (347), dimethylsulfoxide (224,347), dimethylformamide (169,254), tetramethylammonium chloride (169), tetramethylene sulfoxide (254), and dimethylacetamide (254). The known coordination compounds are listed in Table XI-15.

The stoichiometry of these coordination compounds is dependent on the nature of the ligand, the organolead moiety, and the acid anion. Thus, diphenyllead dichloride and diphenyllead dibromide form one-to-two (lead-to-ligand) complexes with dimethyl sulfoxide (169), but dimethyllead diperchlorate is reported to form a one-to-four complex (347). Pyridine is reported to form a one-to-one complex with trimethyllead perchlorate, a one-to-two complex with dimethyllead diperchlorate (347). The stoichiometry of most of these complexes is consistent with a six coordinate mononuclear lead atom with an octahedral distribution of ligands, or a four coordinate lead atom having a tetrahedral configuration. However, Matwiyoff and Drago (254) have shown that triethyllead chloride forms a one-to-one adduct with tetramethylene sulfoxide, the structure of which necessitates a coordination number of five. The infrared and NMR spectral data are consistent with a trigonal bipyramidal structure; similar structures were proposed for the analogous dimethylacetamide and dimethylformamide complexes.

The infrared spectrum of the solid complex $Me_3PbClO_4{\cdot}pyr$ showed a weak absorption band at 460 cm^{-1} which disappeared when the solid complex was dissolved in pyridine. This phenomenon has been interpreted by Shier and Drago (347) to be indicative of the presence of a nonplanar trimethyllead species in the solid complex which becomes planar in pyridine solution as a result of coordination of a second molecule of pyridine to the trimethyllead moiety; the 460 cm^{-1} band, which was assigned to the lead–carbon symmetric stretching vibration, persisted when the solid complex was dissolved in methylene chloride. The absence of a band in the 900–1200 cm^{-1} region in the spectra of both the pyridine and hexamethylphosphoramide complexes of trimethyllead perchlorate was interpreted as evidence that the perchlorate ion is not coordinated to the lead atom; the infrared spectrum of the hexamethylphosphoramide complex is consistent with the presence of a planar trimethyllead species (no 460 cm^{-1} band), the hexamethylphosphoramide being coordinated to lead through the phosphoryl oxygen.

Shier and Drago (347) also showed that the $^{207}Pb{-}CH_3$ coupling constant of trimethyllead chloride in various solvents increased as the donor character of the solvent increased (benzene < nitromethane < acetonitrile < acetone < methanol < pyridine < dimethyl sulfoxide < dimethylacetamide < dimethylformamide < hexamethylphosphoramide). They concluded that the solvent effect cannot be attributed solely to an increasing planarity of the trimethyllead species, since the coupling constant of trimethyllead perchlorate in hexamethylphosphoramide was appreciably greater than that of trimethyllead tetrafluoborate in water and this latter increase would be independent of any apparent configurational change in the geometry of the planar trimethyllead species. A similar solvent effect has been observed for solutions of dimethylthallium salts in various

Table XI-15 Coordination Compounds of Organolead Salts

COMPOUND	MELTING POINT, °C	REFERENCES
	A. R_3PbX *Derivatives*	
$Me_3PbCl{\cdot}TMSO$[a]		254
$Me_3PbCl{\cdot}DMA$[b]		254
$Me_3PbCl{\cdot}DMF$[c]		254
$Me_3PbClO_4{\cdot}2HMPA$[d]	86	347
$Me_3PbClO_4{\cdot}HMPA$[d]	96	347
$Me_3PbClO_4{\cdot}Pyr$[e]	135	347
Me_3PbDz[f]		38,180
$Et_3PbCl{\cdot}TMSO$		254
$Et_3PbCl{\cdot}DMA$		254
$Et_3PbCl{\cdot}DMF$		254
$Ph_3PbCl{\cdot}xNH_3$		97
$Ph_3PbCl{\cdot}Me_4NCl$	dec. 298	169
Ph_3PbDz		180
Ph_3Pb Oxine[g]	108–11	35,91,92
	B. R_2PbX_2 *Derivatives*	
$Me_2Pb(ClO_4)_2{\cdot}4DMSO$	66	347
$Me_2Pb(ClO_4)_2{\cdot}2Pyr$	115	347
$Me_2Pb(OCOCF_3)$ (Oxine)	>190	172
$Me_2Pb(Acac)_2$[h]	163–3.5	185,368
$Et_2PbBr_2{\cdot}Phen$[i]	106 (dec.)	174
$Et_2PbBr_2{\cdot}Bipyr$[j]	118 (dec.)	174
$Et_2Pb(Br)$ (Oxine)	110	172
$(Bu)_2PbCl_2{\cdot}Phen$	132 (dec.)	174
$(Bu)_2PbCl_2{\cdot}Bipyr$	130 (dec.)	174
Bu_2Pb $(Oxine)_2$	dec. >120	173
$Bu_2Pb(Br)$ (Oxine)	dec. >110	173
$Bu_2Pb(OAc)$ (Oxine)	dec. >180	172,173
$Bu_2Pb(Cl)$ (Oxine)	dec. >122	172
$Ph_2PbCl_2{\cdot}2DMSO$	dec. 168–70	169,224
$Ph_2PbCl_2{\cdot}Phen$	275 (dec.)	169,174
$Ph_2PbCl_2{\cdot}2Pyr$	>310	169,174
$Ph_2PbCl_2{\cdot}4Pyr$		291
$Ph_2PbBr_2{\cdot}4Pyr$		291
$Ph_2PbBr_2{\cdot}2NH_3$		291
$Ph_2PbCl_2{\cdot}2DMF$	dec. 76–80	169
$Ph_2PbCl_2{\cdot}Bipyr$	dec. 250; >300	92a,169

(continued)

Table XI-15 (continued)

COMPOUND	MELTING POINT, °C	REFERENCES
$Ph_2Pb(OBz)_2 \cdot DMF$	~233–40	386
$Ph_2Pb(NO_3)_2 \cdot 4Pyr$		291
$Ph_2Pb(OAc)_2 \cdot Pyr$	dec. 185–93	174
$Ph_2PbBr_2 \cdot Bipyr$	250(dec.)	92a
$Ph_2PbI_2 \cdot Bipyr$	180(dec.)	92a
$Ph_2PbCl_2 \cdot Terpyr$[k]	253–5	92a
$Ph_2PbBr_2 \cdot Terpyr$	233–4	92a
$Ph_2PbI_2 \cdot Terpyr$	150	92a
$Ph_2Pb\ (Oxine)_2$	174–7	91,173
$Ph_2Pb(Cl)(Oxine)$	dec. >187	172,173
$Ph_2Pb(Br)(Oxine)$	dec. >195	173
$Ph_2Pb(NO_3)(Oxine)$	dec. >228	172,173
$Ph_2Pb(OAc)(Oxine)$	dec. >200	173
$Ph_2Pb(OOCC_2H_5)(Oxine)$	dec. >212	172,173
$Ph_2Pb(OOCH_2Cl)(Oxine)$	dec. >204	172,173
$Ph_2Pb(OOCCF_3)(Oxine)$	dec. >240	172
$Ph_2Pb \cdot EDTAH_2$[l]		223
C. $RPbX_3$ *Derivatives*		
$PhPb(OAc)(Oxine)_2$	dec. ~158°	175
$PhPb(OEt)(Oxine)_2$	152.5	175
$PhPb(OMe)(Oxine)_2$	148	175
$PhPb(Oxine)_3$	dec. >120	175
D. *Other Derivatives*		
$Et_4Pb \cdot 2N_2O_4$, $[Et_4Pb(NO)_2]^{+2}(NO_3)_2^{-2}$	104–5 (dec.)	167,168
$Pr_4Pb \cdot 2N_2O_4$, $[Pr_4Pb(NO)_2]^{+2}(NO_3)_2^{-2}$	110–1 (dec.)	167,168

[a] Tetramethylene sulfoxide.
[b] *N,N*-Dimethylacetamide.
[c] *N,N*-Dimethylformamide.
[d] Hexamethylphosphoramide.
[e] Pyridine.
[f] Dithizonate.
[g] 8-Hydroxyquinolinate.
[h] Acetylacetonate.
[i] 1,10-Phenanthroline.
[j] Bipyridine.
[k] Terpyridine.
Ethylenediaminetetraacetic acid dianion.

solvents and attributed to increasing interaction between the metal atom and the donor solvent (346). A similar explanation probably applies to NMR effects observed with organolead salts in various coordinating solvents.

A number of complexes of organolead salts with chelating ligands are known; these include such ligands as dithizone (38,180), acetylacetone (185,368), ethylenediaminetetraacetic acid (223), 8-hydroxyquinoline (36,91,92,173), and 1-(2-pyridylazo)-2-naphthol (292). On the basis of the infrared and Raman spectra, dimethyllead bisacetylacetonate is proposed to have a *trans* configuration, the two-to-one stoichiometry being consistent with a six coordinate lead atom having an octahedral configuration. Triphenyllead salts form monooxinate(8-hydroxyquinolinate) complexes, while diphenyllead salts form dioxinates (91). The infrared spectra of diphenyllead dioxinate and diphenyltin dioxinate are consistent with a chelate structure in which lead and tin have a coordination number of six. The monooxinate derivatives of triphenyllead and triphenyltin can be either simple coordination compounds in which the oxine acts as a monodentate ligand, or chelate compounds, dependent on the solvent in which they are dissolved. From the ultraviolet spectra of triphenyllead oxinate in water, methanol, and ethanol, a simple, nonchelated coordination compound is indicated in which the lead atom has a coordination number of four (35,92). On the other hand, the ultraviolet spectrum of triphenyllead oxinate in benzene solution or in the solid state is consistent with a chelate complex in which the oxine is bidentate (91,92). This difference in behavior has been attributed to solvolysis of the lead–nitrogen bond of triphenyllead oxinate in methanol or ethanol solution; solvolysis occurs in ethanol solution as evidenced by the fact that a solution of triphenyllead oxinate in ethanol conducts electricity. The structures of the dithizone derivatives of di- and triorganolead halides are postulated to be very similar to those of the oxinates (35,317).

By means of ion exchange techniques, anionic complexes of the type $R_3PbCl_2^-$ and $R_2PbCl_3^-$ have been detected for triethyllead chloride or diethyllead dichloride in aqueous solutions containing hydrochloric acid or lithium chloride. The complexes are strongly adsorbed on Amberlite IRA-410 and SB-2 anion exchange paper and show bell-shaped distribution curves between the resin and the aqueous solution. Their formation has also been established by electrophoresis (129). The electrophoresis results indicate that the completely coordinated complexes have the composition $(C_2H_5)_3PbCl_3^{2-}$ and $(C_2H_5)_2PbCl_4^{2-}$; on the other hand, from the partition coefficients of triethyllead chloride between carbon tetrachloride and water, and the solubility of triethyllead chloride in aqueous lithium chloride and lithium perchlorate solutions, it was concluded that $(C_2H_5)_3PbCl_2^-$ is the highest complex formed by triethyllead chloride

(356). An anionic complex of the type $(R_4N^+)(Ph_3PbCl_2^-)$ has been postulated to account for the one-to-one coordination compound formed between triphenyllead chloride and tetramethylammonium chloride (169).

From the available data, it is obvious that the coordination chemistry of organolead salts is very similar to that of lead(II) halides and other divalent lead salts but further investigation is needed to elucidate the critical interrelationships in this area of organolead chemistry. Little information is available on the coordination chemistry of aryllead tricarboxylates. Huber and Haupt (175) recently reported the synthesis of a series of monophenyllead oxinates containing two or three oxinate groups. These compounds were prepared by treatment of phenyllead triacetate with 8-hydroxyquinoline. In benzene solvent, phenyllead acetate dioxinate, $PhPb(OAc)(Oxine)_2$, was obtained; in ethanol solvent, the ethoxide was formed, $PhPb(OEt)(Oxine)_2$, instead of the acetate. Alkoxide derivatives were also formed when $PhPb(OAc)(Oxine)_2$ was heated in methanol or ethanol solvent. Treatment of phenyllead triacetate with aqueous ammonia at 60° gave the trioxinate, $PhPb(Oxine)_3$. The coordination chemistry of the lead atom in these compounds was not elucidated. The compounds were very stable; all melted or decomposed above 120°.

As might be expected, coordination compounds are very useful in the analysis of organolead compounds. Electrophoresis of the anionic halide complexes is reported to permit a rapid and accurate separation of diethyllead dichloride and triethyllead chloride. Spectrophotometric methods of analysis for triethyllead and diethyllead salts have been developed based on the dithizone complexes and a recent publication describes the spectrophotometric determination of $(C_2H_5)_2Pb^{2+}$ by means of its complex with 4-(2-pyridylazo)resorcinol (293). Organolead complexes are discussed in recent review articles (76,103,145).

8. ORGANOLEAD–METAL COMPOUNDS, R_3PbM, R_2PbM_2

A. General

A limited number of compounds is known containing an organolead moiety in combination with an alkali or alkaline earth metal and having the composition R_3PbM. However, this class of organolead compounds is not well characterized; they are presumed to be ionic in nature, with the lead-to-metal bond exhibiting a high degree of ionic character. This type of ionic organolead compound is grossly different from the organolead salts of the type R_3PbX discussed above in that the organolead moiety in R_3PbM derivatives is anonic in character so that the charge distribution is best represented by the designation, $R_3Pb^{\delta-}M^{\delta+}$. Their composition was deduced from their utility in the synthesis of unsymmetrical tetraorganolead compounds by reaction with an organic halide:

$$R_3PbM + R'X \rightarrow R_3PbR' + MX$$

The first synthesis of an organoleadmetal compound was reported almost simultaneously by Gilman and Bailie (106) and Foster, Dix, and Gruntfest (96) in 1939 by reaction of triaryllead chloride or hexaaryldilead with sodium metal in liquid ammonia. The formation of a triarylplumbylsodium derivative was deduced by its conversion to benzyltriaryllead upon the addition of benzyl chloride to the reaction mixture. Triphenylplumbyllithium was subsequently prepared by Gilman, Summers, and Leeper (123) by reaction of lead(II) chloride with three moles of phenyllithium in diethyl ether, and triphenylplumbylmetal derivatives of other alkali and alkaline earth metals were presumed to be formed in reactions of hexaphenyldilead with the respective metal in liquid ammonia (113).

Compounds of the type R_3PbMgX are proposed to be important intermediates in reactions of lead chloride with Grignard reagents for syntheses of R_6Pb_2 and R_4Pb compounds. Triarylplumbylsilver compounds are postulated to be formed in low temperature reactions of hexaaryldilead compounds with alcoholic silver nitrate (220) and the product of electrolysis of hexaphenyldilead in tetrabutylammonium perchlorate is concluded to be the triphenylplumbyl anion in association with the tetrabutylammonium gegenion (83). However, neither these nor any of the other types of R_3PbM derivatives mentioned above have been isolated in the pure state. Their existence has been deduced from the reactions they undergo and the products formed.

Cursory evidence has been reported for the possible existence of R_2PbM_2 compounds, which is based on the formation of an R_2PbR_2' derivative by reaction of a diorganolead dihalide with lithium metal and organic halide. These compounds, if they do exist, probably exist only in solution and are probably much more reactive and less stable than their R_3PbM analogs.

Since triphenylplumbylmetal compounds can be prepared directly by the reaction of lead(II) halide with three moles of phenyllithium, R_3PbM compounds are best represented as being complexes of a diorganolead compound with an organometallic derivative of an alkali metal, analogous to the complexes of lead(II) halides with alkali metal halides. This conclusion is consistent with the finding that triphenylplumbyllithium is completely hydrolyzed by water to lead oxide (386), as are diaryllead compounds, e.g., diphenyllead; with few exceptions, derivatives of tetravalent lead, i.e., hexaorganodilead and tetraorganolead compounds, are very stable to hydrolysis.

Very little effort has been made to isolate any R_3PbM compounds in their pure state, and no physical properties measurements have been reported. For many years, the only interest in these compounds evolved from their utility in the synthesis of unsymmetrical tetraorganolead compounds. However, the recent work of Willemsens and van der Kerk (386) has shown that triphenylplumbyllithium undergoes facile reaction with a wide variety of compounds; these reactions have permitted the successful synthesis of many new and unusual organolead derivatives.

B. Methods of synthesis

(*1*) *Triorganoplumbylsodium*, R_3PbNa

A number of R_3PbNa derivatives has been prepared by reaction of triphenyllead chloride, hexaaryldilead or tetraorganolead with sodium in liquid ammonia (43,84,96,106,108,395). These reactions proceed according to the equations:

$$R_3PbX + 2Na \longrightarrow R_3PbNa + NaX$$

$$R_6Pb_2 + 2Na \longrightarrow 2R_3PbNa$$

$$R_4Pb + 2Na + NH_3 \longrightarrow R_3PbNa + NaNH_2 + RH$$

The mechanism of the last reaction above has been investigated using tetramethyllead (171) and was discussed in Chapter VI. Both tetra-alkyl- and tetraaryllead compounds have been shown to react with sodium metal to form the corresponding R_3PbNa derivatives.

The reaction of hexaphenyldilead and sodium in liquid ammonia produces a faintly yellow saturated solution from which an intense lemon-yellow solid crystallizes upon evaporation of the ammonia; the solid becomes cream colored upon final removal of the ammonia (96). The temperature of the system was maintained below $-34°$ until all of the ammonia had been removed in order to avoid solvolysis of the Ph_3PbNa.

Gilman and Bindschadler (108) found that a mixed solvent of diethyl ether–ammonia is a better solvent than ammonia alone for the R_4Pb reaction; diethyl ether alone is reported to be unsatisfactory. However, Hein and Nebe (154) prepared tricyclohexylplumbylsodium by shaking a slurry consisting of finely divided sodium metal in ethereal hexacyclo-hexyldilead. Also, triphenylplumbylsodium has been prepared by reaction of hexaphenyldilead and sodium metal in tetrahydrofuran (382).

Limited evidence exists for the existence of unsymmetrical derivatives of the type $R_2R'PbNa$. Phenyldiethylplumbylsodium was believed to be the product from reaction of phenyltriethyllead with sodium (43).

However, an attempt to prepare unsymmetrical triorganoplumbylsodium derivatives containing a benzyl or allyl group was unsuccessful; the resulting products were unstable (43).

(2) *Triorganoplumbyllithium*, Ph_3PbLi, (Ph_2PbLi_2)

The best known triorganoplumbyllithium compound is the triphenyl derivative which was prepared originally (123) by the reaction of phenyllithium and lead halide in diethyl ether at $-10°$:

$$PbCl_2 + 3PhLi \longrightarrow Ph_3PbLi + 2LiCl$$

The reaction is usually carried out in diethyl ether or tetrahydrofuran at $-10°$ or below in order to avoid decomposition of the intermediate diphenyllead to hexaphenyldilead and lead metal (123). The reaction can be conducted using stepwise addition of the phenyllithium. At a mole ratio of $2PhLi/PbCl_2$, a bright yellow solution is obtained which has been assumed to be diphenyllead or possibly bis(triphenylplumbyl)lead; at this stoichiometry, the yellow solution becomes darkened at higher temperatures and lead metal precipitates. Addition of a third mole of phenyllithium to the yellow solution at low temperature causes the color to disappear; in diethyl ether solvent a grayish white solid is formed, suspended in a faintly yellow solution. The resultant reaction mixture is stable at higher temperature, i.e., no decomposition accompanied by formation of lead metal occurs.

Several other methods of preparation of triphenylplumbyllithium are available. It is easily prepared by reaction of hexaphenyldilead and lithium metal in tetrahydrofuran (357). Since hexaphenyldilead is easily obtained as a pure compound, its reaction with lithium metal represents a convenient method of preparing triphenylplumbyllithium free of major impurities:

$$Ph_6Pb_2 + 2Li \xrightarrow{\text{THF}} 2Ph_3PbLi$$

Triphenylplumbyllithium has also been prepared from triphenyllead chloride and lithium metal in tetrahydrofuran (114,115,357):

$$Ph_3PbCl + 2Li \xrightarrow{\text{THF}} Ph_3PbLi + LiCl$$

Undoubtedly, this reaction proceeds via intermediate formation of hexaphenyldilead as evidenced by the appearance of a white cloudiness at an intermediate stage of the reaction.

These reactions involving lithium metal in tetrahydrofuran are carried out at a temperature of $-20°$ in order to avoid side reactions which occur at higher temperatures. Also, a large excess of lithium must be avoided, because lithium metal will react further with triphenylplumbyllithium with precipitation of lead metal. Triphenylplumbyllithium is soluble in tetrahydrofuran, the color of the reaction mixture varying from green to brown immediately after reaction; upon standing for several days, a dark brown sludge settles out of the reaction mixture leaving a clear yellow solution of triphenylplumbyllithium (386).

Triphenylplumbyllithium was formed, along with tetraphenyllead, by the reaction of hexaphenyldilead and phenyllithium (44):

$$Ph_6Pb_2 + PhLi \rightarrow Ph_3PbLi + Ph_4Pb$$

However, this reaction offers little attraction as a method of synthesis, because reaction of Ph_6Pb_2 with lithium metal is cleaner and is not complicated by the formation of by-products.

Finally, the preparation of triphenylplumbyllithium has been accomplished by reaction of hexaphenyldilead and lithium metal in liquid ammonia (113). High yields (~70%) of benzyltriphenyllead were obtained upon addition of benzyl chloride to the resultant reaction mixture.

Attempts to prepare diphenyllead by the reaction of diphenyllead dichloride and two moles of lithium metal in liquid ammonia gave lead(II) chloride and hexaphenyldilead. However, reaction of diphenyllead dichloride in liquid ammonia with four moles of lithium produced a red solution along with hexaphenyldilead and lead(II) chloride (16). Reaction of this red solution with ethyl bromide gave diphenyldiethyllead (60%) as the major product, along with ethyltriphenyllead (25%), phenyltriethyllead (8%), and tetraethyllead (4%). The formation of diphenyldiethyllead as the major product is indicative of the formation of some diphenylplumbyldilithium in this system.

Little is known about trialkylplumbyllithium compounds since no attempt has been made to synthesize them. Since trialkylplumbylsodium derivatives are known, it is reasonable to expect that the lithium derivatives can be prepared.

(3) *Triorganolead derivatives of other metals*

Gilman and Leeper (113) have shown that hexaphenyldilead undergoes facile reaction with potassium (56%), rubidium (52%), calcium (80%), strontium (70%), and barium (58%) in liquid ammonia; the percentages shown in parentheses correspond to the yield of benzyltriphenyllead obtained upon addition of benzyl chloride to the reaction mixture. From the yields of benzyltriphenyllead obtained and the stoichiometry employed, it was concluded that the alkaline earth metals function as monovalent ions in this system.

The existence of a compound of the type R_3PbMgX was postulated many years ago. Gilman and Bailie (106) originally proposed it to be an intermediate in the reaction of triethyllead chloride and magnesium metal to form tetraethyllead; Willemsens and van der Kerk (386) have recently postulated that triphenylplumbylmagnesium bromide is a reactive intermediate in the formation of tetraphenyllead from lead(II) chloride and phenylmagnesium bromide in tetrahydrofuran or pyridine solvent. Also, Glocking and co-workers (130) postulated trimesitylplumbylmagnesium bromide to be an intermediate in the reaction of lead(II) bromide with mesitylmagnesium bromide. No attempt has been made to isolate any of these magnesium derivatives, although bis-(triphenylstannyl)magnesium, $(Ph_3Sn)_2Mg$, has been isolated (358).

Dessy and co-workers (83) have demonstrated that hexaphenyldilead undergoes electrochemical reduction to form the triphenylplumbyl anion. The electrochemical reduction was carried out in ethylene glycol dimethyl ether solution using tetrabutylammonium perchlorate as electrolyte, so that the triphenylplumbyl moiety was probably present in solution as $Bu_4N^+PbPh_3^-$. The electrolysis of hexaphenyldilead in the presence of silver nitrate gave a green coloration which was presumed to result from the formation of triphenylplumbylsilver. Triorganoplumbylsilver compounds have also been postulated to be the source of the green coloration formed in the reaction of hexaorganodilead compounds with alcoholic silver nitrate solution at Dry Ice temperature (220).

The reaction of diphenyllead dichloride and difluoride with calcium metal has also been investigated (113); hexaphenyldilead was obtained as the major product, along with some tetraphenyllead.

C. Physical and chemical properties

No physical properties measurements have been reported for any of the organolead metal compounds. The triorganoplumbylsodium derivatives are soluble in liquid ammonia and probably exist as solvated species; triphenylplumbylsodium gives a yellow solution in liquid ammonia. Triphenylplumbyllithium is soluble in tetrahydrofuran, but insoluble in diethyl ether; tricyclohexylplumbylsodium is reported to be soluble in diethyl ether.

Little information is available on the thermal stability of the pure triorganoplumbyl metal compounds. The aryl derivatives appear to be more stable than the alkyl derivatives and the stability of both the alkyl and aryl derivatives in solution is enhanced with increasing basicity of the solvent. Triethylplumbylsodium is reported to undergo decomposition upon removal of the ether-ammonia solvent in which it is prepared; it is reported to be stable in liquid ammonia or ether–ammonia but unstable in diethyl ether (84). Triphenylplumbyllithium has been shown to be stable in tetrahydrofuran solution to 115° (384); removal of the tetrahydrofuran gave a yellow oil which did not crystallize even after prolonged standing. Attempts to remove the last traces of tetrahydrofuran caused decomposition as evidenced by lead metal formation (385). Yet, triphenylplumbyllithium precipitates as a stable, gray-white solid from diethyl ether medium (123,386).

Triorganoplumbyllithium and -sodium compounds are rapidly oxidized by air and hydrolyzed by water. Solid triphenylplumbyllithium tends to be pyrophoric; slow oxidation of the solid compound with dry air gave hexaphenyldilead in 92% yield, according to the stoichiometry (378):

$$2Ph_3PbLi + \tfrac{1}{2}O_2 \longrightarrow Ph_6Pb_2 + Li_2O$$

Oxidation with wet air gave triphenyllead hydroxide in nearly 50% yield along with some hexaphenyldilead.

Hydrolysis of triphenylplumbyllithium occurs with the intermediate formation of a transient orange-red coloration (386). A similar orange-red color was observed in the hydrolysis of tricyclohexylplumbylsodium which Hein and Nebe (154) ascribed to the intermediate formation of tricyclohexyllead hydride or dicyclohexyllead. However, hydride formation has been questioned by Willemsens and van der Kerk (386) since no hydrogen is formed in the hydrolysis of triphenylplumbyllithium, and triphenyllead hydride would be expected to give hydrogen upon decomposition. Instead, Willemsens and van der Kerk postulated that the hydrolysis of triphenylplumbyllithium involves the initial formation of diphenyllead, which is hydrolyzed further to lead oxide and benzene:

$$Ph_3PbLi + H_2O \longrightarrow [Ph_2Pb] + LiOH + PhH$$

$$[Ph_2Pb] + H_2O \longrightarrow PbO + 2PhH$$

Hydrolysis of triphenylplumbyllithium in the absence of air gave hexaphenyldilead in small amounts; lead oxide (386) or basic lead bromide (130) were formed as the major products. Similar results were obtained with triphenylplumbylsodium (106). By conducting the hydrolysis of triphenylplumbyllithium in the presence of an oxidizing agent such as hydrogen peroxide, Willemsens and van der Kerk were able to sustain the red color and demonstrated that the source of the coloration was the novel compound, tetrakis(triphenylplumbyl)lead. The synthesis of this and related compounds is discussed in Chapter XV.

Triphenylplumbyllithium reacts with absolute ethanol to form lead ethoxide; decomposition is accompanied by the formation of a yellow to red color. Triphenylplumbyllithium also reacts with acetone, but not with benzophenone. A clear red solution was obtained in the acetone reaction in tetrahydrofuran, which appeared to be stable for several days. However, no well-defined products could be isolated from the reaction mixture (385). With benzophenone, the color of the solution changed slightly from yellow to reddish, but no change was observed by infrared in the CO peak of the benzophenone.

Triphenylplumbyllithium reacts readily with sulfur, selenium, and tellurium in tetrahydrofuran solution to form Ph_3PbSLi, $Ph_3PbSeLi$, and $Ph_3PbTeLi$ as orange-red solutions in tetrahydrofuran (340). These compounds have been used for the synthesis of such compounds as bis(triphenyllead) sulfide, selenide, and telluride, as well as mixed metal derivatives of the type $Ph_3PbSSnPh_3$, by reaction with Ph_3PbX or Ph_3SnX (see Chapter XI.6).

One puzzling aspect of the chemistry of triphenylplumbyllithium has been the question of its tendency to dissociate according to the following equation:

$$Ph_3PbLi \leftrightharpoons Ph_2Pb + PhLi$$

In their original synthesis of Ph_3PbLi from lead chloride and phenyllithium, Gilman, Summers and Leeper (123,354) concluded that a dissociation does occur but that dissociation was less than 20% in diethyl ether solution. This was deduced from the observation that some benzoic acid was formed upon carboxylation-hydrolysis of a solution of triphenylplumbyllithium prepared from lead chloride and phenyllithium in diethyl ether:

$$Ph_3PbLi \leftrightharpoons Ph_2Pb + PhLi \xrightarrow{CO_2} PhCOOLi \xrightarrow{H_2O} PhCOOH$$

Also, d'Ans and his colleagues (15) obtained fluorene-9-carboxylic acid in 52% yield by reaction of triphenylplumbyllithium (prepared from lead chloride and phenyllithium in diethyl ether) with fluorene, followed by carboxylation-hydrolysis; triphenylstannyllithium gave a 21% yield of fluorene-9-carboxylic acid by similar treatment. The formation of fluorene-9-carboxylic acid was attributed to transmetallation of the fluorene by the phenyllithium formed via dissociation of triphenylplumbyllithium (or the tin analog).

Subsequently, Gilman and co-workers (114) found that no benzoic acid was formed in the carboxylation of triphenylplumbyllithium prepared from triphenyllead chloride and lithium metal in tetrahydrofuran; also, fluorene-9-carboxylic acid was obtained in only 0.5% yield from the carboxylation of the mixture obtained from reaction of fluorene with triphenylplumbyllithium prepared in a similar manner (115). Thus, triphenylplumbyllithium prepared from triphenyllead chloride and lithium metal in tetrahydrofuran is much less reactive to carbon dioxide and fluorene than Ph_3PbLi prepared from lead chloride and phenyllithium in diethyl ether; in tetrahydrofuran, little or no "dissociation" of Ph_3PbLi to Ph_2Pb and PhLi is indicated. Willemsens and van der Kerk (388) have reported that triphenylplumbyllithium reacted sluggishly with carbon dioxide to form lead oxide and benzoic acid. However, no indication was given of the method used to prepare the triphenylplumbyllithium or the solvent used in the carboxylation reaction.

Additional indirect evidence against gross dissociation of triphenylplumbyllithium has been reported by Glockling and co-workers (130) who obtained benzyltriphenyllead in 28% yield by addition of benzyl chloride to a reaction mixture containing *equimolar* amounts of lead bromide and phenyllithium in tetrahydrofuran at −40°. Presumably, the benzyltriphenyllead arose from reaction of triphenylplumbyllithium and benzyl chloride. This high yield of benzyltriphenyllead from a reaction mixture containing equimolar amounts of $PbBr_2$ and PhLi is further evidence that if dissociation of triphenylplumbyllithium does occur in tetrahydrofuran, the equilibrium must lie far to the left in favor of triphenylplumbyllithium.

The dissociation of triphenylplumbyllithium in diethyl ether has been questioned by Willemsens and van der Kerk (386) on the grounds that the diphenyllead in the equilibrium mixture should disproportionate to tetraphenyllead and lead metal so that the equilibrium would shift to the right; also, if dissociation to phenyllithium occurs, the triphenylplumbyllithium solution should give a positive Gilman test, but it did not. Hence, the preponderance of data indicates that far less dissociation occurs in tetrahydrofuran than in diethyl ether, and dissociation in diethyl ether is open to serious question. The relatively "high" degree of dissociation in diethyl ether may actually be associated with the method of preparation used; any unreacted phenyllithium in the $PbCl_2$–PhLi system would be carboxylated to form benzoic acid upon hydrolysis. The whole question requires further elucidation.

The reactivities of the Ph_3MLi derivatives of the Group IVb metals have been shown by Cartledge (67) to decrease in the order $Ph_3SiLi > Ph_3GeLi > Ph_3SnLi > Ph_3PbLi$, on the basis of their relative reactivities with *n*-butyl chloride. A similar order was obtained by Dessy and co-workers (82), who measured the rates of reaction of the triphenylplumbyl, -stannyl, and -germyl anions with various alkyl bromides and iodides by polarographic methods. On the basis of the kinetic data it was concluded that the nucleophilicity of the triphenylmetal anion decreases with increasing atomic weight of the Group IVb metal. For the triphenylplumbyl anion, the second-order rate constants for its reactions with the alkyl halides decreased in the relative order $CH_3I = 8167 > C_2H_5I = 44 > C_2H_5Br = 1$.

Trialkylplumbylsodium compounds appear to be more reactive with organic halides than the triaryl analogs. Thus, triphenylplumbylsodium is reported to react with alkyl halides but not with aryl halides; on the other hand, triethylplumbylsodium has been shown to react with bromobenzene as well as various alkyl halides (108). The reaction of triethylplumbylsodium with alkenylethynyl bromides has been used to prepare derivatives of the type $(C_2H_5)_3PbC{\equiv}C{-}CH{=}CH_2$ (395). Triphenylplumbyllithium did not react with hexachloroethane or *p*-dichlorobenzene except under drastic conditions; no definite products were isolated (386). However, reactions of triphenylplumbylsodium and -lithium with other dihalo- and polyhaloalkanes have been used to synthesize organolead derivatives containing up to four lead atoms in the same molecule. Similarly, the reaction of triphenylplumbyllead with triphenylmetal halides and metal(IV) halides has been used to synthesize compounds containing two and five metal atoms, respectively, in the same molecule. These polymetallic compounds are discussed in Chapter XV.

Triphenylplumbyllithium undergoes lithium-halogen exchange with alpha-halo esters. Thus, reaction with ethyl bromoacetate gave 3% hexaphenyldilead and 67% triphenyllead bromide, while ethyl chloroacetate gave 78% hexaphenyldilead and 9% triphenyllead chloride (386). With methyl 3-chloropropionate, only lead oxide was isolated after hydrolysis of the reaction mixture. The following sequence was proposed to account for this result:

$$Ph_3PbLi + ClCH_2CH_2COOMe \rightarrow$$

$$[Ph_2Pb] + CH_2 = CHCOOMe + C_6H_6 + LiCl$$

$$[Ph_2Pb] + H_2O \rightarrow PbO + 2C_6H_6$$

On the other hand, Willemsens and van der Kerk (388) have found that the reaction of triphenylplumbyllithium with ethyl chloroformate proceeded according to the equation:

$$Ph_3PbLi + ClCOOEt \rightarrow Ph_3PbCOOEt + LiCl$$

This novel organolead derivative was more stable than expected but decomposed at elevated temperature to form tetraphenyllead, lead metal, carbon monoxide, and diethyl carbonate.

The reaction of Ph_3MLi compounds with small ring heterocycles has been extended recently to the last member of the Group IVb series, triphenylplumbyllithium. Reaction of triphenylplumbyllithium with three- and four-membered heterocycles has been demonstrated by Willemsens and van der Kerk (384,386). No reaction was obtained with tetrahydrofuran; pyrrolidone did react, but the reaction involved reduction of the carbonyl group and formation of lead metal. Since tetrahydrofuran does not react with triphenylplumbyllithium, reactions with small ring heterocycles are conveniently carried out in this solvent. Reactions of triphenylplumbyllithium with heterocyclic compounds are discussed further in Chapter XII.

The reaction of triethylplumbylsodium with bis(cyclopentadienyl)-titanium dichloride was investigated by Dickson and West (84). In ammonia–diethyl ether solution, a blue-green solid was obtained which was speculated to be $(C_5H_5)_2TiNH_2Pb(C_2H_5)_3$. Removal of ammonia from these solids caused a transformation to buff-colored solids, accompanied by liberation of gas. However, the exact composition of the products was not established nor was the reaction mechanism defined.

Dickson and West (84) also reported that the reaction of triethylplumbylsodium with ammonium chloride in liquid ammonia gave green solids accompanied by gas evolution. Removal of the ammonia at $-78°$ led to decomposition of the green solids to gaseous products and a gray residue from which tetraethyllead and sodium chloride were isolated. The green solids were speculated to contain triethyllead hydride. The following reaction sequence was proposed to explain the products and observations:

$$NaPb(C_2H_5)_3 + NH_4Cl \rightarrow NH_4Pb(C_2H_5)_3 + NaCl$$

$$NH_4Pb(C_2H_5)_3 = NH_3[HPb(C_2H_5)_3] \rightarrow NH_3 + C_2H_6 + (C_2H_5)_2Pb$$

$$2(C_2H_5)_2Pb \rightarrow Pb + (C_2H_5)_4Pb$$

Triethylplumbylsodium reacted readily with ammonium bromide; hexaethyldilead and ethane were isolated from the reaction mixture (106,108). Triarylplumbylsodium compounds also react with ammonium bromide in liquid ammonia, but they are less reactive than the alkyl analogs (108). This reaction was investigated by Gilman (106) and Foster (96) and co-workers in attempts to prepare an organolead hydride (96,106). In all cases Gilman and Bailie (106) obtained decomposition products consisting of a mixture of Ar_6Pb_2 and ArH. Thus, tri-*p*-methoxyphenylplumbylsodium gave anisole and hexa-*p*-methoxyphenyldilead, the tri-*p*-tolyl and tri-*o*-tolyl compounds gave toluene and the corresponding hexatolyldilead derivatives and the tri-*p*-ethoxyphenyl compound gave phenetole and hexa-*p*-ethoxyphenyldilead (106). Similarly, triphenylplumbylsodium and ammonium bromide were found to give hexaphenyldilead and benzene. On the other hand, Foster and co-workers (96) obtained no reaction between triphenylplumbylsodium and ammonium bromide except in the presence of a large excess of the ammonium bromide, under which conditions complete decomposition occurred with formation of benzene and lead metal. They also concluded that, in the presence of a slight excess of ammonium bromide, triphenylplumbylammonium was formed, in admixture with triphenylplumbylsodium, and that the ammonium compound was very soluble in liquid ammonia. Foster used ammonium bromide pellets in his reactions, which may explain the poor reactivity observed.

Tricyclohexylplumbylsodium is reported to react with mercury metal to form hexacyclohexyldilead and sodium amalgam (154); undoubtedly, the formation of sodium amalgam is the driving force in this reaction. Reaction of tricyclohexylplumbylsodium and iodine in ether, pyridine or ethanol gave dicyclohexyllead diiodide and lead iodide, and no tricyclohexyllead iodide; in benzene, a red solution was obtained which was attributed to the formation of dicyclohexyllead (154). On the other hand, the reaction of an ice cold ether solution of triphenylplumbyllithium with an equimolar amount of bromine in chloroform gave a 25% yield of triphenyllead bromide along with hexaphenyldilead in at least 60% yield (378).

Triphenylplumbyllithium has been shown to react with certain reactive alkynes to form organolead products containing a functional group. These reactions are discussed in Chapter XII.

Dessy and Weissman (83) have shown that polarographic reduction of a solution of hexaphen ldilead and triphenyllead acetate in the dimethyl ether of ethylene glycol gives a half-wave potential of -2.2 V (versus a Ag/Ag^+ electrode) instead of -2.0 V, the latter being the half wave potential for reduction of hexaphenyldilead to the triphenylplumbyl anion. The -2.2 V wave gradually diminished in height and was ultimately replaced by a wave at -3.2 V; concomitantly, an intense yellow color appeared and then disappeared. Since tetraphenyllead exhibits a half-wave potential at -3.2 V, it was concluded that it was the ultimate product of the electrochemical reaction and the formation of triphenyllead hydride was proposed to explain the wave at -2.2 V. A number of triphenylplumbyl derivatives of transition metal carbonyls have been synthesized via electrochemical generation of the triphenylplumbyl anion (Chapter XV).

The reactivity of triphenylplumbyllithium follows the trend established by the other members of the Group IVb series, namely that of decreasing reactivity of the Ph_3MLi compound with increasing atomic weight of the Group IVb metal. The reactions of triorganoplumbylmetal compounds, particularly triphenylplumbyllithium are similar in many respects to the reactions which organolithium and organosodium compounds undergo. They are generally less reactive than the simple organoalkali metal analogs.

Thus, triphenylplumbyllithium does not tend to cleave ethers as readily as phenyllithium and is less reactive to ammonia. The extensive studies of Willemsens and van der Kerk have strikingly demonstrated the broad scope of the reactions of triphenylplumbyllithium and its utility for the synthesis of a wide variety of novel organolead derivatives.

REFERENCES

1. Abel, E. W., D. A. Armitage, and D. B. Brady, *Trans. Faraday Soc.*, **62,** 3459 (1966).
2. Abel, E. W., and D. A. Armitage, "Organosulfur Derivatives of Silicon, Germanium, Tin, and Lead," in *Advances in Organometallic Chemistry*, Vol. 5, F. G. A. Stone and R. West, Eds., Academic Press, N.Y., 1967, pp. 2–92.
3. Abel, E. W., and D. B. Brady, *J. Chem. Soc.*, **1965,** 1192; *J. Organometal Chem.*, **11,** 145 (1968).
4. Aleksandrov, Yu. A., T. G. Brilkina, and V. A. Shushunov, *Khim. Perekisnykh Soedin., Akad. Nauk SSSR, Inst. Obshch. i Neorgan. Khim.*, **1963,** 291; through *C.A.*, **60,** 12040 (1964).
5. Aleksandrov, Yu. A., T. G. Brilkina, and V. A. Shushunov, *Tr. po Khim. i Khim. Tekhnol.*, **2,** 623 (1959); through *C.A.*, **56,** 14314 (1962).
6. Aleksandrov, Yu. A., T. G. Brilkina, and V. A. Shushunov, *Dokl. Akad. Nauk SSSR*, **136,** 89 (1961); *C.A.*, **55,** 27027 (1961).
7. Aleksandrov, Yu. A., T. G. Brilkina, and V. A. Shushunov, *Tr. po Khim. i Khim. Tekhnol.*, **3,** 381 (1960); *C.A.*, **55,** 27023 (1961).
8. Aleksandrov, Yu. A., T. G. Brilkina, and V. A. Shushunov, *Tr. po Khim. i Khim. Tekhnol.*, **4,** 3 (1961); *C.A.*, **56,** 492 (1962).
9. Aleksandrov, Yu. A., and T. I. Mokeeva, *Tr. po Khim. i Khim. Tekhnol.*, **4,** 365 (1961); through *C.A.*, **56,** 493 (1962).
10. Altamura, M. S., U.S. Pat. 2,493,213 (to Socony-Vacuum Oil Co.), Jan. 3, 1950; *C.A.*, **44,** 2229 (1950).
11. Amberger, E., *Angew. Chem.*, **72,** 494 (1960).
12. Amberger, E., and R. Hönigschmid-Grossich, *Chem. Ber.*, **98,** 3795 (1965).
13. Anon., *Chem. Eng. News*, p. 50, Sept. 20, 1965.
14. d'Ans, J., H. Zimmer, and M. v. Brauchitsch, *Chem. Ber.*, **88,** 1507 (1955).
15. d'Ans, J., H. Zimmer, E. Endrulat, and K. Lübke, *Naturwiss.*, **39,** 450 (1952).
16. Apperson, L. C., *Iowa St. Coll. J. Sci.*, **16,** 7 (1941); through *C.A.*, **36,** 4476 (1942).
17. Arbuzov, B. A., and A. N. Pudovik, *Zh. Obshch. Khim.*, **17,** 2158 (1947); through *C.A.*, **42,** 4522 (1948).
18. Austin, P. R., *J. Am. Chem. Soc.*, **55,** 2948 (1933).
19. Austin, P. R., *J. Am. Chem. Soc.*, **53,** 1548 (1931).
20. Austin, P. R., *J. Am. Chem. Soc.*, **53,** 3514 (1931).
21. Austin, P. R., *J. Am. Chem. Soc.*, **54,** 3287 (1932).
22. Austin, P. R., *J. Am. Chem. Soc.*, **54,** 3726 (1932).
23. Aynsley, E. E., N. N. Greenwood, G. Hunter, and M. J. Sprague, *J. Chem. Soc.* **A, 1966,** 1344.
24. Baer, M., U.S. Pat. 2,561,044 (to Monsanto Chemical Co.), July 17, 1951; *C.A.*, **45,** 8807 (1951).
25. Bähr, G., *Z. anorg. allgem. Chem.*, **253,** 330 (1947).

26. Ballinger, P., U.S. Pat. 3,073, 852 (to California Research Corp.), Jan. 15, 1963; *C.A.*, **58,** 12599 (1963).
27. Ballinger, P., U.S. Pat. 3,073,853 (to California Research Corp.), Jan. 15, 1963; *C.A.*, **58,** 12599 (1963).
28. Ballinger, P., U.S. Pat. 3,073,854 (to California Research Corp.), Jan. 15, 1963; *C.A.*, **58,** 12599 (1963).
29. Ballinger, P., U.S. Pat. 3,081,325 (to California Research Corp.), March 12, 1963; *C.A*, **59,** 6440 (1963)
30. Ballinger, P., U.S. Pat. 3,116,127 (to California Research Corp.), Dec. 31, 1963; *C.A.*, **60**, 6684 (1964).
31. Ballinger, P., U.S. Pat. 3,143,399 (to California Research Corp.), Aug. 4, 1964; *C.A.*, **61**, 10520 (1964).
32. Ballinger, P., U.S. Pat. 3,143,400 (to California Research Corp.), Aug. 4, 1964; *C.A.*, **61,** 10520 (1964).
33. Barbieri, R., U. Belluco, and G. Tagliavini, *Ann. Chim.* (*Rome*), **48,** 940 (1958); through *C.A.*, **53,** 9901 (1959).
34. Barbieri, R., U. Belluco, and G. Tagliavini, *Ric. sci.*, **30,** 1671 (1960); through *C.A.*, **55,** 14175 (1961).
35. Barbieri, R., G. Faraglia, M. Giustiniani, and L. Roncucci, *J. Inorg. Nucl. Chem.*, **26,** 203 (1964).
36. Barbieri, R., G. Faraglia, and M. Giustiniani, *Ric. sci. Rend.*, *Sez.* **A4,** 109 (1964); through *C.A.*, **61,** 4388 (1964).
37. Barbieri, R., M. Giustiniani, G. Faraglia, and G. Tagliavini, *Ric. sci. Rend.*, *Sez.* **A3,** 975 (1963); through *C.A.*, **61,** 1884 (1964).
38. Barbieri, R., G. Tagliavini, and U. Belluco, *Ric. sci.*, **30,** 1963 (1960); through *C.A.*, **55,** 14159 (1961).
39. Barry, A. J., H. N. Beck, D. C. Bradley, R. K. Ingham, and H. Gilman, in *Inorganic Polymers*, F. G. A. Stone and W. A. G. Graham, Eds., Academic Press, New York, 1961, pp. 285–445.
40. Beck, W., and E. Schuierer, *Chem. Ber.*, **97,** 3517 (1964).
41. Becker, W. E., and S. E. Cook, *J. Am. Chem. Soc.*, **82,** 6264 (1960).
42. Belluco, U., G. Tagliavini, and P. Favero, *Ric. sci. Rend.*, *Sez.* **A2,** 98 (1962); through *C.A.*, **57,** 13786 (1962).
43. Bindschadler, E., *Iowa St. Coll. J. Sci.*, **16,** 33 (1941); through *C.A.*, **36,** 4476 (1942).
44. Bindschadler, E., and H. Gilman, *Proc. Iowa Acad. Sci.*, **48,** 273 (1941); through *C.A.*, **36,** 1595 (1942).
45. Bock, R., and H. Deister, *Naturwiss.*, **50,** 496 (1963); *Chem. Zent.*, **135,** 535 (1964).
46. Bogomolov, S. G., I. A. Veselkova, and V. I. Lodochnikova, *Tr. Komis. po Spektroskopii*, *Akad. Nauk SSSR*, **1964,** 475; through *C.A.*, **63,** 13029 (1965).
47. Braun, D., S. Bong Chang, and M. Thallmaier, *Gummi Asbest. Kunstat.*, **19,** 1353 (1966); through *C.A.*, **66,** 46885h (1967).
48. Brilkina, T. G., M. K. Safonova, and V. A. Shushunov, *Zh. Obshch. Khim.*, **32,** 2684 (1962); *C.A.*, **58,** 9112 (1963).
49. Brilkina, T. G., M. K. Safonova, and N. A. Sokolov, *Zh. Obshch. Khim.*, **36,** 2202 (1966); *C.A.*, **66,** 76106x (1967).
50. Brilkina, T. G., and V. A. Shushunov, *Tr. po Khim. i Khim. Tekhnol.*, **3,** 505 (1960); *C.A.*, **55,** 26993 (1961).
51. Browne, O. H., and E. E. Reid, *J. Am. Chem. Soc.*, **49,** 830 (1927).

52. Buck, J. S., and D. M. Kumro, *J. Pharmacol.*, **38,** 161 (1930); through *C.A.*, **24,** 2806 (1930).
53. Buckton, G. B., *Ann.*, **112,** 220 (1859).
54. Buckton, G. B., *Chem. Gazette*, **1859,** 276.
55. Buckton, G. B., *Ann.*, **109,** 218 (1859).
56. Bullard, R. H., and F. R. Holden, *J. Am. Chem. Soc.*, **53,** 3150 (1931).
57. Burrows, L. A., W. F. Filbert, and E. E. Reid, U.S. Pat. 2,105,635 (to E. I. du Pont de Nemours & Co.), Jan. 18, 1938; through *C.A.*, **32,** 2357 (1938).
58. Cahours, A., *Ann.*, **122,** 48 (1862).
59. Calingaert, G., *Chem. Rev.*, **2,** 43 (1925).
60. Calingaert, G., and H. A. Beatty, *J. Am. Chem. Soc.*, **61,** 2748 (1939).
61. Calingaert, G., F. J. Dykstra, and H. Shapiro, *J. Am. Chem. Soc.*, **67,** 190 (1945).
62. Calingaert, G., H. Shapiro, F. J. Dykstra, and L. Hess, *J. Am. Chem. Soc.*, **70,** 3902 (1948).
63. Calingaert, G., and H. Soroos, *J. Org. Chem.*, **2,** 535 (1938).
64. Calingaert, G., H. Soroos, and H. Shapiro, *J. Am. Chem. Soc.*, **64,** 462 (1942).
65. Calingaert, G., H. Soroos, and H. Shapiro, *J. Am. Chem. Soc.*, **62,** 1104 (1940).
66. Carothers, W. H., U.S. Pat. 2,008,003 (to E. I. du Pont de Nemours & Co.), July 16, 1935; *C.A.*, **29,** 5862 (1935).
67. Cartledge, F., *Diss. Abs.*, **25,** 6226 (1965).
68. Challenger, F., and E. Rothstein, *J. Chem. Soc.*, **1934,** 1258.
69. Clark, H. C., and R. J. O'Brien, *Proc. Chem. Soc.*, **1963,** 113.
70. Collier, W. A., *Z. Hyg. Infektionskrankh.*, **110,** 169 (1929).
71. Collier, H. E., Jr., U.S. 3,308,061 (to E. I. du Pont de Nemours & Co.), March 7, 1967; *C.A.*, **66,** 97295y (1967).
71a. Cornell, J. H., H. Gorth, and M. C. Henry, *J. Inorg. Nucl. Chem.*, **29,** 1411 (1967).
72. Costa, G., *Ann. Chim. (Rome)*, **40,** 541 (1950); through *C.A.*, **45,** 6507 (1951).
73. Costa, G., *Ann. Chim. (Rome)*, **41,** 207 (1951); through *C.A.*, **45,** 10094 (1951).
74. Creemers, H. M. J. C., A. J. Leusink, J. G. Noltes, and G. J. M. van der Kerk, *Tetrahedron Letters*, **1966,** 3167.
75. Criegee, R., P. Dimroth, and R. Schempf, *Ber.*, **90,** 1337 (1957).
76. Croatto, U., and R. Barbieri, *Ric. sci. Rend., Sez.* **A8,** 441 (1965); through *C.A.*, **64,** 4563 (1966).
77. Danzer, R., *Monatsh.*, **46,** 241 (1925); through *C.A.*, **20,** 1589 (1926).
78. Davidson, W. E., K. Hills, and M. C. Henry, *J. Organometal. Chem.*, **3,** 285 (1965).
79. Davies, A. G., and R. J. Puddephatt, *J. Organometal. Chem.*, **5,** 590 (1966).
80. DePree, D. O., U.S. Pat. 2,893,857 (to Ethyl Corporation), July 7, 1959; *C.A.*, **53,** 18372 (1959).
81. Dessy, R. E., W. Kitching, and T. Chivers, *J. Am. Chem. Soc.*, **88,** 453 (1966).
82. Dessy, R. E., R. L. Pohland, and R. B. King, *J. Am. Chem. Soc.*, **88,** 5121 (1966).
83. Dessy, R. E., and P. M. Weissman, *J. Am. Chem. Soc.*, **88,** 5124 (1966).
84. Dickson, R. S., and B. O. West, *Australian J. Chem.*, **14,** 555 (1961).
85. Druce, J., *Chem. News*, **120,** 229 (1920).
86. Duffy, R., J. Feeney, and A. K. Holliday, *J. Chem. Soc.*, **1962,** 1144.
87. Duffy, R., and A. K. Holliday, *Proc. Chem. Soc.*, **1959**, 124.
88. Duffy, R., and A. K. Holliday, *J. Chem. Soc.*, **1961,** 1679.
89. Emeléus, H. J., and P. R. Evans, *J. Chem. Soc.*, **1964,** 510.
90. Evans, D. P., *J. Chem. Soc.*, **1938,** 1466.

91. Faraglia, G., L. Roncucci, and R. Barbieri, *Ric. sci. Rend., Sez.* **A8,** 205 (1965); through *C.A.*, **63,** 12654 (1965).
92. Faraglia, G., L. Roncucci, and R. Barbieri, *J. Organometal. Chem.*, **6,** 2078 (1966).
92a. Fergusson, J. E., W. R. Roper, and J. C. Wilkins, *J. Chem. Soc.*, **1965,** 3716.
93. Finholt, A. E., A. C. Bond, K. E. Wilzbach, and H. I. Schlesinger, *J. Am. Chem. Soc.*, **69,** 2692 (1947).
94. Flitcroft, N., and H. D. Kaesz, *J. Am. Chem. Soc.*, **85,** 1377 (1963).
95. Flood, E. A., and L. Horvitz, *J. Am. Chem. Soc.*, **55,** 2534 (1933).
96. Foster, L. S., W. M. Dix, and I. J. Gruntfest, *J. Am. Chem. Soc.*, **61,** 1685 (1939).
97. Foster, L. S., I. J. Gruntfest, and L. A. Fluck, *J. Am. Chem. Soc.*, **61,** 1687 (1939).
98. Freidline, C. E., and R. S. Tobias, *Inorg. Chem.*, **5,** 354 (1966).
99. Fritz, H. P., and K. E. Schwarzhans, *Chem. Ber.*, **97,** 1390 (1964).
100. Gelius, R., *Z. anorg. allgem. Chem.*, **349,** 22 (1966).
100a. Geluis, R., and R. Muller, *Z. anorg. allgem. Chem.*, **351,** 42 (1967).
101. George, M. V., P. B. Talukdar, and H. Gilman, *J. Organometal. Chem.*, **5,** 397 (1966).
102. Gershbein, L. L., and V. N. Ipatieff, *J. Am. Chem. Soc.*, **74,** 1540 (1952).
103. Gielen, M., and N. Sprecher, *Organometal. Chem. Rev.*, **1,** 455 (1966).
104. Gilman, H., and R. K. Abbott, Unpublished Data; ref. 95 in Leeper, R. W., L. Summers, and H. Gilman, *Chem. Rev.*, **54,** 101 (1954).
105. Gilman, H., and L. D. Apperson, *J. Org. Chem.*, **4,** 162 (1939).
106. Gilman, H., and J. C. Bailie, *J. Am. Chem. Soc.*, **61,** 731 (1939).
107. Gilman, H., and M. M. Barnett, *Rec. trav. chim.*, **55,** 563 (1936).
108. Gilman, H., and E. Bindschadler, *J. Org. Chem.*, **18,** 1675 (1953).
109. Gilman, H., W. G. Bywater, and P. T. Parker, *J. Am. Chem. Soc.*, **57,** 885 (1935).
110. Gilman, H., W. A. Gregory, and S. M. Spatz, *J. Org. Chem.*, **16,** 1788 (1951).
111. Gilman, H., and O. M. Gruzhit, *J. Pharmacol.*, **41,** 1 (1931).
112. Gilman, H., and R. G. Jones, *J. Am. Chem. Soc.*, **72,** 1760 (1950).
113. Gilman, H., and R. W. Leeper, *J. Org. Chem.*, **16,** 466 (1951).
114. Gilman, H., O. L. Marrs, and S. Y. Sim, *J. Org. Chem.*, **27,** 4232 (1962).
115. Gilman, H., O. L. Marrs, W. Trepka, and J. W. Diehl, *J. Org. Chem.*, **27,** 1260 (1962).
116. Gilman, H., and D. S. Melstrom, *J. Am. Chem. Soc.*, **72,** 2953 (1950).
117. Gilman, H., and J. Robinson, *Rec. trav. chim.*, **49,** 766 (1930).
118. Gilman, H., and J. Robinson, *J. Am. Chem. Soc.*, **52,** 1975 (1930).
119. Gilman, H., and J. Robinson, *J. Am. Chem. Soc.*, **51,** 3112 (1929).
120. Gilman, H., S. M. Spatz, and M. J. Kolbezen, *J. Org. Chem.*, **18,** 1341 (1953).
121. Gilman, H., and C. G. Stuckwisch, *J. Am. Chem. Soc.*, **72,** 4553 (1950).
122. Gilman, H., and L. Summers, *J. Am. Chem. Soc.*, **74,** 5924 (1952).
123. Gilman, H., L. Summers, and R. W. Leeper, *J. Org. Chem.*, **17,** 630 (1952).
124. Gilman, H., and E. B. Towne, *Rec. trav. chim.*, **51,** 1054 (1932).
125. Gilman, H., and E. B. Towne, *J. Am. Chem. Soc.*, **61,** 739 (1939).
126. Gilman, H., E. B. Towne, and H. L. Jones, *J. Am. Chem. Soc.*, **55,** 4689 (1933).
127. Gilman, H., and L. A. Woods, *J. Am. Chem. Soc.*, **65,** 435 (1943).
128. Giraitis, A. P., P. Kobetz, and H. Shapiro, U.S. Pat. 3,147,293 (to Ethyl Corp.), Sept. 1, 1964; *C.A.*, **61,** 13345 (1964).
129. Giustiniani, M., G. Faraglia, and R. Barbieri, *J. Chromatog.*, **15,** 207 (1964).
130. Glockling, F., K. Hooton, and D. Kingston, *J. Chem. Soc.*, **1961,** 4405.

131. Goddard, A. E., *J. Chem. Soc.*, **123,** 1161 (1923).
132. Goddard, A. E., J. N. Ashley, and R. B. Evans, *J. Chem. Soc.*, **121,** 978 (1922).
133. Goddard, A. E., and D. Goddard, *J. Chem. Soc.*, **121,** 482 (1922).
134. Goddard, D. and A. E. Goddard, *J. Chem. Soc.*, **121,** 256 (1922).
135. Goldach, A., *Helv. Chim. Acta*, **14,** 1436 (1931).
136. Gorsich, R. D., U.S. Pat. 3,261,806 (to Ethyl Corp.), July 19, 1966; *C.A.*, **65,** 17152 (1966).
137. Grüttner, G., *Ber.*, **47,** 3257 (1914).
138. Grüttner, G., *Ber.*, **51,** 1298 (1918).
139. Grüttner, G., and E. Krause, *Ber.*, **49,** 1415 (1916).
140. Grüttner, G., and E. Krause, *Ber.*, **49,** 1546 (1916).
141. Grüttner, G., and E. Krause, *Ber.*, **49,** 2666 (1916).
142. Grüttner, G., and E. Krause, *Ber.*, **50,** 202 (1917).
143. Grüttner, G., and E. Krause, *Ber.*, **50,** 278 (1917).
144. Grüttner, G., and E. Krause, *Ber.*, **50,** 574 (1917).
145. Hagihara, N., *Yukagaku*, **14,** 462 (1965); through *C.A.*, **64,** 270 (1966).
146. Hartle, R. J., U.S. Pat. 3,055,748 (to Gulf Research & Dev. Co.), Sept. 25, 1962; *C.A.*, **58,** 7776 (1963).
147. Hartle, R. J., U.S. Pat. 3,055,925 (to Gulf Research & Dev. Co.), Sept. 25, 1962; *C.A.*, **59,** 1681 (1963).
148. Harvey, D. R., and R. O. C. Norman, *J. Chem. Soc.*, **1964,** 4860.
149. Hathaway, B. J., and D. E. Webster, *Proc. Chem. Soc.*, **1963,** 14.
150. Heap, R., and B. C. Saunders, *J. Chem. Soc.*, **1949,** 2983.
151. Heap, R., B. C. Saunders, and G. J. Stacey, *J. Chem. Soc.*, **1951,** 658.
152. Hein, F., and E. Heuser, *Z. anorg. allgem. Chem.*, **255,** 125 (1947).
153. Hein, F., and A. Klein, *Ber.*, **71,** 2381 (1938).
154. Hein, F., and E. Nebe, *Ber.*, **75,** 1744 (1942).
155. Hein, F., E. Nebe, and W. Reimann, *Z. anorg. allgem. Chem.*, **251,** 125 (1943).
156. Hein, F., and H. Pobloth, *Z. anorg. allgem. Chem.*, **248,** 84 (1941).
157. Henry, M. C., U.S. Dept. Comm., Office Tech. Serv., AD 286,653 (1962); from U.S. Gov't. Res. Rept., **38**(2), 17 (1963); *C.A.*, **60,** 6859 (1964).
158. Henry, M. C., *Inorg. Chem.*, **1,** 917 (1962).
159. Henry, M. C., U.S. Dept. Comm., Office Tech. Serv., AD611,432 (1964); *C.A.*, **63,** 14893 (1965).
160. Henry, M. C., and H. Gorth, Quarterly Report No. 16, Project LC-28, Intern. Lead Zinc Res. Organ., New York, April–June, 1965.
161. Henry, M. C., and H. Görth, Quarterly Report No. 18, Project LC-28, Intern. Lead Zinc. Res. Organ., New York, October–December, 1965.
162. Henry, M. C., and K. Hills, Quarterly Report No. 11 Project LC-28, Intern. Lead Zinc. Res. Organ., New York, March 3, 1964.
163. Henry, M. C., and K. Hills, Summary Report No. 14A, Project LC-28, Intern. Lead Zinc Res. Organ., New York, Dec. 31, 1964.
164. Henry, M. C., and A. W. Krebs, Quarterly Report No. 2, Project No. 7-99-01-001, Lead Industries Association, New York, Dec. 31, 1961.
165. Henry, M. C., and A. W. Krebs, *J. Org. Chem.*, **28,** 225 (1963).
166. Henry, M. C., J. G. Noltes, et al. U.S. Dept. Comm., Office Tech. Serv., AD 286, 653 (1962); through *C.A.*, **60,** 6859 (1964).
167. Hetnarski, B., and T. Urbanski, *Roczniki Chem.*, **37,** 1073 (1963); through *C.A.*, **60,** 5533 (1964).
168. Hetnarski, B., and T. Urbanski, *Tetrahedron*, **19,** 1319 (1963).

169. Hills, K., and M. C. Henry, *J. Organometal. Chem.*, **3,** 159 (1965).
170. Holliday, A. K., and W. Jeffers, *J. Inorg. Nucl. Chem.*, **6,** 134 (1958).
171. Holliday, A. K., and G. Pass, *J. Chem. Soc.*, **1958,** 3485.
172. Huber, F., *Angew. Chem.*, **77,** 1084 (1965).
173. Huber, F., and M. Enders, *Z. Naturforsch.*, **20b,** 601 (1965).
174. Huber, F., M. Enders, and R. Kaiser, *Z. Naturforsch.*, **21b,** 83 (1966).
175. Huber, F., and H. J. Haupt, *Z. Naturforsch.*, **21b,** 808 (1966).
175a. Huber, F., H. Horn, and H. J. Haupt, *Z. Naturforsch.*, **22b,** 918 (1967).
176. Huber, F., and F. J. Padberg, 19th Intern. Congr. Pure Appl. Chem., London, Div. A, **1963,** 196.
177. Hurd, C. D., and P. R. Austin, *J. Am. Chem. Soc.*, **53,** 1543 (1931).
178. Imura, S., and K. Fukutaka, *Bunseki Kagaku*, **14,** 1167 (1965); through *C.A.*, **64,** 14952 (1966).
179. Ipatieff, V. N., G. A. Razuvaev, and I. F. Bogdanov, *J. Russ. Phys. Chem. Soc.*, **61,** 1791 (1929); through *C.A.*, **24,** 2660 (1930).
180. Irving, H., and J. J. Cox, *J. Chem. Soc.*, **1961,** 1470.
181. Janssen, M. J., J. G. A. Luijten, and G. J. M. van der Kerk, *Rec. trav. chim.*, **82,** 90 (1963).
182. Jones, W. J., D. P. Evans, T. Gulwell, and D. C. Griffiths, *J. Chem. Soc.*, **1935,** 39.
183. Juenge, E. C., and S. E. Cook, *J. Am. Chem. Soc.*, **81,** 3578 (1959).
184. Juenge, E. C., and R. D. Houser, *J. Org. Chem.*, **29,** 2040 (1964).
185. Kawasaki, Y., T. Tanaka, and R. Okawara, *Bull. Chem. Soc. Japan*, **37,** 903 (1964); through *C.A.*, **61,** 9047 (1964).
186. Kiseleva, T. M., M. M. Koton, and G. M. Chetyrkina, *Izv. Akad. Nauk SSSR, Otd. Khim. Nauk*, **1962,** 1798; through *C.A.*, **58,** 7962 (1963).
187. Klippel, J., *Jahresber.*, **1860,** 383.
188. Klippel, J., *J. prakt. Chem.*, **81,** 287 (1860).
189. Kocheshkov, K. A., and A. P. Aleksandrov, *Ber.*, **67,** 527 (1934).
190. Kocheshkov, K. A., and A. P. Aleksandrov, *Zh. Obshch. Khim.*, **7,** 93 (1937); through *C.A.*, **31,** 4291 (1937).
191. Kocheshkov, K. A., and G. M. Borodina, *Bull. acad. sci. URSS, Classe sci. math. nat., Ser. chim.*, **1937,** 569; through *C.A.*, **32,** 2095 (1938).
192. Kocheshkov, K. A., and R. Kh. Freidlina, *Izv. Akad. Nauk SSSR*, Otd. *Khim. Nauk*, **1950,** 203; through *C.A.*, **44,** 9342 (1950).
193. Kocheshkov, K. A., and R. Kh. Freidlina, *Uchenye Zapiski Moskov. Gosudarst. Univ. M. V. Lomonosova*, **No. 132,** *Org. Khim.*, **No. 7,** 144 (1950); through *C.A.*, **50,** 7728 (1956).
194. Kocheshkov, K. A., and A. N. Nesmeyanov, *Zh. Obshch. Khim.*, **4,** 1102 (1934); *C.A.*, **29,** 3993 (1935).
195. Kocheshkov, K. A., A. N. Nesmeyanov, and N. K. Gipp, *Zh. Obshch. Khim.*, **6,** 172 (1936); through *C.A.*, **30,** 4834 (1936).
196 Kocheshkov, K. A., and E. M. Panov, *Izv. Akad. Nauk SSSR, Otd. Khim. Nauk*, **1955,** 711; through *C.A.*, **50,** 7075 (1956).
197. Kocheshkov, K. A., and E. M. Panov, *Izv. Akad. Nauk SSSR, Otd. Khim. Nauk*, **1955,** 718; through *C.A.*, **50,** 7076 (1956).
198. Kocheshkov, K. A., E. M. Panov, and N. N. Zemlyanskii, *Izv. Akad. Nauk SSSR, Otd. Khim. Nauk*, **1961,** 2255; *C.A.*, **57,** 7294 (1962).
199. Kochkin, D. A., *Dokl. Akad. Nauk SSSR*, **135,** 857 (1960); *C.A.*, **55,** 11341 (1961).

200. Kochkin, D. A., and I. B. Chekmareva, *Zh. Obshch. Khim.*, **31,** 3010 (1961); *C.A.*, **56,** 15532 (1962).
201. Kochkin, D. A., and I. B. Chekmareva, *Tr. Vses. Nauchn.-Issl. Vitamin. Inst.*, **7,** 37 (1961); through *C.A.*, **59,** 6427 (1963).
202. Kochkin, D. A., L. V. Luk'yanova, and E. B. Reznikova, *Zh. Obshch. Khim.*, **33,** 1945 (1963); *C.A.*, **59,** 11551 (1963).
203. Kochkin, D. A., Yu. P. Novichenko, G. I. Kuznetsova, and L. V. Laine, USSR Pat. 133,224, Nov. 10, 1960; through *C.A.*, **55,** 11923 (1961).
204. Kochkin, D. A., and S. G. Verenkina, *Tr. Vses. Nauchn.-Issl. Vitamin. Inst.*, **8,** 39 (1961); through *C.A.*, **58,** 6851 (1963).
205. Kochkin, D. A., S. G. Verenkina, and I. B. Chekmareva, *Dokl. Akad. Nauk SSSR*, **139,** 1375 (1961); *C.A.*, **56,** 7343 (1962).
206. Korshunov, I. A., and N. I. Malyugina, *Zh. Obshch. Khim.*, **31,** 1062 (1961); *C.A.*, **55,** 23320 (1961).
207. Koton, M. M., *Zh. Obshch. Khim.*, **9,** 2283 (1939); *C.A.*, **34,** 5049 (1940).
208. Koton, M. M., *Zh. Obshch. Khim.*, **11,** 376 (1941); *C.A.*, **35,** 5870 (1941).
209. Koton, M. M., and F. S. Florinskii, *Zh. Obshch. Khim.*, **32,** 3057 (1962); *C.A.*, **58,** 9111 (1963).
210. Koton, M. M., and T. M. Kiseleva, *Izv. Akad. Nauk SSSR, Otd. Khim. Nauk*, **1961,** 1783; through *C.A.*, **56,** 8740 (1962).
211. Koton, M. M., T. M. Kiseleva, and F. S. Florinskii, *Mezhdunarod. Simpozium po Makromol. Khim., Doklady Moscow 1960 Sektsiya*, **1,** 167; through *C.A.*, **55,** 7272 (1961).
212. Koton, M. M., T. M. Kiseleva, and F. S. Florinskii, *Vysokomolekulyarnye Soedineniya*, **2,** 1639 (1960); through *C.A.*, **55,** 26522 (1961).
213. Koyama, K., Japan. Pat. 8530, Sept. 24, 1958; through *C.A.*, **54,** 6111 (1960).
214. Krause, E., *Ber.*, **54,** 2060 (1921).
215. Krause, E., *Ber.*, **62,** 135 (1929).
216. Krause, E., and E. Pohland, *Ber.*, **55,** 1282 (1922).
217. Krause, E., and G. G. Reissaus, *Ber.*, **55,** 888 (1922).
218. Krause, E., and O. Schlöttig, *Ber.*, **58,** 427 (1925).
219. Krause, E., and O. Schlöttig, *Ber.*, **63,** 1381 (1930).
219a. Krause, E., and N. Schmitz, *Ber.*, **52,** 2165 (1919).
220. Krause, E., and A. von Grosse, *Die Chemie der metall-organischen Verbindungen*, Borntraeger, Berlin, 1937, pp. 372–429.
221. Krebs, A. W., and M. C. Henry, Quarterly Report No. 1, Project No. 7-99-01-001, Lead Industries Association, Sept. 30, 1961.
222. Krebs, A. W., and M. C. Henry, *J. Org. Chem.*, **28,** 1911 (1963).
223. Langer, H. G., U.S. Pat. 3,120,550 (to Dow Chemical Co.), Feb. 4, 1964; through *C.A.*, **60,** 12051 (1964).
223a. Langer, H. G., *Tetrahedron Letters*, **1967,** 43.
224. Langer, H. G., and A. H. Blut, *J. Organometal. Chem.*, **5,** 291 (1966).
225. Langkammerer, C. M., U.S. Pat. 2,253,128 (to E. I. du Pont de Nemours & Co.), Aug. 19, 1941; *C.A.*, **35,** 8151 (1941).
226. Lappert, M. F., and H. Psyzora, "Pseudohalides of Groups IIIB and IVB Elements," in *Advances in Inorganic Chemistry and Radio-Chemistry*, Vol. 9, H. J. Emeleus and E. G. Sharpe, Eds., Academic Press, N.Y., 1966, pp. 163–184.
227. Leeper, R. W., L. Summers, and H. Gilman, *Chem. Rev.*, **54,** 101 (1954).
228. Lesbre, M., *Compt. rend.*, **200,** 559 (1935).
229. Lesbre, M., *Compt. rend.*, **204,** 1822 (1937).

230. Lesbre, M., *Compt. rend.*, **206,** 1481 (1938).
231. Lesbre, M., *Compt. rend.*, **210,** 535 (1940).
232. Leusink, A. J., J. W. Marsman, and H. A. Budding, *Rec. trav. chim.*, **84,** 689 (1965).
233. Leusink, A. J., and G. J. M. van der Kerk, *Rec. trav. chim.*, **84,** 1617 (1965).
234. Lewis, G. L., P. F. Oesper, and C. P. Smyth, *J. Am. Chem. Soc.*, **62,** 3243 (1940).
235. Lieber, E., and F. M. Keane, *Chem. Ind.* (*London*), **1961,** 747.
236. Lieber, E., and F. M. Keane, *Inorg. Syn.*, **8,** 56 (1966).
237. Lieber, E., C. N. R. Rao, and F. M. Keane, *J. Inorg. Nucl. Chem.*, **25,** 631 (1963).
238. Ligett, W. B., R. D. Closson, and C. N. Wolf, U.S. Pat. 2,595,798 (to Ethyl Corp.), May 6, 1952; *C.A.*, **46,** 7701 (1952).
239. Ligett, W. B., R. D. Closson, and C. N. Wolf, U.S. Pat. 2,640,006 (to Ethyl Corp.), May 26, 1953; *C.A.*, **47,** 8307 (1953).
240. Lile, W. J., and R. C. Menzies, *J. Chem. Soc.*, **1950,** 617.
241. Lodochnikova, V. I., E. M. Panov, and K. A. Kocheshkov, *Izv. Akad. Nauk SSSR, Otd. Khim. Nauk*, **1957,** 1484; through *C.A.*, **52,** 7245 (1958).
242. Lodochnikova, V. I., E. M. Panov, and K. A. Kocheshkov, *Zh. Obshch. Khim.*, **29,** 2253 (1959); *C.A.*, **54,** 10967 (1960).
243. Lodochnikova, V. I., E. M. Panov, and K. A. Kocheshkov, *Zh. Obshch. Khim.*, **33,** 1199 (1963); *C.A.*, **59,** 10101 (1963).
244. Lodochnikova, V. I., E. M. Panov, and K. A. Kocheshkov, *Zh. Obshch. Khim.*, **34,** 946 (1964); *C.A.*, **60,** 15905 (1964).
245. Lodochnikova, V. I., E. M. Panov, and K. A. Kocheshkov, *Zh. Obshch. Khim.*, **34,** 4022 (1964); *C.A.*, **62,** 9164 (1965).
246. Löwig, C., *Ann.*, **88,** 318 (1853).
247. Löwig, C., *J. prakt. Chem.*, **60,** 304 (1853).
248. Luijten, J. G. A., F. Rijkens, and G. J. M. van der Kerk, "Organometallic Nitrogen Compounds of Germanium, Tin and Lead," in *Advances in Organometallic Chemistry*, Vol. 3, F. G. A. Stone and R. West, Eds., Academic Press, New York, 1965, pp. 397–443.
249. Makarova, L. G. and A. N. Nesmeyanov, *Zh. Obshch. Khim.*, **9,** 771 (1939); through *C.A.*, **34,** 391 (1940).
250. Malatesta, L., and R. Pizzotti, *Gazz. chim. ital.*, **73,** 349 (1943); through *C.A.*, **41,** 1137 (1947).
251. Manulkin, Z. M., *Zh. Obshch. Khim.*, **16,** 235 (1946); through *C.A.*, **41,** 90 (1947).
252. Manulkin, Z. M., *Zh. Obshch. Khim.*, **18,** 299 (1948); through *C.A.*, **42,** 6742 (1948).
253. Manulkin, Z. M., *Zh. Obshch. Khim.*, **20,** 2004 (1950); through *C.A.*, **45,** 5611 (1951).
254. Matwiyoff, N. A., and R. S. Drago, *Inorg. Chem.*, **3,** 337 (1964).
255. Matwiyoff, N. A., and R. S. Drago, *J. Organometal. Chem.*, **3,** 393 (1965).
256. McCombie, H., and B. C. Saunders, *Nature*, **158,** 382 (1946).
257. McCombie, H., and B. C. Saunders, *Nature*, **159,** 491 (1947).
258. McDyer, T., and R. D. Closson, U.S. Pat. 2,571,987 (to Ethyl Corp.), Oct. 16, 1951; *C.A.*, **46,** 3556 (1952).
259. Meals, R. N., *J. Org. Chem.*, **9,** 211 (1944).
260. Midgley, T., Jr., C. A. Hochwalt, and G. Calingaert, *J. Am. Chem. Soc.*, **45,** 1821 (1923).
261. Moedritzer, K., *Organometal. Chem. Rev.*, **1,** 179 (1966).
262. Möller, S., and P. Pfeiffer, *Ber.*, **49,** 2441 (1916).

263. Monserrat, M. P., *Combustibles (Zaragoza)*, **5,** No. 25/6, 15, No. 27/8, 67 (1945); through *C.A.*, **42,** 8451 (1948).
264. Nadj, M. M., and K. A. Kocheshkov, *Zh. Obshch. Khim.*, **12,** 409 (1942); through *C.A.*, **37,** 3068 (1943).
265. Nederlandse Centrale Organisatie voor Toegepast-Natuurwetenschappelijk Onderzoek, Belg. Pat. 660,428, March 1, 1965; *C.A.*, **64,** 3792 (1966).
266. Nederlandse Centrale Organisatie voor Toegepast-Natuurwetenschappelijk Onderzoek, Neth. Appl. 299,409, Aug. 25, 1965; through *C.A.*, **64,** 6853 (1966).
267. Nesmeyanov, A. N., and A. E. Borisov, *Dokl. Akad. Nauk SSSR*, **60,** 67 (1948); through *C.A.*, **43,** 560 (1949).
268. Nesmeyanov, A. N., R. Kh. Freidlina, and K. A. Kocheshkov, *Izv. Akad. Nauk SSSR, Otd. Khim. Nauk*, **1948,** 127; through *C.A.*, **43,** 1716 (1949).
269. Nesmeyanov, A. N., K. A. Kocheshkov, W. A. Klimova, and N. K. Gipp, *Ber.*, **68,** 1877 (1935).
270. Neumann, W. P., and K. Kühlein, *Angew. Chem.*, **77,** 808 (1965).
271. Neumann, W. P., paper presented at Second International Symposium on Organometallic Chemistry, Madison, Wis., Aug. 30, 1965.
272. Neumann, W. P., and K. Kühlein, *Tetrahedron Letters*, **1966,** 3415.
273. Neumann, W. P., and K. Kühlein, *Tetrahedron Letters*, **1966,** 3419.
274. Neumann, W. P., and K. Kühlein, *Tetrahedron Letters*, **1966,** 3423.
275. Okawara, R., *Proc. Chem. Soc.*, **1963,** 13.
276. Okawara, R., and H. Sato, *J. Inorg. Nucl. Chem.*, **16,** 204 (1961).
277. Okawara, R., D. E. Webster, and E. G. Rochow, *J. Am. Chem. Soc.*, **82,** 3287 (1960).
278. Overmars, H. G. J., and G. M. van der Want, *Chimia*, **19,** 126 (1965).
279. Padberg, F. J., *Dissertation*, *Univ. of Aachen*, 1965.
280. Panov, E. M., and K. A. Kocheshkov, *Dokl. Akad. Nauk SSSR*, **85,** 1037 (1952); *C.A.*, **47,** 6365 (1953).
281. Panov, E. M., and K. A. Kocheshkov, *Dokl. Akad. Nauk SSSR*, **85,** 1293 (1952); *C.A.*, **47,** 6887 (1953).
282. Panov, E. M., and K. A. Kocheshkov, *Zh. Obshch. Khim.*, **25,** 489 (1955); *C.A.*, **50,** 3271 (1956).
283. Panov, E. M., and K. A. Kocheshkov, *Dokl. Akad. Nauk SSSR*, **123,** 295 (1958); *C.A.*, **53,** 7133 (1959).
284. Panov, E. M., V. I. Lodochnikova, and K. A. Kocheshkov, *Dokl. Akad. Nauk SSSR*, **111,** 1042 (1956); *C.A.*, **51,** 9512 (1957).
285. Panov, E. M., N. N. Zemlyanskii, and K. A. Kocheshkov, *Dokl. Akad. Nauk SSSR*, **143,** 603 (1962); *C.A.*, **57,** 12521 (1962).
286. Peach, M. E., and T. C. Waddington, *J. Chem. Soc.*, **1961,** 1238.
287. Pedinelli, M., *Chim. Ind. (Milan)*, **44,** 651 (1962); through *C.A.*, **60,** 14303 (1964).
288. Pedinelli, M., R. Magri, and M. Randi, *Chim. Ind. (Milan)*, **48,** 144 (1966); *C.A.*, **64,** 12719 (1966).
289. Perilstein, W. L., U.S. Pat. 3,287,265 (to Intern. Lead Zinc Res. Organ.), Nov. 22, 1966; *C.A.*, **66,** 48102z (1967).
290. Pfeiffer, P., and P. Truskier, *Ber.*, **37,** 1125 (1904).
291. Pfeiffer, P., P. Truskier, and P. Disselkamp, *Ber.*, **49,** 2445 (1916).
292. Pilloni, G., *Anal. Chim. Acta*, **37,** 497 (1967).
293. Pilloni, G., and G. Plazzogna, *Anal. Chim. Acta*, **35,** 325 (1966); **37,** 260 (1967).
294. Podall, H. E., H. E. Petree, and J. R. Zietz, *J. Org. Chem.*, **24,** 1222 (1959).

295. Polees, B. M., A. B. Bruker, and L. Z. Soborovskii, USSR Pat. 148,404, July 13, 1962; through *C.A.*, **58,** 7974 (1963).
296. Polis, A., *Ber.*, **20,** 716 (1887).
297. Polis, A., *Ber.*, **20,** 3331 (1887).
298. Polis, A., *Ber.*, **21,** 3424 (1888).
299. Preuss, Fr. R., and I. Janshen, *Arch. Pharm.*, **293,** 933 (1960); through *C.A.*, **55,** 5396 (1961).
300. Preuss, Fr. R., and I. Janshen, *Arch. Pharm.*, **295,** 284 (1962); through *C.A.*, **58,** 2465 (1963).
301. Ramsden, H. E., Brit. Pat. 824,944 (to Metal and Thermit Corp.), Dec. 9, 1959; through *C.A.*, **54,** 17238 (1960).
302. Razuvaev, G. A., Yu. I. Dergunov, and N. S. Vyazankin, *Zh. Obshch. Khim.*, **31,** 998 (1961); *C.A.*, **55,** 23321 (1961).
303. Razuvaev, G. A., O. S. D'yachkovskaya, N. S. Vyazankin, and O. A. Shchepetkova, *Dokl. Akad. Nauk SSSR*, **137,** 618 (1961); *C.A.*, **55,** 20925 (1961).
304. Reichle, W. T., *Inorg. Chem.*, **1,** 650 (1962).
305. Reichle, W. T., *Inorg. Chem.*, **3,** 402 (1964).
306. Renger, G., *Ber.*, **44,** 337 (1911).
307. Riccoboni, L., *Gazz. chim. ital.*, **71,** 696 (1941); through *C.A.*, **36,** 6902 (1942).
308. Riccoboni, L., *Gazz. chim. ital.*, **72,** 47 (1942); through *C.A.*, **37,** 574 (1943).
309. Richard, W. R., U.S. Pat. 2,477,349 (to Monsanto Chemical Co.), July 26, 1949; *C.A.*, **44,** 4718 (1950).
310. Richardson, W. L., U.S. Pat. 3,010,980 (to California Research Corp.), Nov. 28, 1961; *C.A.*, **56,** 11620 (1962).
311. Richardson, W. L., U.S. Pat. 3,116,126 (to California Research Corp.), Dec. 31, 1963; *C.A.*, **60,** 6686 (1964).
312. Rieche, A., *XVII Internationaler Kongress für reine u. ang. Chemie*, Band 1, 14, Verlag Chemie, GMBH, Weinheim, 1959.
313. Rieche, A., *Wiss. Z. Friedrich Schiller Univ., Jena Math.-Naturwiss. Reihe*, **9,** 35 (1959/60) (Pub. 1960); *C.A.*, **55,** 21945 (1961).
314. Rieche, A., and J. Dahlmann, *Monatsber. deut. Akad. Wiss. Berlin*, **1,** 491 (1959); *C.A.*, **55,** 18640 (1961).
315. Rieche, A., and J. Dahlmann, *Ann.*, **675,** 19 (1964); *C.A.*, **61,** 10698 (1964).
316. Robinson, G. C., Ethyl Corporation, Unpublished Data.
317. Roncucci, L., G. Faraglia, and R. Barbieri, *J. Organometal. Chem.*, **1,** 427 (1964).
318. Rzaev, Z. M., D. A. Kochkin, and P. I. Zubov, *Dokl. Akad. Nauk SSSR*, **172,** 364 (1967); *C.A.*, **66,** 95146 (1967).
319. Saunders, B. C., *J. Chem. Soc.*, **1950,** 684.
320. Saunders, B. C., and G. J. Stacey, *J. Chem. Soc.*, **1948,** 1773.
321. Saunders, B. C., and G. J. Stacey, *J. Chem. Soc.*, **1949,** 919.
322. Saunders, B. C., G. J. Stacey, F. Wild, and I. G. E. Wilding, *J. Chem. Soc.*, **1948,** 699.
323. Schafer, H., *Z. anal. Chem.*, **180,** 15 (1961).
324. Scherer, O. J., and M. Schmidt, *J. Organometal. Chem.*, **1,** 490 (1964).
325. Scherer, O. J., and M. Schmidt, *J. Organometal. Chem.*, **3,** 156 (1965).
326. Schlemper, E. O., and D. Britton, Ref. 108 in Tobias, R.S., *Organometal Chem. Rev.*, **1,** 93 (1966).
327. Schmidbaur, H., and H. Hussek, *J. Organometal. Chem.*, **1,** 235 (1964); **1,** 244 (1964).
328. Schmidbaur, H., and H. Hussek, *J. Organometal. Chem.*, **1,** 257 (1964).
329. Schmidbaur, H., and M. Schmidt, *J. Am. Chem. Soc.*, **83,** 2963 (1961).

330. Schmidbaur, H., and M. Schmidt, *Chem. Ber.*, **94,** 1138 (1961).
331. Schmidbaur, H., and M. Schmidt, *Angew. Chem.*, **73,** 408 (1961).
332. Schmidbaur, H., and M. Schmidt, *Angew. Chem.*, **73,** 655 (1961).
333. Schmidbaur, H., and M. Schmidt, *Angew. Chem.*, **74,** 328 (1962).
334. Schmidt, H., *Medicine in Its Chemical Aspects*, Vol. III, Bayer, Leverkusen, Germany, 1938, pp. 394–404.
335. Schumann, H., A. Roth, O. Stelzer, and M. Schmidt, *Inorg. Nucl. Chem. Letters*, **2,** 311 (1966).
336. Schumann, H., and M. Schmidt, *J. Organometal. Chem.*, **3,** 485 (1965).
337. Schumann, H., and M. Schmidt, *Inorg. Nucl. Chem. Letters*, **1,** 1 (1965).
338. Schumann, H., P. Schwabe, and M. Schmidt, *J. Organometal. Chem.*, **1,** 366 (1964); *Inorg. Nucl. Chem. Letters*, **2,** 309 (1966).
339. Schumann, H., P. Schwabe, and M. Schmidt, *Inorg. Nucl. Chem. Letters*, **2,** 313 (1966).
340. Schumann, H., K. -F. Thom, and M. Schmidt, *J. Organometal. Chem.*, **4,** 28 (1965).
341. Schweitzer, G. K., and S. W. McCarty, *J. Inorg. Nucl. Chem.*, **27,** 191 (1965).
342. Semion, V. A., and D. N. Kravtsov, *Zh. Strukt. Khim.*, **7,** 814 (1966); through *C.A.*, **66,** 23141 (1967).
343. Setzer, W. C., R. W. Leeper, and H. Gilman, *J. Am. Chem. Soc.*, **61,** 1609 (1931).
344. Shapiro, H., V. F. Hnizda, and G. Calingaert, U.S. Pat. 2,674,519 (to Ethyl Corp.), April 6, 1954; *C.A.*, **48,** 10308 (1954).
345. Shapiro, H., V. F. Hnizda, and G. Calingaert, U.S. Pat. 2,674,610 (to Ethyl Corp.), April 6, 1954; *C.A.*, **49,** 4706 (1955).
346. Shier, G. D., and R. S. Drago, *J. Organometal. Chem.*, **5,** 330 (1966).
347. Shier, G. D., and R. S. Drago, *J. Organometal. Chem.*, **6,** 359 (1966).
348. Shushunov, V. A., and T. G. Brilkina, *Dokl. Akad. Nauk SSSR*, **141,** 1391 (1961); *C.A.*, **56,** 12921 (1962).
349. Singh, G., *J. Org. Chem.*, **31,** 949 (1966); *J. Organometal. Chem.*, **11,** 133 (1968)
350. Sijpesteijn, A. K., F. Rijkens, J. G. A. Luijten, and L. C. Willemsens, *Antonie van Leeuwenhoek, J. Microbiol. Serol.*, **28,** 346 (1962); through *C.A.*, **58,** 7308 (1963).
351. Smyth, C. P., *J. Org. Chem.*, **6,** 421 (1941).
352. Smyth, C. P., *J. Am. Chem. Soc.*, **63,** 57 (1941).
353. Sosnovsky, G., and J. H. Brown, *Chem. Rev.*, **66,** 529 (1966).
354. Summers, L., *Iowa St. Coll. J. Sci.*, **26,** 292 (1952); through *C.A.*, **47,** 8673 (1953).
355. Tafel, J., *Ber.*, **44,** 323 (1911).
356. Tagliavini, G., *Ric. sci. Rend., Sez.* **A8,** 1533 (1965); through *C.A.*, **65,** 3071 (1966).
357. Tamborski, C., F. E. Ford, W. L. Lehn, G. J. Moore, and E. J. Soloski, *J. Org. Chem.*, **27,** 619 (1962).
358. Tamborski, C., and E. J. Soloski, *J. Am. Chem. Soc.*, **83,** 3734 (1961).
359. Taylor, H. S., and W. H. Jones, *J. Am. Chem. Soc.*, **52,** 1111 (1930).
360. Thayer, J. S., *Organometal. Chem. Rev.*, **1,** 157 (1966).
361. Thayer, J. S., and D. P. Strommen, *J. Organometal. Chem.*, **5,** 383 (1966).
362. Thayer, J. S., and R. West, *Inorg. Chem.*, **3,** 406 (1964).
363. Thayer, J. S., and R. West, "Organometallic Pseudohalides" in *Advances in Organometallic Chemistry*, Vol. 5, F. G. A. Stone and R. West, Eds., Academic Press, N.Y., 1967, pp. 169–224.

364. Thomas, A. B., and E. G. Rochow, *J. Am. Chem. Soc.*, **79,** 1843 (1957).
365. Thomas, A. B., and E. G. Rochow, *J. Inorg. Nucl. Chem.*, **4,** 205 (1957).
366. Tobias, R. S., *Organometal. Chem. Rev.*, **1,** 93 (1966).
367. Torpova, V. F., and M. K. Saikina, *Sb. Statei Obshch. Khim. Akad. Nauk SSSR*, **1,** 210 (1953); through *C.A.*, **48,** 12579 (1954).
368. Ueeda, R., Y. Kawasaki, T. Tanaka, and R. Okawara, *J. Organometal. Chem.*, **5,** 194 (1966).
369. Van der Kelen, G. P., L. Verdonck, and D. van de Vondel, *Bull. Soc. Chim. Belges*, **73,** 733 (1964).
370. von Hevesy, G., and L. Zechmeister, *Ber.*, **53,** 410 (1920).
371. Vorlander, D., *Ber.*, **58,** 1893 (1925).
372. Vyshinskii, N. N., Yu. A. Aleksandrov, and N. K. Rudnevskii, *Izv. Akad. Nauk SSSR, Ser. Fiz.*, **26,** 1285 (1962); through *C.A.*, **58,** 7517 (1963).
372a. Vyshinskii, N. N. and N. K. Rudnevskii, *Optkica i Spekroskopiya*, **10,** 797 (1961); through *C.A.*, **58,** 4059 (1963).
373. Vyshinskii, N. N., and N. K. Rudnevskii, *Spektroskopiya, Metody i Primenenie, Akad. Nauk SSSR, Sibirsk. Otd.*, **1964,** 115; through *C.A.*, **62,** 3533 (1965).
374. Wang, C-H., P. L. Levins, and H. E. Pars, *Tetrahedron Letters*, **1964,** 687.
375. Wartik, T., and H. I. Schlesinger, *J. Am. Chem. Soc.*, **75,** 835 (1953).
376. Werner, A., and P. Pfeiffer, *Z. anorg. allgem. Chem.*, **17,** 82 (1898).
377. West, R., and R. H. Baney, *J. Phys. Chem.*, **64,** 822 (1960).
378. Willemsens, L. C., Quarterly Report No. 10, Project LC-18, to Intern. Lead Zinc Res. Organ., New York, April–June, 1962.
379. Willemsens, L. C., Quarterly Report No. 16, Project No. LC-18, to Intern. Lead Zinc Res. Organ., New York, July–Dec., 1963.
380. Willemsens, L. C., Quarterly Report No. 18, Project No. LC-18, to Intern. Lead Zinc Res. Organ., New York, Jan–March, 1964.
381. Willemsens, L. C., and G. J. M. van der Kerk, *J. Organometal. Chem.*, **2,** 260 (1964).
382. Willemsens, L. C., and G. J. M. van der Kerk, Quarterly Report No. 19, Project LC-18, to Intern. Lead Zinc Res. Organ., New York, July–October, 1964.
383. Willemsens, L. C., and G. J. M. van der Kerk, *Rec. trav. chim.*, **84,** 43 (1965).
384. Willemsens, L. C., and G. J. M. van der Kerk, *J. Organometal. Chem.*, **4,** 34 (1965).
385. Willemsens, L. C., and G. J. M. van der Kerk, Quarterly Report No. 24, Project LC-18, to Intern. Lead Zinc Res. Organ., New York, January–December, 1965.
386. Willemsens, L. C., and G. J. M. van der Kerk, *Investigations in the Field of Organolead Chemistry*, Intern. Lead Zinc Res. Organ., New York, 1965.
387. Willemsens, L. C., and G. J. M. van der Kerk, *J. Organometal. Chem.*, **4,** 241 (1965).
388. Willemsens, L. C., and G. J. M. van der Kerk, Quarterly Report No. 25, Project LC-18, to Intern. Lead Zinc Res. Organ., New York, January–March, 1966.
389. Witherspoon, S. C., P.B. Report No. 25847, *Bib. Sci. Ind. Repts.*, **2,** 79 (1946).
390. Wu, C-L., C. Y. Ch'en, Cg-W. Weng, and S-M. Chang, *Yuan Tzu Neng*, **1,** 27 (1965); through *C.A.*, **63,** 14318 (1965).
391. Yakubovich, A. Ya., S. P. Makarov, V. A. Ginsburg, G. I. Gavrilov, and E. N. Merkulova, *Dokl. Akad. Nauk SSSR*, **72,** 69 (1950); *C.A.*, **45,** 2856 (1951).
392. Yakubovich, A. Ya, E. N. Merkulova, S. P. Makarov, and G. I. Gavrilov, *Zh. Obshch. Khim.*, **22,** 2060 (1952); *C.A.*, **47,** 9257 (1953).
393. Yakubovich, A. I., and I. Petrov, *J. prakt. Chem.*, **144,** 67 (1935).

394. Yngve, V., U.S. Pat. 2,307,090 (to Carbide and Carbon Chemicals Corp.), Jan. 5, 1943; through *C.A.*, **37,** 3532 (1943).
395. Zavgorodnii, S. V., and A. A. Petrov, *Dokl. Akad. Nauk SSSR*, **143,** 855 (1962); *C.A.*, **57,** 3466 (1962).
396. Zartman, W. H., and H. Adkins, *J. Am. Chem. Soc.*, **54,** 3398 (1932).
397. Zechmeister, L., and J. Csabay, *Ber.*, **60,** 1617 (1927).
398. Zemlyanskii, N. N., V. N. Lodochnikova, E. M. Panov, and K. A. Kocheshkov, *Zh. Obschch. Khim.*, **35,** 843 (1965); *C.A.*, **63,** 7031 (1965).
399. Zimmer, H., and E. Endrulat, Abstract, p. 61N, 129th Meeting Amer. Chem. Soc., Dallas, Texas, April, 1956.
400. Zimmer, H., and O. A. Homberg, *J. Org. Chem.*, **31,** 947 (1966).

XII Organofunctional lead compounds

Organofunctional lead compounds, for the present purpose, may be defined as those compounds in which one or more functional atoms or groups are present in an organic substituent carbon-bonded to the lead atom.

Until very recently, the organofunctional compounds were a comparatively small group of lead compounds as compared with the unsubstituted tetraorganolead compounds. The most common method of attaching an organic group to a lead atom in the past depended on the reaction of a lead compound with a Grignard reagent or an organolithium compound, and since such reactions lead to functional group reactions as well, the number of organofunctional lead compounds was understandably small until recently. Furthermore, attempts to introduce functionality into an organic group already carbon-bonded to lead usually resulted in cleavage of the carbon–lead bond instead.

Since about 1962, several methods of introducing functionality into organolead compounds have been developed or extended. Willemsens of the T.N.O. group at Utrecht found that some triphenyllead compounds react with ketene to introduce a β C=O functionality [$=PbCH_2C(O)^-$]. Willemsens extended the scope of a reaction first investigated by Gilman, that of the ring opening of small ring heterocycles with triphenylplumbyllithium (described below), and he has also shown that triphenylplumbyllithium will add to some compounds containing reactive triple bonds to yield organofunctional lead compounds containing double bond unsaturation; this was discussed in Chapter X. Willemsens and van der Kerk have also extended Kocheshkov and Aleksandrov's decarboxylation of organolead salts to form two new compounds, as described below. Creemers, Leusink, Noltes, and van der Kerk have shown that organofunctionally-substituted lead compounds can be prepared by reaction of trialkyllead hydrides (generated *in situ*) with acetylenic functional compounds. Perhaps more importantly, the simple hydroplumbation of unsaturated organic compounds with alkyllead hydrides has been demonstrated by Leusink and van der Kerk, and also by Henry, who found that alkyllead hydrides will add to reactive acetylenic compounds to yield trialkylplumbyl-substituted olefin derivatives. The versatility of the hydroplumbation of olefinic compounds has been demonstrated, especially by Neumann and Kühlein. This reaction seems to be a mild, quite general method for introducing functionality into organolead molecules. Lukevits and Voronkov (18) have recently published a book on the organic insertion reactions of the Group IVb elements, including the hydroplumbation of unsaturated compounds. Finally, Davies and Puddephatt (5) have shown that triorganolead oxides and alkoxides add to certain unsaturated compounds in a manner analogous to the trialkyllead hydrides in the hydroplumbation reaction.

Compounds resulting from organolead hydride reactions are discussed in Chapter XI, compounds in which there is unsaturation are discussed in Chapter X; and in addition, Chapters XIII and XIV on "Heterocyclic Lead Compounds" and "Optical Activity," respectively, should be consulted for the specific organofunctional lead compounds pertinent to these subjects.

Most of the progress prior to 1960 in the area of organofunctional lead chemistry was due primarily to the investigations of Austin, Krause and his co-workers, and Kocheshkov and Aleksandrov, but especially of Gilman and his students, who prepared various aryl and some alkyl compounds containing halogen, amino, hydroxy, carboxy, methoxy, and other groups. These compounds were prepared by such methods as various halogen-metal interconversion reactions, and through the use of triphenylplumbyllithium. Compounds prepared before 1953 have been tabulated in Leeper, Summers, and Gilman's review (17). Compounds prepared since 1952 are listed in Table XII-1.

1. COMPOUNDS WITH SUBSTITUENTS BEARING HALOGEN

As indicated above, lead compounds with organic substituents bearing halogen atoms cannot usually be prepared with the facility of the analogous compounds of silicon, germanium, or even tin. Since the lead–carbon bond is cleaved so easily, only indirect halogenation of alkyl side-chains is possible. (In this discussion only halogen-substituted alkyllead compounds are covered; halogen-substituted aromatic compounds are included in Chapter VI.)

An elegant method of preparing chloromethyl derivatives of the Group IVb metals via diazomethane methylenation was discovered by Yakubovich and his associates (26,38,39) in 1950. Ethyllead chlorides were found to react with diazomethane in ether solution at 0–15° in the presence of copper bronze catalyst. Triethyllead chloride gave triethylchloromethyllead in 73% yield, but in admixture with tetraethyllead, from which separation by careful distillation was required. Diethyllead dichloride gave diethyl-bis(chloromethyl)lead; however, the latter was obtained in poor yield, and was unstable at room temperature even in the absence of light. Similarly, α-chloroethyltriethyllead was prepared from diazoethane, but was too unstable to isolate due to a tendency to disproportionate into lead chloride and tetraethyllead on removal of the ether solvent.

The triphenylhalomethyllead compounds have been prepared by the addition of triphenylplumbyllithium to carbon tetrachloride or chloroform, and by the addition of triphenyllead methoxide to hexachloroacetone or bromal, as described below.

Perfluoroalkyl compounds of lead (and tin) have been prepared by Kaesz, Phillips, and Stone (13,14,29), who utilized the ability of tetraalkyllead and perfluoroalkyl iodide compounds to undergo free radical reactions.

$$(CH_3)_4Pb + C_nF_{2n+1}I \xrightarrow[\text{or uv}]{\text{heat}} (CH_3)_3PbC_nF_{2n+1} + CH_3I \qquad (n = 1 \text{ or } 2)$$

In a similar manner, trimethylperfluoroethyllead was prepared from hexamethyldilead and pentafluoroiodoethane. Yields were relatively low, with considerable amounts of fluoroform and pentafluoroethane being produced, but the products were stable compounds. For example, trimethylperfluoroethyllead, obtained in 28% yield, survived repeated vapor phase chromatography. The infrared spectra of the compounds were described, and some experiments were made on cleavage of the alkyl groups; base cleaved the perfluoroalkyl group quantitatively. The perfluoroalkyllead compounds were much less stable to hydrolysis than normal tetraalkyllead compounds. The cleavage of the perfluoroalkyl group by acids and bases is discussed in Chapter VI.

Tetraalkylsilicon and tetraalkyltin compounds do not react with perfluoroalkyl iodides in a similar fashion. Presumably, higher temperatures would be required to achieve the homolytic fission of the C—Si and C—Sn bonds, and under these conditions, hydrogen is abstracted from the alkyl groups, yielding perfluoroalkanes instead. Thus lead is unique among the Group IVb metals in this respect. As Treichel and Stone (29) note, there is now no reason why a wide variety of new fluorocarbon–lead compounds cannot be prepared, since a large number of perfluoroalkyl lithium and Grignard compounds can now be synthesized.

Table XII-1 New Organofunctional Lead Compounds

COMPOUND	PHYSICAL PROPERTIES	REFERENCES
$(C_2H_5)_3PbCH_2Cl$	b.p.$_3$ 65.5–6.5°, d_{20} 1.7917, n_D^{20} 1.5434	17,38,39
$(C_2H_5)_2Pb(CH_2Cl)_2$	b.p.$_2$ 96°, d_{20} 1.9890	17,38,39
$(C_2H_5)_3PbCH_2CH_2Cl$		17,38,39
$(C_2H_5)_3PbCH_2CH_2CH_2CH_2CH_2Br_5$		12
$(CH_3)_3PbCF_3$		13,14,29
$(CF_3)_4Pb$		4
$(CH_3)_3PbCF_2CF_3$	b.p. 138°	13,14,29
$(F_2C{=}CF)_4Pb$	b.p.$_8$ 51–2°, d_{18} 2.4020, n_D^{18} 1.4193	27
$(C_6H_5)_3PbCHCl_2$	m.p. 124–5°	37
$(C_6H_5)_3PbCCl_3$	m.p. 172–3°	37
$[(C_6H_5)_3Pb]_2CCl_2$	m.p. 207–8°	37
$(C_6H_5)_3PbCBr_3$	m.p. 135–40°	5
$(C_6H_5)_3PbCH_2CH_2CH_2Br$	m.p. 62–6°	11
$(C_6H_5)_3PbCH_2(CH_2)_2CH_2Br$	m.p. 61–3°	11
$(C_6H_5)_3PbCH_2(CH_2)_3CH_2Br$	m.p. 61–2°	11
$(C_6H_5)_2Pb(p\text{-}C_3H_5C_6H_5)(C_6H_5CH_2CHBrCH_2Br)$		25
$(o\text{-}FC_6H_4CH_2)_4Pb$		3
$(o\text{-}ClC_6H_4CH_2)_4Pb$		3
$(o\text{-}BrC_6H_4CH_2)_4Pb$		3
$[(ClC_6H_4CH_2)_3Pb]_2$		3
$[(BrC_6H_4CH_2)_3Pb]_2$		3
$(C_6H_5)_3PbCH_2C_6H_4\text{-}p\text{-}Br$		10,17,30
$(N{\equiv}CCH_2CH_2)_4Pb$	d_{20} 1.7480, n_D^{20} 1.5489	28
$(C_4H_9)_3PbCH_2CH_2CN$	b.p.$_{0.01}$ 116–9°	1,21,22
$(C_4H_9)_3PbCH_2CH_2COOCH_3$	b.p.$_{10^{-4}}$ 94–7°	1,21,22
$(C_4H_9)_3PbCH(C_6H_5)C(H)(CO_2R)(CN)$		23
$(C_4H_9)_3PbC(CO_2R)(CN)\text{—}CH_2C_6H_5$		23
trans-$C_6H_5CH{=}CHPb(C_4H_9)_3$	dec. on distillation	1,21,22
$(C_6H_5)_3PbCH_2CH_2OH$	m.p. 72°	24,32,33,35
$(C_6H_5)_3PbCH_2CH_2CH_2OH$	m.p. 99°	32,33,35
$(C_6H_5)_3PbC_4H_8OH$		32,33,35

(continued)

Table XII-1 (continued)

COMPOUND	PHYSICAL PROPERTIES	REFERENCES
$(C_6H_5)_3Pb(C_6H_4CH_2OH)$-*p*	m.p. 98–100°	9,17
$(C_6H_5)_3PbC_6H_4CHO$	m.p. 110.5–2°	8
$(C_6H_5)_3PbC_6H_4CH_2OOCC_6H_5$		8
$(C_6H_5)_3PbC_6H_4CH(C_8H_{10}O_2)_2$	m.p. 161.5–3°, with dec.	8
$(C_6H_5)_3PbC_6H_4CH_2OOCC_{18}H_{18}O$	m.p. 152.5–3.5°	8
$(C_6H_5)_3PbCH_2CH_2SH$	m.p. 104–6°	32,33,35
$(C_6H_5)_3PbCHOHCH_2Cl$	m.p. 90–1°	32,33,35
$(C_6H_5)_3PbCH_2COOCH_2CH_3$	m.p. 60°	31,36
$[(C_6H_5)_3PbCH_2CO]_2O$	dec. 97–103°	31,36
$(C_6H_5)_3PbCH_2COOC_2H_5$	m.p. 59–60°	15–17
$(C_6H_5)_3PbCOOEt$		34
$(C_6H_5)_3PbC(CH_2C_6H_5)COOC_2H_5$	m.p. 82–4°	15–17
$(C_6H_5)_3PbCH_2CN$		37
$(C_6H_5)_3PbCH_2CH_2N_3$		11
$(C_6H_5)_3PbCH_2(CH_2)_2CH_2N_3$		11
$(C_6H_5)_3Pb(CH_2)_3NCH{=}C[C(CH_3)_2OH]N{=}N$ (ring: N…N)		11
$(C_6H_5)_3Pb(CH_2)_4NCH{=}C[C(CH_3)_2OH]N{=}N$ (ring: N…N)		11
$(C_6H_5)_3PbCH_2CH_2NHCOCH_3$	m.p. 113°	37
$(C_6H_5)_3PbCH_2CH_2NHCOC_6H_5$	m.p. 136–8°	32,33,35,37
$(C_6H_5)_3PbCH_2CH_2COOCH_3$	m.p. 52.5°	35,37
$HO[(C_6H_5)_2Pb^{+}CH_2CH_2COO^{-}]_3H$	m.p. >240°	35,37
$(C_6H_5)_3PbCH_2CHOHCH_2OH$	m.p. 124–5°	25
$(CH_3)_2BrPbC(C_6H_5){=}C(C_6H_5)C(C_6H_5){=}C(C_6H_5)Br$		5
$(p\text{-}HOCC_6H_4)_4Pb$	m.p. 188°	6
$\left(\begin{smallmatrix}H_2C-O\\ \vert \\ H_2C-O\end{smallmatrix}\!>\!\overset{H}{C}-\text{⬡}\right)_4Pb$	m.p. 268–70°	6
$(\text{⬡})_3Pb-\overset{H}{C}(COOH)-CH_2-\text{⬡}$ (H–C–H)	165–7	40

To date, no tetraperfluoroalkyllead compound has been prepared and identified unequivocally. However, Bell, Pullman, and West (4) pyrolyzed hexafluoroacetone and passed the CF_3 free radicals over a lead mirror. They obtained a colorless liquid, characterized only by infrared spectrum, which was probably tetrakis(trifluoromethyl)lead.

In 1961, Sterlin et al. (27) prepared tetrakis(perfluorovinyl)lead from lead(II) chloride and perfluorovinylmagnesium iodide. The stability of the perfluoro compound was compared with the other similar Group IV metal derivatives by treatment with alkali. The order of stability was found to be Si > Sn > Pb > Ge; this order was concluded to be consistent with increasing corrected values of Pauling's electronegativity scale.

In 1955, Bähr and Zoche (3) synthesized several tetrakis(*o*-halobenzyl)-lead compounds by a similar type of Grignard reaction. The *o*-fluoro, *o*-chloro, and *o*-bromo compounds so prepared were all low-melting solids. Surprisingly, as compared with the usual colorless tetraalkyl- or tetraaryllead compounds, the tetrakis-*o*-halobenzyl compounds were yellow in color. The *o*-chloro and *o*-bromo hexabenzyldilead compounds were also prepared; these exhibited orange to red colors (see Chapter VII). Also, in 1955, Wallen (30) described the preparation of triphenyl-*p*-bromobenzyllead, a compound previously reported by Gilman and Summers (19), according to the following equation:

$$LiPb(C_6H_5)_3 + ClCH_2C_6H_4Br \rightarrow (C_6H_5)_3PbCH_2C_6H_4Br + LiCl$$

Gorth and Henry (11) synthesized a series of bromoalkyllead derivatives in the course of a study of the reactivity of functional groups as a function of the distance from the central lead atom of the bromine atom. When the functional group was five or more carbon atoms away, its reactivity was apparently normal. However, when the functionality was in the β position, the reactivity of the carbon–halogen bond was drastically changed. The triphenyllead bromoalkanes were prepared in approximately 70% yields by reacting triphenylplumbyllithium with a five-fold molar excess of 1,3-dibromopropane, 1,4-dibromobutane, or 1,5-dibromopentane, according to the general method originated by Grüttner and Krause (12):

$$LiPb(C_6H_5)_3 + Br(CH_2)_{3-5}Br \rightarrow (C_6H_5)_3Pb(CH_2)_{3-5}Br + LiBr$$

However, with dibromoethane the coupling product was not obtained; instead, tetraphenyllead was the principal product, presumably resulting from β-elimination in the expected product. With dibromomethane, the main product of the reaction was bis(triphenyllead)methane. The triphenyllead bromoalkanes were converted in good yields to the corresponding azido compounds with excess sodium azide in dimethyl sulfoxide at 90°. The azido compounds were in turn reacted with 2-methyl-butyne-3-ol-2 to yield the expected triazole derivatives:

$$(C_6H_5)_3Pb(CH_2)_{3-5}Br + NaN_3 \rightarrow (C_6H_5)_3Pb(CH_2)_{3-5}N_3 + NaBr$$

$$(C_6H_5)_3Pb(CH_2)_{3-5}N_3 + (CH_3)_2C(OH)C{\equiv}CH \rightarrow$$

$$(C_6H_5)_3Pb(CH_2)_{3-5}\underline{NCH{=}C[C(CH_3)_2OH]N{=}N}$$

Willemsens and van der Kerk (37) prepared three novel chloro compounds by means of the addition of triphenylplumbyllithium to carbon tetrachloride or chloroform. At −60°, addition to excess carbon tetrachloride gave $(C_6H_5)_3PbCCl_3$ in 92% yield, while at room temperature the disubstituted product, $[(C_6H_5)_3Pb]_2CCl_2$, was obtained in 80% yield. The $(C_6H_5)_3PbCCl_3$ decomposed at 80° to dichlorocarbene and triphenyllead chloride. Addition of triphenylplumbyllithium to chloroform at −60° yielded $(C_6H_5)_3PbCHCl_2$ in 75% yield. Attempts to prepare $[(C_6H_5)_3Pb]_3CCl$ and $[(C_6H_5)_3Pb]_2CHCl$ were unsuccessful because of further reaction of these products with triphenylplumbyllithium.

Freedman (7) has claimed the following ring-opening reaction in a 1963 patent:

$$\text{1,1-(CH}_3)_2\text{-2,3,4,5-Ph}_4\text{-plumbole} + Br_2 \longrightarrow (CH_3)_2Pb(Br){-}C(Ph){=}C(Ph){-}C(Ph){=}C(Ph){-}Br$$

However, the only example described pertains to the preparation of (4-bromo-1,2,3,4-tetraphenyl-1,3-*cis*-butadienyl)dimethyltin bromide from 1,1-dimethyl-2,3,4,5-tetraphenylstannole and bromine.

2. COMPOUNDS WITH "WATER-SOLUBILIZING" GROUPS

Organolead compounds containing a variety of organofunctional groups other than halogen have been synthesized. These other groups have been sometimes referred to generically for convenience as "water-solubilizing" groups.

Much of the older work on "water-solubilizing" groups is due to Gilman and his co-workers. For example, Gilman and Melstrom (9) employed the reaction of triphenyllead chloride and the dilithium compound prepared from *n*-butyllithium and *p*-bromobenzyl alcohol to synthesize [*p*-(hydroxymethyl)phenyl]triphenyllead in fair yield. Recently, Gerrard and Green

(8) have oxidized this alcohol to *p*-(triphenylplumbyl)benzaldehyde by shaking it with "precipitated" manganese dioxide. An attempt to oxidize the aldehyde with silver oxide failed. However, the aldehyde was successfully converted to the 5,5-dimethyl-1,3-cyclohexanedione derivative, and also esterified with benzoyl chloride and a steroid acid chloride, (±)-*cis*-7-methylbisdehydrodoisynolic acid chloride; two different crystalline benzoates were isolated.

Willemsens and van der Kerk (32,33,35,37) investigated the scope of Gilman's ring-opening reaction of triphenylplumbyllithium with small ring heterocycles to produce a number of functionally-substituted compounds, according to the general equation:

$$\mathrm{LiPb(C_6H_5)_3} + \overbrace{\mathrm{(CH_2)_nX}} \longrightarrow \mathrm{(C_6H_5)_3Pb(CH_2)_nXLi} \xrightarrow{\mathrm{H_2O}} \mathrm{(C_6H_5)_3Pb(CH_2)_nXH}.$$

In a typical reaction, addition of an equimolar quantity of ethylene oxide to triphenylplumbyllithium at $-60°$ in an inert atmosphere followed by hydrolytic workup gave (2-hydroxyethyl)triphenyllead in 83% yield. Similar reactions were carried out with trimethylene oxide, ethylene sulfide, epichlorohydrin, and benzoylethylene imine to give the corresponding hydroxy, mercapto, chlorohydroxy, and benzoamido compounds, respectively. Unsubstituted ethylene imine, *N*-butylethylene imine, and tetrahydrofuran did not react with triphenylplumbyllithium. Trimethylene sulfide and pyrrolidone reacted with triphenylplumbyllithium at elevated temperatures, but no organic products could be isolated.

A Russian patent (24) claims the preparation of β-hydroxyalkyltrialkyllead compounds by reaction of trialkyllead chloride with an olefin oxide in liquid ammonia, followed by hydrolytic decomposition of the alcoholate thus formed. The ethylene oxide reaction is described as an example.

Willemsens and van der Kerk (31,36,37) also reported the reaction of triphenyllead hydroxide and triphenyllead acetate with ketene. This reaction is an extension of the reaction of organomercury and organotin compounds with ketene, discovered by Lutsenko et al. (19,20).

$$2(\mathrm{C_6H_5})_3\mathrm{PbOH} + 4\mathrm{H_2C{=}CO} \rightarrow [(\mathrm{C_6H_5})_3\mathrm{PbCH_2CO}]_2\mathrm{O} + \mathrm{H_3CCOOH}$$

The preparation of α-triphenylplumbylacetic anhydride in 85% yield was carried out by passing ketene into a solution of the triphenyllead hydroxide in diethyl ether at 10°. The ketene insertion reaction failed with a number of other lead salts, including $\mathrm{Et_3PbOAc}$, $\mathrm{Ph_3PbCl}$, $\mathrm{Ph_2Pb(OAc)_2}$, and

Ph_2PbO. However, triphenyllead acetate reacted with ketene in ethanol to give a 75% yield of ethyl-α-triphenylplumbylacetate.

$$(C_6H_5)_3PbOOCCH_3 + H_2C{=}CO + C_2H_5OH \rightarrow (C_6H_5)_3PbCH_2COOCH_2CH_3 + H_3CCOOH$$

Kocheshkov and Aleksandrov (15,16) showed that a decarboxylation reaction of the type

$$(C_6H_5)_3PbOOCCH_2COOC_2H_5 \rightarrow (C_6H_5)_3PbCH_2COOC_2H_5 + CO_2$$

proceeded easily at 160–180° with the monoethyl esters of both malonic and benzylmalonic acids. The starting triphenyllead esters were prepared easily from triphenyllead chloride and the potassium malonate esters. The method has an inherent limitation since organolead compounds ordinarily undergo homolytic decomposition at high temperatures. However, Willemsens and van der Kerk (37) have been able to extend the reaction to two other cases, decomposing triphenyllead cyanoacetate to cyanomethyltriphenyllead, $(C_6H_5)_3PbCH_2CN$, in 48% yield, and triphenyllead phenylpropiolate to phenylethynyltriphenyllead, $(C_6H_5)_3PbC{\equiv}CC_6H_5$, in 80% yield. Both products were relatively unstable. Attempts to decarboxylate triphenyllead acetoacetate resulted in easy decomposition, but the desired acetonyltriphenyllead was apparently even more unstable and could not be isolated.

Davies and Puddephatt (5) demonstrated recently that triphenyllead and tributyllead methoxides, and bis(triphenyllead) oxide, can be reacted with several unsaturated compounds, as the analogous organotin compounds have been shown to do. This type of reaction is similar to hydroplumbation reactions of organolead hydrides.

$$R_3PbOR' + A{=}B \rightarrow R_3PbABOR'$$

These reactions were rapid and exothermic at room temperature, and it was shown that the Pb—O bond is clearly more reactive than the Sn—O bond. The adducts containing the tributyl group were thermally and photolytically unstable, and blackened in a few days. On the other hand, the triphenyllead-containing products were stable under nitrogen and sufficiently stable toward hydrolysis to allow molecular weight determinations by vapor pressure osmometry. Adducts of triphenyllead methoxide were isolated from reactions with 1-naphthyl isocyanate, phenyl isothiocyanate, carbon disulfide, chloral, trichloroacetonitrile, and di-1-naphthylcarbodiimide, while a tributyllead methoxide adduct was isolated only from the 1-naphthyl isocyanate reaction.

Davies and Puddephatt showed also that this type of reaction could be employed to prepare trihalomethyllead compounds by decomposition of intermediate adducts. For example, trichloromethyltriphenyllead was prepared from hexachloroacetone in 60% yield by the reaction:

$$Ph_3PbOMe + OC(CCl_3)_2 \longrightarrow \left[\begin{array}{c} CCl_3 \\ | \\ Ph_3PbOCOMe \\ | \\ CCl_3 \end{array} \right] \longrightarrow Ph_3PbCCl_3 + Cl_3CC(O)OMe$$

Similarly, tribromomethyltriphenyllead was synthesized from bromal.

$$Ph_3PbOMe + OCHCBr_3 \longrightarrow \left[\begin{array}{c} Ph_3PbOCHOMe \\ | \\ CBr_3 \end{array} \right] \longrightarrow Ph_3PbCBr_3 + OCHOMe$$

The authors point out that the intermediate adducts can be used as catalysts in reactions similar to those which have been established for the corresponding organotin compounds. Thus, triphenyllead methoxide and tributyllead methoxide were shown to catalyze the addition of methanol to butyl isocyanate and to di-1-naphthylcarbodiimide, and the trimerization of 1-naphthyl isocyanate.

A functional compound of lead has recently been prepared by an electrolytic method. Tomilov and his associates (28) electrolyzed 3-iodopropionitrile in either 0.5*N* sulfuric acid or aqueous sodium sulfate solution, using lead electrodes, and obtained $Pb(CH_2CH_2CN)_4$ in 13% yield. The mercury compounds, $Hg(CH_2CH_2CN)_2$ and $IHgCH_2CH_2CN$, were synthesized in a similar manner.

REFERENCES

1. Anon., *Chem. Eng. News*, p. 50, Sept. 20, 1965.
2. Austin, P. R., *J. Am. Chem. Soc.*, **53,** 3514 (1931).
3. Bähr, G., and G. Zoche, *Chem. Ber.*, **88,** 542 (1955).
4. Bell, T. N., B. J. Pullman, and B. O. West, *Australian J. Chem.*, **16,** 722 (1963); through *C.A.*, **59,** 7549 (1963).
5. Davies, A. G., and R. J. Puddephatt, *J. Organometal. Chem.*, **5,** 590 (1966).
6. Drehfahl, G., and D. Lorenz, *J. prakt. Chem.*, **24,** 106 (1964).
7. Freedman, H. H., U.S. Pat. 3,090,797 (to Dow Chemical Co.), May 21, 1963; *C.A.*, **59,** 11560 (1963).
8. Gerrard, W., and D. B. Green, *J. Organometal. Chem.*, **7,** 91 (1967).
9. Gilman, H., and D. S. Melstrom, *J. Am. Chem. Soc.*, **72,** 2953 (1950).
10. Gilman, H., and L. Summers, *J. Am. Chem. Soc.*, **74,** 5924 (1952).
11. Gorth, H., and M. C. Henry, *J. Organometal. Chem.*, **9,** 117 (1967).
12. Grüttner, G., and E. Krause, *Ber.*, **49,** 2666 (1916).
13. Kaesz, H. D., J. R. Phillips, and F. G. A. Stone, *Chem. Ind.* (London), **1959,** 1409.

14. Kaesz, H. D., J. R. Phillips, and F. G. A. Stone, *J. Am. Chem. Soc.*, **82,** 6228 (1960).
15. Kocheshkov, K. A., and A. P. Aleksandrov, *Ber.*, **67,** 527 (1934); *C.A.*, **28,** 4044 (1934).
16. Kocheshkov, K. A., and A. P. Aleksandrov, *J. Gen. Chem. USSR*, **7,** 93 (1937); *C.A.*, **31,** 4291 (1937).
17. Leeper, R. W., L. Summers, and H. Gilman, *Chem. Rev.*, **54,** 101 (1954).
18. Lukevits, E. Ya., and M. G. Voronkov, *Organic Insertion Reactions of Group IV Elements*, translated by M. J. Newlands, Consultants Bureau, Plenum Press, N.Y., 1966.
19. Lutsenko, I. F., V. L. Foss, and N. L. Ivanova, *Dokl. Akad. Nauk SSSR*, **141,** 1107 (1961); through *C.A.*, **56,** 12920 (1962).
20. Lutsenko, I. F., and L. Ponomarev, *Zh. Obshch. Khim.*, **31,** 2025 (1961); through *C.A.*, **55,** 27024 (1961).
21. Neumann, W. P., paper presented at Second Intern. Symp. on Organometallic Chemistry, Madison, Wis., Aug. 30, 1965.
22. Neumann, W. P., and K. Kühlein, *Angew. Chem.*, **77,** 808 (1965).
23. Neumann, W. P., and K. Kühlein, *Tetrahedron Letters*, **1966,** 3419.
24. Polees, B. M., A. B. Bruker, and L. Z. Soborovskii, USSR Pat. 148,404, July 13, 1962; through *C.A.*, **58,** 7974 (1963).
25. Puchinyan, E. A., and Z. M. Manulkin, *Dokl. Akad. Nauk., Uz. SSR*, **19,** 47 (1962); through *C.A.*, **57,** 13788 (1962).
26. Seyferth, D., *Chem. Rev.*, **55,** 1155 (1955).
27. Sterlin, R. N., S. S. Dubov, Wei-Kang Li, L. P. Vakhomchik, and I. L. Knunyants, *Zh. Vsesoyuz. Khim. Obshchestva im. D.I. Mendeleeva*, **6,** 110 (1961); through *C.A.*, **55,** 15336 (1961).
28. Tomilov, A. P., Yu. D. Smirnov, and S. L. Varshavskii, *Zh. Obshch. Khim.*, **35,** 391 (1965); through *C.A.*, **63,** 5238 (1965).
29. Treichel, P. M., and F. G. A. Stone, "Fluorocarbon Derivatives of Metals," in *Advances in Organometallic Chemistry*, Vol. I, F. G. A. Stone and R. West, Eds., Academic Press, New York, 1964.
30. Wallen, L. L., *Iowa State Coll. J. Sci.*, **29,** 526 (1955); through *C.A.*, **49,** 10418 (1955).
31. Willemsens, L. C., Quarterly Report No. 11, Project LC-18, to Intern. Lead Zinc Res. Organ., New York, Nov. 22, 1962.
32. Willemsens, L. C., Quarterly Report No. 16, Project LC-18, to Intern. Lead Zinc Res. Organ., New York, March, 1964.
33. Willemsens, L. C., Quarterly Report No. 18, Project LC-18, to Intern. Lead Zinc Res. Organ., New York, May, 1964.
34. Willemsens, L. C., Quarterly Report No. 25, Project LC-18, to Intern. Lead Zinc Res. Organ., New York, March, 1966.
35. Willemsens, L. C., and G. J. M. van der Kerk, *J. Organometal. Chem.*, **4,** 34 (1965).
36. Willemsens, L. C., and G. J. M. van der Kerk, *J. Organometal. Chem.*, **4,** 241 (1965).
37. Willemsens, L. C., and G. J. M. van der Kerk, *Investigations in the Field of Organolead Chemistry*, Intern. Lead Zinc Res. Organ., New York, 1965.
38. Yakubovich, A. Ya., S. P. Makarov, V. A. Ginsburg, G. I. Gavrilov, and E. N. Merkulova, *Dokl. Akad. Nauk SSSR*, **72,** 69 (1950); through *C.A.*, **45,** 2856 (1951).
39. Yakubovich, A. Ya., E. N. Merkulova, S. P. Makarov, and G. I. Gavrilov, *Zh. Obsch. Khim.*, **22,** 2060 (1952); *C.A.*, **47,** 9257 (1953).
40. Zimmer, H., and E. Endrulat, *Angew. Chem.*, **68,** 390 (1956).

XIII Heterocyclic lead compounds

1. COMPOUNDS WITH LEAD IN HETEROCYCLIC RING

Only a few compounds have been synthesized in which the lead atom has been shown unequivocally to form part of a heterocyclic ring. In making this statement, we exclude compounds of undefined constitution from consideration, such as possible Pb—O bonded trimers of the organoplumbonic acid type.

In 1916, Grüttner and Krause (10) prepared diethylcyclopentamethylenelead,

```
C2H5       CH2—CH2
    \     /       \
     Pb            CH2
    /     \       /
C2H5       CH2—CH2
```

by the reaction between diethyllead dichloride and the di-Grignard derivative prepared from 1,5-dibromopentane:

$$Et_2PbCl_2 + BrMg(CH_2)_5MgBr \rightarrow Et_2Pb(CH_2)_5$$

(The analogous tin heterocycle has also been prepared, using a similar method.) The liquid cyclopentamethylenelead compound was stable in the absence of air, but since the ring is probably strained, the compound was slowly oxidized when exposed to air, yielding a brown oxygen-containing resin. Bromine cleaved the ring at $-75°$ to give diethyl-5-bromoamyllead bromide, which was reacted with ethyl Grignard reagent to yield triethyl-5-bromoamyllead. In further proof of structure, chlorine reacted with the diethylcyclopentamethylenelead under more drastic conditions to produce ethyl-5-chloroamyllead dichloride.

Diphenylcyclopentamethylenelead was prepared in 1962 by Bajer and Post (1) from diphenyllead dichloride and 1,5-dilithiopentane in ether, but it proved to be quite sensitive to light and air, forming a colored resinous solid. The infrared spectra of this compound and its silicon, germanium, and tin analogs showed a series of three absorption bands at 2650, 990–965, and 910 cm^{-1}, which were ascribed to the heterocyclic, six-membered structural unit. Using essentially the method of Grüttner and Krause, Juenge and Gray (13) recently synthesized diethylcyclotetramethylenelead and diphenylcyclotetramethylenelead from the 1,4-dibromobutane di-Grignard derivative. Both compounds oxidize slowly in air, but the aromatic compound is more stable than the aliphatic derivative. Cleavage of the diethylcyclotetramethylenelead with acids demonstrated that the proportion of ring opening rises with increasing acid strength in the series from acetic acid to trichloroacetic acid. Treatment of the diethyl compound with halogens indicated initial formation of trialkyllead salt, but no salt could be isolated as in the case of Grüttner and Krause's compound containing a six-membered ring.

More stable compounds have resulted from the presence of benzo groups in the molecule. In 1959 Gilman et al. (8,9) prepared 5-ethyl-10,10-dihydro-10,10-diphenyl-5*H*-phenazaplumbine, (**1**), an analog of

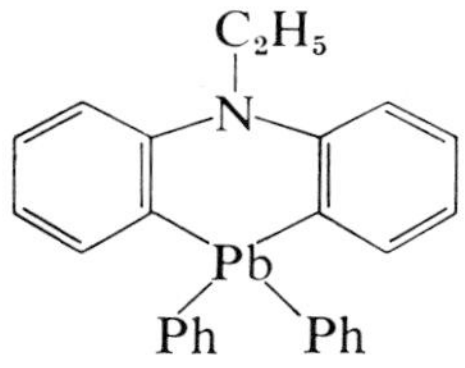

C_2H_5 N Pb Ph Ph

(1)

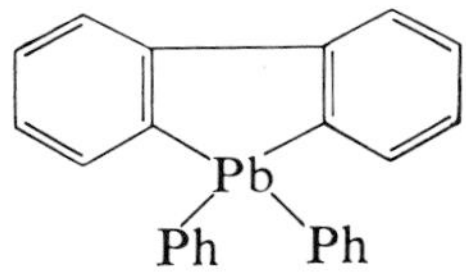

Pb Ph Ph

(2)

5,10-dihydroacridine. 2,2′-Dibromodiphenylamine was converted to the desired product in 26% yield by means of the following series of reactions:

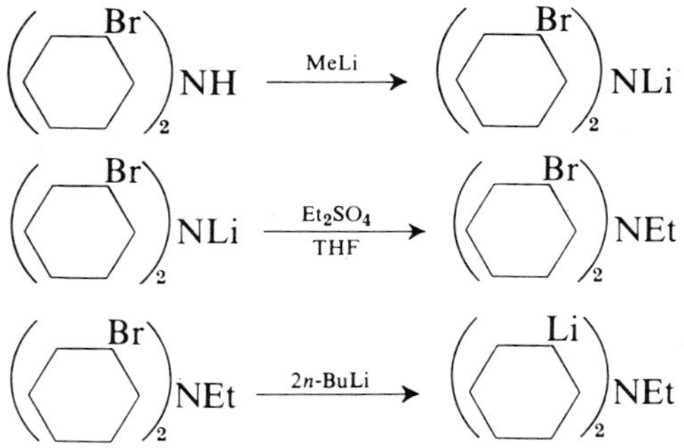

$$(BrC_6H_4)_2NH \xrightarrow{MeLi} (BrC_6H_4)_2NLi$$

$$(BrC_6H_4)_2NLi \xrightarrow[THF]{Et_2SO_4} (BrC_6H_4)_2NEt$$

$$(BrC_6H_4)_2NEt \xrightarrow{2n\text{-BuLi}} (LiC_6H_4)_2NEt$$

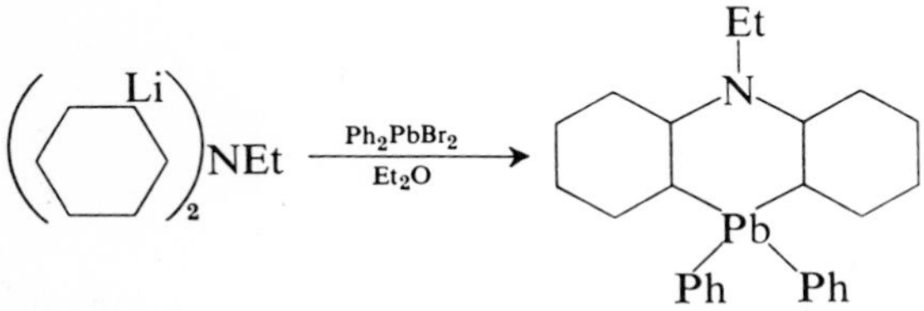

$$\left(\text{C}_6\text{H}_4\text{Li}\right)_2\text{NEt} \xrightarrow[\text{Et}_2\text{O}]{\text{Ph}_2\text{PbBr}_2} \text{(10-ethyl-5,5-diphenylphenazaplumbine: Et–N, Pb(Ph)}_2\text{)}$$

Gelius (4) prepared 9,9-diphenyldibenzoplumbole (**2**) in 30% yield, by the reaction of triphenyllead chloride (not diphenyllead dichloride as erroneously reported in some review literature) and 2,2′-dilithiobiphenyl. He could not obtain 2,2′-bis(triphenylplumbyl)biphenyl.

In 1947, Closson and Shapiro (2) reacted the intermetallic compound dimagnesium-lead (Mg_2Pb) with 1,4-dibromobutane or 1,4-dichlorobutane, using an iodide–ether catalyst, to prepare air-stable, lead-containing oils. By analogy with the Grignard reaction, heterocyclic lead compounds containing four carbon links were expected. The oils were separated by high-vacuum distillation to yield fractions which analyzed correctly for the following formulas:

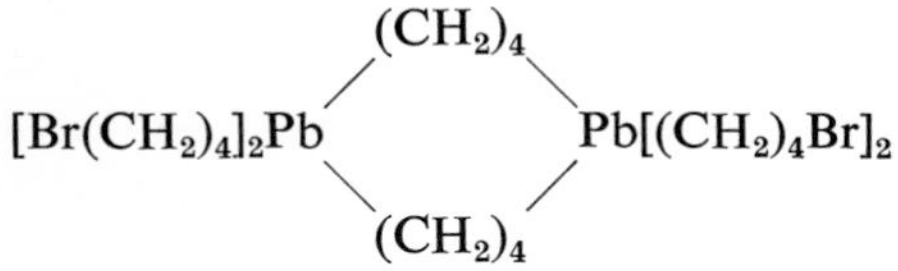

$$[Br(CH_2)_4]_2Pb\begin{matrix}(CH_2)_4\\ \\(CH_2)_4\end{matrix}Pb[(CH_2)_4Br]_2$$

(3)

$$Br(CH_2)_4Pb\begin{matrix}(CH_2)_4\\ -(CH_2)_4-\\(CH_2)_4\end{matrix}Pb(CH_2)_4Br$$

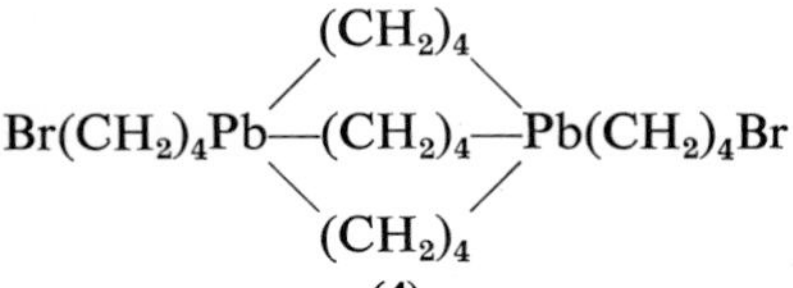

(4)

However, the constitution of the fractions was not definitely established.

Davidson, Hills, and Henry (3) prepared a novel heterocyclic compound in which the lead atom is bonded to two sulfur atoms. By reacting ethanedithiol with diphenyllead dihalide in benzene in the presence of an excess of triethylamine they obtained

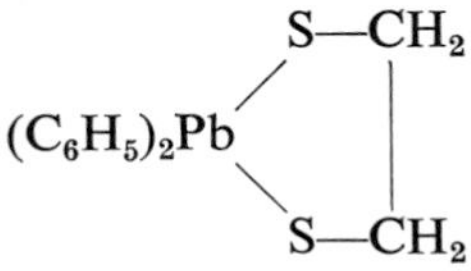

$$(C_6H_5)_2Pb\begin{matrix}S-CH_2\\ |\\ S-CH_2\end{matrix}$$

The properties of this compound were not described. The somewhat

similar polymeric compound

$$\left[(C_6H_5)_2Pb \begin{matrix} \diagup S-CH_2 \\ \quad\quad | \\ \diagdown O-C{=}O \end{matrix} \right]_n$$

was prepared by the same group (12) by the reaction of diphenyllead dichloride (or triphenyllead chloride) with β-mercaptoacetic acid.

2. HETEROCYCLIC RING-BONDED COMPOUNDS

Among the compounds in which heterocyclic rings could conceivably be bonded to lead, the only compounds isolated to date are those containing either oxygen, nitrogen, sulfur, or selenium.

Gilman and Towne (6) prepared tetra-2-furyllead and tetra-2-thienyllead, the latter compound having been prepared earlier by Krause and Renwanz (14,15) from the corresponding Grignard reagent. These compounds are fairly stable solids from which unsymmetrical derivatives can be easily synthesized by treatment with halogen, followed by reaction with an appropriate Grignard reagent. Gilman and Towne (6), using this type of procedure, prepared di-2-furyldiphenyllead, 2-furyltriphenyllead, di-2-furyl-di-2-thienyllead, di-2-thienyldiphenyllead, 2-thienyltriphenyllead, and di-2-thienyllead dichloride. These investigators found the thienyl and furyl groups were cleaved by halogen preferentially, as compared with the phenyl group. On this basis they regarded both heterocyclic groups as more electronegative than the phenyl group, or "superaromatic" in nature (see Chapter VI).

Gilman and his students (5) prepared the 2-, 3-, and 4-dibenzofuryltriphenyllead compounds by analogous methods from triphenyllead chloride and either 4-dibenzofuryllithium or the appropriate Grignard reagent. Gilman and Towne (7) prepared 2-furyl-tri-*p*-methoxyphenyllead and di-2-furyl-di-*p*-methoxyphenyllead from furylmagnesium iodide and the corresponding *p*-methoxyphenyllead chloride.

Panov and Kocheshkov (22) prepared di-2-thienyllead diisobutyrate. This ester was prepared by two reactions; first, from the reaction of tetra-2-thienyllead with isobutyric acid at 60–70°; and second, by the reaction of lead tetraisobutyrate with thiophene at room temperature over a period of several days. On treatment at room temperature with a solution of chloroacetic acid in benzene, the diisobutyrate underwent an acid interchange reaction to give a 68% yield of the di-2-thienyllead bis(chloroacetate).

Yur'ev et al. (26) have described the synthesis of di-2-thienyl-di-2-selenieyllead. This compound was prepared by adding 2-iodoselenophene in ether to magnesium and mercuric bromide under ether, followed by refluxing and addition of di-2-thienyllead dichloride. After 12 hr at room temperature, workup gave a 6% yield of product. Treatment of the di-2-thienyl-di-2-selenieyllead with dry hydrogen chloride dissolved in heptane–ether resulted in cleavage to both thiophene and selenophene in 81% combined yield.

A few nitrogen-bonded heterocyclic lead compounds have been known for a number of years. Saunders and his associates (11,23) synthesized several trialkyllead sulfonamides and phthalimides as potential military sternutators during World War II. Two types of preparation were used: the reaction of triethyl- or tripropyllead chloride with the sodium salt of a sulfonamide, and the reaction of triethyl- or tripropyllead hydroxide with a sulfonamide. Compounds of this type are stable to hydrolysis.

Substituted imides of the *N*-triethyllead phthalimide type have also been patented as fungicides by Ligett et al. (18–20). The triethyllead phthalimides were synthesized from triethyllead hydroxide and either phthalimide or tetrachlorophthalimide in ethanol. *N*-triethyllead phthalohydrazide was also prepared in aqueous medium, employing the reaction between triethyllead chloride, sodium hydroxide, and phthalohydrazide.

In 1964 and 1965, Willemsens and van der Kerk (24,25) reported on the preparation of a number of nitrogen-bonded heterocyclic lead compounds, for example *N*-(triethyllead)imidazole. The compounds were easily prepared in good yield by mixing a triorganolead hydroxide with the appropriate heterocycle in an organic solvent such as ether, for example:

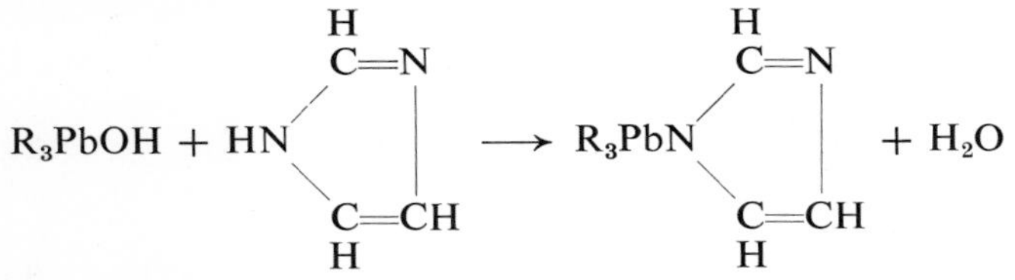

$$R_3PbOH + HN\begin{matrix} \overset{H}{C}=N \\ | \\ \underset{H}{C}=CH \end{matrix} \longrightarrow R_3PbN\begin{matrix} \overset{H}{C}=N \\ | \\ \underset{H}{C}=CH \end{matrix} + H_2O$$

These nitrogen-bonded heterocyclic lead compounds are quite stable to hydrolysis. This stability has been attributed to a polymeric pentacoordinated structure of the type:

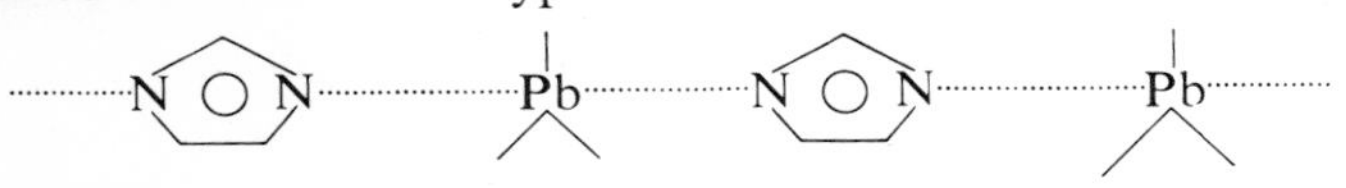

$$\cdots N\bigcirc N \cdots Pb \cdots N\bigcirc N \cdots Pb \cdots$$

Such structures, in which the nitrogen atom occurs in a five-membered aromatic heterocycle containing at least one other nitrogen atom in the 3-position (e.g., imidazole, benzimidazole, 1,2,4-triazole), had been previously shown by Luijten and van der Kerk (21) to be remarkably stable to hydrolysis when tin was the metal present.

Heterocyclic lead compounds of the two types discussed above are listed in Table XIII-1.

Table XIII-1   Heterocyclic Lead Compounds

COMPOUND	MELTING POINT, °C	REFERENCES
Lead in Heterocyclic Ring		
$(C_2H_5)_2PbCH_2CH_2CH_2CH_2CH_2$ (ring closed Pb to terminal CH_2)		10,17
$(C_2H_5)_2PbCH_2CH_2CH_2CH_2$ (ring closed Pb to terminal CH_2)		13
$(C_6H_5)_2PbCH_2CH_2CH_2CH_2CH_2$ (ring closed Pb to terminal CH_2)		1
$(C_6H_5)_2PbCH_2CH_2CH_2CH_2$ (ring closed Pb to terminal CH_2)		13
C_2H_5; N; Pb; Ph, Ph (two benzene rings bridged by N and Pb)	121.5–3	8,9
Pb; Ph, Ph (two directly bonded benzene rings bridged by Pb)	136–7.5	4
$[Br(CH_2)_4]_2Pb$ bridged by two $(CH_2)_4$ groups to $Pb[(CH_2)_4Br]_2$		2
$Br(CH_2)_4Pb—(CH_2)_4—Pb(CH_2)_4Br$, further bridged by two $(CH_2)_4$ groups		2
$(C_6H_5)_2PbSCH_2CH_2S$ (ring closed Pb to S)		3
$[(C_6H_5)_2PbSCH_2C(O)O]_n$ (ring closed Pb to O)		12
Heterocyclic Ring Bonded Lead Compounds		
$\left(\text{HC=CH–CH=C(–)–O (2-furyl ring: H H / C—C / C— / C—O / H)}\right)_4Pb$	b.p.$_{53}$52	6,17

(continued)

COMPOUND	MELTING POINT, °C	REFERENCES
$(C_4H_3S)_4Pb$	152	6,14,16,17
$(C_4H_3O)Pb(C_6H_5)_2$	118	6,17
$C_4H_3O{-}Pb(C_6H_5)_3$	166–7	6,17
$(C_4H_3S)_2Pb(C_4H_3Se)_2$	147	26
$(C_4H_3O)Pb(C_4H_3S)_2$	117–9	6,17
$(C_4H_3S)_2Pb(C_6H_5)_2$	185	6,17

(*continued*)

Table XIII-1 (continued)

COMPOUND	MELTING POINT, °C	REFERENCES
H H C—C C—$Pb(C_6H_5)_3$ C—S H		6,12,15,17
(H H C—C C— C—S H)$_2$ $PbCl_2$		15,17
$(C_6H_5)_3Pb(2\text{-}C_{12}H_7O)$		5,17
$(C_6H_5)_3Pb(3\text{-}C_{12}H_7O)$		5,17
$(C_6H_5)_3Pb(4\text{-}C_{12}H_7O)$		5,17
H H C—C C—$Pb(C_6H_4OCH_3\text{—}p)_3$ C—O H		7.17
(H H C—C C— C—O H)$_2$ $Pb(C_6H_4OCH_3-p)_2$		7,17
(H H C—C C— C—S H)$_2$ $Pb[OOCCH(CH_3)_2]_2$	dec. 192	22
(H H C—C C— C—S H)$_2$ $Pb(OOCCH_2Cl)_2$	dec. 174	22

(continued)

Table XIII-1 (*continued*)

COMPOUND	MELTING POINT, °C	REFERENCE
H C $(C_2H_5)_3Pb$—N N C=C H H	dec. 100	25
H C $(C_4H_9)_3Pb$—N N C=C H H	48–50	25
H C $(i\text{-}C_4H_9)_3Pb$—N N C=C H H	167	25
H C $(C_6H_5)_3Pb$—N N C=C H H	dec. 285	25
N $(C_2H_5)_3Pb$—N CH C=N H		25
N $(C_6H_5)_3Pb$—N CH C=N H	dec. 265	25

(*continued*)

COMPOUND	MELTING POINT, °C	REFERENCES
N $(C_2H_5)_3$Pb—N C—NH_2 C═N H	178–81	25
N $(C_4H_9)_3$Pb—N C—NH_2 C═N H	116–8	25
N $(C_6H_5)_3$Pb—N C—NH_2 C═N H	248–56	25
H C $(C_2H_5)_3$Pb—N N C═C HC CH C—CH H	189–91	25
H C $(C_4H_9)_3$Pb—N N C═C HC CH C—C H H	104–5	25

(*continued*)

Table XIII-1 (*continued*)

COMPOUND	MELTING POINT, °C	REFERENCES
H C $(C_6H_5)_3Pb$—N N C=C HC CH C—C H H	dec. 270	25
N $(C_4H_9)_3Pb$—N N C=C HC CH C—C H H	103	25
N $(C_6H_5)_3Pb$—N N C=C HC CH C—C H H	275–6	25
SH C $(C_4H_9)_3Pb$—N N C=C HC CH C—C H H	67	25
C(O) $(C_2H_5)_3PbN$ C(O)	131	11,17,19

(*continued*)

Table XIII-1 (continued)

COMPOUND	MELTING POINT, °C	REFERENCES
$(C_2H_5)_3PbN$–C(O)–C_6H_4–C(O)–NH–N (cyclic: PbN, C(O), HN—C(O), benzene ring)	135	17,18,20,23
$(C_2H_5)_3PbN$(C(O))$_2$$C_6Cl_4$ (cyclic: C(O), C(O), ring with Cl, Cl, Cl, Cl)		19
$(C_2H_5)_3PbN$(C(O))(SO_2)C_6H_4 (cyclic: C(O), SO_2, benzene ring)		18
$(n\text{-}C_3H_7)_3PbN$(C(O))$_2$$C_6H_4$ (cyclic: C(O), C(O), benzene ring)		11,17
$(n\text{-}C_3H_7)_3PbN$(C(O))(SO_2)C_6H_4 (cyclic: C(O), SO_2, benzene ring)	130	17,23
Unidentified 1:2 adduct from tributyllead hydroxide and 2-mercaptobenzimidazole	dec. 106	25

REFERENCES

1. Bajer, F. J., and H. W. Post, *J. Org. Chem.*, **27,** 1422 (1962).
2. Closson, R. D. and H. Shapiro, Ethyl Corp., 1947, Unpublished Data.
3. Davidson, W. E., K. Hills, and M. C. Henry, *J. Organometal. Chem.*, **3,** 285 (1965).
4. Gelius, R., *Angew. Chem.*, **72,** 322 (1960).
5. Gilman, H., W. G. Bywater, and P. T. Parker, *J. Am. Chem. Soc.*, **57,** 885 (1935).
6. Gilman, H., and E. B. Towne, *Rec. trav. chim.*, **51,** 1054 (1932).
7. Gilman, H., and E. B. Towne, *J. Am. Chem. Soc.*, **61,** 739 (1939).
8. Gilman, H., W. J. Trepka, B. J. Gaj, O. L. Marrs, G. Schwebke, *WADC Technical Report 53–426, Part VIII*, Wright Air Development Division, Air Research Development Command, U.S. Air Force, Wright-Patterson Air Force Base, Ohio.
9. Gilman, H., and E. A. Zuech, *J. Am. Chem. Soc.*, **82,** 2522 (1960).
10. Grüttner, G., and E. Krause, *Ber.*, **49,** 2666 (1916).
11. Heap, R., and B. C. Saunders, *J. Chem. Soc.*, **1949,** 2983.
12. Henry, M. C., and K. Hills, Quarterly Report No. 11, Project LC-28, to Intern. Lead Zinc Res. Organ., New York, March 31, 1964.
13. Juenge, E. C., and S. Gray, *J. Organometal. Chem.*, **10,** 465 (1967).

14. Krause, E., and G. Renwanz, *Chem. Ber.*, **60,** 1582 (1927).
15. Krause, E., and G. Renwanz, *Chem. Ber.*, **62,** 1710 (1929).
16. Krause, E., and G. Renwanz, *Chem. Ber.*, **65,** 777 (1932).
17. Leeper, R. W., L. Summers and H. Gilman, *Chem. Rev.*, **54,** 101 (1954).
18. Ligett, W. B., R. D. Closson, and C. N. Wolf, U.S. Pat. 2,595,798 (to Ethyl Corp.), May 6, 1952; *C.A.*, **46,** 7701 (1952).
19. Ligett, W. B., R. D. Closson, and C. N. Wolf, U.S. Pat. 2,640,006 (to Ethyl Corp.), May 26, 1953; *C.A.*, **47,** 8307 (1953).
20. Ligett, W. B., R. D. Closson, and C. N. Wolf, U.S. Pat. 2,654,689 (to Ethyl Corp.), Oct. 6, 1953; *C.A.*, **48,** 942 (1954).
21. Luijten, J. G. A., and G. J. M. van der Kerk, *Rec. trav. chim.*, **82,** 1181 (1963).
22. Panov, E. M., and K. A. Kocheshkov, *Dokl. Akad. Nauk SSSR*, **123,** 295 (1958); *C. A.*, **53,** 7133 (1959).
23. Saunders, B. C., *J. Chem. Soc.*, **1950,** 684.
24. Willemsens, L. C., *Organolead Chemistry*, Intern. Lead Zinc. Res. Organ., New York, 1964, p. 54.
25. Willemsens, L. C., and G. J. M. van der Kerk, *Investigations in the Field of Organolead Chemistry*, Intern. Lead Zinc. Res. Organ., New York, 1965.
26. Yur'ev, Yu. K., M. A. Gal'bershtam, and I. I. Kandror, *Zh. Obshch. Khim.*, **34,** 4116 (1964); through *C. A.*, **62,** 9163 (1965).

XIV Lead compounds with optical activity

Optically active compounds containing asymmetric Group IVb metal atoms have been known for many years. However, the demonstration of optical activity has been confined to the elements silicon, germanium, and tin; lead has been conspicuous by its absence from the list. It is well known that the configuration about all the Group IVb elements is tetrahedral in four-covalent bonded systems (see Chapter II). Accordingly, attempts have been made to resolve several asymmetric lead compounds, but these efforts have been without success. There would seem to be no theoretical reason why the lead derivatives should be singularly resistant to resolution, and it is hoped that further research in this direction will eventually lead to clarification of the problem.

As early as 1917, several asymmetric lead alkyls with four different groups were prepared by Grüttner and Krause (3,4), and their resolution into optically active isomers was attempted using *d*-bromocamphorsulfonic acid as the reagent. These asymmetric lead alkyls, for example, methylethylpropylbutyllead, were always obtained as impure oils. This was not surprising, since unsymmetrical lead alkyls, as compared with symmetrical lead alkyls, are in general lower melting, more unstable to heat, more soluble in organic solvents, and more reactive to oxygen and other common reactants. In Grüttner and Krause's resolution work, ethylpropylisobutyllead *d*-bromocamphorsulfonate (**1**) was obtained as an uncrystallizable oil. In contrast, the compound *n*-propyl-*i*-amyllead bromo-*d*-bromocamphorsulfonate (**2**) was fractionally crystallized, but no optical activity could be shown.

In 1933, Austin (1) attempted the resolution of lead compounds with nonpolar substitutents only, since such compounds should have little, if any, tendency to racemize. In a further effort to simplify the problem,

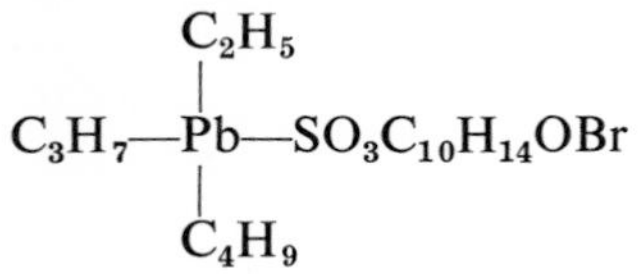

$$C_3H_7\text{—}\underset{\underset{C_4H_9}{|}}{\overset{\overset{C_2H_5}{|}}{Pb}}\text{—}SO_3C_{10}H_{14}OBr \qquad C_3H_7\underset{\underset{Br}{|}}{\overset{\overset{C_5H_{11}}{|}}{Pb}}\text{—}SO_3C_{10}H_{14}OBr$$

(1) 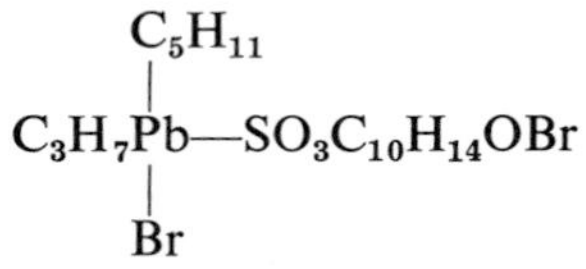 (2)

Austin introduced an optically active carbon atom into one of the substituent groups in order to be able to produce two stereoisomers of the tetracovalent lead compound. In order to assure crystallinity insofar as possible, aromatic derivatives were used. The optically active carbon atom was introduced by the reaction of *p*-lithiophenyl-*sec*-octyl ether with phenyl-*n*-propyl-*o*-tolyllead chloride to yield the active phenyl-*n*-propyl-*o*-tolyllead-*p*-(*sec*-octylphenyl ether) (**4**).

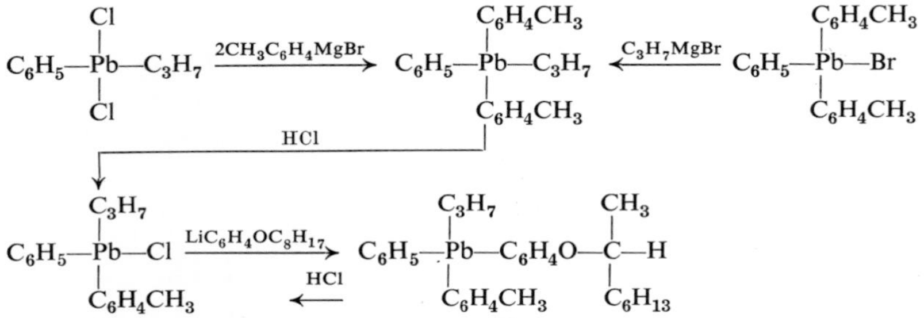

$$C_6H_5\text{—}\underset{\underset{Cl}{|}}{\overset{\overset{Cl}{|}}{Pb}}\text{—}C_3H_7 \xrightarrow{2CH_3C_6H_4MgBr} C_6H_5\text{—}\underset{\underset{C_6H_4CH_3}{|}}{\overset{\overset{C_6H_4CH_3}{|}}{Pb}}\text{—}C_3H_7 \xleftarrow{C_3H_7MgBr} C_6H_5\text{—}\underset{\underset{C_6H_4CH_3}{|}}{\overset{\overset{C_6H_4CH_3}{|}}{Pb}}\text{—}Br$$

$$\xrightarrow{HCl}$$

$$C_6H_5\text{—}\underset{\underset{C_6H_4CH_3}{|}}{\overset{\overset{C_3H_7}{|}}{Pb}}\text{—}Cl \underset{HCl}{\overset{LiC_6H_4OC_8H_{17}}{\rightleftarrows}} C_6H_5\text{—}\underset{\underset{C_6H_4CH_3}{|}}{\overset{\overset{C_3H_7}{|}}{Pb}}\text{—}C_6H_4O\text{—}\underset{\underset{C_6H_{13}}{|}}{\overset{\overset{CH_3}{|}}{C}}\text{—}H$$

(3) (4)

Obviously, the final lead compound (**4**) should exist in two separable forms if the lead is the center of asymmetry. However, although all of the intermediate compounds were solids, the final lead derivative (**4**) was an oil which exhibited optical activity, but could not be separated into its two stereoisomers. As Austin pointed out, this was not surprising, since in the absence of crystalline products the resolution of optically active compounds has invariably failed, even when stereoisomers were available. Cleavage of various fractions of Austin's asymmetric lead compound by means of hydrogen chloride resulted in the elimination of the active carbon center, and the phenyl-*n*-propyl-*o*-tolyllead chloride (**3**) was always inactive. It appears that a repetition of Austin's method using other aromatic groups might well be successful, since a tendency toward a slightly higher melting point could be critical.

Gilman and Summers (2) attempted to prepare optically active lead compounds from diphenyl-γ-diethylaminopropyllead chloride as the hydrochloride by reaction with ethyl or benzyl Grignard reagent, followed by hydrogen chloride treatment. However, the products were uncrystallizable oils.

In the most recent attempt, Zimmer and Endrulat (5) prepared triphenyllead benzylacetic acid, which has an asymmetric carbon atom, according to the equation:

$$\mathrm{EtOOC{-}\underset{\underset{\underset{Ph}{|}}{HCH}}{\overset{H}{\overset{|}{C}}}{-}COOPbPh_3} \xrightarrow[\text{hydrolysis}]{\text{decarboxylation}} \mathrm{Ph_3Pb{-}\underset{\underset{\underset{Ph}{|}}{HCH}}{\overset{H}{\overset{|}{C}}}{-}COOH}$$

The product was a crystalline material, m.p. 165–7°, but only very weakly acidic, so that attempts to resolve it into optically active isomers by means of salt formation with *d*-bornylamine failed.

REFERENCES

1. Austin, P. R., *J. Am. Chem. Soc.*, **55,** 2948 (1933).
2. Gilman, H., and L. Summers, *J. Am. Chem. Soc.*, **74,** 5924 (1952).
3. Grüttner, G., and E. Krause, *Ber.*, **50,** 202 (1917).
4. Grüttner, G., and E. Krause, *Ann.*, **415,** 338 (1918).
5. Zimmer, H., and E. Endrulat, *Angew. Chem.*, **68,** 390 (1956).

XV Organolead–organometal compounds

In addition to the hexaorganodilead and triorganoplumbylmetal compounds discussed in preceding chapters, other types of organolead derivatives are known in which more than one metal atom is present in the same molecule. These include (*1*) compounds in which the lead atom is bound directly to a second metal atom, and (*2*) compounds in which the lead atom is bonded to the second metal atom through one or more carbon atoms or through a phenylene group. Organolead derivatives are now known which contain five metal atoms in a neopentane configuration. These and other organolead compounds containing two or more metal atoms in the same molecule are discussed below.

1. TETRAKIS(TRIPHENYLPLUMBYL)LEAD AND RELATED COMPOUNDS

In 1964, Willemsens and van der Kerk (53,54) published two papers describing the synthesis of a series of novel compounds of the type $(R_3M)_4M'$, in which both M and M′ are Group IVb metals and R is phenyl. These compounds were the first organolead compounds to be prepared containing more than two lead atoms in a chain or of a stable organolead derivative in which lead is bound directly to a second Group IVb metal. The parent compound, tetrakis(triphenylplumbyl)lead, is believed to be the red crystalline product described earlier by Krause and Reissaus as diphenyllead. A number of mixed metal derivatives was also prepared in which M and M′ were lead, tin, or germanium. However, compounds containing germanium as the central metal atom were unstable and could not be isolated in the pure state.

Evidence has been obtained for the preparation of bis(triphenylplumbyl)diphenyllead, $(Ph_3Pb)_2PbPh_2$ (or octaphenyltrilead), by the reaction of diphenyllead dichloride and triphenylplumbyllithium, but the product was too unstable to be isolated (56). However, a germanium analog, $(Ph_3Pb)_2GePh_2$, was successfully isolated by Neumann and Kühlein (40) who prepared it by the reaction of diphenylplumbylbis(diethylamine) with diphenylgermane:

$$x\mathrm{Ph_2Pb(NEt_2)_2} + y\mathrm{Ph_2GeH_2} \longrightarrow z\mathrm{(Ph_3Pb)_2GePh_2}$$

Mixing these two reactants at −70° gave a yellow product from which lead metal was deposited upon warming to 20°; colorless, crystalline bis(triphenylplumbyl)diphenylgermane, melting at 154° with decomposition, was subsequently isolated from the reaction mixture.

No alkyllead compounds are known analogous to the above phenyllead derivatives. However, they have been postulated to be possible intermediates in redistribution reactions of unsymmetrical tetraorganolead compounds (44).

A. Synthesis of $(Ph_3M)_4M'$ compounds

Tetrakis(triphenylplumbyl)lead was discovered by Willemsens and van der Kerk (54,56) from the chance observation that a red surface coloration is formed when solid triphenylplumbyllithium is exposed to air. It was subsequently found that the red coloration could be intensified by hydrolysis of solid, ether-free triphenylplumbyllithium with an ice–salt mixture containing a small amount of hydrogen peroxide. Extraction of the hydrolysate with chloroform, followed by evaporation of the chloroform at low temperature, gave a red, crystalline product which was shown to be tetrakis(triphenylplumbyl)lead. The same product was also obtained by oxidative hydrolysis of the solid reaction mixture prepared by reaction of lead(II) chloride and two equivalents of phenylmagnesium bromide in diethyl ether at −10°, followed by evaporation of the ether. The absence of ether is essential in the isolation of the tetrakis compound because it is unstable in ether, as well as other polar solvents.

The mechanism of formation of $(Ph_3Pb)_4Pb$ via oxidative hydrolysis of triphenylplumbyllithium has not been defined. Willemsens and van der Kerk have speculated that diphenyllead and/or triphenyllead hydride may be intermediates, the triphenyllead hydride being formed by hydrolysis of the triphenylplumbyllithium. Another possible intermediate which has been suggested is bis(triphenylplumbyl)lead, $(Ph_3Pb)_2Pb$ (from triphenylplumbyllithium and lead(II) chloride), which might then disproportionate into $(Ph_3Pb)_4Pb$ and lead metal; it has been suggested that the yellow color obtained upon reaction of lead(II) chloride with two equivalents of phenyllithium at $-10°$ in diethyl ether may arise from the presence of bis(triphenylplumbyl)lead (56).

Tetrakis(tri-*p*-tolylplumbyl)lead was also prepared by the oxidative hydrolysis of tri-*p*-tolylplumbyllithium (54,56); however, the resultant product was contaminated with hexa-*p*-tolyldilead. Similarly, Gilman and co-workers (10) obtained a bright orange-red solution upon hydrolysis of an ether solution of tri-*p*-dimethylaminophenylplumbyllithium with "ice water" under a nitrogen atmosphere. The lead content of the isolated solids was 49.3% (this was the only analysis reported); recrystallization of the product from benzene-petroleum ether gave a deep brick-red, crystalline solid but the lead content was decreased to 39.2%. Theory for di-*p*-dimethylaminophenyllead, which was considered to be a likely composition, is 46.3% Pb. In view of the successful synthesis of tetrakis(triphenylplumbyl)lead, the dimethylaminophenyl derivative merits further investigation, especially since the orange-red product was formed in the presence of ether.

Attempts to prepare tetrakis(triphenylplumbyl)lead by the reaction of triphenylplumbyllithium and lead(II) chloride were unsuccessful, probably because of its instability in ethereal media. However, Willemsens and van der Kerk (53) were able to prepare a series of mixed metal derivatives using the metathesis reactions shown in the equations below.

$$4Ph_3MLi + M'Cl_4 \longrightarrow (Ph_3M)_4M' + 4LiCl$$

where $M' =$ Ge or Sn and $M =$ Ge, Sn, or Pb;

$$4Ph_3MLi + 2PbCl_2 \longrightarrow (Ph_3M)_4Pb + Pb + 4LiCl$$

where $M =$ Ge or Sn. These reactions were conducted by adding the metal chloride to a solution of triphenylplumbyllithium or Ph_3MLi at about $-50°$ in tetrahydrofuran or diethyl ether. It was concluded on the basis of the yields obtained that the formation of the $(Ph_3M)_4M'$ derivative proceeded more easily the larger the central metal atom M'; the same effect also was true for the surrounding metal M, but to a lesser extent. Some Ph_6M_2 was also formed in these reactions, which probably resulted from lithium–chlorine exchange reactions. No pure $(Ph_3Ge)_4M'$ derivative could be isolated; however, Willemsens and van der Kerk concluded that they did synthesize $(Ph_3Ge)_4Pb$, but that it could not be separated from by-product hexaphenyldigermane. Also, just as the metal halide reaction could not be used to prepare tetrakis(triphenylplumbyl)lead, the oxidative hydrolysis reaction was not satisfactory for the preparation of the other $(Ph_3M)_4M'$ derivatives, although oxidative hydrolysis of triphenylstannyllithium did give low yields of $(Ph_3Sn)_4Sn$.

The $(Ph_3M)_4M'$ compounds prepared by Willemsens and van der Kerk are summarized in Table XV-1.

Table XV-1 $(Ph_3M)_4M'$ *Compounds*

COMPOUND	YIELD, %	COLOR	APPROXIMATE DECOMPOSITION POINT, °C
$(Ph_3Pb)_4Pb$	30	Red	
$(Ph_3Pb)_4Sn$	54	Yellow	160
$(Ph_3Pb)_4Ge$	16	Yellow	210
$(Ph_3Sn)_4Pb$	55	Yellow	200
$(Ph_3Sn)_4Sn$	22	White	280
$(Ph_3Sn)_4Ge$	4	White	324
$(Ph_3Ge)_4Pb$[a]	23	Yellow	

[a] Could not be isolated in the pure state.
Taken from Willemsens and van der Kerk, *J. Organometal. Chem.*, **2**, 260 (1964).

B. Physical and chemical properties

The red all-lead compound, $(Ph_3Pb)_4Pb$, is a highly reactive crystalline solid. Although relatively unstable by normal organometallic standards, it is surprisingly stable for an organolead derivative containing four lead–lead bonds. It is insoluble in alcohol and petroleum ether, but readily soluble in chloroform and benzene. The solid compound decomposes in air within a few days to form substantial amounts of hexaphenyldilead and lead oxide. It is more stable in a nitrogen atmosphere, but still undergoes slow decomposition to form hexaphenyldilead and lead metal. Tetrakis(triphenylplumbyl)lead is much less stable in solution, especially in polar solvents; in acetone, its red color disappears within a few seconds. It is also unstable in ether; hence, ether must be completely removed from the triphenylplumbyllithium before oxidative hydrolysis is effected.

Tetrakis(triphenylplumbyl)lead reacts readily with iodine at low temperature; the stoichiometry of this reaction provided the basis for its structure determination. At $-60°$, one mole reacts with 6–8 atoms of iodine to form triphenyllead iodide and lead iodide:

$$(Ph_3Pb)_4Pb + 3I_2 \longrightarrow 4Ph_3PbI + PbI_2$$

The triphenyllead iodide and lead iodide were isolated in 92 and 93% yields, respectively, based on the above stoichiometry; no diphenyllead diiodide was detected. This stoichiometry and the resultant products are more consistent with the neopentane type structure $(Ph_3Pb)_4Pb$, than with a linear structure of the type $Ph_3Pb(PbPh_2)_3PbPh_3$. The linear structure would consume 8 atoms of iodine per mole, but would be expected to yield two moles of triphenyllead iodide and three moles of diphenyllead diiodide per mole of $Ph_{12}Pb_5$.

The ultraviolet absorption spectra of tetrakis(triphenylplumbyl)-lead and tetrakis(tri-*p*-tolyl)lead show absorption maxima at 358 and 365 mμ, respectively, which are analogous to the absorption band at 294 mμ found in the spectrum of hexaphenyldilead. These bands, therefore, are assigned to vibrations associated with the lead–lead bonds (6,54). Similar absorption bands are found in the ultraviolet spectra of the other $(Ph_3M)_4M'$ derivatives, but at different wavelengths. In addition, both compounds show a second absorption band at higher wavelengths; in tetrakis(triphenylplumbyl)lead this band occurs at 444 mμ, in tetrakis-(tri-*p*-tolylplumbyl)lead at 448 mμ. A definite explanation of this second band has not been offered. Drenth and co-workers (6) suggested that it may result from the symmetry properties of the molecule or, alternatively, from an interaction between the filled 5*d*-orbital of the central lead atom and empty orbitals of the surrounding lead atoms. Although an analogous second band was not detected in those compounds in which tin is the central metal atom, it was speculated that this absorption may be shifted to lower wavelengths in the tin compounds and thus was not detectable.

X-ray analysis of $(Ph_3M)_4M'$ compounds containing two different Group IVb metals showed several of them to be isomorphous with tetrakis(triphenylplumbyl)lead (53). All of the compounds which were successfully prepared and isolated were found to be stable at room temperature as solids. In solution, the derivatives containing lead as the central metal atom were unstable and developed a rapid turbidity. However, compounds in which lead was the surrounding metal, $(Ph_3Pb)_4M'$, exhibited a much higher stability in solution. In general, the thermal stability of $(Ph_3Pb)_4M'$ and $(Ph_3Sn)_4M'$ compounds decreased in the order $M' = Ge > Sn > Pb$.

Finally, it is interesting to note that those compounds containing lead as the central or surrounding metal atom are colored (yellow in most cases), whereas those containing tin or germanium are white, e.g., $(Ph_3Sn)_4Sn$ and $(Ph_3Sn)_4Ge$. Bis(triphenylplumbyl)diphenylgermane is also white (40).

2. DERIVATIVES OF THE TYPE $R_3Pb(CH_2)_nPbR_3$, $(R_3Pb)_nCH_{4-n}$ AND RELATED COMPOUNDS

A second class of organolead compounds containing two or more lead atoms, or a lead atom and a second Group IVb metal, is that in which the two metal atoms are separated by one or more carbon atoms, a phenylene group, or a $—C_6H_4—CH_2—CH_2—$ moiety. Grüttner, Krause, and Wiernik (19) reported the successful synthesis of 1,5-bis(trimethylplumbyl)-pentane, $Me_3Pb(CH_2)_5PbMe_3$, in 1917 from the reaction of trimethyllead bromide with an ethereal solution of the di-Grignard reagent of 1,5-dichloropentane.

$$2Me_3PbBr + ClMg(CH_2)_5MgCl \rightarrow Me_3Pb(CH_2)_5PbMe_3 + 2Mg(Br)Cl$$

The product was a colorless oil, which could be distilled at reduced pressure without decomposition. The mixed tin–lead analog was reported at the same time. It was obtained as a colorless oil by the reaction:

$$Me_3PbCl + ClMg(CH_2)_5SnMe_3 \rightarrow Me_3Pb(CH_2)_5SnMe_3$$

Some thirty years later Wiczer (52) patented the synthesis of such compounds as bis(trimethylplumbyl)methane by the reaction of methylene bromide and hexamethyldilead in the presence of aluminum chloride. The patent also claims the preparation of 1,2-bis(trimethylplumbyl)ethane by the reaction of sodium–lead alloy with a mixture of methyl chloride and 1,2-dichloroethane. Bindschadler (1a) had earlier reported an unsuccessful attempt to prepare 1,2-bis(triphenylplumbyl)ethane, the phenyl analog of one of Wiczer's products, by the reaction of triphenylplumbyllithium and 1,2-dibromoethane in diethyl ether; however, Bindschadler obtained only hexaphenyldilead. A reaction similar to Bindschadler's was attempted in 1966 by Henry and Gorth (28); these workers obtained tetraphenyllead instead of hexaphenyldilead, but speculated that the tetraphenyllead may have been formed from the disproportionation of hexaphenyldilead. Henry and Gorth did prepare bis(triphenylplumbyl)-methane by the reaction of triphenylplumbyllithium and methylene bromide, although they had actually hoped to prepare (bromomethyl)-triphenyllead; bis(triphenylplumbyl)methane was the only product isolated, even when a large excess of methylene bromide was employed. On the other hand, reaction of triphenylplumbyllithium with dibromopropane, -butane, and -pentane gave the desired $Ph_3Pb(CH_2)_nBr$ compounds. Foster and co-workers (7) isolated a soft waxy solid from the reaction of triphenylplumbylsodium and excess methylene chloride. The product was soluble in petroleum ether and had a molecular weight of 640; it was speculated to contain a triphenyllead moiety but was not further characterized. Within the past few years a series of poly(triphenylplumbyl)methane derivatives of the type $(Ph_3Pb)_nCH_{4-n}$ has been synthesized. These surprisingly stable compounds are discussed below.

In 1952, Gilman and Summers (9) described the preparation of 1,3-bis(triphenylplumbyl)propane and the 1,4-butane analog by the reaction of triphenylplumbyllithium with 1,3 dichloropropane and 1-chloro-4-bromobutane, respectively. These products were obtained as colorless, crystalline solids which melted without decomposition. A compound having the structure $Et_3PbCH(CH_3)CH(CH_3)PbEt_3$ has been isolated by Hedden (20) as a heavy residue from tetraethyllead prepared via the sodium–lead alloy–ethyl chloride reaction using a continuous process (46). Presumably, this product arises from dimerization of a free radical having the structure $Et_3Pb\dot{C}HCH_3$, which in turn probably arises from a free radical attack on tetraethyllead. It decomposes at slightly elevated temperature in the presence of aluminum chloride to form tetraethyllead. Except for the latter compound and the compounds described by Wiczer, no other bis(triorganoplumbyl)ethane compounds are known. Furthermore, the efficacy of the reactions described by Wiczer is open to serious question because these particular reactions should yield complex reaction mixtures; also, none of the products was isolated in a pure state. Several ethynyl derivatives of the type $R_3PbC{\equiv}CPbR_3$ are known; these are discussed in Chapter X.

In 1965, Willemsens and van der Kerk (32,55,56) reported the preparation of a series of poly(triphenylplumbyl)methane derivatives by the reaction of triphenylplumbyllithium with various chloromethanes:

$$4Ph_3PbLi + CCl_4 \rightarrow (Ph_3Pb)_4C + 4LiCl$$
$$3Ph_3PbLi + HCCl_3 \rightarrow (Ph_3Pb)_3CH + 3LiCl$$
$$2Ph_3PbLi + H_2CCl_2 \rightarrow (Ph_3Pb)_2CH_2 + 2LiCl$$

These compounds were prepared by the addition of the chloromethane reactant to a solution of triphenylplumbyllithium in tetrahydrofuran at $-60°$. Addition of triphenylplumbyllithium to carbon tetrachloride at room temperature gave $(Ph_3Pb)_2CCl_2$; a dibromo analog, $(Ph_3Pb)_2CBr_2$, was prepared from carbon tetrabromide (57). Attempts to prepare $(Ph_3Pb)_2CHCl$ by the reaction of triphenylplumbyllithium and excess chloroform gave only bis(triphenylplumbyl)methane. However, several chloromethyl derivatives were prepared containing a single lead atom, e.g., Ph_3PbCCl_3 and $Ph_3PbCHCl_2$; these latter compounds are discussed in Chapter XII.

The poly(triphenylplumbyl)methane compounds are crystalline solids of remarkable thermal stability and low reactivity. They melt sharply without decomposition; tetrakis(triphenylplumbyl)methane did not melt up to 294–6°, where it decomposed sharply. Also, it did not react with glacial acetic acid at temperatures up to 150° (32). It crystallized from chloroform with four molecules of solvation, $(Ph_3Pb)_4C{\cdot}4CHCl_3$ (55). Surprisingly, it reacted with four moles of bromine at $-40°$ in chloroform solution with cleavage of four phenyl groups; the molecular weight and chemical composition of the resultant yellow crystalline product agreed with the formula $(Ph_2PbBr)_4C$ (56).

$$(Ph_3Pb)_4C + 4Br_2 \rightarrow (Ph_2PbBr)_4C + 4PhBr$$

A few mixed-metal organolead compounds are known in which silicon is present as the second metal and is separated from lead by a single carbon atom. Thus, Seyferth and Freyer (50) prepared tetrakis(trimethylsilylmethyl)lead, $(Me_3SiCH_2)_4Pb$, by the reaction of Me_3SiCH_2MgCl (from Me_3SiCH_2Cl + Mg) and lead(II) chloride in tetrahydrofuran. The liquid product was distillable at reduced pressure without decomposition and reacted with an excess of phosphorus trichloride at reflux to form the monochloride, $(Me_3SiCH_2)_3PbCl$, and $Me_3SiCH_2PCl_2$.

Schmidbaur and Waldmann (47,48) prepared a series of mixed metal derivatives of the type $Me_3MCH_2SiMe_3$ and $Me_3MCH_2Si(H)Me_2$, where M was Si, Ge, Sn, or Pb. These liquid compounds were synthesized by the general scheme:

$$Me_3MX + Me_3SiCH_2MgX \rightarrow Me_3MCH_2SiMe_3 + MgX_2 \qquad (M = Si, Ge, Sn, or Pb)$$

The nmr and infrared spectra were determined; the absorption band of the Si—H bond in the $CH_2Si(H)Me_2$ derivatives was found to shift to higher wavelengths with increasing atomic weight of the heteroatom.

Besides the above compounds in which a lead atom is bound to a second metal atom via an alkyl chain, a few compounds are also known in which a phenyl group is present as part of the chain. In 1917, Grüttner and Krause (18) reported the first such compound, (*p*-trimethylplumbylphenyl)triethylsilane, which they prepared by the reaction of trimethyllead bromide and *p*-$Et_3SiC_6H_4MgBr$ in ether. Gilman and Melstrom (8) later prepared *p*-bis(triphenylplumbyl)benzene, *p*-$Ph_3PbC_6H_4PbPh_3$, by the reaction of *p*-dilithiobenzene and triphenyllead halide in diethyl ether; the white crystalline product melted at 285–288° without decomposition. Its infrared spectrum showed absorption bands at 1062, 1017, 997 and 726 cm^{-1}, which were also present in the spectrum of tetraphenyllead (37); bands at 1075, 1008 and 794 cm^{-1} were attributed to vibrations associated with the *p*-phenylene group.

A series of mixed lead–tin derivatives was prepared by Noltes and van der Kerk (41) in which the metal atoms are separated by a —CH_2—CH_2—C_6H_4— grouping. These derivatives were synthesized by addition of an Sn–H moiety to the olefinic bond in *p*-styrenyllead compounds.

Typical reactions are:

$$Ph_3SnH + p\text{-}CH_2{=}CHC_6H_4PbPh_3 \rightarrow Ph_3SnCH_2CH_2C_6H_4PbPh_3$$

$$Ph_2SnH_2 + 2p\text{-}CH_2{=}CHC_6H_4PbPh_3 \rightarrow Ph_2Sn(CH_2CH_2C_6H_4PbPh_3)_2$$

$$2Ph_3SnH + (p\text{-}CH_2{=}CHC_6H_4)_2PbPh_2 \rightarrow (Ph_3SnCH_2CH_2C_6H_4)_2PbPh_2$$

$$Ph_2SnH_2 + (p\text{-}CH_2{=}CHC_6H_4)_2PbPh_2 \rightarrow (\text{—}C_6H_4CH_2CH_2(Ph_2)SnCH_2CH_2C_6H_4Pb(Ph_2)\text{—})_n$$

The products ranged from crystalline solids with sharp melting points to amorphous products having wide melting points to glass-like solids. In the case of the last equation above, the product was a glass-like solid with a softening point of 70–5°, a molecular weight of about 14,000, and a value of $n = 18$; dipropyl tin dihydride gave a tough, rubbery, slightly crosslinked polymer. Similar derivatives were prepared in which tin and germanium were substituted for lead.

Using a somewhat similar procedure, Leusink, Noltes and co-workers (36) prepared mixed methyltin–phenyllead derivatives by the reaction of *p*-phenylenebis(dimethyltin hydride) with *p*-styrenyllead compounds.

$$p\text{-}H(Me)_2SnC_6H_4Sn(Me)_2H + p\text{-}CH_2{=}CHC_6H_4PbPh_3 \rightarrow Ph_3PbC_6H_4CH_2CH_2Sn(Me)_2C_6H_4Sn(Me)_2CH_2CH_2C_6H_4PbPh_3$$

$$p\text{-}H(Me)_2SnC_6H_4Sn(Me)_2H + (p\text{-}CH_2{=}CHC_6H_4)_2PbPh_2 \rightarrow (-(Me)_2SnC_6H_4Sn(Me)_2CH_2CH_2C_6H_4Pb(Ph)_2C_6H_4CH_2CH_2\text{—})_n$$

The last product above was a glass-like, hard, and brittle polymer, with a softening point at about 85–95°. The other product was a colorless, amorphous solid which melted over the broad range 115–30°, even after repeated crystallizations.

Henry and Noltes (29) attempted to prepare a mixed tin–lead derivative of ethane by the addition of a tin–hydride moiety to a vinyl group bound to lead, according to:

$$Ph_3SnH + CH_2{=}CHPbPh_3 \not\rightarrow Ph_3SnCH_2CH_2PbPh_3$$

Only lead metal and hexaphenylditin were formed, although this reaction worked well when a vinyl–tin or vinyl–germanium derivative was used instead of vinyl-lead. However, Korshak and co-workers (35) obtained a polymeric material in which oxygen atoms are present in the backbone of the polymer by the addition of a silicon–hydrogen bond to the vinyl groups in diethyldivinyllead:

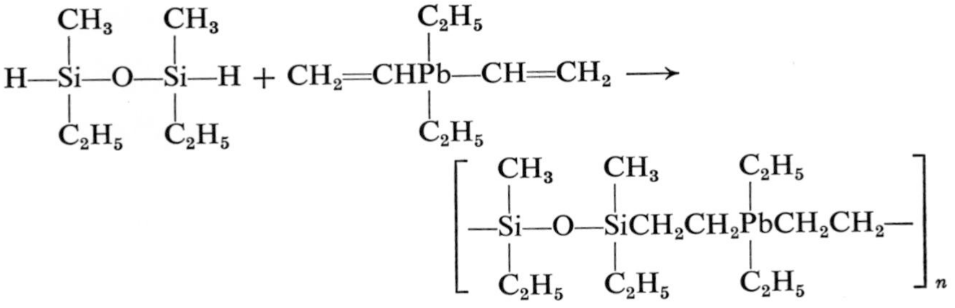

$$\begin{array}{c}
CH_3 \quad\;\; CH_3 \qquad\qquad\qquad\qquad C_2H_5 \\
H—\underset{|}{\overset{|}{Si}}—O—\underset{|}{\overset{|}{Si}}—H + CH_2{=}CH\underset{|}{\overset{|}{Pb}}—CH{=}CH_2 \longrightarrow \\
C_2H_5 \quad C_2H_5 \qquad\qquad\qquad\qquad C_2H_5
\end{array}$$

$$\left[\begin{array}{c}
CH_3 \qquad CH_3 \qquad\quad C_2H_5 \\
—\underset{|}{\overset{|}{Si}}—O—\underset{|}{\overset{|}{Si}}CH_2CH_2\underset{|}{\overset{|}{Pb}}CH_2CH_2— \\
C_2H_5 \quad\; C_2H_5 \qquad\quad C_2H_5
\end{array}\right]_n$$

The homopolymerization and copolymerization of styrenyllead compounds have been accomplished, leading to polymetallic molecules. The conditions used and the nature of the products are discussed in Chapter XII. Many of the compounds discussed above are included in a review of organopolymers of the Group IVb elements (30). Compounds of the types discussed above are listed in Table XV-2.

3. ORGANOLEAD DERIVATIVES OF METAL CARBONYLS

A third class of organolead compounds containing a second metal atom in the molecule is that of the type $R_3PbM(CO)_n$ or $R_2Pb(M(CO)_n)_2$ in which the lead atom is bonded to a metal from Groups VIb, VIIb, or VIII in the periodic table, the latter metals being present as a metal carbonyl moiety. The first such derivative was reported in 1941 by Hein and Pobloth (26) who described the preparation of diethyllead iron tetracarbonyl by the reaction of triethyllead hydroxide with an aqueous solution containing the iron tetracarbonyl hydride anion. The reaction required addition of the iron carbonyl hydride in the presence of an excess of the triethyllead hydroxide at all times and was run under a layer of ether into which the product was extracted. The product $Et_2PbFe(CO)_4$ was isolated as red platelets or needles which slowly darkened at about 110°, gave off yellow vapors at about 138° and subsequently deflagrated. The product was soluble in alcohol, acetone, ether, and hexane. Using a similar procedure, Hein and co-workers (22,24) also prepared the analogous methyl, propyl, butyl, and phenyl derivatives, as well as bis(tricyclohexyllead) iron tetracarbonyl and bis(triphenyllead) iron tetracarbonyl; bis(triethyllead) iron tetracarbonyl was prepared recently by Kahn and Bigorgne (31). The formation of the dialkyllead iron tetracarbonyl derivatives from trialkyllead hydroxide was attributed to a disproportionation of the bis(trialkyllead) iron tetracarbonyl formed as the

Table XV-2 $(R_3Pb)_2(CH_2)_n$, $(R_3Pb)_nCH_{4-n}$ *and Related Compounds*

COMPOUND	MELTING POINT, °C	REFERENCES
$Me_3Pb(CH_2)_5PbMe_3$	b.p., $_{14}$166.5	19
$Me_3Pb(CH_2)_5SnMe_3$	b.p., $_{17\cdot5}$162	19
$Me_3PbCH_2PbMe_3$, $(Me_3Pb)_2CH_2$(?)		52
$Me_3PbCH_2CH_2PbMe_3$(?)		52
$Ph_3Pb(CH_2)_3PbPh_3$	94–5	9,51
$Ph_3Pb(CH_2)_4PbPh_3$	134–6	9,51
$Et_3PbCH(CH_3)CH(CH_3)PbEt_3$	(oil)	20
$(Ph_3Pb)_4C$	dec. 292–4	32,55,56
$(Ph_3Pb)_3CH$	166; 167–8	55,56
$(Ph_3Pb)_2CH_2$	94–5; 97–7.5	55,56
	98–9.5	28
$(Ph_3Pb)_2CCl_2$	207–8	55,56
$(Ph_3Pb)_2CBr_2$	175–9	57
$(Ph_2PbBr)_4C$	dec. 192–4	56
$(Me_3SiCH_2)_4Pb$	b.p., $_{0\cdot01}$104–5	50
$(Me_3SiCH_2)_3PbCl$	214–6	50
$Me_3PbCH_2SiMe_3$	b.p., $_{1\cdot5}$36; $_{740}$179	47
$Me_3PbCH_2Si(H)Me_2$	b.p., $_1$26–7	48
$Me_3PbC_6H_4SiEt_3$-*p*	b.p., $_{17}$190–2	18
$Ph_3PbC_6H_4PbPh_3$-*p*	285–8	8
$Ph_3SnCH_2CH_2C_6H_4PbPh_3$-*p*	177–9	41
$Ph_2Sn(CH_2CH_2C_6H_4PbPh_3)_2$-*p*	~70	41
$(Ph_3SnCH_2CH_2C_6H_4)_2PbPh_2$-*p*	180–2	41
$(—Ph_2SnCH_2CH_2C_6H_4PbPh_2—)_n$-*p*	70–5 (softening)	41
$(—Pr_2SnCH_2CH_2C_6H_4PbPh_2—)_n$-*p*		41
$Ph_3PbC_6H_4CH_2CH_2Sn(Me)_2C_6H_4Sn(Me)_2CH_2CH_2C_6H_4PbPh_3$-*p*	115–30	36
$(—p(Me)_2SnC_6H_4Sn(Me)_2CH_2CH_2C_6H_4Pb(Ph)_2C_6H_4CH_2CH_2—)_n$	85–95 (softening)	36

initial product, according to (23):

$$2R_3PbOH + H_2Fe(CO)_4 \rightarrow (R_3Pb)_2Fe(CO)_4 + 2H_2O$$

$$(R_3Pb)_2Fe(CO)_4 \rightarrow R_4Pb + R_2PbFe(CO)_4$$

Support for this explanation was found in the successful isolation of tetrapropyllead from the reaction of tripropyllead hydroxide with iron tetracarbonyl hydride. The successful preparation of the bis(triphenyllead) and bis(tricyclohexyllead) iron carbonyls, in a similar system, then results presumably from the greater stability of these compounds to disproportionation. When bis(triphenyllead) iron tetracarbonyl was heated in refluxing toluene, no change in appearance occurred; however, in refluxing xylene complete decomposition occurred with formation of a black color. Diphenyllead iron tetracarbonyl was subsequently prepared by the reaction of triphenyllead bromide and an aqueous solution of iron tetracarbonyl hydride (24,25). The red, crystalline product decomposed at 152° and was found to be much less soluble in organic solvents than the dialkyllead analogs.

Attempts to prepare organolead iron carbonyl derivatives by the reaction of tetramethyllead and tetraphenyllead with an aqueous solution of iron carbonyl hydride were unsuccessful (22). Similarly, no reaction occurred between iron pentacarbonyl and a mixture of tetraethyllead and hexaethyldilead in ether. Dialkyltin iron tetracarbonyl derivatives have been prepared by the reaction of tetraorganotin compounds with iron pentacarbonyl, but only from unsymmetrical tetraorganotin compounds containing an sp_2-hybridized carbon atom bound to lead, such as is present in dimethyldivinyltin or dimethyldiphenyltin; in these latter reactions, the sp_2-hybridized carbon atom is preferentially cleaved (34).

Organolead iron tetracarbonyl compounds are relatively unstable and decompose at fairly low temperatures. Diethyllead iron tetracarbonyl is decomposed by concentrated nitric acid; it is unstable in air and slowly changes in color from red to brown when exposed to air in the presence of light (22). Bis(triphenyllead) iron tetracarbonyl reacted with mercuric chloride or bromide in ether or acetone to form the diphenyllead dihalide and mercury iron tetracarbonyl (27). The following sequence was proposed to account for the products:

$$(Ph_3Pb)_2Fe(CO)_4 + HgCl_2 \rightarrow 2Ph_3PbCl + HgFe(CO)_4$$

$$2Ph_3PbCl + 2HgCl_2 \rightarrow 2Ph_2PbCl_2 + 2PhHgCl$$

Bis(triphenyllead) iron tetracarbonyl was also decomposed by cadmium iodide, copper(II) chloride dihydrate, cobalt(II) chloride hexahydrate, and bismuth tribromide; carbon monoxide was evolved in varying amounts but no metal iron tetracarbonyl derivatives could be isolated from the reaction mixture. Heating bis(triphenyllead) iron tetracarbonyl with *o*-phenanthroline caused the evolution of one mole of carbon monoxide per mole of bis(triphenyllead) iron tetracarbonyl; a brick-red crystalline product was formed but it was not characterized (25).

The infrared spectra of diethyllead iron tetracarbonyl and bis(triethyllead) iron tetracarbonyl have been examined recently by Kahn and Bigorgne (31). On the basis of the frequencies observed for the four carbonyl groups, the bis(triethyllead) compound was concluded to have an octahedral structure with the triethyllead moieties in a *cis* configuration. The structure of the diethyl compound was proposed to be dimeric with the diethyllead moieties occupying the bridging positions:

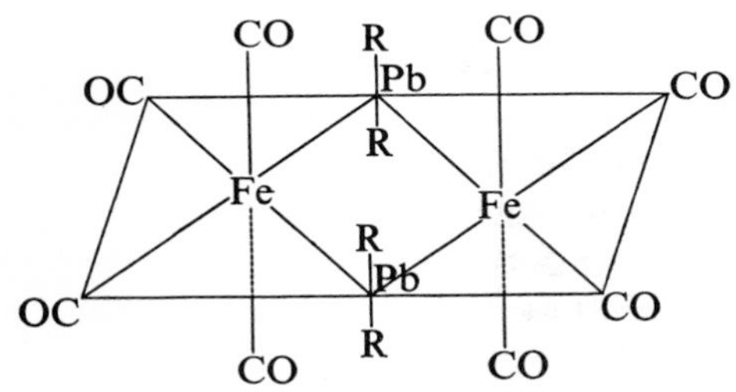

The dimeric structure for the diethyllead compound is consistent with the finding that it is dimeric in benzene; this structure was also proposed originally by Hein and Heuser (22). The covalent character of the lead–iron bond in both the diethyllead and bis(triethyllead) iron tetracarbonyls was concluded to be comparable to that of the lead–carbon bond.

Organolead derivatives of cobalt tetracarbonyl were also prepared by Hein and co-workers (24,25) by the reaction of an organolead hydroxide or halide with the sodium salt of cobalt tetracarbonyl. Triphenyllead cobalt tetracarbonyl, tricyclohexyllead cobalt tetracarbonyl, and diphenyllead bis(cobalt tetracarbonyl) were prepared in this fashion. The products ranged from dark red to yellow in color and decomposed without melting. These, and the phenyllead iron tetracarbonyl derivatives, were shown to be diamagnetic (24).

In recent years, a number of other organolead derivatives of metal carbonyls have been prepared. Gorsich (11) prepared a series of derivatives of manganese pentacarbonyl of the type $R_3MMn(CO)_5$ and $R_2M[Mn(CO)_5]_2$, where R was methyl, ethyl, or phenyl and M was tin or lead. Except for triethyllead manganese pentacarbonyl, which was a yellow liquid, all of the products were solids; the color of the lead derivatives ranged from yellow to orange. They were found to be soluble in common organic solvents such as methylene chloride, ether, acetone, and tetrahydrofuran, decomposed at varying rates in solution and were oxidized by air, particularly in the presence of light. The $R_3MMn(CO)_5$ compounds were found to be more stable to air than the $R_2M[Mn(CO)_5]_2$ analogs.

All of the compounds were prepared by the reaction of sodium manganese pentacarbonyl and the corresponding organolead or organotin halide in tetrahydrofuran:

$$R_3PbCl + NaMn(CO)_5 \xrightarrow{THF} R_3PbMn(CO)_5 + NaCl$$

In the case of trimethyllead chloride, both trimethyllead manganese pentacarbonyl and dimethyllead bis(manganese pentacarbonyl) were formed, even though no dimethyllead dichloride was detected in the starting trimethyllead chloride. Heating trimethyllead manganese pentacarbonyl gave none of the dimethyllead derivative, so that the source of formation of the $Me_2Pb[Mn(CO)_5]_2$ was not defined.

The infrared spectra of all the manganese pentacarbonyl derivatives exhibited no absorption bands characteristic of CO bridging groups. The compounds reacted readily with halogens and halogen acids. In the case of the lead compounds, cleavage of the lead–manganese bond always occurred. Thus, reaction of hydrogen chloride with $Ph_3PbMn(CO)_5$ at 25° gave $ClMn(CO)_5$, benzene or chlorobenzene, and unidentified lead species. Similarly, reaction of diethyllead bis(manganese pentacarbonyl) with chlorine at 5° gave $ClMn(CO)_5$ and a solid containing both diethyl- and triethyllead groupings. On the other hand, the tin–manganese bond in $R_3SnMn(CO)_5$ and $R_2Sn[Mn(CO)_5]_2$ compounds is much more stable to halogens and halogen acids. Thus, in carbon tetrachloride at 25° excess chlorine cleaved only the three phenyl groups from $Ph_3SnMn(CO)_5$ to give $Cl_3SnMn(CO)_5$; similarly, $Ph_2Sn[Mn(CO)_5]_2$ reacted with chlorine to form $Cl_2Sn[Mn(CO)_5]_2$, which reacted further with chlorine to give $Cl_3SnMn(CO)_5$. The preparation of such chlorotin derivatives of manganese carbonyl attests to the remarkable stability of the tin–manganese bond in these compounds. The organolead and organotin derivatives of manganese pentacarbonyl are described in a number of patents (12–17).

Tin and lead derivatives of rhenium carbonyl have been prepared by Nesmeyanov and co-workers (38) via a reaction procedure similar to that used by Gorsich. Triphenyllead rhenium pentacarbonyl was obtained as a white crystalline product from reaction of triphenyllead chloride with sodium rhenium pentacarbonyl in tetrahydrofuran; triphenyltin rhenium pentacarbonyl was similarly prepared. Like Gorsich, Nesmeyanov found that reaction of the tin–rhenium derivatives with chlorine, bromine, or hydrogen chloride at 5° resulted in cleavage of all three phenyl groups to yield the trihalo compound, $X_3SnRe(CO)_5$.

Nesmeyanov's group (39) also prepared by similar methods phenyllead and phenyltin derivatives of cyclopentadienylmolybdenum tricarbonyl and cyclopentadienyltungsten tricarbonyl of the type $Ph_3MM'(CO)_3C_5H_5$, in which M was tin or lead and M′ was molybdenum or tungsten. These compounds, as well as a chromium derivative, were synthesized independently by Patil and Graham (42), who also prepared the trimethyllead

$$R_3MX + NaM'(CO)_3C_5H_5 \rightarrow R_3MM'(CO)_3C_5H_5$$

derivative of cyclopentadienylmolybdenum tricarbonyl. The molybdenum and tungsten derivatives were described as beautifully crystalline compounds, stable indefinitely in closed containers and stable in air for several days without appreciable oxidation. However, the chromium derivatives were less stable and decomposed over a period of several weeks at room temperature, even in an inert atmosphere. The phenyl derivatives were generally more stable than the methyl derivatives, and the lead compounds were less stable than their tin (and germanium) analogs.

The infrared spectra of the $R_3MM'(CO)_3C_5H_5$ derivatives showed three strong terminal CO bands, while the $R_2M[M'(CO)_3C_5H_5]$ compounds, such as $(CH_3)_2Sn[Mo(CO)_3C_5H_5]_2$, showed up to six terminal CO bands. By X-ray diffraction analysis, the tin–molybdenum bond distance in $Ph_3SnMo(CO)_3C_5H_5$ was found to be 2.85 ± 0.02 Å, which is only 0.08 Å greater than the sum of the atomic radii of these two metals, so that the metal–metal bonds in these compounds are concluded to be covalent (39). Their covalent character is also indicated by their solubility in hexane and other hydrocarbons. Hydrogen chloride cleaved the phenyl groups from $Ph_3SnMo(CO)_3C_5H_5$ and $Ph_2SnMo(CO)_3C_5H_5$ and the tungsten analogs without cleaving the metal–metal bonds (39).

The NMR spectra of the methyl compounds showed that the position of the methyl resonance was shifted to lower field in $Me_3SnM(CO)_3C_5H_5$ and $Me_2Sn[M(CO)_3C_5H_5]_2$, compared to tetramethyltin; also, the proton-tin coupling constants were smaller than in tetramethyltin (42). This was interpreted to be indicative of enhanced *s* character in the hybrid tin orbitals bonding to the transition metal. A similar effect was observed in $Me_3PbMo(CO)_3C_5H_5$. The NMR data for the lead compounds are given in Table XV-3.

In an investigation of the electrochemical behavior of organometallic compounds, Dessy and co-workers (2–5) investigated the electrolysis of a number of triorganolead transition metal carbonyls and found that these compounds tend to undergo one of two common electrochemical reduction reactions, typified by the following general equations (m and m′ being

Table XV-3 NMR Data for Triphenyllead Cyclopentadienylmetal Tricarbonyls

COMPOUND	SOLVENT	$\tau_{C_5H_5}$	$\tau_{C_6H_5}$	τ_{CH_3}	$J_{^{207}PbCH_3}$
$(C_6H_5)_3Pb—Cr(CO)_3C_5H_5$	THF	4.95	2.5		
$(C_6H_5)_3Pb—Mo(CO)_3C_5H_5$	$CDCl_3$	4.74	2.62		
$(CH_3)_3Pb—Mo(CO)_3C_5H_5$	$CHCl_3$	4.72		8.83	44.5
$(C_6H_5)_3Pb—W(CO)_3C_5H_5$	$CDCl_3$	4.57	2.5		

Taken from Patil and Graham, *Inorg. Chem.*, **5,** 1401 (1966).

organometal moieties):

$$m \cdot m' \xrightarrow{2e} m:^- + m':^-$$

$$m \cdot m' \xrightarrow{1e} m:^- + m'\cdot$$

In the first equation, a two-electron process is involved which results in the conversion of both of the organometallic moieties to anionic species. In the second system, a one-electron reduction occurs which results in the formation of an anionic species of one organometallic moiety, while the second organometallic moiety is released as a radical which can couple with a second of its kind (e.g., $2R_3Pb\cdot \rightarrow R_6Pb_2$). The observed half-wave potentials and the nature of the reduced species for a number of tri-organolead derivatives are summarized in Table XV-4.

In the case of $Et_3PbMn(CO)_5$ and $Ph_3PbRe(CO)_5$ (where $n = 1$) a second wave was observed; the second wave occurred at $E_{1/2}$ values of -2.9 and -2.0 V, respectively. The half-wave value of -2.0 for the triphenyllead compounds corresponds to the half wave potential observed

Table XV-4 Electrochemical Data for Triorganolead Metal Carbonyls

COMPOUND	$-E_{1/2}$[a]	n[b]	PRODUCTS
$Ph_3PbMo(CO)_3C_5H_5$	2.2	2	mo:$^-$ + pb:$^-$
$Et_3PbMn(CO)_5$	1.8	1	mn:$^-$ + pb·
$Ph_3PbMn(CO)_5$	2.1	2	mn:$^-$ + pb:$^-$
$Ph_3PbRe(CO)_5$	2.4	1	unidentified
$Ph_3PbFe(CO)_2C_5H_5$	2.1	2	fe:$^-$ + pb:$^-$

[a] Versus 10^{-3} *M* Ag^+/Ag electrode.
[b] Number of electrons involved.
Taken from Dessy, Weissman, and Pohl, *J. Am. Chem. Soc.*, **88,** 5117 (1966).

under identical conditions for the electrochemical reduction of hexaphenyldilead. The triethyllead derivatives were reduced at potentials more anodic than their triphenyl analogs; organotin derivatives showed a similar effect. It was concluded that electron-withdrawing ligands (such as a phenyl group) reduce nonbonding *d*-electron interaction and thereby increase the stability of the metal–metal bond. From the observed half wave potentials, the thermodynamic stability of the tin–metal bond was concluded to be stronger than the lead–metal bond in the triorganometal-metal carbonyls (5); this is consistent with the results obtained from the reactions of the tin and lead compounds with chlorine.

Dessy and Weissman (3) have demonstrated the electrochemical synthesis of several mixed-metal organometallics. Triphenyllead cyclopentadienyliron dicarbonyl was prepared as pale orange needles by the exhaustive electrolysis of hexaphenyldilead at -2.1 V in a solution of dimethyl ether of ethylene glycol containing tetrabutylammonium perchlorate (to reduce the hexaphenyldilead to triphenyllead anion), followed by addition of cyclopentadienyliron dicarbonyl iodide. This reaction was found to be rather clean and appears to offer promise for other reactions involving organolead anions. Addition of triphenyllead acetate to a solution containing the electrolytically-generated cyclopentadienyliron dicarbonyl anion gave the same lead–iron derivative as above; a lead–manganese derivative was similarly obtained by addition of triphenyllead acetate to electrolytically-derived manganese pentacarbonyl anion [from $Mn_2(CO)_{10}$]. On the other hand, no lead–manganese product was obtained when manganese pentacarbonyl chloride was added to a solution containing the electrolytically-derived triphenyllead anion. Attempts to prepare triphenyllead cyclopentadienylmolybdenum tricarbonyl using either electrolytically-derived triphenyllead anions or cyclopentadienylmolybdenum tricarbonyl anions were also unsuccessful. The former system [$Ph_3Pb^- + C_5H_5Mo(CO)_3Cl$] gave hexaphenyldilead, while the latter reaction gave an unidentified product which was not the desired lead–molybdenum compound.

Some of the known organolead derivatives of metal carbonyls are included in a review of the alkali metal derivatives of metal carbonyls (33); organolead derivatives of iron tetracarbonyl and other organometallic derivatives of iron tetracarbonyl have been reviewed by Hein (21), and the antiknock properties of organolead iron tetracarbonyls have been discussed by Pobloth (43). The properties of the various organolead carbonyls are given in Table XV-5.

In addition to the numerous organolead metal carbonyls, one other type of organolead derivative is known in which lead is bonded to a transition metal. Schrauzer and Kratel (49) prepared a series of triphenyl

Table XV-5 Organolead Derivatives of Metal Carbonyls

COMPOUND	COLOR	MELTING OR DECOMPOSITION POINT, °C	REFERENCES
$Me_2PbFe(CO)_4$	Reddish brown	dec. 120–30	22
$Et_2PbFe(CO)_4$	Red	dec. 110	26,31
$Pr_2PbFe(CO)_4$	Bright red	98	22,23
$Bu_2PbFe(CO)_4$	Orange	87	22
$(Et_3Pb)_2Fe(CO)_4$			31
$(Ph_3Pb)_2Fe(CO)_4$	Orange-yellow	dec. 135–40	22–24,27
$[(C_6H_{11})_3Pb]_2Fe(CO)_4$	Yellow	dec. 140	22
$Ph_2PbFe(CO)_4$	Red	dec. 152	24,25
$Ph_2Pb[Co(CO)_4]_2$	Dark red	dec. 103–6	24
$Ph_3PbCo(CO)_4$	Yellow	dec. 98; 100–2	25,42a
$(C_6H_{11})_3PbCo(CO)_4$	Orange-yellow		24
$Ph_3PbMn(CO)_5$	Yellow	dec. 146–8	11
	Pale yellow	133; 142–4	3,5,30a
$Et_2Pb[Mn(CO)_5)]_2$	Orange	77–9	11
$Me_3PbMn(CO)_5$	Yellow	30–1	11
$Et_3PbMn(CO)_5$	Yellow	b.p., $_{0.1}$70–3	5,11
$Me_2Pb[Mn(CO)_5]_2$	Orange	108–10	11,30a
$Ph_3PbRe(CO)_5$	White	131–3	5,38
	Pale yellow	133–6	30a
$Ph_3PbFe(CO)_2C_5H_5$	Pale orange	134–7	3,5
$Ph_3PbCr(CO)_3C_5H_5$	Yellow	195–7 (dec.)	42
$Ph_3PbMo(CO)_3C_5H_5$	Yellow	200 (dec.)	42
	Pale yellow	dec. 201	3,5
	Yellow	205	39
$Me_3PbMo(CO)_3C_5H_5$	Pale Orange	93–5	42
$Ph_3PbW(CO)_3C_5H_5$	Yellow	214–5 (dec.)	42
	Yellow	210–1.5	39

metal derivatives containing the bis(dimethylglyoxime)cobalt(III) anion and having the general formula $Ph_3MCo(DMG)_2$·Base, where M is a Group IVb metal, DMG is the dimethylglyoxime dianion, and Base is a Lewis base such as pyridine or tributylphosphine. Both the pyridine and tributylphosphine adducts of triphenyllead cobalt(III) bis(dimethylglyoximate) were prepared. These compounds are believed to contain lead–cobalt bonds. They are soluble in such organic solvents as methylene chloride, chloroform, benzene and acetone; they are stable to alkali, but are decomposed by strong acids. Their stability increases with increasing atomic weight of the Group IVb metal M.

An unstable lead–platinum compound has been reported recently (1). The white, solid compound had the composition, $PtCl(PbPh_3)(PPh_3)_2$, and was prepared by the reaction of triphenyllead nitrate with *trans* $PtHCl$-$(PPh_3)_2$. It could not be obtained pure but disproportionated to

$PtPhCl\text{-}(PPh_3)_2$. The formation of this product was attributed to phenyl group migration and formation of diphenyllead. The pure tin analog was also prepared; it was more stable but also disproportionated to the phenylplatinum chloride bis (triphenylphosphine).

Finally, a brief report by Rosenberg (45) describes the synthesis of "tetraferrocenyl" derivatives of silicon, germanium, tin and lead by the reaction of ferrocenyllithium with metal(IV) chlorides or lead(II) chloride. The lead(II) chloride reaction is postulated to involve the intermediate formation of hexaferrocenyldilead. The tetraferrocenyl derivatives are surprisingly stable, the silicon compound being stable to above 450°. The existence of an organolead compound of this type is surprising, especially since all attempts to prepare tetracyclopentadienyllead have been unsuccessful, and also since the simple cyclopentadienyltrialkyllead compounds are unstable. The molecular configuration and the nature of the bonding in these unusual compounds is of special interest. In all probability, these compounds will prove not to contain metal–metal bonds.

REFERENCES

1. Baird, M. C., *J. Inorg. Nucl. Chem.*, **29,** 367 (1967).
1a. Bindschadler, E., *Iowa St. Coll. J. Sci.*, **16,** 33 (1941); through *C.A.*, **36,** 4476 (1942).
2. Dessy, R. E., R. L. Pohl, and R. B. King, *J. Am. Chem. Soc.*, **88,** 5121 (1966).
3. Dessy, R. E., and P. M. Weissman, *J. Am. Chem. Soc.*, **88,** 5124 (1966).
4. Dessy, R. E., and P. M. Weissman, *J. Am. Chem. Soc.*, **88,** 5129 (1966).
5. Dessy, R. E., P. M. Weissman, and R. L. Pohl, *J. Am. Chem. Soc.*, **88,** 5117 (1966).
6. Drenth, W., M. J. Janssen, G. J. M. van der Kerk, and J. A. Vliegenthart, *J. Organometal. Chem.*, **2,** 265 (1964).
7. Foster, L. S., W. M. Dix, and I. J. Gruntfest, *J. Am. Chem. Soc.*, **61,** 1685 (1939).
8. Gilman, H., and D. S. Melstrom, *J. Am. Chem. Soc.*, **72,** 2953 (1950).
9. Gilman, H., and L. Summers, *J. Am. Chem. Soc.*, **74,** 5924 (1952).
10. Gilman, H., L. Summers, and R. W. Leeper, *J. Org. Chem.*, **17,** 630 (1952).
11. Gorsich, R. D., *J. Am. Chem. Soc.*, **84,** 2486 (1962).
12. Gorsich, R. D., U.S. Pat. 3,030,396–7 (to Ethyl Corp.), Apr. 17, 1962, *C.A.*, **57,** 7309 (1962).
13. Gorsich, R. D., U.S. Pat. 3,033,885 (to Ethyl Corp.), May 8, 1962, *C.A.*, **57,** 13803 (1962).
14. Gorsich, R. D., U.S. Pat. 3,050,537 (to Ethyl Corp.), Aug. 21, 1962, *C.A.*, **57,** 16657 (1962).
15. Gorsich, R. D., U.S. Pat. 3,069,445 (to Ethyl Corp.), Dec. 18, 1962; *C.A.*, **58,** 10237 (1963).
16. Gorsich, R. D., U.S. Pat. 3,069,449 (to Ethyl Corp.), Dec. 18, 1962; *C.A.*, **58,** 10241 (1963).
17. Gorsich, R. D., U.S. Pat. 3,099,667 (to Ethyl Corp.), July 30, 1963, *C.A.*, **60,** 549 (1964).
18. Grüttner, G. and E. Krause, *Ber.*, **50,** 1559 (1917).

19. Grüttner, G., E. Krause, and M. Wiernik, *Ber.*, **50,** 1549 (1917).
20. Hedden, G. D., U.S. Pat. 3,110,719 (to E. I. duPont de Nemours & Co.), Nov. 12, 1963, *C.A.*, **60,** 3006 (1964).
21. Hein, Fr., *Angew. Chem.*, **62,** 205 (1950).
22. Hein, Fr., and E. Heuser, *Z. anorg. allgem. Chem.*, **254,** 138 (1947).
23. Hein, Fr., and E. Heuser, *Z. anorg. allgem. Chem.*, **255,** 125 (1947).
24. Hein, Fr., and W. Jehn, *Ann. Chem.*, **684,** 4 (1965).
25. Hein, Fr., P. Kleinert and W. Jehn, *Naturwiss.*, **44,** 34 (1957).
26. Hein, Fr., and H. Pobloth, *Z. anorg. allgem. Chem.*, **248,** 84 (1941).
27. Hein, Fr., and H. Scheiter, *Z. anorg. allgem. Chem.*, **259,** 183 (1949).
28. Henry, M. C., and H. Gorth, Quarterly Report No. 19, Proj. LC-28, to Intern. Lead Zinc Res. Organ., New York, March 1966.
29. Henry, M. C. and J. G. Noltes, *J. Am. Chem. Soc.*, **82,** 558 (1960).
30. Ingham, R. K., and H. Gilman, "Organopolymers of Silicon, Germanium Tin and Lead", in *Inorganic Polymers*, F. G. A. Stone and W. A. G. Graham, Eds., Academic Press, New York, 1962, pp. 321–446.
30a. Jetz, W., P. B. Simons, J. A. J. Thompson, and W. A. G. Graham, *Inorg. Chem.*, **5,** 2217 (1966).
31. Kahn, O., and M. Bigorgne, *Compt. rend.*, **261,** 2483 (1965).
32. Kerk, G. J. M. van der, *Ind. Eng. Chem.*, **58,** 29 (1966).
33. King, R. B., "Alkali Metal Derivatives of Metal Carbonyls" in *Advances in Organometallic Chemistry*, Vol. 2, F. G. A. Stone and R. West, Eds., Academic Press, New York, 1964, pp. 157–256.
34. King, R. B., and F. G. A. Stone, *J. Am. Chem. Soc.*, **82,** 3833 (1960).
35. Korshak, V. V., A. M. Polyakova, and M. D. Suchkova, *Vysokomolekulyarye Soedineniya*, **2,** 13 (1960); through *C.A.*, **55,** 361 (1961).
36. Leusink, A. J., J. G. Noltes, H. A. Budding, and G. J. M. van der Kerk, *Rec. trav. chim.*, **83,** 609 (1964).
37. Leusink, A. J., J. G. Noltes, H. A. Budding, and G. J. M. van der Kerk, *Rec. trav. chim.*, **83,** 844 (1964).
38. Nesmeyanov, A. N., K. N. Anisimov, N. E. Kolobova, and V. N. Khandozhko, *Dokl. Akad. Nauk SSSR*, **156,** 383 (1964); *C.A.*, **61,** 7041 (1964).
39. Nesmeyanov, A. N., K. N. Anisimov, N. E. Kolobova, and M. Ya. Zakharova, *Dokl. Akad. Nauk SSSR*, **156,** 612 (1964); *C.A.*, **61,** 5173 (1964).
40. Neumann, W. P., and K. Kühlein, *Tetrahedron Letters*, **1966** 3419.
41. Noltes, J. G., and G. J. M. van der Kerk, *Rec. trav. chim.*, **80,** 623 (1961); *Chimia*, **16,** 122 (1962).
42. Patil, H. R. H., and W. A. G. Graham, *Inorg. Chem.*, **5,** 1401 (1966).
42a. Patmore, D. J., and W. A. G. Graham, *Inorg. Chem.*, **6,** 981 (1967).
43. Pobloth, H., PB Report No. 95601, **1943,** *Bib. of Sc. and Ind. Repts.*, **11,** 487 (1949).
44. Razuvaev, G. A., N. S. Vyazankin, Yu. I. Dergunov. and O. S. D'yachkovskaya, *Dokl. Akad. Nauk SSSR*, **132,** 364 (1960); *C.A.* **54,** 20937 (1960).
45. Rosenberg, H., *RTD Technology Briefs*, **2**(8), 14 (1964), Headquarters, Res. Tech. Div., Bolling Air Force Base, Washington, D.C.
46. Schlaudecker, G. F., U.S. Pat. 2, 891,977 (to E. I. du Pont de Nemours & Co.), June 23, 1959; *C.A.*, **53,** 19881 (1959).
47. Schmidbaur, H., *Chem. Ber.*, **97,** 270 (1964).
48. Schmidbaur, H., and S. Waldmann, *Chem. Ber.*, **97,** 3381 (1964).
49. Schrauzer, G. N., and G. Kratel, *Angew. Chem.*, **77,** 130 (1965).

50. Seyferth, D., and W. Freyer, *J. Org. Chem.*, **26,** 2604 (1961).
51. Summers, L., *Iowa St. Coll. J. Sci.* **26,** 292 (1952); through *C.A.*, **47,** 8673 (1953).
52. Wiczer, S. B.. U.S. Pat. 2,447,926, Aug. 24, 1948; *C.A.*, **42,** 7975 (1948).
53. Willemsens, L. C., and G. J. M. van der Kerk, *J. Organometal. Chem.*, **2,** 260 (1964).
54. Willemsens, L. C., and G. J. M. van der Kerk, *J. Organometal. Chem.*, **2,** 271 (1964).
55. Willemsens, L. C., and G. J. M. van der Kerk, *Rec. trav. chim.*, **84,** 43 (1965).
56. Willemsens, L. C., and G. J. M. van der Kerk, *Investigations in the Field of Organolead Chemistry*, Internat. Lead Zinc Res. Organ., New York, 1965.
57. Willemsens, L. C., and G. J. M. van der Kerk, Quarterly Report No. 21, Project LC-18, to Intern. Lead Zinc Res. Organ. March 1965.

XVI Analysis of organolead compounds

1. GENERAL

There is an extensive literature on the analytical determination of organolead compounds because of their commercial significance and importance in industrial hygiene.

The qualitative detection of lead in its organic compounds or in their concentrated solutions is relatively simple. On the other hand, the quantitative lead analysis of dilute solutions such as gasolines or oils is a somewhat more complex problem, and one of considerable industrial importance. This is especially true of gasolines, since the lead content is closely related to the octane number value, which is economically important.

Organolead compounds may be easily converted to inorganic lead forms such as lead metal, lead oxide, or a lead salt (depending on the particular compound originally present) by heating, burning, or by acids or halogens. These reactions usually suffice for qualitative analysis, and a simple ashing microtest has been described by Spialter and Ballester (146).

In a similar manner, quantitative analytical treatment of organolead compounds is usually started by treatment with nitric, hydrochloric, or sulfuric acid, or mixtures of these acids, or by decomposition with free halogen. For the volatile alkyllead compounds, decomposition with bromine in the cold, followed by heating with hydrobromic acid to give lead bromide, is an excellent method. If desired, the lead may then be determined gravimetrically, as originally described by Grüttner and Krause (67). Alternatively, the organolead compound in benzene solution may be treated with aqueous iodine solution with potassium iodide in slight excess, using heat and light as necessary, and the excess iodine may be backtitrated with standard sodium thiosulfate solution, as described in Calingaert's review of organolead chemistry (35), and by Hein, Klein, and Mesée (70). A method based on the use of excess alcoholic potassium iodide solution has also been described (31,70).

The method of decomposition with bromine or iodine does not give satisfactory results with aryllead compounds because they are more stable than the alkyllead compounds. For the aryllead compounds, a more applicable method of determination depends on the acid treatment mentioned above. Such lead determinations may be completed either gravimetrically by weighing the lead sulfate, according to Gilman and Robinson (61), or volumetrically by lead chromate titration, as described by Saunders and Stacey (134).

When organolead compounds are to be analyzed for elements other than lead, the usual combustion tube techniques or sodium peroxide fusion are employed. Difficulties are occasionally encountered in using such methods for organometallic compounds; for a discussion and references to various modified methods, see the review on "Organotin Compounds" by Ingham Rosenberg, and Gilman (78).

Analytical methods for lead with particular reference to medical and biological specimens were discussed by Cholak (41) at a 1963 symposium.

2. ANALYSIS OF COMMERCIAL ANTIKNOCK FLUIDS

A technical problem which arises occasionally is the analysis of commercial antiknock fluids or mixtures for lead. Such concentrated fluids are sometimes analyzed for both tetraalkyllead, ammonia-soluble alkyllead salts, and also for bromine and chlorine (34,150). The standard control assay of fluids for tetraethyllead (TEL) is based on titration with iodine.

In the past few years, several papers have appeared on the determination of small amounts of hexaethyldilead in antiknock fluids or in solutions of TEL by a high frequency permanganimetric titration, a technique used by Goldshtein and co-workers (63,93), by use of absorption spectra as described by Rudnevskii and Vyshinskii (129), by coulorimetric iodination favored by Tagliavini, Belluco et al. (22,151), or by a polarographic method discussed by DeVries et al. (47).

3. ANALYSIS OF GASOLINES AND OTHER DILUTE SOLUTIONS

As indicated above, the determination of tetraalkyllead compounds in gasolines and other dilute solutions is a matter of considerable practical importance, and hundreds of publications have appeared on this subject. Until recently, TEL was the only alkyl lead compound for which analyses were made routinely. This compound has now been joined for use in gasoline, however, by tetramethyllead (TML) and various mixed methylethyllead compounds. Undoubtedly, additional reports will appear in the future on the determination of mixed lead alkyls in gasoline; at this writing, however, only a few publications have appeared on this subject. Parker and co-workers (118,119) have described an accurate, rapid method utilizing gas chromatography to separate the lead alkyls, and a dithizone procedure to measure them (see also "Other Methods" below for additional discussion of gas chromatography). However, a cooperative ASTM Committee program by ten laboratories on blends of TEL, TML, and mixed alkyllead compounds in several types of base stocks has shown that reproducibility in such analyses is not as precise as that for TEL in the standard ASTM Method D526-56 (126).

Paper chromatography has been employed recently by Pedinelli (120) and Schafer (135) to identify components of mixed methylethyl lead mixtures. Paper chromatography has also been found satisfactory by Barbieri et al. (15–19) for the separation of R_4Pb, R_3PbCl, and R_2PbCl_2 compounds prior to analysis. Paper electrophoresis has been used by the same group (62) to separate Et_3Pb^+ ions from Et_2Pb^{++} ions in chloride solutions.

Abel et al. (1) showed that vapor phase chromatography could be used to separate tetramethyllead from other Group IVb organometallic compounds with complete separation and sharp elution curves.

Henneberg and Schomberg (73) reported in 1965 on the use of mass spectroscopy as a vapor phase chromatography detector for detecting as little as 10 ppm of TEL.

Broadly speaking, the methods of analysis for lead in gasoline may be classified as either chemical or instrumental. These classifications may be further subdivided into various categories, such as gravimetric, volumetric, X-ray absorption, X-ray fluorescence, polarographic, flame photometric, electrolytic, colorimetric, and turbidimetric. Critical reviews of some of these methods have been written by Chaudhuri (40) and Tarjan and Knezevic (152), and more recently by Nagypataki and Tamasi (108) and Gelius and Preussner (57). Ethyl Corporation has compiled a collection of analytical methods (3).

A. Chemical methods

In general, the chemical methods are quite accurate but they require excessive attention and thus they are too time consuming. These methods have been multiplied almost ad infinitum within the past 20 years by slight alterations of existing methods, but this extensive development work has not produced any substantial improvement in either rapidity or economy.

One special precaution that must be taken in the analysis of dilute solutions for lead is not to volatilize the lead compound while concentrating the solution. The standard ASTM chemical procedure D526-61 (4) prevents such loss by chemical reaction of the lead while it is still in dilute solution. This standard procedure is based on the method of Calingaert and Gambrill (36), in which the TEL is decomposed quantitatively by refluxing the gasoline solution with concentrated hydrochloric acid, and then the lead chloride is extracted with water. The aqueous solution of inorganic lead is usually oxidized with nitric acid, and the lead is then determined gravimetrically by the sulfate or chromate method, or volumetrically by the molybdate method.

Several Russian workers have suggested modifications of the standard hydrochloric acid–chromate method, involving smaller sample size and other changes (85,131,153). Recently, the standard ASTM D526 method has been compared with the Russian procedure GOST-63, by American and Italian investigators. The ASTM procedure appears to be somewhat more precise for catalytically cracked gasolines containing TEL, but time requirements for the methods are similar (6). Siniramed and Renzanigo (137) reported in 1965 that a minor modification of the ASTM method proposed by the U.S. delegation to the International Standards Association is generally the most accurate method of all.

Among the chemical methods, considerable interest has been displayed within the past few years in titration methods based on complexometric titration of the lead in order to speed the analysis. Most complexometric methods are based on chelation with the disodium salt of ethylenediaminetetraacetic acid, according to the general method of Schwarzenbach (7,26,30,44,48,51,54,66,90,103,105,130,133,154). A good deal of work has been done on various indicators and techniques for this method, principally to improve the end point, but the working time per determination is still at least 20 min. The precision compares favorably with the chemical ASTM Method D526-61.

B. Colorimetric methods

Colorimetric methods have a special usefulness in the determination of lead in gasoline in micro quantities, when perhaps only a limited supply of sample is available. Also, colorimetric methods of analysis are significant for refinery operations, because trace amounts of lead are capable of poisoning platinum refining catalysts which are in widespread use. The colorimetric methods afford a means of determining lead in concentrations below 0.01 ppm.

Excellent methods of this type using dithizone chelation of the lead have been established by Levine (94), Henderson and Snyder (72), and Griffing et al. (65). These methods usually are based on the decomposition of the organolead compound to ionic form by treatment with iodine or bromine in chloroform solution. Henderson and Snyder, cited above, describe a simple and rapid method—based on complexes colored yellow, orange, and red—for the simultaneous determination of triethyllead, diethyllead, and inorganic lead ions, respectively, since TEL may be converted to these forms by halogen. A similar method has been described by Barbieri et al. (20). Ethyl Corporation has published a rapid field method for the analysis of gasoline based on dithizone, using the Hellige Color Comparator Kit (8). Also, a microdetermination method based on the precipitation of lead chromate has been described by Gordon and Burdett (64), and a method using color developed by sodium sulfide has been reported by Lorone and Paul (99).

Although the dithizone reagent is widely used, and the stoichiometry of some simple dithizonates has been established, the structures of the dithizonates are not yet clearly established. Irving and Cox (79) have recently discussed the absorption spectra and possible structures for dithizone complexes of Pb(II), $Me_3Pb(IV)$, and $Ph_3Pb(IV)$, as well as other heavy metal ions.

Colorimetry has also been applied to macro methods of determining organic lead. For example, rapid determination of TEL in gasoline has been described by Smith et al. (140), based on the turbidimetric reaction of alcoholic silver nitrate with the TEL to yield ethylsilver, which decomposes to metallic silver; the colloidal silver is then determined by using a photoelectric colorimeter. This method is excellent for aviation gasolines, but unsuitable in the presence of the free sulfur or polysulfides usually found in "doctor-sweetened" motor gasoline stocks. An automatic process based on ultraviolet light decomposition of TEL, followed by photometric determination of the lead on a paper tape, also has been patented by Obraztsova and Novikov (111).

C. Polarographic methods

As a result of the increasing demand for speed in analysis in recent years, there has been a growing interest in new instrumental methods to replace the existing chemical methods. For example, due to the pioneering work of Frediani and Bass (55) in 1940, direct polarographic methods have come into extensive use to determine lead by a Calingaert-Gambrill type hydrochloric acid–Cellosolve extract of the gasoline without further treatment. This saves considerable time required in the tedious steps of oxidation and evaporation, while still retaining an accuracy generally within ±3% (13,29,68,76,112,113,114,124,139,149,157–159,164–6). Because of the minimum of manipulation and lack of complex chemical operations, as many as 40 determinations per man-day may be made by this method. A commercial instrument named the TEL-Meter is available, designed specifically for use with the cadmium pilot-ion polarographic method currently proposed by the ASTM as a standard method (D1269-61) (5,9). Also, a method of titration has been described based on a combination of bromination and mercurization (122).

Nightingale et al. (110) have used triethanolamine complexes to determine both lead and manganese polarographically in the same gasoline sample.

Dessy and associates (46) made a recent study of the polarographic behavior of several Group IVb organometallic derivatives (see also Chapter XV). With respect to lead, Riccoboni (127) had previously published on the polarographic, one-electron reduction of triethyllead chloride to "triethyllead" in aqueous potassium chloride, and Korshunov and Malyugina (89) have investigated the reduction of triethyllead hydroxide.

Plazzogna and Pilloni (123) have recently described amperometric titration methods for organolead ions such as $(C_2H_5)_2Pb^{++}$, $(CH_3)_2Pb^{++}$, and $(C_2H_5)_3Pb^{+}$ using a dropping mercury electrode and titration with $K_4Fe(CN)_6$ in the case of the dialkyllead ions; trialkyllead ions do not interfere. The triethyllead ion was determined by precipitation with $NaB(C_6H_5)_4$, and backtitration of the excess $NaB(C_6H_5)_4$ with $TlNO_3$; in this instance, dialkyllead ions do not interfere.

D. X-ray methods

X-ray absorption methods, first investigated as early as 1929 by Aborn and Brown (2,32,37,53,77,83,95–97,148,159) have also come into vogue within the past decade, owing principally to their rapidity (5–10 min per sample) and their nondestructive character. However, the equipment is expensive, and these methods are sensitive to other elements in the gasoline, especially heavy elements. Corrections are ordinarily required even for sulfur and halogens, which are present in practically all fuels, as well as for the base stock carbon–hydrogen ratio. The accuracy is comparable to that of other standard methods of analysis when the kind of antiknock compound present is known and the corrections mentioned above are applied. The method has also been applied by Oikawa and Okuda (115,116) to the analysis of tetraphenyllead.

The X-ray fluorescence method (25,39,92) is somewhat similar to the X-ray absorption method, but it is unique for lead and also relatively insensitive to the base stock and the other interferences from which the absorption method suffers. The chief drawback is the increased expense of the equipment. In 1957, as a further refinement, Kannuna (82) published an investigation of the use of X-rays generated from β particles from tritium (tritium bremsstrahlung) for determining TEL. Hasegawa et al. (69) have recently applied fluorescent X-ray spectroscopy to the determination of both lead and bromine in gasoline, using the Pb Lα and Br Kα lines. The accuracy of the method is reported to be about $\pm 1\%$. Mahr and Stork (102) have recently suggested addition of the test compound in X-ray fluorescence to reduce the background and increase the accuracy.

X-ray spectrochemical analysis was reviewed by Birks (24) in 1959, and a new book on reflectance spectroscopy by Wendlandt and Hecht (156) appeared in 1966.

E. Flame photometry methods

Within the past several years, a flame photometric method (10,33,60, 81,98,104,117,132) based on burning the gasoline sample in an oxy-hydrogen flame and measuring the lead emission in a spectrophotometer has become perhaps the most popular instrumental method. This is because flame photometry has the speed advantage of the X-ray methods while it utilizes relatively inexpensive equipment. Flame photometry is also comparatively insensitive to halogens and sulfur. A correction for the base stock is usually made by the incremental addition of a standard solution of lead compound. Combined with distillation, the method is suitable for mixtures of lead alkyls (132). Chemical analysis by flame photometry has been reviewed by Herrmann and Alkemade (74).

F. Other instrumental methods

It should be pointed out that considerable effort has been applied in the past few years to other kinds of instrumental methods for determining lead in gasolines. Nuclear radiation and electron absorption techniques are under study, as well as other spectrographic procedures. In 1961, Robinson (128) published a method based on atomic absorption spectroscopy. This procedure is relatively inexpensive, rapid, and remarkably free from interferences both by other trace elements present, and by variations in the base stock, including sulfur and nitrogen. The method, based on an earlier demonstration by Walsh (155), utilizes a hollow lead cathode as the source and a modified Perkin–Elmer Model 13 Spectrophotometer as the detector.

In 1963, Barrall and Ballinger (21) described a method suitable for separation and analysis of several lead alkyls in gasoline based on gas chromatographic analysis with electron capture detectors. Similar methods have been published by Bonnelli and Hartmann (28), Dawson (45), and Kramer (91). Bonelli and Hartmann, for example used a 10 ft × 1/8 inch stainless steel column packed with 10% 1,2,3-tris-(2-cyanoethoxy)propane on 80–100 mesh Chromosorb W treated with hexamethyldisilizane and operated at 72°. This column was followed by a small column to absorb column material and prevent contamination of the electron capture detector. Kramer (91) used a similar Chromosorb W column coated with 10% Apiezon L, employing bromobenzene in hexane as an internal standard, with reanalysis after the addition of a pre-column to remove the internal standard and the scavengers. The results from this system are said to agree with those obtained by the ASTM Method D-1949-64 or IP188/64T. The analysis requires about 45 min and is stated to detect 0.002 g of lead alkyl per liter of gasoline with an average error of 2.5% for TEL and 1% for TML.

Lovelock and Zlatkis (100,101) also have described an analytical method based on gas phase electron absorption. The gasoline sample is separated into its major components by passing the vapor through a chromatographic column at 90°. The lead compounds are thus separated from the scavenger halides. The detector consists of two parallel electrodes; the cathode is stainless steel coated on one side with titanium tritide as a source of soft β radiation. The density of the thermal electrons produced in the carrier sample gas (usually argon) by the β radiation is measured by applying dc pulses to the anode, while the concentration of vapor in the gas stream is measured by a photoionization detector connected in parallel to the electron detector chamber. This method is quite rapid, but more expensive than conventional analytical tools. It has potential applicability to many other areas of analysis and detection, wherever small quantities of electron-absorbing materials are available and separable. The method has been described further by Farkas and Csanyi (52).

Mass spectrometry has been used by Howard et al. (75) for determining TEL and TML separately in gasoline.

Ashbel et al. (14) have discussed a β ray method suitable for both organic lead and tin compounds. Using a Sr^{90} source, they found that the concentration of organometallic compound in solution is proportional to the intensity of the back-scattered β radiation. Analyses were reported to require only 10–15 min, with relative accuracy of 0.05–3%, and relative sensitivity of 0.2%.

In 1966, Soulages (145) described a gas chromatographic method for the simultaneous determination of lead alkyls and halide scavengers in gasoline. In Soulages' method, the compounds are separated in a partition column using polypropylene glycol 400 as the stationary phase. Following this separation, the lead alkyls and the halides are catalytically hydrogenated over nickel at 140°, and the resulting methane and/or ethane is separated in a charcoal-packed chromatographic column for final detection by flame ionization. The method appears to be suitable for the analysis of all methylethyllead alkyl mixtures in a continuous system. Also, Kolb and associates (88) described a gas chromatographic method combined with atomic absorption spectroscopy for the determination of lead alkyls in gasoline.

Undoubtedly, such systems as described above will be improved further, and further progress in rapid and inexpensive instrumental analysis may be anticipated in the next few years.

4. ANALYSIS FOR ORGANOLEAD COMPOUNDS IN AIR

As was discussed in Chapter V on Physiological and Toxicological Properties, the chief danger from alkyllead compounds in both industry and laboratory lies in continued exposure to small doses. For this reason there has been a great deal of interest in the determination of small amounts of tetraalkyllead compounds in air. Micro methods have been devised for this analysis; a kit is now in general industrial use for TEL in air which passes an air sample through a scrubber containing an iodine solution, permitting the colorimetric determination of the ionic lead produced by means of dithizone solution (59,107,141). This standard kit is not quantitative for TML and mixed lead alkyls. However, Snyder and Henderson (142,143) have recently introduced a suitable quantitative kit based on a solid elemental iodine absorber. Khrustaleva (86) has used plumbone to determine the lead colorimetrically after the usual type of collection from air with iodine. As might be expected, iodine has also been used by Zemskov et al. (161) to purify air of lead.

In a modification of Snyder and Henderson's method, Moss and Browett (106) have used iodine monochloride solution instead of iodine to collect TEL or TML. After removing inorganic lead dust by prior filtration and collecting the organic lead in iodine monochloride solution as the dialkyllead ion, the lead is determined colorimetrically as the dithizonate. This is described both as a manual procedure, and as an automatic procedure using the Technicon Auto-Analyzer.

Cholak (41,42) has discussed the analysis of lead in air from the sampling standpoint and also the concentration of lead in urban atmospheres in a 1964 symposium.

The electron capture detector has not only been used in the analysis of lead in gasoline, but it was investigated in 1965 for sensitivity to both lead and chlorinated compounds in air, using several different commercial chromatographs (27). It was concluded that the electron capture detector is quite useful in air analysis, not because it is unusually sensitive to lead, but because it discriminates lead compounds from unsubstituted hydrocarbons very well. For heavily chlorinated compounds, the electron capture detector is reported to be from 100 to 1000 times more sensitive than ionization detectors.

Khrustaleva (87) has disclosed an analytical method for lead compounds in air based on the aspiration of the air sample through an ashless filter, followed by conversion to $PbSO_4$ with H_2SO_4, ashing at 500°, and nephelometric determination as the chromate. For such analyses, adsorption of lead in air on activated carbon is described as suitable by Zemskov and Sidelnikova (163), as is absorption in wet nitric acid described by Dulinski and Köhsling (49).

A continuous recorder for determining traces of lead or other contaminants in air has been described by Strange et al. (147), the so-called M.S.A. Billion-Aire. This is based on gas ionization and pyrolysis or ultraviolet degradation of the organolead vapor. An ionizing ray method for TEL in air (or gasoline) has also been patented in Germany by Hendee and Fine (71). In 1966, Skalicka and Cejka (138) compared several methods for the determination of TEL in air and concluded that colorimetry is most suitable.

Air containing TEL vapor is claimed to be purified by treatment with ozonized air in a Russian patent by Emelyanov et al. (50). Furthermore, it is claimed in a Russian patent issued to Penkov et al. (121) that TEL is completely removed from air by passing a mixture containing ozone through a layer of activated carbon. A fluidized bed of activated carbon has also been employed to remove TEL from air by Zemskov (160,162).

5. DELEADING OF GASOLINE

Sometimes it is necessary to remove antiknock compounds from their dilute solution in gasoline in order to obtain an unleaded fuel. This problem is of some importance when it is desired to burn gasoline in lanterns, stoves, etc., in the field, or to purify a refinery charge stock when it is to be treated with platinum catalyst, since lead poisons the catalyst. The usual chemical analytical reagents for TEL cannot be used in a lead removal procedure since they react with the fuel as well as the TEL.

The chemical approach generally employed involves a mild reaction, such as reacting the TEL with a metal halide. Soroos and Shapiro (38,144) solved the problem in this way during World War II. They devised a field procedure based on interaction of the TEL with anhydrous tin(IV) chloride released by breaking an ampoule. The chemicals react within a few minutes to give principally the dialkyl salts:

$$Et_4Pb + SnCl_4 \rightarrow Et_2PbCl_2 + Et_2SnCl_2$$

The reaction products are easily extracted from the gasoline by shaking with water, and the fuel is separated by decantation or filtration. The resulting fuel is of sufficiently high quality to be usable in motor vehicles in an emergency.

Friedman (56) and Kharasch (84) later patented quite similar methods employing tin(IV) chloride. For example, Kharasch's method involved stannic chloride treatment of the gasoline, followed by shaking the mixture with tertiary amine-treated charcoal to adsorb the reaction products and settle them out, followed by filtration of the fuel.

Other such chemical methods could obviously be devised, based on the known reactions of TEL with other metal salts. For example, in 1961 Choudhuri et al. (43) proposed to use anhydrous ferric chloride in ether solution. This method is apparently also effective, but requires good filtration.

In still another type of chemical method, Shapiro and Olson (136) patented the use of sulfur dioxide as a reactant at 125–140°F. It is claimed that lead sulfate is precipitated from the fuel, and the lead-free fuel is then filtered. Reactions of R_4Pb compounds with sulfur dioxide and sulfur trioxide were discussed in Chapter VI.

Nonchemical methods of TEL removal from gasoline have also been devised. Rasmussen (125) patented heating the gasoline in an apparatus filled with iron particles to vaporize the hydrocarbons and thermally decompose the TEL. The vaporized fuel was then passed through a mat of fine metal fibers to retain the finely divided lead. Wilkinson (11,12,158) developed this method to a practicable field procedure for Army stoves during World War II.

In another nonchemical method, German and Degering (58) patented a TEL removal method based on irradiation of the fuel. They used a Van de Graff electron accelerator to precipitate the lead.

Using an adsorption procedure, Neef (109) obtained a patent on the removal of TEL with an activated clay, such as Filtrol, which reacts strongly with TEL, but the reactions involved are not yet understood. Also, Jezl and Mills (80) patented a method of removing TEL involving contact of the gasoline with silica gel at 200–450°F, and separation of the gasoline. The lead is retained in the silica gel.

In a process probably involving reactions related to the deleading procedures, it has been shown by Berezniak and Zimoch (23) that 0.1% TEL added to gasoline being gelled by aluminum salts or naphthenates has the effect of increasing the viscosity twofold and shortening the ripening time by 50%.

REFERENCES

1. Abel, E. W., G. Nickless, and F. H. Pollard, *Proc. Chem. Soc.*, **1960,** 288.
2. Aborn, R. H., and R. H. Brown, *Ind. Eng. Chem., Anal. Ed.*, **1,** 26 (1929).
3. Anon., *Ethyl Analytical Methods*, Ethyl Corp. Research and Development Dept., Ferndale, Michigan.
4. Anon., *ASTM Standards on Petroleum Products and Lubricants*, Method D526-61 (1961), Gravimetric Chromate Method.
5. Anon., *ASTM Standards on Petroleum Products and Lubricants*, Method D1269-61 (1961), Polarographic Method.
6. Anon., *ASTM Bull. No. 250*, 13 (1960); through *C.A.*, **55,** 6831 (1961).
7. Anon., *Erdöl Kohle*, **12,** 987 (1959).
8. Anon., *EAM 77*, *Ethyl Analytical Methods*, Ethyl Corp. Research and Development Department, Ferndale, Michigan.
9. Anon., Proposed Method of Test for Tetraethyllead in Gasoline by Direct Reading Polarograph, Appendix VII, *ASTM Standards on Petroleum Products and Lubricants*, November, 1955.
10. Anon., *EAM 90*, *Ethyl Analytical Methods*, Ethyl Corp. Research and Development Dept., Ferndale, Michigan.
11. Anon., Ethyl Corporation, "Instructions for the Operation and Care of Thermal Deleading Unit," PB Report 107,045; *Bib. of Tech. Repts.*, **18,** (1952).
12. Anon., U.S. Quartermaster Board, "Report of Test of Outfit, Deleading Gasoline, Project T-260, QMB Test 1316," PB Report 28796, Feb., 1944; *Bib. of Sci. and Ind. Repts.*, **2,** 801 (1946).
13. Anzaldi, O., *Bol. inform. Petrol. (Buenos Aires)*, **312,** 170 (1959); through *C.A.*, **53,** 19361 (1959).
14. Ashbel, F. B., A. M. Parshina, M. S. Goizman, L. I. Zhizhina, and K. M. Kuptsova, *Zavodsk. Lab.*, **31** (9), 1062 (1965); through *C.A.*, **64,** 11 (1966).
15. Barbieri, R., U. Belluco, and G. Tagliavini, *Ann. Chim. (Rome)*, **48,** 940 (1958); through *C.A.*, **53,** 9901 (1959).
16. Barbieri, R., U. Belluco, and G. Tagliavini, *Ric. sci.*, **30,** 1671 (1960); through *C.A.*, **55,** 14175 (1961).
17. Barbieri, R., G. Faraglia, M. Giustiniani, and L. Roncucci, *J. Inorg. Nucl. Chem.*, **26,** 203 (1964).
18. Barbieri, R., G. Faraglia, and M. Giustiniani, *Ric. sci.*, *Rend. Sez. A***4** (2), 109 (1964); through *C.A.*, **61,** 4388 (1964).
19. Barbieri, R., M. Giustiniani, G. Faraglia, and G. Tagliavini, *Ric. sci.*, *Rend. Sez. A***3** (7), 975 (1963); through *C.A.*, **61,** 1884 (1964).
20. Barbieri, R., G. Tagliavini, and U. Belluco, *Ric. sci.*, **30,** 1963 (1960); through *C.A.*, **55,** 14159 (1961).
21. Barrall, E. M., II, and P. R. Ballinger, *J. Gas Chromatog.* **1,** 7 (1963); *C.A.*, **59,** 13739 (1963).
22. Belluco, U., G. Tagliavini, and R. Barbieri, *Ric. sci.*, **30,** 1675 (1960); through *C.A.*, **55,** 14175 (1961).

23. Berezniak, A., and T. Zimoch, *Biul. Wojskowej Akad. Tech. Prace Chem.*, **7,** 62 (1958); through *C.A.*, **53,** 3672 (1959).
24. Birks, L. S., *X-Ray Spectrochemical Analysis*, Interscience, N.Y., 1959.
25. Birks, L. S., E. J. Brooks, H. Friedman, and R. M. Roe, *Anal. Chem.*, **22,** 1258 (1950).
26. Blumenthal, A., *Mitt. Gebiete Lebensm. Hyg.*, **51,** 159 (1960); through *C.A.*, **55,** 2071 (1961).
27. Boettner, E. A., and F. C. Dallos, *J. Gas Chromatog.*, **1,** 190 (1965).
28. Bonelli, E. J., and H. Hartmann, *Anal. Chem.*, **35,** 1980 (1963).
29. Borup, R., and H. Levin, *Am. Soc. Testing Materials*, **Proc. 47,** 1010 (1947); *C.A.*, **42,** 7661 (1948).
30. Brandt, M., and R. H. Vandenberg, *Anal. Chem.*, **31,** 1921 (1959).
31. Brown, R., *Mitt. Lebens. Hyg.*, **41,** 90 (1950); through *C.A.*, **44,** 7521 (1950).
32. Brown, J. F., and R. J. Weir, *J. Sci. Instr.*, **33,** 222 (1956); *C.A.*, **50,** 13417 (1960).
33. Buell, B. E., *Am. Soc. Testing Materials*, *Spec. Techn. Publ.* **No. 269,** 157 (1960); *C.A.*, **55,** 4935 (1961).
34. Burriel, F., and L. Garcia Escolar, *Anales fis. y. quim.* (*Madrid*), **42,** 777; through *C.A.*, **41,** 4735 (1947).
35. Calingaert, G., *Chem. Rev.*, **2,** 43 (1925).
36. Galingaert, G., and C. M. Gambrill, *Ind. Eng. Chem., Anal. Ed.*, **11,** 324 (1939).
37. Calingaert, G., F. W. Lamb, H. L. Miller, and G. E. Noakes, *Anal. Chem.*, **22,** 1238 (1950).
38. Calingaert, G., H. Soroos, and H. Shapiro, U.S. Pat. 2,390,988 (to United States of America), Dec. 18, 1945; *C.A.*, **48,** 1025 (1946).
39. Campo, C. *Chim. Anal.* (*Paris*), **45,** 343 (1963).
40. Chaudhuri, J. C., *J. Indian Chem. Soc., Ind. & News Ed.*, **16,** 207 (1953); *C.A.*, **48,** 12616 (1954).
41. Cholak, J., *Arch. Environ. Health*, **8,** 222 (1964).
42. Cholak, J., *Arch. Environ. Health*, **8,** 314 (1964).
43. Choudhuri, B. K., C. R. Viswanathan, S. S. Vats, and A. R. Aiyar, *Defence Sci. J.*, **2,** 34 (1961); through *C.A.*, **56,** 628 (1962).
44. Cipriano, J. M., *Rev. Port. Quim.*, **5** (3), 129 (1963); through *C.A.*, **61,** 9337 (1964).
45. Dawson, H. J., Jr., *Anal. Chem.*, **35,** 542 (1963).
46. Dessy, R. E., W. Kitching, and T. Chivers, *J. Am. Chem. Soc.*, **88,** 453 (1966).
47. DeVries, J. E., A. Lauw-Zecha, and A. Pellecer, *Anal. Chem.*, **31,** 1995 (1959).
48. Dmitrievskii, V. S., *Khim. i Tekhnol. Topliv i Masel*, **3**(11), 59 (1958); through *C.A.*, **53,** 3671 (1959).
49. Dulinski, W., and Z. Köhsling, *Gaz. Woda i Tech. Sanit.*, **34,** 258 (1960); *C.A.*, **55,** 6747 (1961).
50. Emelyanov, B. V., Z. N. Shemyakina, and M. N. Khalyovin, USSR Pat. 127,263, March 25, 1960; *C.A.*, **54,** 18842 (1960).
51. Escolar, L. Garcia, and M. Paz Castro, *Inform. Quim. Anal.* (*Madrid*), **18** (3), 66 (1964); *C.A.*, **61,** 9337 (1964).
52. Farkas, M., and F. P. Csanyi, *Magy. Kem. Folyoirat.*, **69** (9), 407 (1963); through *C.A.*, **59,** 13742 (1963).
53. Ferro, A., and C. P. Galotto, *Ann. Chim.* (*Rome*), **45,** 1234 (1956); *C.A.*, **50,** 17401 (1956).
54. Flaschka, H., *Mikrochemie ver. Mikrochim. Acta*, **39,** 315 (1952); through *C.A.*, **46,** 9463 (1952).

55. Frediani, H. A., and L. A. Bass, *Oil Gas J.*, **39,** 51 (1940).
56. Friedman, M., U.S. Pat. 2,392,846 (to United States of America), Jan. 15, 1946. *C.A.*, **40,** 1308 (1946).
57. Gelius, R., and K. R. Preussner, *Chem. Tech. (Berlin)*, **15** (5), 290 (1963); *C.A.*, **59,** 13357 (1963).
58. German, J. F., and E. F. Degering, U.S. Pat. 2,867,572 (to United States of America), Jan. 6, 1959; *C.A.*, **54,** 1841 (1960).
59. Gernet, E. V., *Nauch. Raboty Khim. Lav. Gor'kovsk, Nauch.-Issledovatel Inst. Gigieny Truda i Professional. Boleznei, Sbornik*, **1957** (6), 5; through *C.A.*, **54,** 6391 (1960).
60. Gilbert, P. T., Jr., *Am. Soc. Testing Materials, Spec. Tech.* Publ. **No. 116,** 77 (1951); *C.A.*, **47,** 7766 (1953).
61. Gilman H., and J. Robinson, *J. Am. Chem. Soc.*, **50,** 1714 (1928).
62. Giustiniani, M., G. Faraglia, and R. Barbieri, *J. Chromotog.*, **15,** 207 (1964); through *C.A.*, **61,** 13916 (1964).
63. Goldshtein, A. L., N. P. Lapisova, and L. M. Shtifman, *Zhur. Analit. Khim.*, **17,** (1), 143 (1962); through *C.A.*, **57,** 5709 (1962).
64. Gordon, G. E., and R. A. Burdett, *Anal. Chem.*, **19,** 137 (1947).
65. Griffing, M. E., A. Rozek, L. J. Snyder, and S. R. Henderson, *Anal. Chem.*, **29,** 190 (1957).
66. Grünwald, A., *Erdöl Kohle*, **6,** 550 (1953); *C.A.*, **48,** 986 (1954).
67. Grüttner, G., and E. Krause, *Chem. Ber.*, **49,** 1125 (1916).
68. Hansen, K. A., T. D. Parks, and L. Lykken, *Anal. Chem.*, **22,** 1232 (1950).
69. Hasegawa, K., M. Kajikawa, and N. Okamoto, *Bunseki Kagaku*, **14** (8), 717 (1965); *C.A.*, **63,** 11207 (1965).
70. Hein, F., A. Klein, and H. J. Mesée, *Z. Anal. Chem.*, **115,** 177 (1939).
71. Hendee, C. F., and S. Fine, Ger. Pat. 1,074,293 (to N.V. Phillips Gloeilampenfabriecken), Jan. 28, 1960; *C.A.*, **55,** 17125 (1961).
72. Henderson, S. R., and L. J. Snyder, *Anal. Chem.*, **33,** 1172 (1961).
73. Henneberg, H., and G. Schomburg, *Z. Anal. Chem.*, **215,** 424 (1965).
74. Hermann, R., and C. T. J. Alkemade, *Chemical Analysis by Flame Photometry*, Second Revised Edition, translated by P. T. Gilbert, Jr., Interscience, N.Y., 1963.
75. Howard, H. E., W. C. Ferguson, and L. R. Snyder, *Anal. Chem.*, **32,** 1814 (1960).
76. Hubis, W., and R. O. Clark, *Anal. Chem.*, **27,** 1009 (1955).
77. Hughes, H. K., and F. P. Hochgesang, *Anal. Chem.*, **22,** 1248 (1950).
78. Ingham, R. K., S. D. Rosenberg, and H. Gilman, *Chem. Rev.*, **60,** 459 (1960).
79. Irving, H., and J. J. Cox, *J. Chem. Soc.*, **1961,** 1470.
80. Jezl, J. L., and I. W. Mills, U.S. Pat. 2,745,793 (to Sun Oil Co.), May 15, 1956; *C.A.*, **50,** 13423 (1956).
81. Jordan, J. H., Jr., *Petroleum Refiner*, **32,** 139 (1953); *C.A.*, **50,** 8186 (1956).
82. Kannuna, M. M., *J. Inst. Petrol.*, **43,** 198 (1957); *C.A.*, **51,** 14247 (1957).
83. Kellog, H. W., E. F. Mahlke, and J. T. Jones, *Petroleum Processing*, **7,** 1430 (1952).
84. Kharasch, M. S., U.S. Pat. 2,453,138 (to United States of America) (1948); *C.A.*, **44,** 2746 (1950); U.S. Pat. 2,504,134 (to United States of America), Apr. 18, 1950; *C.A.*, **44,** 5581 (1950).
85. Khaikin, M. O., *Zh. Prikl. Khim.*, **14,** 652 (1941); through *C.A.*, **36,** 3663 (1942).
86. Khrustaleva, V. A., *Metody Opred. Vredn. Veshehestv v Vozdukhe, Moscow Sb.*, **1961,** 105; through *C.A.*, **58,** 8403 (1963).
87. Khrustaleva, V. A., *Gigiena i Sanit.*, **25,** 57 (1960); through *C.A.*, **54,** 25385 (1960).

88. Kolb, B., G. Kemmner, F. H. Schleser, and E. Wiedeking, *Z. Anal. Chem.*, **221,** 166 (1966); through *C.A.*, **66,** 12664v (1967).
89. Korshunov, I. A., and N. I. Malyugina, *Zh. Obshch. Khim.*, **31,** 1062 (1961); *C.A.*, **55,** 23320 (1961).
90. Koyama, K., Y. Taguchi, and S. Eguchi, *Bunzeki Kagaku*, **12,** 435 (1963); *C.A.*, **59,** 7281 (1963).
91. Kramer, K., *Erdöl Kohle*, **19,** 182 (1966); *C.A.*, **64,** 15640 (1966).
92. Lamb, F. W., L. M. Niebylski, and E. W. Kiefer, *Anal. Chem.*, **27,** 129 (1955); *C. A.*, **49,** 4977 (1955).
93. Lapisova, N. P., and A. L. Goldshtein, *Zhur. Analit. Khim.*, **16,** 508 (1961); through *C.A.*, **56,** 8012 (1962).
94. Levine, W. S., *Petroleum Engr.*, **25,** C37-44 (1953); *C.A.*, **47,** 7766 (1953).
95. Levine, W. S., and A. H. Okamoto, *Anal. Chem.*, **23,** 1293 (1951); *C.A.*, **46,** 1237 (1952).
96. Liebhafsky, H. A., *Anal. Chem.*, **24,** 16 (1952).
97. Liebhafsky, H. A., and E. H. Winslow, *ASTM Bull.* **167,** 67 (1950); *C.A.*, **44,** 11073 (1950).
98. Linne, W., and H. D. Wulfken, *Erdöl Kohle*, **10,** 757 (1957); *C.A.*, **52,** 4157 (1958).
99. Lorone, D., and G. Paul, *Rev. Inst. Franc. Petrole Am. Combust. Liquides*, **17,** 830 (1962); through *C.A.*, **57,** 11452 (1962).
100. Lovelock, J. E., and N. L. Gregory, *Gas Chromatog., Intern. Symp.*, **3,** 219 (1961, publ. 1962); through *C.A.*, **58,** 3865 (1963).
101. Lovelock, J. E., and A. Zlatkis, *Anal. Chem.*, **33,** 1958 (1961).
102. Mahr, C., and G. Stork, *Z. Anal. Chem.*, **221,** 1 (1966); through *C.A.*, **65,** 16062c (1966).
103. Manns, T. J., N. U. Reschovsky, and A. J. Certa, *Anal. Chem.*, **24,** 908 (1952).
104. Meine, W., *Erdöl Kohle*, **8,** 711 (1955); *C.A.*, **50,** 8186 (1956).
105. Milner, O. I., and G. F. Shipmen, *Anal. Chem.*, **26,** 1222 (1954).
106. Moss, R., and E. V. Browett, *Analyst*, **91,** 428 (1966).
107. Nagasawa, S., and M. Funakubo, *J. Chem. Soc. Japan*, Pure Chem. Sect., **69,** 16 (1948); through *C.A.*, **46,** 8568 (1952).
108. Nagypataki, G., and Z. Tamasi, *Magy. Kem. Lapja*, **17,** 140 (1962); through *C.A.*, **57,** 2493 (1962).
109. Neef, F. E., Jr., U.S. Pat. 2,368,261, Jan. 30, 1945; *C.A.*, **39,** 4470 (1945).
110. Nightingale, E. R., G. W. Wilcox, and A. D. Zielinski, *Anal. Chem.*, **32,** 625 (1960).
111. Obraztsova, E. N., and V. A. Novikov, USSR Pat. 154,431, July 24, 1963; through *C.A.*, **60,** 1124 (1964).
112. Offutt, E. B., and L. V. Sorg, *Anal. Chem.*, **22,** 1234 (1950).
113. Offutt, E. B., and L. V. Sorg, U.S. Pat. 2,773,020 (to Standard Oil Co. of Indiana), Dec. 4, 1956; *C.A.*, **51,** 3984 (1957).
114. Offutt, E. B., and L. V. Sorg, U.S. Pat. 2,773,021 (to Standard Oil Co. of Indiana), Dec. 4, 1956; *C.A.*, **51,** 3984 (1957).
115. Oikawa, H., and T. Okuda, *Tanken*, **1960,** 74; through *C.A.*, **55,** 15884 (1961).
116. Oikawa, H., and T. Okuda, *Kogyo Kagaku Zasshi*, **64,** 34 (1961); through *C.A.*, **57,** 2853 (1962).
117. Okada, T., T. Ueda, and T. Kohzuma, *Bunko Kenkyu*, **4** (1), 30 (1956); through *C.A.*, **53,** 6586 (1959).
118. Parker, W. W., and R. L. Hudson, *Anal. Chem.*, **35,** 1334 (1963).

119. Parker, W. W., G. Z. Smith, and R. L. Hudson, *Anal. Chem.*, **33,** 1170 (1961).
120. Pedinelli, M., *Chim. Ind.* (*Milan*), **44** (6), 651 (1962); through *C.A.*, **60,** 14303 (1964).
121. Penkov, N. I., M. N. Khalyavin, B. V. Emelyanov, Z. N. Shemyakina, V. I. Tsvetkov, G. V. Kudryavtseva, and V. N. Shvarov, USSR Pat. 154,382, July 24, 1963; through *C.A.*, **60,** 13786 (1964).
122. Pilloni, G., and G. Plazzogna, *Ric. sci., Rend. Sez. A***4** (1), 27 (1964); through *C.A.*, **61,** 1278 (1964).
123. Plazzogna, G., and G. Pilloni, *Anal. Chim. Acta*, **37,** 260 (1967); through *C.A.*, **66,** 82188s (1967).
124. Ptashinskii, I. A., and M. K. Frolova, *Vsesoyuz. Nauch.-Issledovatel. Inst. po Pererabotke Nefti i Gaza i Poluchen. Zhidkogo Topliva*, **1957** (6), 181; through *C.A.*, **53,** 15538 (1959).
125. Rasmussen, S., U.S. Pat. 2,448,235 (to American Gas Machine Co.), Aug. 21, 1948; *C.A.*, **44,** 834 (1950).
126. Rather, J. B., Jr., *Materials Research and Standards*, **2,** 494 (1962).
127. Riccoboni, L., *Gazz. chim. ital.*, **72,** 47 (1942); through *C.A.*, **37,** 574 (1943).
128. Robinson, J. W., *Anal. Chim. Acta*, **24,** 451 (1961).
129. Rudnevskii, N. K., and N. N. Vyshinskii, *Izvest. Akad. Nauk. SSSR, Ser. Fiz.*, **23,** 1228 (1959); through *C.A.*, **54,** 6392 (1960).
130. Russ, J. J., and W. Reeder, *Anal. Chem.*, **29,** 1331 (1957).
131. Ryabova, A. S., S. R. Belova, V. I. Sharapov, *Neftepererabotka i Neftekhim., Nauchn-Tekhn. Sb.*, **1963** (2), 11; through *C.A.*, **59,** 7281 (1963).
132. Rysselberg, J. van, and R. Leysen, *Nature*, **189,** 478 (1961).
133. Saori, H., *Shoseki Giho*, **2,** 182 (1958); through *C.A.*, **53,** 15538 (1959).
134. Saunders, B. C., and G. J. Stacey, *J. Chem. Soc.*, **1949,** 919.
135. Schafer, H., *Z. Anal. Chem.*, **180,** 15 (1961); through *C.A.*, **55,** 17374 (1961).
136. Shapiro, A., and D. A. H. Olsen, U.S. Pat. 2,969,320 (to Socony Mobil Oil Co.), Jan. 24, 1961; *C.A.*, **55,** 14898 (1961).
137. Siniramed, C., and F. Renzanigo, *Riv. Combust.*, **19** (7–8), 351 (1965); *C.A.*, **64,** 4828 (1966).
138. Skalicka, B., and M. Cejka, *Ropa Uhlie* **8**, 246, (1966); through *C.A.*, **66,** 31814k (1967).
139. Smelik, J., *J. prakt. Chem.*, **5,** 9 (1954); *C.A.*, **48,** 6679 (1954).
140. Smith, V. A., W. E. Delaney, W. J. Tancig, and J. C. Bailie, *Anal. Chem.*, **22,** 1230 (1950).
141. Snyder, L. J., W. R. Barnes, and J. V. Tokos, *Anal. Chem.*, **20,** 772 (1948).
142. Snyder, L. J., and S. R. Henderson, *Anal. Chem.*, **33,** 1175 (1961).
143. Snyder, L. J., and S. R. Henderson, U.S. Pat. 3,071,446 (to Ethyl Corp.), Jan. 1, 1963; *C.A.*, **58,** 6200 (1963).
144. Soroos, H., and H. Shapiro, *Development of a Chemical Process for Deleading Gasoline*, OSRD Report 3018; PB Report 32222, Dec., 1943; *Bib. of Sci. and Ind. Repts.*, **2,** 801 (1946).
145. Soulages, N. C., *Anal. Chem.*, **38,** 28 (1966).
146. Spialter, L., and M. Ballester, *Anal. Chem.*, **34,** 1183 (1962).
147. Strange, J. P., K. E. Ball, and D. O. Barnes, *J. Air Pollution Control Assoc.*, **10,** 423 (1960).
148. Sullivan, M. V., and H. Friedman, *Ind. Eng. Chem., Anal. Ed.*, **18,** 304 (1946).
149. Swanson, B. W., and P. H. Daniels, *J. Inst. Petroleum*, **39,** 487 (1953).
150. Tagliavini, G., *Chim. Ind.* (*Milan*), **39,** 902 (1957); through *C.A.*, **52,** 7672 (1958).

151. Tagliavini, G., U. Belluco, and L. Riccoboni, *Ric. sci.*, *Rend. Sez. A***1** (2), 338 (1961); through *C.A.*, **57,** 5306 (1963).
152. Tarjan, D., and E. Knezevic, *Nafta* (*Yugoslavia*), **5,** 14 (1953); through *C.A.*, **48,** 7287 (1954).
153. Titov, N. G., *Izvest. Akad. Nauk SSR Otdel. Tekhnick. Nauk*, **1944,** 690.
154. Uvarova, E. I., and N. M. Vanyarkina, *Zavodskaya Lab.*, **26,** 1907 (1960); through *C.A.*, **56,** 10443 (1962).
155. Walsh, A., *Spectrochim. Acta*, **7,** 108 (1955).
156. Wendlandt, W. W., and H. G. Hecht, *Reflectance Spectroscopy*, Interscience, N.Y., 1966.
157. West, P. W., and C. H. Hale, *Petroleum Refiner*, **29** (6), 109 (1950).
158. Wilkinson, W. R., *Thermal Deleading of Gasoline*, PB Report 107,044, Aug., 1944; *Bib. of Tech. Repts.*, **18,** No. 1 (1952).
159. Zemany, P. D., E. H. Winslow, G. S. Poellmitz, and H. A. Liebhafsky, *Anal. Chem.*, **21,** 493 (1949); *C.A.*, **43,** 5693 (1949).
160. Zemskov, I. F., *Khim. Prom.*, **1961,** 290; through *C.A.*, **55,** 18206 (1961).
161. Zemskov, I. F., and M. N. Khalyavin, *Khim. Prom.*, **1958,** 500; through *C.A.*, **54,** 7020 (1960).
162. Zemskov, I. F., and M. N. Khalyavin, *Khim. Prom.*, **1961,** 135; through *C.A.*, **55,** 19356 (1961).
163. Zemskov, I. F., and G. I. Sidelnikova, *Tr. po Khim. i Khim. Tekhnol.*, **1960** (3), 20; through *C.A.*, **56,** 9448 (1962).
164. Zuliani, G., *Metano* (*Padua*), **14,** 31 (1960); through *C.A.*, **62,** 51112 (1965).
165. Zuliani, G., and P. Dal Pozzo, *Metano* (*Padua*), **14,** 111 (1960); through *C.A.*, **62,** 5112 (1965).
166. Zuliani, G., and P. Dal Pozzo, *Metano* (*Padua*), **14,** 207 (1960); through *C.A.*, **62,** 5112 (1965).

XVII Uses of organolead compounds

1. COMMERCIAL ANTIKNOCK AGENTS

A. Tetraethyllead (TEL)

The principal commercial application of organolead compounds is their use as antiknock agents, a discovery due to Midgley and Boyd (240), following World War I. Antiknock agents are additives for aviation and motor gasolines which are incorporated in order to suppress the reactions that produce the undesirable phenomenon of knock in the internal combustion engine. Knock frequently limits the power output of an engine, and if it is sufficiently severe, it may damage the engine. It may also preclude the use of higher compression ratios, which are more efficient thermodynamically and thus allow greater power output or greater economy, as desired. The phenomenon of engine knock, as well as a description of the engine cycle in the internal combustion engine, are described in the standard reference work by Lewis and von Elbe (227). It should be recognized that other gasoline additives are frequently used together with TEL to accomplish various non-antiknock purposes; engine fuel additives in general have been discussed recently by Barusch and Macpherson (45).

Although the mechanism of knock is relatively little understood, the technology of its control is well developed. Many alkyllead compounds have been proposed as antiknocks; however, the chief antiknock in general use in gasoline-type fuels for the past 45 years has been tetraethyllead. Either this compound or a similar alkyllead compound or mixture is used in over 99% of all gasoline consumed. It is always combined with corrective agents, or "scavengers"—of which many have been used experimentally, but principally ethylene bromide and ethylene chloride are used commercially—and blended into gasoline at levels up to 4 ml/gal. Special halogen-free fuel compositions containing TEL have been patented by Kerley and Felt (197), but are not being used commercially. The theory of scavenging has been discussed recently by Buck and Ryason (65). Average U.S. TEL contents in gasoline in October, 1965, were 2.88 g Pb/gal in premium gasoline and 2.35 g Pb/gal in regular gasoline, while in October, 1966, premium gasoline averaged 100.0 research octane number and regular gasoline 94.1 research octane number (10,14).

Only a few organic chemicals outrank TEL in their value of production, namely, nylon fiber, GSR rubber, polyethylene, and ethylene monomer. The total production of all lead alkyls in the U.S. in 1962 was placed at 513 MM lb (9), while in 1963 production was estimated at 540 MM lb (13). TEL production in 1964 was between 540 MM lb and 587 MM lb (9). A forecast has been made for 1969 at 720 MM lb (13), and a slightly older projection is available for 1972 at 599 MM lb (304); the latter older forecast is by now in error. Such forecasts are obviously not too reliable, although U.S. demand has been estimated to have a future growth of about 3% annually for the next five years (15). The annual rate of growth for Western Europe has been assessed at 6.8%, for the Middle East at 8.4%, and for Japan at the impressive rate of 10.2%. The present production figures for lead alkyls outside the U.S. are not available. Current U.S. plant capacity for lead alkyls is more than ample for years to come; it was estimated in 1964 to be at least 850 MM lb (13). A history of the development of the TEL industry as a product of American chemical research has been written by Nickerson (251).

The commercial production of TEL was undertaken for a short time by the Standard Oil Company of New Jersey in 1922, and gasoline containing TEL as an antiknock agent was initially sold by General Motors Development Company at Dayton, Ohio, in February, 1923. In this same year, commercial production was begun by E. I. du Pont de Nemours & Company, at Deepwater, Delaware, for the then Ethyl Gasoline Corporation account. By 1945, Ethyl Corporation had entered into manufacture at Baton Rouge, La., and in 1948, du Pont began to sell TEL for its own account. Others have since followed. At present, lead alkyl antiknock agents are manufactured in several locations in the U.S. by Ethyl Corporation, du Pont, Houston Chemical Co., and Nalco Chemical Co. Manufacture in the remainder of the Free World is on a significantly smaller scale, with plants operating in Canada, Mexico, Greece, England, France, Italy, and a new plant in Germany (5,7). No plants exist at present in South America or Asia. Nalco Chemical Co. at present manufactures only tetramethyllead, using a new Grignard electrolysis method, represented by the equation:

$$2CH_3MgCl + 2CH_3Cl + Pb \rightarrow (CH_3)_4Pb + 2MgCl_2$$

TEL can also be prepared by this process, according to the patent claims.

TEL is commonly manufactured in either batch or continuous reactors by the reaction of ethyl chloride with lead–sodium alloy, according to the equation:

$$4NaPb + 4C_2H_5Cl \rightarrow (C_2H_5)_4Pb + 3Pb + 4NaCl$$

The process has undergone appreciable improvement in the past 40 years, but the basic reaction is still the same. The reaction is complicated, and commercial yields are in the range of 85–90% based on sodium. For a thorough discussion of the preparation and use of TEL, see the papers by Edgar (115,116), the review of preparative methods by Shapiro (314), and the article on "Organolead Compounds," by Shapiro and Frey (316) in the *Encyclopedia of Chemical Technology*.

B. Tetramethyllead (TML), mixed lead alkyls, and other antiknocks

Hundreds of compounds have been tested as antiknock agents in competition with TEL, and many papers and patents (both important and commercially irrelevant) have been published. Until recently, no new compound could match TEL in the economical production of octane numbers in gasolines. However, in 1960 Standard Oil of California and Mobil Oil Company inaugurated the marketing of gasoline containing tetramethyllead as an antiknock (16,17,69,71). The Ethyl Corporation began the manufacture of TML on a commercial scale early that year (18,20), followed closely by a second U.S. supplier of TEL, E. I. du Pont de Nemours & Co. Shortly thereafter, both corporations started to supply mixed methylethyllead alkyl compounds to the petroleum companies (55,70). Various methylethyllead mixtures have been patented as antiknocks (113,125). These mixtures are manufactured by a redistribution reaction between TML and TEL in the presence of a suitable catalyst, for example, boron trifluoride etherate, methylaluminum sesquichloride, activated clay, zeolites, and activated alumina (28,89, 331,344). (See Chapter VI for a discussion of this method of preparation.)

As previously mentioned, TML is manufactured by a new Grignard electrolysis process by the Nalco Chemical Co. This process, described in some detail by Guccione (153), Bott (54), Shapiro and Frey (316), and Seyferth and King (313), depends on the electrolysis of lead pellets as the anode in a membrane-separated, refrigerant-cooled electrolytic cell, with the reactor walls as the cathode, and the electrolyte consisting of separately prepared methylmagnesium chloride dissolved in a high-boiling ether.

Except for the manufacture by Nalco, TML is manufactured by a batch NaPb alloy process based on methyl chloride:

$$4NaPb + 4CH_3Cl \rightarrow (CH_3)_4Pb + 3Pb + 4NaCl$$

Approximately 5 million pounds of TML and the methylethyllead alkyl mixes were consumed in 1960, whereas by 1962 the production of TML had risen to 18 million pounds (12,21). Current production levels are not available.

It has been known for many years that the greater volatility of methyl-containing lead compounds as compared with TEL is beneficial in internal combustion engines with induction systems that maldistribute the fuel from cylinder to cylinder. Indeed, for this reason, the preparation of mixed methylethyllead compounds by the redistribution reaction was extensively studied two decades ago (73). Also, in recent years it has become recognized that TML, although its intrinsic antiknock effect is less than that of TEL in many gasolines (146), is particularly effective in gasolines containing large amounts of aromatic compounds (126). This fact is especially important in California, where the olefin content of gasolines is limited by state law in order to help alleviate the smog problem. In general, the comparative effectiveness of TML and TML-containing mixtures increases with higher lead content, higher octane number, lower sulfur content, and lower TEL susceptibility; thus, there are made available a variety of antiknock agents that can be tailored for specific gasoline stocks. TML and mixtures containing methyl groups increase "Motor" method ratings more than "Research" method ratings, and they exert their greatest measurable differential effect in automobile road ratings at low speeds, where it is most important (22,146–148,209,243,265,279,288,289, 329). Furthermore, the occurrence of the phenomenon called "rumble," like knock a deterrent to the use of high compression ratios, can be greatly reduced with highly aromatic gasolines by the addition of both TML and a phosphorus compound to the gasoline (266).

Because of its volatility, TML has been shown to be useful in increasing the antiknock quality of LPG (liquefied petroleum gases) (133). This finding has considerable commercial implication for the future.

Recently, lead antiknocks have been shown to be supplemented by new compounds that are synergistic with organolead compounds in increasing octane numbers. One of these compounds is an organometallic that has been introduced commercially by Ethyl Corporation, methylcyclopentadienylmanganese tricarbonyl, or "AK-33X" (61,62). Mixtures of TEL and AK-33X have been designated commercially as Motor 33 Mix (23). Other organic manganese compounds and nickel carbonyls and nitrosyls have been shown to be synergistic when added with lead alkyls (58,60,90). Ferrocene and iron pentacarbonyl have also been described as supplements (59,155,156), as has hexamethylbenzene molybdenum tricarbonyl (25).

Nonmetallic supplement compounds have also been researched extensively in recent years. This development is exemplified by *tert*-butyl acetate and other esters and fatty acids (26,28,152,250,290,291). Since the evaluation of these materials requires taking into consideration the base fuel and the nature of the other antiknocks and synergists present, it appears that specialized areas of usefulness may eventually be found for several potentially available nonlead antiknock agents (232). The use of many acids and amines has been patented (114). Monomethylaniline has also been suggested as a supplement for alkyllead antiknock agents by Laznowski (224). Ethyl alcohol has been suggested as an extender for gasoline in numerous papers, but it does not act as a supplement for lead antiknocks.

To explain the action of *tert*-butyl acetate, it has been postulated that it and TEL decompose to acetic acid and lead oxide, and these intermediates react to give lead acetate (182). Later in the combustion cycle the lead acetate decomposes to give active lead oxide. Organic acids are postulated to behave in the same manner.

Many additional organolead compounds have some degree of antiknock activity. Tetravinyllead has recently been patented as a gasoline additive (72,124), as have redistributed mixtures of vinylalkyllead compounds (127), bis(trimethyllead) sulfide (37,42), trimethyllead methylmercaptide (38,39,41), trimethyllead methylthioglycolate (39), trialkyllead derivatives of thioglycolamides (40), trialkyllead dialkyl phosphates (161,162), and trialkyllead selenides (286,287).

A short review of new developments in organometallic compounds with special reference to organolead antiknock agents was published in 1963 by Gelius (141), and another appeared in 1965 by Nottes and Cordes (256), with greater stress on new cyclopentadienyl and aromatic metal antiknocks, including their synergism with organolead antiknocks.

C. Mechanism of antiknock action

Since the time TEL was first introduced as an antiknock agent for internal combustion engines, considerable effort has been devoted to an understanding of the mechanism of its antiknock action. A review of the various theories proposed to explain this action was given by Beatty and Edgar (47), and more recently by Chamberlain, Hoare, and Walsh (82), Ross and Rifkin (303), Lewis and von Elbe (227), and Barusch and Macpherson (45). Unfortunately, direct proof of the mechanism is lacking, and quite possibly, both homogeneous and heterogeneous mechanisms may be operative. In the light of accumulated experimental data, however, certain deductions are possible.

It has been established that TEL must decompose to be effective (77,295), and also that decomposition is likely in the environmental conditions in an engine (295). Walsh and his co-workers (83,84,110,346) have evidence which supports the belief that the effective agent after decomposition is a fog of solid lead monoxide particles. In contrast, however, Norrish and his collaborators (74,75,120,121,255) indicate from their studies that the agent is gaseous lead oxide. In both cases it has been most generally postulated that the action is one of inhibition of active oxygen atoms which, if not controlled, would enhance a chain reaction mechanism resulting in knock.

The concept of metal oxide as the (homogeneous) effective agent is supported by Egerton (117,118), who has indicated that the effective metals are those which can exist in a number of different oxidation states. Agnew (2) has also supported the homogeneous theory and Rifkin (293) cites evidence which indicates that lead oxide particles varying in size between 15 and 500 Å do exist in an engine prior to knock, based on measurements of the particles which emerge in the exhaust of a "motored" engine. This latter study indicates that particulate yellow lead monoxide deactivates intermediates such as hydroperoxides which would otherwise lead to chain branching reactions and knock. Cartlidge and Tipper (79) came to very much the same conclusion. On a theoretical basis, Wright (359) has shown that particles of the dimensions found in the "motored" engine exhaust would indeed be formed in the engine, and Oosterhoff (259) had concluded previously that particles of such dimensions would be appropriate for a knock inhibitor to be effective. Finally, it has been shown by Downs et al. (108) that there is far more scattering of light in an engine running with a lead antiknock than in one running without an antiknock. This also suggests that solid particles must be formed in the engine when TEL is introduced into the gasoline.

Some investigations have been made on flame speed and cool flame limits and intensity, using TEL as an additive. Like many other investigators, Curry (95) believed knock to be caused by a rapid rate of pressure rise associated with an accelerating flame front, and he showed that the addition of TEL resulted in a distinct lowering of the flame speed.

Downs et al. (109) have shown that TEL reduces the cool flame intensity, but has no effect on the cool flame limit. The antiknock action is not due to changes in the spontaneous ignition temperature of the fuel (332). Thus, TEL appears to exert its major antiknock effect in the reactions occurring during the passage of the cool flame and prior to the hot flame. Some other antiknocks, such as iron carbonyl, for example, act on the reactions prior to the passage of the cool flame, since they raise the cool flame limit. Other cool flame studies on lead oxide-coated vessels have been made by Chernesky and Bardwell (43,86), and on lead oxide-coated engines by Walsh and his co-workers (84). The chemical reactions which occur in an engine prior to the occurrence of knock have been examined systematically by Pipenberg et al. (277) using several kinds of hydrocarbons and TEL as an antiknock.

Salooja (305) has studied the combustion of hydrocarbons containing TEL in glass. The lead compounds formed were mainly orthorhombic and tetragonal polymorphs of PbO. Combustion inhibition appeared to be due to both forms, but the tetragonal form has the greater effect, destroying peroxides, decreasing the concentration of carbonyl compounds, etc. Mechanisms were proposed to account for the effects.

The reactions leading to knock are extremely rapid, as discussed by Gluckstein et al. (143). Thus, the difficulties in obtaining a more precise analysis in an engine relate to the establishment of physical and chemical entities at concentrations of about one part per quarter million and in a split millisecond of time. Obviously, improved high-speed techniques would facilitate further research in this field. For a recent discussion of the kinetics and analysis of very fast chemical reactions, see the 1965 review by Norrish (255).

D. Oxidation-related uses outside of the gasoline engine

There are a number of proposed uses for organolead compounds that are related to their oxidation-inhibiting ability. Indeed, as indicated in the preceding discussion, it has been shown that TEL inhibits many oxidation reactions outside of the gasoline engine at relatively low temperatures. For example, TEL inhibits the oxidation of tetraborane (234), the oxidation of butane (43), the shock-induced oxidation of acetylene (157), and it reduces the propagation velocity of laminar flames (223).

Some of these oxidation-inhibiting effects may have practical implications. For example, Howard (180) has shown that the addition of TEL increases the flame-induction time during the partial combustion of methane to acetylene. The flame-induction period was almost doubled by the addition of only 0.07 vol % TEL. Similarly, Berl (50) has claimed that TEL in an amount of about 0.5 vol % of the reactants serves as an antidetonating agent in the oxidation of ethylene to ethylene oxide with gaseous oxygen.

About 1% of TEL in a mixture of tetranitromethane and 10–20% benzene lowers the burning rate to the point where this mixture can be employed successfully as a propellant, according to a patent by Hannum (160). On the other hand, Wyler (361) claims that a mixture of TEL and tetranitromethane constitutes an explosive of unusual strength and high rate of detonation. These two results are not contradictory: in the former instance, TEL apparently exerts its characteristic antioxidant action; while in the latter case, the TEL present in larger amount acts as a fuel with a high heat of combustion per unit volume.

Mortar propellants of improved combustion regularity have been patented, based on the incorporation of 1–4% of an organic lead compound, 2–12% of *p*-phenylenediamine, and ≥25% of nitroglycerine (184).

Two British patents (122,123) claim that the combustion properties of a diesel fuel oil are improved by the addition of a mixture of a small amount of a Group IV metal compound, such as a lead alkyl, and a larger amount of a detergent; both horsepower increases and fuel consumption decreases are claimed. Also, in a recent patent Riegert and Siganos (292) claim that the same type of mixture, for example, TEL plus calcium sulfonate, reduces smoke emission from a diesel engine. Alkyl lead compounds have also been patented by Hesselberg (172) as additives for propane-burning diesel engines.

In an interesting patent on the promotion of oxidation, Hochwalt (174) has patented the use of TEL as a co-catalyst with metal oxides or salts at temperatures above 140° in the liquid-phase oxidation of toluene and other alkylbenzene compounds to acids. Dialkylbenzenes produce dicarboxylic acids in this process. Supposedly, lead metal is formed by the thermal decomposition of the TEL. Moreover, it has been patented by Scott (312) that fuel mixtures containing as little as 16% TEL are hypergolic with either white or red fuming nitric acid.

The use of 0.05–1 wt % of either tetraethyllead or tetraphenyllead in combination with an organic bromide has been patented as a fire-retardent additive for plastics (183).

Finally, it should be mentioned that a novel space-age use for tetramethyllead has recently been developed for research in the meteorological field. Hord and Tolefson (176) report the use of tetramethyllead vapor reaction with atmospheric oxygen at altitudes above 90 km to form a chemiluminescent gas trail which can be used for the measurement of upper atmosphere wind velocities at night.

2. OTHER USES OF ORGANOLEAD COMPOUNDS

The various present and potential uses of the organolead compounds outside of the antiknock field have been reviewed briefly by Harwood (163), Azerbaev and Kochkin (32), Willemsens (351), Klema (200), and van der Kerk (196).

As compared with the application of organolead compounds as antiknock agents, the other commercial uses of these compounds are quite limited in importance at the present time. Undoubtedly, new uses will be developed, particularly due to the continuing interest of the International Lead Zinc Research Organization (ILZRO) in this field. Currently, this organization is sponsoring research by several institutions on the preparation and utilization of lead compounds, and also is engaged in the dissemination of information on this research work, as summarized in the ILZRO Research Digest (19). On May 2, 1967, ILZRO and the U.S. Army Natick Laboratories sponsored an international conference on organolead chemistry, with emphasis on the potential uses of these compounds as preservative agents for cotton, wood, and paints, lubricant additives, polyurethan foam catalysts, and molluscicides; the conference proceedings will probably be available from ILZRO. The most significant types of application research being done are in the fields of catalysis and biological effects, as discussed below. The primary sponsored fundamental and application research is under way at the Organisch Chemisch Instituut T.N.O. at Utrecht, The Netherlands (19,353,354). Other examples of sponsored work are fundamental work at the U.S. Army Natick Laboratories, Natick, Massachusetts (19,170); organolead antiwear compounds at Ethyl Corporation, Detroit, Michigan (19); the application of organolead compounds to impart special properties to cotton at the U.S. Department of Agriculture, Cotton Finishes Laboratory (19); the use of organolead compounds as wood impregnants against marine borers at the British Columbia Research Council (3,8,19); and antifouling paint formulations at Battelle Memorial Institute (19).

Under ILZRO sponsorship, the T.N.O. group and the U.S. Army Natick Laboratories distribute gratis samples of lead compounds for testing by many other research organizations. In 1965 and 1966, the T.N.O. group distributed about 1100 samples of about 100 different organolead compounds to approximately 200 institutions (196).

The second largest commercial use for organolead compounds is as an alkylating agent in the manufacture of mercurial fungicides, which is described below. Other uses of organolead compounds today take place primarily on the laboratory scale. As implied by the reference to ILZRO work above, however, a great many additional uses for organolead compounds have been suggested, and interest is increasing in these potential uses. Sometimes these suggestions, both in the technical literature and the patent literature, have been based simply by analogy to the uses for which the compounds of the other Group IVb metals have been suggested or employed.

A. Manufacture of mercury fungicides

The largest commercial use for organolead compounds outside of the antiknock field is relatively small: it is for the production of organomercury fungicides by alkylation. The manufacture of mercurials from organolead compounds is based on a method patented by Kharasch (198,199) in 1930. The preparation of alkyl mercurial compounds with TEL was reviewed in 1957 by Whelen (347) and is the subject of a number of patents (111,119,134,175,244).

The mercurial alkylation reaction is a general one, and is illustrated by the following representative equation:

$$2HgCl_2 + Et_4Pb \rightarrow Et_2PbCl_2 + 2EtHgCl$$

In commercial practice, the mercuric chloride is generally ball-milled with talc as an inert ingredient, while the TEL is added gradually. The product mass has an ethylmercuric chloride content of 2%, and is usually used in this form. Du Pont is a principal manufacturer of ethylmercuric compounds according to this method.

Other typical mercury compounds used as fungicides are ethylmercuric phosphate, ethylmercuric sulfate, phenylmercuric acetate, and mercury compounds of various substituted ureas and phenols. They are used mainly as fungicides in the disinfection of seeds and grains, such as wheat, corn, and cotton.

B. Other uses as alkylating agents

No commercial use is made of organolead compounds as alkylating agents, except for the preparation of mercury compounds. However, it has been well known for many years that TEL may be used as an alkylating agent, since many metal salts will cleave this tetraalkyllead compound. The general subject of the reactions of R_4Pb compounds with metal salts is discussed in Chapter VI, and the use of metal salts to delead gasolines is described in Chapter XVI.

C. Organolead polymerization catalysts

It has been known for many years that organolead compounds, like many other organometallic compounds, are thermally unstable and will pyrolyze to free radicals and metal. (See Chapters VI and VII for discussion.) The free radicals produced by the thermal decomposition of organolead compounds can initiate chain reactions, and therefore lead compounds have long been regarded as potentially useful catalysts for promoting a variety of reactions. In 1930 Taylor and Jones (333) demonstrated that the dissociation of TEL at 250°–350° causes ethylene gas to polymerize to a high-boiling liquid.

Little, if any, industrial use is made today of the catalytic polymerization properties of organolead compounds. However, as polymerization technology matures, lead compounds may well find some specialized commercial future as polymerization catalysts. For example, recent tests in polyurethan foams have been very promising, as described below.

Numerous reviews of polymerization using organometallic catalysts have been published in the past several years, including the excellent book by Gaylord and Mark (140) on linear and stereoregular addition polymers. Other recent reviews include those by Ziegler (368), Furukawa (137), Lefebvre (225) on stereospecific polymerization, Furukawa and Tsuruta (138) on vinyl polymerization, Nowakowska (258) on ethylene copolymerization, and Tsuruta (337) on vinyl polymerization.

The general theory of organometallic catalysts for polymerization is discussed in the reviews cited above. In addition to free radical initiation, complex coordinated catalyst structures may be involved for stereospecific polymerization, as well as active intermediate compounds such as the TEL–$TiCl_4$ reaction product, ethyltitanium trichloride. Thus, TEL may exert a catalytic influence on polymerization in more than one way.

Specific references to the theory of catalysis involving organolead compounds have been made recently by Huff and Perry (181), who indicated that there is a correlation between the polarizability of the metal compound and the intrinsic reactivity of the metal compound within a given group of the periodic table. Nazarova and Aleksandrova (246) have shown that a relationship exists between the catalyst activity of a phenyl metal catalyst pair containing lead, and the ability of the two metals involved to exchange phenyl groups.

As might be expected, most of the literature references to the use of organolead compounds as polymerization catalysts are concerned specifically with TEL. Only a comparatively few references are made, however, to the use of TEL or any other organolead compound as the sole catalytic agent.

With reference to TEL as the sole catalyst, Cramer (94) reported the polymerization of ethylene with TEL at about 250° in 1934, and Beeck and Rust (48) and van Peski (267) have reported that propylene polymerized in this manner at higher temperatures. The yields and molecular weights were low for ethylene, and still lower for propylene. Gol'dshtein et al. (144) have more recently patented the polymerization of ethylene with TEL catalyst in the USSR, and Sargent and Hanford (309), and Roedel (297–299) have done the same in the U.S.

In 1955, Caldwell and Wellman (68) obtained a patent on the use of TEL as a condensation catalyst for the polymerization of polyesters. For example, a resin was easily produced from ethylene glycol and methyl terephthalate using as little as 0.03% TEL. No water or acid was present, and the TEL was not decomposed quickly under the conditions employed.

Thus, the nature of the actual catalyst was not at all evident. Very recently the Dutch fibers firm AKU has drawn attention to the catalytic properties of several alkyllead acylates in promoting transesterification reactions, as for example, in producing ethylene terephthalate (353). Chas. Pfizer, Inc., has also noted the activity of these compounds in transesterification reactions (353). Further work on these transesterification reactions would appear to be justified both by the industrial importance of the reactions and the unresolved questions regarding the nature of the catalyst and the mechanism of the reactions.

As early as 1930, Young and Douglas (365) patented the polymerization of vinyl acetate and vinyl chloride in the presence of TML, tetrapropyllead, and other lead compounds at 80–100°. Similarly, the polymerization of vinyl acetate has been shown by Koton (210) to be strongly affected by tetracyclohexyllead. In still other references Faragher (130,131) patented the use of TEL as a catalyst for the polymerization of butadiene or butadiene–styrene mixtures. TEL has been claimed in several patents to be a catalyst for ethylene telomerization (in the presence of HCl) at 50–150° and 400–1000 atm (158,308,311); greases and waxes of the alkyl chloride type are formed with molecular weights of about 400. The use of TEL has been patented by Breslow (57) as an initiator in the polymerization of various rosin derivatives. Kozlov et al. (211) reported the polymerization of 1-nitro-1-propene by allowing it to stand 15 days with TEL as a catalyst. TEL has also been shown to be an initiator for the polymeric Diels-Alder reactions of olefinic silanes by Woolford (358). Dissolved TEL will dimerize methallyl chloride when it is exposed to light, as has been shown by Wilzbach, Mayo, and Van Meter (357).

The use of organolead compounds in concert with other polymerization catalysts or promoters has received more attention recently than the use of lead compounds as the sole catalysts. The polymerization of olefins will be considered first. TEL and other organolead compounds have been investigated as the co-catalyst in combination with titanium tetrachloride or another transition metal halide as a substitute for the aluminum alkyl in a Ziegler-type catalyst. This sort of work has been done successfully, for example, in the polymerization of ethylene, propylene, and various alpha-olefins (51,64,91,102,132,145,207,208,241,242,271,273). Patents have also been obtained on the use of TEL with both aluminum chloride and ethylaluminum sesquibromide as an "activated aluminum alkyl initiator for polymerizing olefins" (135,136,190,300–302). The products obtained using these catalyst combinations are claimed to have a higher degree of polymerization than products obtained from aluminum alkyl catalysts in the absence of lead.

More recently, Sobue and Saito (322) have studied the polymerization rate and products from propylene, using a TEL–$TiCl_3$ catalyst at 120–140° to produce isotactic polypropylene.

The polymerization of ethylene with a catalyst combination of organolead compound and a cerium chelate was suggested by Stuart (330) in a 1960 patent. Also, olefin polymerizations with tetraalkyllead and either titanium dioxide (168), or a Group I, II, or III organometallic compound (270), or a peroxide, or molecular oxygen (281) have been patented.

Dickson and West (101) have investigated the reaction of triethylplumbylsodium with biscyclopentadienyltitanium dichloride in liquid ammonia in an attempt to prepare superior Ziegler-type homogeneous catalysts for the low pressure polymerization of ethylene, but unstable solid products were formed.

In addition to these two-component catalyst systems, TEL has been employed in various three-component systems for the polymerization of olefins, such as, for example, titanium tetrachloride, aluminum chloride, and TEL. The catalytic activity is claimed to be increased in these three component systems (27,33,63,191,272,325–327,339,367). Both the two- and the three-component systems yield the highly crystalline, high melting, tough polymers characteristic of products prepared with Ziegler-type catalysts. In a fairly typical example of the three-component catalyst systems, workers at Phillips Petroleum (272) described the polymerization of ethylene dissolved in cyclohexane, using TEL, aluminum chloride, and titanium tetrachloride as the catalyst system at 150° for 7 hr. After decomposition of the catalyst, a polymer was obtained with a density of 0.998, m.p. ca. 125°, inherent viscosity 0.924, melt index 10.05, and mol. wt. 23,020 (based on viscosity) and 16,129 (based on melt index).

When conjugated dienes such as butadiene are used rather than ethylene or propylene, the products are rubbery polymers (33). The addition of a little isoprene to the butadiene improves the product, according to a recent British patent (274). Other modifications of the butadiene polymerization catalyst systems containing organolead compounds have been patented recently (275,276).

Badische (34) claims that the addition of carbon tetrachloride as a third component to a TEL–VCl_4 catalyst system yields tough, high-melting polymers from C_2 to C_8 olefins polymerized in isooctane at 30° and 100 atm pressure. Badische chemists (253) have also polymerized propylene to polypropylene melting at 165°, using a metal such as aluminum, magnesium, or zinc as the third catalyst component with titanium tetrachloride and tetraalkyllead. Joyner and Shearer (192) have extended this work to ethylene and other olefins in a 1963 patent.

Ethylene graft copolymers have been prepared by Greber and Egle (151) by adding ethylene at 70–80° and 3–4 atm pressure to titanium tetrachloride and styrene-triphenylstyrenyllead copolymers. Excellent ethylene polymers and copolymers of ethylene with a C_3 to C_8 alpha-olefin also have been prepared by Kluiber and Carrick (78,201,341), according to recent patents. The catalyst mixture claimed typically included three components, such as TEL, aluminum chloride, and a trace of a vanadium salt.

Both TEL and other alkyl lead compounds have been patented in combination with titanium tetrachloride and other titanium sub-group compounds to polymerize styrene and similar aryl-substituted olefins (187,188,237,324,325). In addition to the polymerization of styrene and related molecules, Marvel and Woolford (237) have investigated TEL as a catalyst, either alone or preferably in combination with titanium tetrachloride, for the polymerization of acrylonitrile, vinyl chloride, vinyl acetate, and methyl methacrylate. These polymerizations were carried out in the temperature range 20 to 60°, usually in conjunction with ultraviolet radiation. Actually, ultraviolet light was found to be absolutely necessary for many of the polymerizations when TEL was the sole catalyst. It was noted that the polyacrylonitrile made with these catalysts was somewhat less crystalline, as indicated by X-ray diffraction patterns, than usual.

The catalytic activity and the properties of TEL–$TiCl_4$ and TEL–$TiBr_4$ complexes in the polymerization of styrene and ethylene have been studied by Kashireninov (193), as well as their composition, dipole moments, magnetic susceptibility, and heats of formation.

Some study of the polymerization of vinyl chloride using lead alkyls as co-catalysts with butyllithium has been made by Dynamit-Noble A.G. (66), who have patented this type of process. Solvay (324) has similarly patented the use of TEL as a co-catalyst with titanium tetrachloride for the polymerization of vinyl chloride. Societa Edison (323) has recently applied for a patent in the Netherlands on the combination of TEL or TML (or organotin compounds) with $Ce(NO_3)_4$ for the polymerization of vinyl chloride, and Nicora et al. (252) have described the polymerization of vinyl chloride in methanol at room temperature, using TEL as a co-catalyst with $Cr(NO_3)_3{\cdot}9H_2O$, $Fe(NO_3)_3{\cdot}9H_2O$, $Cu(NO_3)_2{\cdot}3H_2O$, or $(NH_4)_2Ce(NO_3)_6$. The ceric salt gave 32% conversion in one hour. The high catalytic activity of the TEL-ceric salt system was attributed to the strong oxidizing properties of the tetravalent cerium, since it was postulated that the stronger the oxidizer, the more likely that a redox system would initiate polymerization with equal strength reducers.

Wiley (348–350) has successfully polymerized vinylidene chloride using a catalyst system comprising TEL plus a co-catalyst such as dibenzoyl peroxide or copper. Similarly, Reinhardt (284) has patented catalyst mixtures comprised of TEL, a peroxide, polyhalophenols, and electrically deposited spongy copper which is subsequently chemically treated.

Furukawa and his co-workers (139) have described the rapid polymerization of methyl methacrylate and acrylonitrile with binary catalysts composed of TEL and various metal salts, especially $Cu(NO_3)_2{\cdot}3H_2O$, even at room temperature in the dark. Initiation was attributed to the reaction:

$$(C_2H_5)_4Pb + Cu^{++} \rightarrow (C_2H_5)_3Pb^+ + Cu^+ + C_2H_5\cdot$$

This work was followed up by Kawabata, Tsuruta, and Furukawa (194), who polymerized vinyl acetate and also vinyl ethyl oxalate using either TEL alone, or TEL with a co-catalyst such as potassium dichromate, ferrous acetate, or the chloride of either antimony, bismuth, zinc, or tin. The work of the Furukawa group on methyl methacrylate and vinyl esters was preceded by somewhat similar experiments by Bawn and Whitby (46) who obtained rapid, room-temperature polymerization of methyl methacrylate, styrene, and acrylonitrile upon the addition of TEL to an alcoholic solution of the monomer containing cupric nitrate. When silver nitrate was substituted for the copper nitrate, however, polymerization did not occur. Apparently, such polymerizations are not based on free radical processes, but rather the initiation of the process is probably reliant on a redox system based on copper ions.

The production of polyurethans can be accelerated several-thousand-fold by the addition of an organolead compound to the usual amine catalyst, according to a 1962 patent by Hostettler and Cox (179); good flexible and rigid foams are claimed to be produced. More recently, Overmars and Van der Want (19,196,262,353,355) of the T.N.O. group have shown that aryllead triacylate compounds are very active catalysts in the formation of polyurethan foams, promoting both gelation by reaction of the polyether alcohol with a diisocyanate, and formation of carbon dioxide by reaction of the isocyanate groups with water; the carbon dioxide functions as a blowing agent. A patent application on this subject claims the catalyst RPbXYZ, where R is aryl, alkyl, aralkyl, or cycloalkyl, and X, Y, and Z are halogen atoms or groups bound to the lead by oxygen or sulfur linkages (247). The results have been confirmed by Imperial Chemical Industries (353) and Farbenfabriken Bayer A.G. (355).

Also, Marchenko and associates (235) have successfully demonstrated the use of trialkyllead salts of aliphatic monocarboxylic acids as catalysts in polyurethan formation, employing the copolymerization reaction of polydiethylene glycol adipate with 2,4-tolulene diisocyanate. On the other hand, there have been brief, unsuccessful investigations of the possible usefulness of simple trialkyllead salts as catalysts for the production of "one-shot" polyurethan foams from polyethers at A. D. Little, Inc. (231). Organotin compounds, such as stannous dioctoate, are now used commercially as urethane catalysts, and tertiary amines are used to promote the formation of carbon dioxide.

The T.N.O. group has investigated the vulcanization of natural rubber with R_3PbX and R_2PbX_2 compounds (353). These compounds proved to be vulcanization agents, but the product rubbers were of mediocre quality.

Arnold (30) has patented the use of polyalkenylated lead compounds as crosslinking agents for acrylonitrile polymers. The resulting polymers are described as useful for thickening nonpolar solvents.

D. Other organolead catalysis

Organolead compounds have been suggested as catalysts in a number of chemical reactions which do not involve either polymerization or oxidation-type knock inhibition.

Chlorination has received some attention, dating back to 1940. Vaughn and Rust (342,343) studied the chlorination of ethane, propane, and cyclopentane in the vapor phase, and *n*-pentane in the liquid phase, in the dark with TEL present in small amount. In one instance of their study, complete reaction with chlorine occurred at 120° to give the normal thermal reaction products, 80 mole % ethyl chloride and 20 mole % higher chlorides, in contrast to no reaction whatsoever in the absence of TEL. Vaughn and Rust attributed the initiation of chlorination to the formation of a highly reactive species, such as an ethyl radical, during the reaction of chlorine with TEL, although it is not clear how such a radical could be formed in this process.

In other work, Evans, Vaughn, and Rust (128,129) discovered that TEL could be used under ultraviolet light as a free radical initiator for the addition of hydrogen sulfide, mercaptans, and hydrogen bromide to propylene. The normal propyl derivative is formed, contrary to Markownikoff's rule, due to ethyl radicals from the photolysis of the TEL which initiates the abnormal addition.

Langenbeck et al. (220) have indicated that TEL, in common with peroxides and some other organometallic compounds, is a catalyst for the increase of the Gammexane content of crude hexachlorocyclohexane.

Hanford and Harmon (158,159) have patented the use of tetraalkyllead and tetraaryllead compounds to catalyze telomerization reactions between hydrogen chloride and olefinic compounds, for example, the formation of higher alkyl chlorides from ethylene and hydrogen chloride in oxygen at 100 atm and at temperatures above 50° in the presence of water.

It has been shown by Herold and Asinger (171) that the chlorosulfonation of aliphatic hydrocarbons with sulfur dioxide and chlorine can be accelerated by carrying out the reaction in the presence of TEL or TML; this reaction is known to be free-radical initiated.

Chenicek and Bloch (85) have claimed in a 1959 patent that TEL is a catalyst for the alkylation of olefins, allowing the use of lower temperatures and pressures than usual. R_4Pb and R_3PbO_2CR compounds have been described as general esterification catalysts (11).

In still another use of TEL as a catalyst, it was claimed to be a hardening agent for rubber by Twiss and Jones (338), who believed that reinforcing compounds are produced in the rubber through the catalytic decomposition of TEL. The T.N.O. group has more recently attempted the vulcanization of natural rubber by means of organolead compounds. In a similar type of application, Lucas (233) patented the use of organolead compounds as curing agents for polysiloxane resins, and Zherdev et al. (366), have described diethyllead dicaprylate as a curing agent for resins of the average composition $MePh_2(SiO_{1.5})_3$.

Thompson (334) has patented the use of TEL as a catalyst for the reforming of hydrocarbons. 1,2-Diphenylethane, for example, was produced from toluene with TEL as a catalyst, either at 250° or in sunlight at room temperature.

Berezovskaya et al. (49) have suggested that TEL, among other catalysts, can be used to decompose such peroxides as ethyl hydroperoxide, *tert*-butyl hydroperoxide, and trimolecular acetone peroxide at 80–100°.

In an unusual application, Horie (177) has demonstrated that the yield of indole prepared from aniline and acetylene at 400° is raised significantly by the addition of a small amount of TEL to the aniline.

The thermal decomposition of organolead compounds was employed by Wilson (340,356) in 1958 to prepare lead metal catalyst precursors. This procedure has been extended by Novak (257) who proposed that the thermal decomposition of an alkyl lead compound could be used to produce porous hollow lead coatings as catalysts over such substrates as plastics, wax or other substances.

(*1*) PVC *stabilization*

Organolead compounds have also been investigated as potential polyvinyl chloride (PVC) stabilizers. Commercial PVC is especially susceptible during processing to dehydrochlorination, and the onset of dehydrochlorination leads to rapid deterioration. At present, organotin compounds such as dibutyltin maleate are very widely used as heat and light stabilizers in commercial PVC. Inorganic lead compounds, rather than organolead derivatives, are used in considerable amounts for this same purpose. These compounds are such materials as dibasic lead phosphite and lead phthalate, and are used in the concentration range of 2–7%.

It was only natural, therefore, to assume that organic lead compounds and other Group IV metal compounds as well could find a role in commercial PVC technology. Thus, various carbon-bonded lead compounds have been suggested as stabilizers by several investigators, dating back to 1936. However, these lead compounds do not compete commercially with the organotin compounds for this purpose, probably because of their lower effectiveness as stabilizers and because of fear of their toxicity as compared with the tin compounds. The general subject was reviewed by Smith (321) in 1954.

The types of organolead stabilizers suggested in the literature are alkyllead oxides and hydroxides (362), alkyllead esters of organic acids (88,363,364), triethyllead hexylmaleate (285), various organolead compounds (56), alkyllead esters of α,β-unsaturated carboxylic acids (35), trimethyllead cyanide (149), and hexaphenyldilead (150,278). In addition, the Russian work (278) demonstrated that ultraviolet induced degradation is reduced considerably by hexaphenyldilead and hexaphenylditin. The trialkyllead esters have also been proposed as synergists with standard organotin stabilizers.

A number of compounds are being tested in the current research programs in cooperation with the Organisch Chemisch Instituut T.N.O., Utrecht (11,196,353–355). In this work, hexaphenyldilead and organolead–sulfur derivatives have shown some stabilizing effect, but the work to date has not uncovered any potential competitors for the commercial PVC stabilizers.

Sheftel (318) has suggested that PVC stabilized with lead compounds should not be used in water supply pipes because of extraction of the lead by the water.

E. Biological effects

Since an apparently characteristic property of inorganic lead compounds in substantial amounts is their systemic toxicity to living cells, a number of organolead compounds have been investigated for their biological effects apart from their toxicological properties in human beings. The assumed "poisonous" character of organolead compounds for some time tended to hold back investigations in this general area of biological effects. Interest in the area is increasing, however, especially since the powerful pesticidal effects of organotin and organomercury compounds have been demonstrated in the past ten years [see, for example, p. 524 of the "Organotin Compounds" review by Ingham, Rosenberg, and Gilman (189)]. The physiological and toxicological properties of organolead compounds were discussed in Chapter V.

As already indicated, the International Lead Zinc Research Organization has been sponsoring work in the biocidal field for the past seven years at the Organisch Chemisch Instituut T.N.O. at Utrecht, and at some other institutions as well. In recent years, organolead compounds have been tested successfully in such diverse applications, for example, as the rotproofing of cotton, protection of wood against fungal and marine fouling attack, mildewproofing of paint, protection against insects and rodents, and control of human bilharzia transmission through tropical snails. A high degree of specificity has been noted, with regard both to the structure of the lead compounds and the organisms affected.

Actually, the most simple types of trialkyllead and dialkyllead salts were demonstrated some time ago to be potent in a variety of biological applications. For example, both trialkyllead halides and triaryllead halides were reported by Krause (212) as early as 1929 to be effective against mouse carcinomas. The triaryllead halides were shown by Wallen (345) to be active in low concentrations against microorganisms such as *S. cerevisiae*, *L. casei*, *A. suboxydans*, *A. niger*, *C. acetobutylicum*, and *L. delbrueckii*. Additionally, Shapiro, Hnizda, and Calingaert (317) showed that diethyllead dicyanide has strong fungicidal activity against the spores of *Alternaria oleracea* and *Sclerotinia fructicola*.

Another example of the demonstrated biological effects of the simple trialkyllead salts is their use as very effective temporary sternutators toward human beings. Indeed, they were described by a World War II British team as more potent than the corresponding organotin compounds (166,167,238,310). Moreover, in 1934 Carothers (76,112) claimed that both trialkyllead and dialkyllead phenolates act as insecticides, fungicides, and disinfectants (and antioxidants and driers for paints).

Indeed, the simple organolead acetates seem to be unusually effective fungicides and bactericides. In a recent systematic study at Utrecht (196,320), several trialkyllead and triaryllead acetates were compared with the analogous tin and germanium compounds for activity against three fungi and both gram-positive and gram-negative bacteria. The lead compounds were found to be the most effective, with activity peaking in the lead series in the butyl and amyl derivatives; dialkyllead diacetates also were effective against bacteria. The activity pattern was unrelated to the water solubilities of the compounds. In an extension of the fungicidal investigations, the T.N.O. group found that tributyllead laurate displays good protective properties when added to emulsion paints (11,196,215). The T.N.O. group has shown that tripropyllead acetate is extremely effective against the bacterium Desulfovibrio desulfuricans (352). In the test used, this compound was 30 times as active as the standard "Zephirol" compound.

The simple trialkyllead acetates apparently are capable of protecting wood against wood-rotting fungi. Both triethyllead acetate and triethyllead chloride were demonstrated in a recent A. D. Little study (11,150) to be equal to phenylmercuric acetate against a typical wood-rotting fungus. This study has been extended recently by E. W. B. DaCosta of the Commonwealth Scientific and Industrial Research Organization, Melbourne, Australia, according to a T.N.O. source (353). Of four compounds tested (Bu_3PbOAc, Et_3PbOAc, Ph_3PbOAc, Ph_6Pb_2), tributyllead acetate was most active against both white and brown rot fungi.

Work is also being carried out by Donaldson and Drake at the U.S. Department of Agriculture, Southern Utilization Research and Development Division, Cotton Finishes Laboratory, on the rot-proofing of cotton to improve its resistance to mildewing and weather (19,105,196). Representative compounds being tested are triphenyllead laurate, triphenyllead acetate, trimethyllead acetate, hexaphenyldilead, *N*-(tributylplumbyl) imidazole (104), diphenyllead diacetate, triphenyllead thioalkylamides (103), and thioalkyltriphenyllead compounds (8). It has been demonstrated that treated cotton samples buried in the soil can successfully withstand attack by microorganisms over a period of six months, retaining 100% breaking strength. Likewise, tributyllead acetate and dibutyllead diacetate have been shown by the T.N.O. group to impart bacteriostatic and rotproofing properties to cotton fabrics (218,219,353). Also, trialkyllead compounds have been indicated to give some protection to woolen fabrics against attack by larvae of the clothes moth and carpet beetle (214,217).

Several organolead–nitrogen compounds have been the subject of patents on their biological potency, particularly with respect to fungi. Ligett et al. (228,230) found that triethyllead phthalohydrazide is quite active against the spores of the potato blight and peach rot diseases; moreover, they demonstrated this type of fungicidal activity also for *N*-(triethyllead) phthalimide (229). Nagasawa et al. (245) showed that several dimethylglyoxime lead derivatives inhibit the growth of phytopathogenic fungi. Quaternary ammonium compounds containing lead have been suggested as slime control agents in a patent by Sowa and Kenney (328), and van Peski et al. (268) have claimed lead derivatives of aminomethylene aldehydes as general biocides. In a further example of the biological effectiveness of lead–nitrogen compounds, DePree (98) has shown that compounds of the type triethyl(*sec.*-butylamino)lead are good herbicides.

Triethyllead chloride and triethyllead acetate have been suggested as additives to prevent the bacterial and fungal sludges formed at water interfaces in jet fuel on storage (231). However, there is little evidence that these compounds are either effective or acceptable for this use.

Considerable interest has arisen recently in the possibility of using organolead compounds as marine antifouling agents. Marine timbers are frequently destroyed in as little as 5–10 years through attack by limnorial and teredine borers. Again, under International Lead Zinc Research Organization sponsorship, the Organisch Chemisch Instituut T.N.O., Utrecht, has started an investigation of this problem (100,196,353,354). It has been found that the *in vitro* antifouling activities of compounds of the type Bu_3PbX, where X is acetate, caproate, or laurate, are independent of the group X. Previously, triethyllead cyanide was claimed by Tisdale (335) as an effective marine antifouling agent. In confirmation of the T.N.O. data, the University of British Columbia (British Columbia Research Council) has obtained very promising results in ten-month sea-water immersion tests at two locations, using especially tributyllead acetate and dibutyllead diacetate, particularly in combination with creosote (3,8,19,354). This type of investigation is being continued by several organizations. Among the active compounds being investigated further are hexaphenyldilead, dibutyllead diacetate, and a number of tributyl- and triphenyllead derivatives, including tributyllead acetate.

The antifouling activity of organolead compounds, such as tributyllead acetate and triphenyllead acetate, has also been investigated in paint systems, such as vinyl standard, vinyl high rosin, and oleoresinous (196,280). Exposure of test panels at Miami Beach and Pearl Harbor has indicated very promising activity of the lead compounds against barnacles and tube worms. However, Gurevitch and Dolgopol'skaya (154) have found that organolead compounds decompose DDT in PVC- and polyethylene-based antifouling paints; this may be a possible drawback.

The T.N.O. group has demonstrated that 3% of triethyllead acetate acts as an effective rodent repellant for jute in a barrier test separating male and female rats (196,216). Likewise, U.S. Department of the Interior tests on several lead compounds prepared by Hills and Henry (173) showed them to have excellent rodent repellant properties at 1% concentration in white wheat. The unusually effective compounds (100% of rodents repelled) included several triphenyllead mercaptides, methyltriphenyllead, triphenyllead chloride, and several diphenyllead organoarsonates.

The Tropical Products Institute of London and others are testing a series of trialkyllead acetates for bilharzia control in Australorbis glabratus snails (99,296,353,354). Bilharzia is a schistosome-type disease affecting many millions of people through these snails as carriers. Optimum activity has been found in triamyl-, tributyl-, and triphenyllead acetates, with a very favorable snail-to-rice toxicity ratio. Open water tests in the tropics are being carried out by a number of organizations under the auspices of the World Health Organization, since this application appears to be very promising (196,354).

R. I. Dorfman of the Worcester Foundation for Experimental Biology, Shrewsbury, Mass., has reported that tetrabutyllead has a unique and perhaps specific antiandrogenic activity in the testosterone-stimulated castrated mouse (107,352). Dorfman (106) has further reported that thioacetyltriphenyllead, $(C_6H_5)_3PbSC(O)CH_3$, has sufficient antiandrogenic activity to make it a possible candidate for the treatment of acne, baldness, or beard growth inhibition in man; and that the thiobenzyl- and thiophenyltriphenyllead compounds show excellent anti-inflammatory properties, suggesting possible application in treatments of allergies, asthma, and rheumatoid arthritis.

It was only to be expected, after all, that alkyllead compounds would be found to have insecticidal effects, and as long ago as 1936, several simple trialkyllead salts were claimed to be good insecticides by Calcott and associates (67). In further insecticidal investigations, Kochkin et al. (203) have recently demonstrated that monomers such as triphenyllead methacrylate (and many tin compounds as well) have insecticidal activity against a variety of insects; the triphenyllead methacrylate was also found to have a fairly high residual activity. Further, Kochkin et al. (204) demonstrated bactericidal activity as well for these compounds, either alone or admixed with quaternary ammonium salts. Kochkin et al. (205,206) prepared the diphenyllead and triphenyllead (and tin) orotates and uracilacetates and the diphenyllead derivative of nicotinic acid. However, preliminary studies showed that these lead compounds were less toxic than the corresponding tin compounds.

The insecticidal toxicity of organolead compounds was also investigated by Blum (53). Using the housefly as an assay organism, Blum compared the toxicities and knockdown power of several series of analogous compounds of silicon, germanium, tin, and lead. The compounds were administered as acetone solutions by means of a micrometer-driven syringe. The triethyl and diethyl metal chlorides, triphenyl and diphenyl metal chlorides, tetraethyl metal and tetraphenyl metal compounds, and a few monoorganometal compounds were compared. The knockdown activity decreased in the order $Sn > Ge > Pb = Si$. Both triethyltin chloride and triphenyltin chloride were highly insecticidal, in contrast to the R_3MCl compounds of the other metals. The LD_{50}'s for the two tin compounds were 0.28 and 0.18, respectively. The tetraorganometal compounds, including lead, were relatively nontoxic to the housefly. The housefly appears to be incapable of converting these compounds to the more toxic triorganometal forms, since there was no sign of delayed toxicity to these compounds. The Cooper Technical Bureau of Berkhamsted, U.K. (354), has more recently investigated organolead compounds on houseflies and roaches, with similar results.

Dibutyllead diacetate has been shown by Dr. Gras of the University of Montpellier (354) to kill such parasites as *Hymenolepsis fraterna* and *Raillietina cesticillus* in mice and chickens, respectively. It is too early to assess the import of such work, but it should be mentioned that dibutyllead salts have lower mammalian toxicity than the tributyllead salts.

In confirmation of Krause's previously cited work on the treatment of mouse cancers with trialkyllead halides, Marsh and Simpson (236) have shown that TEL has a regressive action on mouse tumors, but it is only very effective when administered in lethal doses.

In a patent devoted mainly to tin compounds, Langer (221) has recently described the use of the diphenyllead chelate of ethylenediaminetetraacetic acid as an insecticide, fungicide, slimicide, and bactericide. As an insecticide, the compound is claimed to be particularly effective against the southern army worm and the spotted spider mite.

Several new organolead compounds have recently been prepared and tested for biocidal activity (without outstanding results) by Willemsens (355). These are believed to have the following structure:

$$\cdots N\langle\rangle N \cdots PbR_3 \cdots N\langle\rangle N \cdots PbR_3 \cdots$$

These pentacoordinated compounds have been prepared by mixing triorganolead hydroxide and the appropriate heterocycle (see Chapter XIII; Heterocyclic Lead Compounds). In general, these compounds appear to be similar to the corresponding triorganolead compounds in their biocidal effects.

F. Lead-containing polymers

In general, fewer organolead polymers have been synthesized than polymers containing other Group IVb metals. Undoubtedly, this is due at least partly to the lower stability of the organolead compounds as well as the instability of the lead–lead bond. The organolead polymers based on the polymerization of unsaturated compounds are discussed in Chapter X. A recent review of Group IVb element organometallic polymers was written by Henry and Davidson (169).

As was described in Chapter X, a number of polymers based on organolead structures have been prepared successfully, using vinyl monomers such as triphenyl-*p*-styrenyllead, triethyl-*p*-styrenyllead, triethyl-*p*-(α-methylstyrenyl)lead, and organolead acrylates and methacrylates. It has also been suggested many times that polymers could be prepared from tetravinyllead and similar vinyl lead compounds, and also from allyltriphenyllead. Up to the present time, however, the homopolymerization of vinyl and allyl groups adjacent to the lead atom has not been accomplished, despite numerous attempts.

Compounds containing long chains of lead atoms are unknown. The absence of literature references to characterized lead compounds containing four or more adjacent lead atoms is due not only to the instability of the lead–carbon bond, but also in even greater measure to the instability of the lead–lead bond, although compounds of the type tetrakis(triphenylplumbyl)lead are known, as well as several hexaalkyl- and hexaaryldileads (see Chapter VII); these disproportionate and decompose readily at relatively low temperatures. Polymers of the type

$$(C_2H_5)_3Pb{+}Pb(C_2H_5)_2]_nPb(C_2H_5)_3$$

have been suggested as intermediates in the disproportionation reactions of hexaethyldilead (282,283).

The R_2Pb compounds are postulated to be polymeric under most conditions (see Chapter VIII). It also appears that diphenyllead oxide is polymeric, although it has not been investigated extensively.

A number of poorly defined polymers containing lead–oxygen bonds are also mentioned in the literature, such as the common lead oxide–glycerol products, and various silanol reaction products with lead oxide or lead chloride. For example, Patnode and Schmidt (263) described the following reaction:

$$2(CH_3)_3SiOH + PbO \longrightarrow [(CH_3)_3SiO]_2Pb + H_2O$$

Adrianov (1) reported the somewhat similar reaction between methylphenyltriethoxysilane and lead acetate in alcohol to yield various organoplumbosiloxanes.

Sidgwick (319) reported the formation of plumboxanes from diaryllead dihydroxides, for example:

$$n(C_6H_5)_2Pb(OH)_2 \longrightarrow \left[\begin{array}{c} C_6H_5 \\ | \\ -Pb-O- \\ | \\ C_6H_5 \end{array}\right]_n + 2nH_2O$$

This type of formation of polydiphenylplumboxane was reported also by Kochkin et al. (202) who prepared a mixture of compounds by heating diphenyllead dichloride at 60° in alcohol and treating with alcoholic potassium hydroxide.

The reaction of thioglycolic acid with diphenyllead dihalide has been shown to produce a high melting (>300°) solid by Davidson, Hills, and Henry (97). This solid, obtained in relatively low yield, was probably polymeric because of the high melting point and insolubility in organic solvents.

Osborn and Yu (260,261) have patented the preparation of polymers from polybutadiene–acrylic acid copolymers and triethyllead hydroxide, and polymers prepared from triethyllead hydroxide and unsaturated hydrocarbons, such as butadiene, isoprene, chloroprene, or 2-cyano-1,3-butadiene. These polymers are said to be useful as films, elastomers, or burning rate modifiers for rocket propellants.

G. Miscellaneous applications

A number of additional areas of potential usefulness of organolead compounds has been explored. For example, it has been known for many years that organolead compounds are thermally unstable and yield metallic lead on pyrolysis. Such dissociation of organolead compounds has usually been employed as a source of alkyl radicals to promote various reactions, but Imhausen (185,186) suggested in a 1921 patent that such compounds could be decomposed thermally to coat iron with lead metal for protection. Other methods of decomposition have been suggested for the deposition of lead metal. Baker and Morris (36) demonstrated that lead films between 50 and 1500 Å in thickness could be deposited from TEL by means of decomposition through the agency of an electron beam gun. Lead alloys have also been deposited by means of thermal decomposition. Norman and Whaley (254) patented the plating of mixed metals by means of thermal decomposition of various bimetallic organometallic compounds; for example, metal films were easily obtained from the decomposition of $Et_2PbFe(CO)_4$ on mild steel. Also, lead telluride films were formed by Cornish (93) by the vapor interaction of TEL and hydrogen telluride at 200–510° directly at the substrate, yielding good mirrors. The mechanism of the reaction was discussed, since the films were not formed by the reaction of the free elements produced by thermal decomposition of the vapors. Lead telluride is of interest because of its semiconductor properties and the discovery by Lalevic (6) that it is a superconductor at 5°K. Bloom (52) pyrolyzed TEL vapor in nitrogen at 800°C in a fused silica tube to form a lead glass "skin."

In a recent patent devoted mainly to tin compounds, Langer (222) suggests that dialkyl- and diaryllead chelates of ethylenediaminetetraacetic acid are useful as metal sources in electroplating and in insoluble dental enamel.

Some promising investigations have been made of the effect of organolead compounds as additives for lubricating oils. TEL, for example, has been shown by Antler (24) in 1962 to be a wear-reducing additive in polydimethylsiloxane oil. TEL has been employed as an antioxidant in paraffinic turbine oils by Hatta (164), with beneficial effects on color, sludge formation, specific gravity, and saponification values; on the other hand, naphthenic oils were adversely affected.

Chakravarty (81) investigated the influence of TEL under ultraviolet decomposition irradiation on the corrosion of lead bearings by lubricating oils. It was found that irradiation reduced the corrosion of lead in contact with such oils prior to oxidation at 160°, whereas in the absence of irradiation there was marked corrosion of the metal. He also showed that a yellow deposit of $(R_2PbO)_2$ was formed on lead immersed in white oil undergoing liquid-phase oxidation in air (80).

In 1964, Hatton and Stark (165) patented the use of tetraphenyllead, triphenyllead chloride, and diphenyllead dichloride as viscosity index improvers for engine oils of the polyphenyl ether type. On the other hand, beneficial results have not been uniformly obtained; it was reported that Gulf Research and Development Co. (335) in recent work tested phenyl lead compounds as oil additives without success. In the latest series of tests, organolead compounds are once more under test as antiwear additives for oils by the T.N.O. group (8,196,353), once again with very promising results. A compound supplied by the T.N.O. Institute to the Ethyl Corporation has displayed unusual properties in reducing wear under load (19). Electron diffraction examination has shown that an unknown crystalline compound results on the wearing surface, forming a thin, durable film. Patent applications have been filed, and an oil additive package has been developed. One patent has issued: Perilstein (264) has shown that compounds of the type $(C_6H_5)_3PbSR$ improve the lubrication properties of oils when added in the concentration of 1 wt % based on lead. However, there has also been reference in the literature to slightly increased engine wear with organolead antiknocks. Aronov et al. (31) found a small effect as measured by incorporating radioactive isotopes in engine parts and measurement of increased radioactivity in the circulating motor oil. It appears that more fundamental research may be required to explain the discrepancies in the oil additive test results.

It was demonstrated some years ago by Shapiro and Capinjola (315) that triethyllead sulfide solutions are sensitive to light and can be used as low speed photographic reproduction chemicals. In 1961, Peters (269) patented the use of tetracyclohexyllead solutions as photosensitive print-out materials. He showed that peroxides are inhibitors of the photodecomposition reactions and that sulfur compounds are activators. With proper mixture preparation, photosensitive layers were obtained with dark-colored image formation on exposure to high-intensity light (PbS deposition), and which remained white on exposure to low-intensity light (PbO deposition). Also, Horizons, Inc. (178) has claimed in a recent Netherlands application that tetraalkyllead compounds, and similar compounds in which the metal is germanium, tin, phosphorus, arsenic, antimony, or bismuth, are useful as antifogging agents in light-sensitive organic systems in which the latent image is developed by exposure to heat radiation, because they prevent fogging due to the heat treatment.

The pentametal compounds of the $(Ph_3Pb)_4M$ type prepared by the T.N.O. group (355) at Utrecht have been investigated briefly for semiconductor properties, but without success.

On a very small scale, tetramethyllead vapor has been used for many years to fill Geiger counters, to which it imparts high efficiency (142, 339,360). Also, Sandler et al. (96,306,307,336) have shown that triphenyl(4-ethylphenyl)lead and the corresponding tin compound might be useful in liquid scintillation counting. Moreover, copolymers containing triphenyl-*p*-styrenyllead or the analogous tin compound were even more effective scintillation quenchers, as were also the analogous ethylbenzene derivative and 2,4-dimethylphenyltriphenyllead. In prior work, Andreeschev et al. (4) and Kropp and Burton (213) studied the quenching effect of incorporation of tetraphenyllead and other phenyl metal compounds into organic scintillation systems.

Keck (195) has shown that the incorporation of lead atoms into plastic films for electron scattering may be accomplished by adding TEL or tetrabutyllead to the film solutions; the distribution of the lead is much more homogeneous with tetrabutyllead than with TEL. Tetraphenyllead and similar organometallic compounds have also been incorporated into polystyrene by Baroni et al. (44) by dissolving the organometallic in the monomer, and then polymerizing the monomer slowly in the absence of air at relatively low temperature.

Leshchenko et al. (226) have patented the use of organolead compounds as stabilizers to prevent the breakdown of polyethylene subjected to ionizing radiation. Cormany et al. (92) patented the use of tetraalkyllead compounds as a stabilizer for methylchloroform to reduce its corrosivity toward steel. Also, Clark (87) has patented the use of organometallic compounds, including TEL, as stabilizers for halogenated aromatic hydrocarbons used as dielectric, insulating, or cooling agents.

Finally, Newland and Tamblyn (248,249) have suggested that organometallic compounds, including R_4Pb compounds, can be used as stabilizers for polyolefins. However, their investigation showed that a given organometallic compound could exhibit a considerable range of stabilizing or antistabilizing activity, depending on its concentration and on the chemical nature of the environment.

REFERENCES

1. Adrianov, K. A., *Usp. Khim.*, **27,** 1257 (1958).
2. Agnew, W. G., *Combust. Flame*, **4,** 29 (1960).
3. Allen, I. V. F., *British Columbia Research Council Progress Report No. 4, Project LC-89*, Semi-Annual Report to July, 1966, to Intern. Lead Zinc Res. Organ., New York.
4. Andreeshchev, E. A., E. E. Baroni, N. S. Kursonova, and I. M. Rosman, *Pribory i Tekhn. Eksperim.*, **6,** 151 (1961); through *C.A.*, **56,** 6764 (1962).
5. Anon., *Chem. Age*, Nov. 26, 1966, p. 977.
6. Anon., *Chem. Eng. News*, **43,** 49 (1965).
7. Anon., *Erdöl Kohle*, **19,** 905 (1966).
8. Anon., "Organolead Job Hunt," *Chem. Wk.*, Oct. 24, 1964, p. 95.
9. Anon., *Stanford Research Institute Chemical Economics Newsletter*, Sept., 1966, p. 2, Stanford Research Institute, Menlo Park, Calif.
10. Anon., Ethyl Corporation Survey and News Release, Nov., 1965.
11. Anon., *Paint Manuf.*, **35,** 36 (1965).
12. Anon., *Stanford Research Institute Chemical Economics Newsletter*, April, 1964, p. 3, Stanford Research Institute, Menlo Park, Calif.
13. Anon., "Chemical Profile," *Oil, Paint & Drug Reporter*, Aug. 3, 1964.
14. Anon., Ethyl Corporation Survey and News Release, Nov., 1966.
15. Anon., Ethyl Corporation Forecast, 1966.
16. Anon., *Oil and Gas. Journal*, **58,** 74 (1960).
17. Anon., *Petr. Wk.*, **10,** 58 (1960).
18. Anon., *Chem. Eng. News*, Aug. 29, 1960, p. 21.
19. Anon., *ILZRO Research Digest No. 18, Part V, Lead Chemistry*, Oct., 1966, Intern. Lead Zinc Res. Organ., New York.
20. Anon., *Chem. Wk.*, Feb. 4, 1961, p. 21.
21. Anon., *Stanford Research Institute Chemical Economics Newsletter*, Oct. 1961, p. 1, Stanford Research Institute, Menlo Park, Calif.
22. Anon., *Oil Daily*, June 14, 1960, p. 3.
23. Anon., *Oil Gas J.*, **57,** 195 (1959).
24. Antler, M., U.S. Pat. 3,058,912 (to Ethyl Corp.), Oct. 16, 1962; *C.A.*, **60,** 13084 (1964).
25. Antonsen, D. H., U.S. Pat. 3,010,978 (to Sun Oil Co.), Nov. 28, 1961; *C.A.*, **56,** 11620 (1962).
26. Aries, R. S., *Petrol. Engr.*, **31,** (12), C15 (1959).
27. Aries, R. S., U.S. Pat. 2,900,374, Aug. 18, 1959; *C.A.*, **53,** 23099 (1959).
28. Aries, R. S., *Rev. inst. franc. petrole et Ann. combustibles liquides*, **15,** 1881 (1960); *C.A.*, **55,** 9848 (1961).
29. Arimoto, F. S., French Pat. 1,328,932 (to E. I. du Pont de Nemours & Co.), June 7, 1963; *C.A.*, **60,** 550 (1964).

30. Arnold, L. F., U.S. Pat. 2,991,276 (to B. F. Goodrich Co.), July 4, 1961; *C.A.*, **55,** 26535 (1961).
31. Aronov, D. I., V. I. Golov, and M. S. Lerner, *Khimi Tekhnol. Topliv. i Masel*, **5,** 43 (1960); through *C.A.*, **54,** 23289 (1960).
32. Azerbaev, I. N., and D. S. Kochkin, *Vestn. Akad. Nauk Kaz. SSR*, **19,** 18 (1963); *C.A.*, **60,** 5534 (1964).
33. Badische Anilin- u. Soda-Fabrik A.G., Belg. Pat. 550,840, Sept. 6, 1956.
34. Badische Anilin- u. Soda-Fabrik A.G., Brit. Pat. 838,227, June 22, 1960; *C.A.*, **55,** 6039 (1961).
35. Baer, M., U.S. Pat. 2,561,044 (to Monsanto Chemical Co.), July 17, 1951; *C.A.*, **45,** 8807 (1951).
36. Baker, A. G., and W. C. Morris, *Rev. Sci. Instr.*, **32,** 458 (1961); *C.A.*, **56,** 126 (1962).
37. Ballinger, P., U.S. Pat. 3,073,852 (to California Research Corp.), Jan. 15, 1963; *C.A.*, **58,** 12599 (1963).
38. Ballinger, P., U.S. Pat. 3,073,853 (to California Research Corp.), Jan. 15, 1963; *C.A.*, **58,** 12599 (1963).
39. Ballinger, P., U.S. Pat. 3,073,854 (to California Research Corp.), Jan. 15, 1963; *C.A.*, **58,** 12599 (1963).
40. Ballinger, P., U.S. Pat. 3,081,325 (to California Research Corp.), March 12, 1963; *C.A.*, **59,** 6440 (1963).
41. Ballinger, P., U.S. Pat. 3,116,127 (to California Research Corp.), Dec. 31, 1963; *C.A.*, **60,** 6684 (1964).
42. Ballinger, P., U.S. Pat. 3,143,399 (to California Research Corp.), Aug. 4, 1964; *C.A.*, **61,** 10520 (1964).
43. Bardwell, J., *Combust. Flame*, **5,** 71 (1961).
44. Baroni, E. E., S. F. Kilin, T. N. Lebsadze, I. M. Rozman, and V. M. Shoniya, *At. Energ.* (*USSR*), **17,** 497 (1964); through *C.A.*, **62,** 10604 (1965).
45. Barusch, M. R., and J. H. Macpherson, "Engine Fuel Additives," in *Advances in Petroleum Chemistry and Refining*, Vol. X, J. J. McKetta, Jr., Ed., Interscience, New York, 1965, Chapt. 10.
46. Bawn, C. E. H., and F. J. Whitby, *Discussions Faraday Soc.*, **1947,** 228.
47. Beatty, H. A., and G. Edgar, "Theory of Knock in Internal Combustion Engines, in *The Science of Petroleum*, Vol. IV, A. E. Dunstan, Ed., Oxford Univ. Press, London, 1938, p. 2927.
48. Beeck, O., and F. F. Rust, *J. Chem. Phys.*, **9,** 480 (1941).
49. Berezovskaya, F. I., E. K. Varfolomeeva, and V. G. Stefanovskaya, *Zh. Fiz. Khim.*, **18,** 321 (1944); *C.A.*, **39,** 2024 (1944).
50. Berl, E., U.S. Pat. 2,270,780 (to Berl Chemical Corp.), Jan. 20, 1942; *C.A.*, **36,** 3191 (1942).
51. Bessant, K. H. C., and S. K. Lachowicz, Brit. Pat. 795,882 (to Distillers Co.), June 4, 1958; *C.A.*, **52,** 21242 (1958).
52. Bloom, M., *J. Am. Ceram. Soc.*, **48,** 649 (1965).
53. Blum, M. S., Louisiana State University private communication, 1964.
54. Bott, L. L., *Hydrocarb. Proc.*, **44,** 115 (1965).
55. Braendle, R. O., Lecture on "New Lead Antiknock Agents," presented before Philadelphia ACS Section, Feb., 1961.
56. Braun, D., S. B. Chang, and M. Thallmaier, *Gummie Asbest. Kunstst.*, **19,** 1353 (1966); through *C.A.*, **66,** 46885h (1967).
57. Breslow, D. S., U.S. Pat. 2,554,810 (to Hercules Powder Co.), May 29, 1951; *C.A.*, **45,** 8809 (1951).

58. Brown, J. E., U.S. Pat. 2,913,413 (to Ethyl Corp.), Nov. 17, 1959; *C.A.*, **54,** 6111 (1960).
59. Brown, J. E., Ger. Pat. 1,101,853 (to Ethyl Corp.), March 9, 1961; *C.A.*, **56,** 7598 (1962).
60. Brown, J. E., E. G. DeWitt, and H. Shapiro, U.S. Pat. 3,086,034 (to Ethyl Corp.), April 16, 1963; *C.A.*, **59,** 8791 (1963).
61. Brown, J. E., and W. G. Lovell, *Ind. Eng. Chem.*, **50,** 1547 (1958).
62. Brown, J. E., H. Shapiro, and E. G. DeWitt, U.S. Pat. 2,818,417 (to Ethyl Corp.), Dec. 31, 1957; *C.A.*, **52,** 9203 (1958).
63. Bruce, Jr., J. M., U.S. Pat. 2,992,190 (to E. I. du Pont de Nemours & Co.), July 11, 1961; *C.A.*, **55,** 27978 (1961).
64. Bua, E., A. Malatesta, and A. Negromanti, Ital. Pat. 531,219 (to "Montecatini"), July 23, 1955; *C.A.*, **54,** 4048 (1960).
65. Buck, R. P., and P. R. Ryason, Preprints, Division of Petroleum Chemistry, April, 1965, A.C.S. Meeting, Detroit, p. D-5, p. D-21.
66. Buening, R., Ger. Pat. 1,144,483 (to Dynamit-Nobel A.G.), Feb. 28, 1963; *C.A.*, **58,** 11481 (1963).
67. Calcott, W. S., W. H. Tisdale, and A. L. Flenner, U.S. Pat. 2,044,934 (to E. I. du Pont de Nemours & Co.), June 23, 1936; *C.A.*, **30,** 5715 (1936).
68. Caldwell, J. R., and J. W. Wellman, U.S. Pat. 2,720,505 (to Eastman Kodak Co.), Oct. 11, 1955; *C.A.*, **50,** 2205 (1956).
69. California Research Corp., Brit. Pat. 853,515, Nov. 9, 1960; *C.A.*, **55,** 14899 (1961).
70. California Research Corp., Brit. Pat. 917,961, Feb. 13, 1963; *C.A.*, **59,** 1426 (1963).
71. California Research Corp., Brit. Pat. 941,742, Nov. 13, 1963; *C.A.*, **60,** 3932 (1964).
72. California Research Corp., Brit. Pat. 949,402, Feb. 12, 1964; *C.A.*, **60,** 14311 (1964).
73. Calingaert, G., and H. A. Beatty, "The Redistribution Reaction," in *Organic Chemistry, An Advanced Treatise*, Vol. 2, 2nd ed., H. Gilman, Ed., Wiley, N.Y., 1943, Chapt. 24.
74. Callear, A. B., and R. G. W. Norrish, *Nature*, **184,** 1794 (1959).
75. Callear, A. B., and R. G. W. Norrish, *Proc. Roy. Soc.* (*London*), **A259,** 304 (1960); through *C.A.*, **55,** 14889 (1960).
76. Carothers, W. H., U.S. Pat. 2,008,003 (to E.I. Du Pont de Nemours and Co.), July 16, 1935; *C.A.*, **29,** 5862 (1935).
77. Carr, M. S., and G. G. Brown, *Ind. Eng. Chem.*, **21,** 1971 (1929).
78. Carrick, W. L., Brit. Pat. 873,499 (to Union Carbide Corp.), Appl. March 25, 1958; through *C.A.*, **56,** 3654 (1962).
79. Cartlidge, J., and C. F. H. Tipper, *Combust. Flame*, **5,** 87 (1961).
80. Chakravarty, N. K., *J. Inst. Petrol*, **48,** 196 (1962); *C.A.*, **57,** 8804 (1962).
81. Chakravarty, N. K., *Indian J. Technol.*, **1,** 181 (1963). through *C.A.*, **59,** 3611 (1963).
82. Chamberlain, G. H. N., D. E. Hoare, and A. D. Walsh, *Discussions Faraday Soc.*, **1953,** 89.
83. Chamberlain, G. H. N., and A. D. Walsh, *Proc. Roy. Soc.* (*London*), **A215,** 175 (1952).
84. Cheaney, D. E., D. A. Davis, A. Davis, D. E. Hoare, J. Protheroe, and A. D. Walsh, *Symp. Intern. Combustion, 7th, London, 1958*, 183 (publ. 1959); *C.A.*, **55,** 22769 (1961).

85. Chenicek, J. A., and H. S. Bloch, U.S. Pat. 2,867,673 (to U.O.P.), Jan. 6, 1959; *C.A.* **53,** 14053 (1959).
86. Chernesky, M., and J. Bardwell, *Can. J. Chem.*, **38,** 482 (1960).
87. Clark, F. M., U.S. Pat. 2,468,544 (to General Electric Co.), April 26, 1949; *C.A.*, **43,** 5887 (1949).
88. Cleverdon, D., and J. J. Straudinger, U.S. Pat. 2,481,086, Sept. 6, 1949; *C.A.*, **44,** 4718 (1950).
89. Closson, R. D., French Pat. 1,362,696 (to Ethyl Corp.), June 5, 1964; *C.A.*, **62,** 4052 (1965).
90. Coffield, T. H., U.S. Pat. 3,200,212 (to Ethyl Corp.), Aug. 6, 1963; *C.A.*, **60,** 549 (1964).
91. Collinson, R. G., and T. T. Jones, Brit, Pat. 767,417 (to Bakelite, Ltd.), Feb. 6, 1957; *C.A.*, **51,** 10124 (1957).
92. Cormany, C. L., W. R. Dial, and B. O. Pray, Brit. Pat. 912,118 (to Pittsburgh Plate Glass Co.), Dec. 5, 1962; *C.A.*, **58,** 13532 (1963).
93. Cornish, E. H., *J. Appl. Chem.* (*London*), **11,** 41 (1961).
94. Cramer, P. L., *J. Am. Chem. Soc.*, **56,** 1234 (1934).
95. Curry, S., *Symp.* (*Intern.*) *Combustion*, *9th*, *Ithaca*, *N.Y.*, **1962,** 1056 (publ. 1963).
96. Dannin, J., S. R. Sandler, and B. Baum, *Intern. J. Appl. Radiation Isotopes*, **16** (10), 589 (1963).
97. Davidson, W. E., K. Hills, and M. C. Henry, *J. Organometal. Chem.*, **3,** 285 (1965).
98. DePree, D. O., U.S. Pat. 2,893,857 (to Ethyl Corp.), July 7, 1959; *C.A.*, **53,** 18372 (1959).
99. Devillers, J. P., and J. G. Mackenzie, *World Health Organization Bilharziasis Research Document*, *Mol. Inf. Bull. No. 13*, p. 63 (1963).
100. deWolf, P., and A. M. van Londen, *T.N.O. Delft Report CL 63-33*, to Intern. Lead Zinc Res. Organ., Aug. 3, 1963.
101. Dickson, R. S. and B. O. West, *Australian J. Chem.*, **14,** 555 (1961); through *C.A.*, **55,** 12133 (1961).
102. Distillers Co., Ltd., Brit. Pat. 841,527, July 20, 1960; *C.A.*, **55,** 1083 (1961).
103. Donaldson, D. J., and G. L. Drake, Jr., *Quarterly Report No. 9*, *Project No. LC-63*, to Intern. Lead Zinc Res. Organ., New York, June 30, 1964.
104. Donaldson, D. J., and G. L. Drake, Jr., *Quarterly Report No. 12*, *Project No. LC-63*, to Intern. Lead Zinc Res. Organ., New York, April 3, 1965.
105. Donaldson, D. J., and G. L. Drake, Jr., *Quarterly Report No. 17*, *Project No. LC-63*, to Intern. Lead Zinc Res. Organ., New York, July 3, 1966.
106. Dorfman, R. I., *Experimental and Biological Test Methods*, Academic Press, N.Y., 1962.
107. Dorfman, R. I., *Proc. Soc. Exptl. Biol. Med.*, **116,** 1055 (1964).
108. Downs, D., S. T. Griffiths, and R. W. Wheeler, *J. Inst. Petrol.*, **47,** 1 (1961).
109. Downs, D., S. T. Griffiths, and R. W. Wheeler, *J. Inst. Petrol.*, **49,** 8 (1963).
110. Downs, D., A. D. Walsh, and R. W. Wheeler, *Trans. Roy. Soc.* (*London*), **A243,** 463 (1951).
111. E. I. du Pont de Nemours & Co., Brit. Pat. 331,494, Jan. 21, 1929; *C.A.*, **25,** 166 (1931).
112. E. I. du Pont de Nemours & Co., Brit. Pat. 408,967, April 18, 1934; *C.A.*, **28,** 5602 (1934).
113. E. I. du Pont de Nemours & Co., Brit. Pat. 948,642, Feb. 5, 1964; *C.A.*, **60,** 14315 (1964).

114. Eckert, G. W., and H. V. Hess, U.S. Pat. 3,009,792 (to Texaco, Inc.), March 7, 1958; *C.A.*, **56,** 7598 (1962).
115. Edgar, G., *Ind. Eng. Chem.*, **31,** 1439 (1939).
116. Edgar, G., Presentation 51–8, Division of Petroleum Chemistry, 120th A.C.S. Meeting, New York, Sept. 1951.
117. Egerton, A. C., *Trans. Faraday Soc.*, **24,** 269 (1928).
118. Egerton, A. C., and S. F. Gates, *J. Inst. Petrol.*, **13,** 244 (1927).
119. Englemann, M., U.S. Pat. 1,783,377 (to E. I. du Pont de Nemours & Co.), Dec. 2, 1930; *C.A.*, **25,** 374 (1931).
120. Erhard, K. H. L., and R. G. W. Norrish, *Proc. Roy. Soc.* (*London*), **A234,** 178 (1956).
121. Erhard, K. H. L., and R. G. W. Norrish, *Proc. Roy. Soc.* (*London*), **A259,** 297 (1960).
122. Esso Res. & Eng. Co., Brit. Pat. 823,839, Nov. 18, 1959; *C.A.*, **54,** 8068 (1960).
123. Esso Res. & Eng. Co., Brit. Pat. 824,555, Dec. 2, 1959; *C.A.*, **54,** 8068 (1960).
124. Esso Res. & Eng. Co., Brit. Pat. 888,456, Jan. 31, 1962; *C.A.*, **58,** 10026 (1963).
125. Ethyl Corp., Brit. Pat. 928,275, June 12, 1963; *C.A.*, **59,** 11168 (1963).
126. Ethyl Corp., Brit. Pat. 961,407, June 24, 1964; *C.A.*, **61,** 6842 (1964).
127. Ethyl Corp., Neth. Appl. 6,505,907, Nov. 12, 1965; *C.A.*, **64,** 14005 (1966).
128. Evans, T. W., W. E. Vaughn, and F. F. Rust, U.S. Pat. 2,376,675 (to Shell Development Co.), May 22, 1945; *C.A*, **39,** 3533 (1945).
129. Evans, T. W., W. E. Vaughn, and F. F. Rust, Brit. Pat. 567,524 (to Shell Development Co.), Feb. 19, 1945; *C.A.*, **41,** 2745 (1947).
130. Faragher, W. F., U.S. Pat. 2,502,444 (to Houdry Process Corp.), April 4, 1950; *C.A.*, **44,** 5632 (1950).
131. Faragher, W. F., U.S. Pat. 2,634,257 (to Houdry Process Corp.), April 7, 1953; *C.A.*, **47,** 7250 (1953).
132. I. G. Farbenindustrie A. G., French Pat. 750,869, June 6, 1933, publ. Aug. 21, 1933.
133. Felt, A. E., and R. V. Kerley, *Hydrocarb. Process. and Petr. Refiner*, **43,** 157 (1964).
134. Flenner, A. L., U.S. Pat. 2,673,869 (to E. I. du Pont de Nemours & Co.), March 30, 1954; *C.A.*, **49,** 5512 (1955).
135. Freimiller, L. R., and C. H. McKeever, U.S. Pat. 2,786,035 (to Rohm and Haas Co.), March 19, 1957; *C.A.*, **51,** 9210 (1957).
136. Freimiller, L. R., and C. H. McKeever, U.S. Pat. 2,786,036 (to Rohm and Haas Co.), March 19, 1957; *C.A.*, **51,** 9210 (1957).
137. Furukawa, J., *Yuki Gosei Kagaku Kyokaishi*, **17,** 649 (1959); through *C.A.* **54,** 1049 (1960).
138. Furukawa, J., and T. Tsuruta, *Bull. Inst. Chem. Research*, *Kyoto Univ.*, **38,** 319 (1960); through *C.A.*, **55,** 18173 (1961).
139. Furukawa, J., T. Tsuruta, and Y. Takeda, *Kogyo Kagaku Zasshi*, **64,** 1307 (1961); through *C.A.*, **57,** 4818 (1962).
140. Gaylord, N. G., and H. Mark, *Linear and Stereoregular Addition Polymers*, Interscience, New York, 1965.
141. Gelius, R., *Freiberger Forschungsh.*, **A264,** 19 (1963). *C.A.* **59,** 6166 (1963).
142. Glassford, H. A., and R. L. Macklin, *PB Report 95754*, July, 1944.
143. Gluckstein, M. E., C. Walcutt, and R. R. Acles, *S.A.E. Preprint 201F* (1960); *C.A.*, **54,** 25721 (1960).
144. Gol'dshtein, A. L., N. P. Lapisova, and N. P. Zorina, USSR Pat. 138,040, Oct. 5, 1959; through *C.A.*, **56,** 4966 (1962).

145. Gol'dshtein, A. L., N. P. Lapisova, and N. P. Zorina, *Plasticheskie Massy, 1960*, **No. 11,** 3; through *C.A.*, **55,** 13900 (1961).
146. Goodacre, C. L., and D. Foord, *Riv. Combust.*, **16,** 340 (1962).
147. Goodacre, C. L., and D. Foord, *Acta Chim. Acad. Sci. Hung.*, **36,** 235 (1963); through *C.A.*, **59,** 4941 (1963).
148. Goodacre, C. L., D. Foord, and M. Hedde, *Bull. Assoc. Franc. Techniciens Petrole.*, **No. 152,** 255 (1962); *C.A.*, **57,** 4932 (1962).
149. Gorsich, R. D., U.S. Pat. 3,262,909 (to Ethyl Corp.), July 26, 1966; *C.A.*, **65,** 15614 (1966).
150. Graham, W. A. G., A. D. Little 1959 Summary Report, *Applications of Organolead Chemicals*, to Lead Industries Association, April 5, 1960.
151. Greber, G., and G. Egle, *Makromol. Chem.*, **64,** 207 (1963); *C.A.*, **59,** 5276 (1964).
152. Griffiths, S. T., and W. D. Pigott, *Erdöl Kohle*, **17,** 997 (1964).
153. Guccione, E., *Chem. Eng.*, **72,** 102 (1965).
154. Gurevitch, E. S., and M. A. Dolgopol'skaya, *Tr. Sevostopol'sk. Biol. St., Akad. Nauk Ukr. SSR*, **15,** 472 (1964); through *C.A.*, **64,** 5293 (1966).
155. Gursky, J., and V. Vesely, *Freiberger Forschungsh.*, **340A,** 303 (1964); through *C.A.*, **63,** 2816 (1965).
156. Gursky, J., and V. Vesely, *Ropa Uhlie*, **7,** 53 (1965); through *C.A.*, **63,** 2816 (1965).
157. Hand, C. W., and G. B. Kistiakowsky, *J. Chem. Phys.*, **37,** 1239 (1962).
158. Hanford, W. E., and J. Harmon, U.S. Pat. 2,418,832 (to E. I. du Pont de Nemours & Co.), April 15, 1947; *C.A.*, **42,** 581 (1948).
159. Hanford, W. E., and J. Harmon, U.S. Pat. 2,440,801 (to E. I. du Pont de Nemours & Co.), May 4, 1948; *C.A.*, **42,** 6839 (1948).
160. Hannum, J. A., U.S. Pat. 2,559,071 (to Borg Warner Corp.), July 3, 1951; *C.A.*, **45,** 10545 (1951).
161. Hartle, R. J., U.S. Pat. 3,055,748 (to Gulf Research & Dev. Co.), Sept. 25, 1962; *C.A.*, **58,** 7776 (1963).
162. Hartle, R. J., U.S. Pat. 3,055,925 (to Gulf Research & Dev. Co.), Sept. 25, 1962; *C.A.*, **59,** 1681 (1963).
163. Harwood, J. H., "Group IVb Section," in *Industrial Applications of the Organometallic Compounds. A Literature Survey*, Chapman and Hall, London, 1963, p. 124.
164. Hatta, S., *J. Soc. Chem. Ind. Japan*, **28,** 1346 (1925); *C.A.*, **20,** 816 (1926).
165. Hatton, R. E., and L. R. Stark, French Pat. 1,356,569 (to Monsanto Chemical Co.), March 27, 1964; *C.A.*, **62,** 3870 (1965).
166. Heap, R., and B. C. Saunders, *J. Chem. Soc.*, **1949,** 2983.
167. Heap, R., B. C. Saunders, and G. J. Stacey, *J. Chem. Soc.*, **1951,** 658.
168. Heiligmann, R. G., and P. B. Stickney, U.S. Pat. 2,765,297 (to Borden Co.), Oct. 12, 1956; *C.A.*, **50,** 17530 (1956).
169. Henry, M. C., and W. E. Davidson, *Annals N.Y. Acad. Sci.*, **125,** 172 (1965).
170. Henry, M. C., and H. Gorth, *Summary Report No. 20-A*, Project LC-28, to Intern. Lead Zinc Res. Organ., New York, June 30, 1966.
171. Herold, P., and F. Asinger, Ger. Pat. 765,790 (to I. G. Farbenindustrie A. G.), June 1, 1953; *C.A.*, **49,** 3238 (1955).
172. Hesselberg, H. E., Ger. Pat. 1,145,434 (to Ethyl Corp.), March 14, 1963; *C.A.*, **58,** 13689 (1963).
173. Hills, K., and M. C. Henry, *Intern. Lead Zinc Res. Organ. Research Digest*, No. 12, p. 13, Oct. 1, 1963.

174. Hochwalt, C. A., U.S. Pat. 2,552,278 (to Monsanto Chemical Co.), May 8, 1951; *C.A.*, **46,** 139 (1952).
175. Holt, L. C., U.S. Pat. 2,344,872 (to E. I. du Pont de Nemours & Co.), March 21, 1944; *C.A.*, **38,** 3776 (1944).
176. Hord, R. A., and H. B. Tolefson, *Virginia J. Sci.*, **16,** 105 (1965); *C.A.*, **63,** 16086 (1965).
177. Horie, S., *Nippon Kagaku Zasshi*, **78,** 1795 (1957); through *C.A.*, **53,** 21868 (1959).
178. Horizons, Inc., Neth. Appl. 6,409,367, Feb. 15, 1965; through *C.A.*, **63,** 181 (1965).
179. Hostettler, F., and E. F. Cox, Brit. Pat. 898,060 (to Union Carbide Corp.), June 6, 1962; *C.A.*, **57,** 11395 (1962).
180. Howard, W. B., U.S. Pat 3,062,906 (to Monsanto Chemical Co.), Nov. 6, 1962; *C.A.*, **58,** 2318 (1963).
181. Huff, T., and E. Perry, *J. Polymer Sci.*, *A.*, **1,** 1553 (1963); through *C.A.*, **59,** 776 (1963).
182. Hughes, F. J., and G. H. Meguerian, *Proc. Am. Petrol. Inst.*, *Sect. III*, **42,** 588 (1962).
183. Ilgemann, R., and R. D. Rauschenbach, French Pat. 1,411,363 (to Badische Anilin- u. Soda-Fabrik A. G.), Sept. 17, 1965; *C.A.*, **65,** 10765 (1966).
184. Ilyana v. Thyssen-Bornemisza, Brit. Pat. 997,367, July 7, 1965; *C.A.*, **63,** 11242 (1965).
185. Imhausen, A., Ger. Pat. 362,814, Sept. 29, 1921; *Soc. Chem. Ind.* (London) **42,** 315 (1923).
186. Imhausen, A., *J. Soc. Chem. Ind.* (*London*), **42,** 315 (1923).
187. Imperial Chemical Industries, Ltd., Brit. Pat. 808,132, Jan. 28, 1959; *C.A.*, **54,** 21850 (1960).
188. Imperial Chemical Industries, Ltd., Brit. Pat. 812,176, April 22, 1959; *C.A.*, **53,** 15649 (1959).
189. Ingham, R. K., S. D. Rosenberg, and H. Gilman, *Chem. Rev.*, **60,** 459 (1960).
190. Isbenjian, C., U.S. Pat. 2,898,330 (to Aries Associates), Aug. 4, 1959; *C.A.*, **54,** 966 (1960).
191. Joyner, F. B., U.S. Pat. 3,072,631 (to Eastman Kodak Co.), Jan. 8, 1963; *C.A.*, **58,** 7006 (1963).
192. Joyner, F. B., and N. H. Shearer, U.S. Pat. 3,086,964 (to Eastman Kodak Co.), April 23, 1963; *C.A.*, **59,** 777 (1963).
193. Kashireninov, O. E., *Materialy 4-oi* (*Chetvertoi*) *Nauchn. Konf. Aspirantov* (Rostov-on-Don: Rostovsk. Univ.) Sb., **1962,** 144; through *C.A.*, **60,** 10792 (1964).
194. Kawabata, N., T. Tsuruta, and J. Furukawa, *Kogyo Kagaku Zasshi*, **65,** 70 (1962); through *C.A.*, **57,** 4850 (1962).
195. Keck, K., *Z. Physik*, **178,** 120 (1964); through *C.A.*, **60,** 13391 (1962).
196. Kerk, G. J. M. van der, *Ind. Eng. Chem.*, **58,** 29 (1966).
197. Kerley, R. V., and A. E. Felt, U.S. Pat. 3,038,792 (to Ethyl Corp.), June 12, 1962; *C.A.*, **57,** 7524 (1962).
198. Kharasch, M. S., U.S. Pat. 1,770,886 (to E. I. du Pont de Nemours & Co.), July 15, 1930; *C.A.*, **25,** 5734 (1931).
199. Kharasch, M. S., U.S. Pat. 1,987,685 (to E. I. du Pont de Nemours & Co.), Jan. 15, 1935; *C.A.*, **29,** 1436 (1935).
200. Klema, F., *Chemiker Ztg.*, **90,** 106 (1966); through *C.A.*, **64,** 17632 (1966).
201. Kluiber, R. W., and W. L. Carrick, Brit. Pat. 856,859 (to Union Carbide Corp.), Dec. 21, 1960; *C.A.*, **55,** 13924 (1961).

202. Kochkin, D. A., L. V. Lukijanova, and E. B. Reznikova, *Zh. Obshch. Khim.*, **33,** 1945 (1963); *C.A.*, **59,** 11551 (1963).

203. Kochkin, D. A., V. I. Vashkov, and V. P. Dermova, *Zh. Obshch. Khim.*, **34,** 325 (1964); through *C.A.*, **60,** 10706 (1964).

204. Kochkin, D. A., V. I. Vashkov, G. V. Kiryutkin, and A. R. Savel'eva, *Zh. Obshch. Khim.*, **34,** 4027 (1964); *C.A.*, **62,** 9042 (1965).

205. Kochkin, D. A., and S. G. Verenikina, *Tr. Vses. Nauchn.-Issled. Vitamin Inst.*, **8,** 39 (1961); through *C.A.*, **58,** 6851 (1963).

206. Kochkin, D. A., S. G. Verenikina, and I. B. Chekmareva, *Dok. Akad. Nauk SSSR*, **139,** 1375 (1961); *C.A.*, **56,** 7343 (1962).

207. Kooijman, P. L., and W. L. Ghijsen, *Rec. trav. chim.*, **66,** 247 (1947).

208. Kooijman, P. L., and W. L. Ghijsen, *Rec. trav. chim.*, **66,** 673 (1947).

209. Korn, T. M., and G. Moss, S.A.E. Preprint, **207D,** 1960; *C.A.*, **57,** 8802 (1962).

210. Koton, M. M., *Dokl. Akad. Nauk SSSR*, **88,** 991 (1953); *C.A.*, **48,** 8727 (1954).

211. Kozlov, L. M., L. S. Drafkina, and V. I. Burmistrov, *Tr. Kazansk. Khim.-Tekhnol. Inst.*, **No. 30,** 109 (1962); through *C.A.*, **60,** 3106 (1964).

212. Krause, E., *Ber.*, **62,** 135 (1929).

213. Kropp, J. C., and M. Burton, *J. Chem. Phys.*, **37,** 1752 (1962).

214. LaBrijn, J., *T.N.O. Delft Report CL 63-56*, to Intern. Lead Zinc Res. Organ., New York, May 15, 1963.

215. LaBrijn, J., *T.N.O.*, *Delft*, *Report CL-63-67*, to Intern. Lead Zinc. Res. Organ., New York, June 12, 1963.

216. LaBrijn, J., *T.N.O. Delft Report CL 63-91*, to Intern. Lead Zinc Res. Organ., New York, Sept. 13, 1963.

217. LaBrijn J., *T.N.O. Delft Report CL 64-15*, to Intern. Lead Zinc Res. Organ., New York, Feb. 10, 1964.

218. LaBrijn, J., *T.N.O. Delft Report CL 64-48*, to Intern. Lead Zinc Res. Organ., New York, Feb. 10, 1964.

219. LaBrijn, J., *T.N.O. Delft Report CL 64-49*, to Intern. Lead Zinc Res. Organ., New York, June 9, 1964.

220. Langenbeck, W., G. Loose, and H. Fürst, *Chem. Tech.* (*Berlin*), **4,** 247 (1952); *C.A.*, **47,** 12726 (1953).

221. Langer, H. G., U.S. Pat. 3,117,147 (to Dow Chemical Co.), Jan. 7, 1964; *C.A.*, **60,** 8061 (1964).

222. Langer, H. G., U.S. Pat. 3,120,550 (to Dow Chemical Co.), Feb. 4, 1964; *C.A.*, **60,** 12051 (1964).

223. Lask, G. W., and H. G. Wagner, *Forsch. Gebiete Ingenieurw.*, **27** (2), 52 (1961); through *C.A.*, **55,** 26614 (1961).

224. Laznowski, I., *Compt. rend. congr. intern. chim. ind.*, *31e*, *Liege*, *1958* [Publ. as *Ind. chim. belge*, *Suppl.* **1,** 552 (1959)]; through *C.A.*, **54,** 3929 (1960).

225. Lefebvre, G., *Rev. inst. franc. petrole et Ann. combustibles liquides*, **15,** 730 (1960); *C.A.*, **55,** 3409 (1961).

226. Leshchenko, S. S., E. E. Finkel, N. I. Sheverdina, L. V. Abramova, and V. C. Karpov, USSR Pat. 138,043, appl. June 21, 1960; through *C.A.*, **56,** 6181 (1962).

227. Lewis, B., and G. von Elbe, *Combustion*, *Flames and Explosions of Gases*, 2nd ed., Academic Press, N. Y., 1961.

228. Ligett, W. B., R. D. Closson, and C. N. Wolf, U.S. Pat. 2,595,798 (to Ethyl Corp), May 6, 1952; *C.A.*, **46,** 7701 (1952).

229. Ligett, W. B., R. D. Closson, and C. N. Wolf, U.S. Pat. 2,640,006 (to Ethyl Corp.), May 26, 1953; *C.A.*, **47,** 8307 (1953).

230. Ligett, W. B., R. D. Closson, and C. N. Wolf, U.S. Pat. 2,654,689 (to Ethyl Corp.), Oct. 6, 1953. *C.A.*, **48,** 941 (1954).
231. A. D. Little, 1960 Summary Report, *Applications of Organic Lead Chemicals*, to Lead Industries Association, April 15, 1961.
232. Lovell, W. G., and E. B. Rifkin, Preprint **A67,** "A Decade of Progress in Petroleum Combustion," presented before Division of Petroleum Chemistry, A.C.S. Meeting, Chicago, Ill, Sept. 3–8, 1961.
233. Lucas, G. R., U.S. Pat. 2,598,403 (to General Electric Co.), May 27, 1952; *C.A.*, **46,** 9342 (1952).
234. Ludlum, K. H., S. E. Wiberley, and W. H. Bauer, *U.S. Dept. Comm., Office Tech. Serv., AD 259,093* (1961); *C.A.*, **58,** 7405 (1963).
235. Marchenko, G. N., S. F. Goldofin, and G. Ya. Smertin, *Vysokomol. Soedin.*, **8,** 2087 (1966); through *C.A.*, **66,** 66095d (1967).
236. Marsh, M. C., and B. T. Simpson, *J. Cancer Research*, **11,** 417 (1927); *C.A.*, **22,** 3696 (1928).
237. Marvel, C. A., and R. G. Woolford, *J. Am. Chem. Soc.*, **80,** 830 (1958).
238. McCombie, H., and B. C. Saunders, *Nature*, **159,** 491 (1947).
239. Meaker, C. L., C. S. Wu, and L. J. Rainwater, *Phys. Rev.*, **73,** 1240 (1947); *C.A.*, **43,** 4095 (1948).
240. Midgley, T., Jr., and T. A. Boyd, *Ind. Eng. Chem.*, **14,** 894 (1922).
241. "Montecatini" (Societa generale l'Industria mineraria e chimica), Belg. Pat. 546,474, filed March 27, 1956.
242. "Montecatini," (Societa generale l'Industria mineraria e chimica) Brit. Pat. 823,236, Nov. 11, 1959; *C.A.*, **54,** 19399 (1960).
243. Morris, W. E., S.A.E. Preprint, **547C,** 1962.
244. Mowery, D. F., U.S. Pat. 2,452,595 (to E. I. du Pont de Nemours & Co.), Nov. 2, 1948; *C.A.*, **43,** 1805 (1949).
245. Nagasawa, M., F. Yamamoto, and T. Maeda, Jap. Pat. 1415 ('64) (to Ihara Agricultural Chemical Co.), Feb. 14, 1964; through *C.A.*, **60,** 11906 (1964).
246. Nazarova, L. M., and G. E. Aleksandrova, *Vysokomolekul. Soedin.*, **3,** 1823 (1961); through *C.A.*, **56,** 14462 (1962).
247. Nederlandse Centrale Organisatie voor Toegepast-Natuurwetenschappelijk Onderzoek., Neth. Appl. 299,409, Aug. 25, 1965; *C.A.*, **64,** 6853 (1966).
248. Newland, G. C., and J. W. Tamblyn, *Proc. Battelle Symp. Thermal Stability Polymers, Columbus, Ohio, 1963*, N1–N7; through *C.A.*, **60,** 10876 (1964).
249. Newland, G. C., and J. W. Tamblyn, *J. Appl. Poly. Sci.*, **9,** 1947 (1965); *C.A.*, **63,** 7169 (1965).
250. Newman, S. R., K. L. Dille, R. W. Heisler, and M. F. Fontaine, *S.A.E. Journal*, **68,** 42 (1960).
251. Nickerson, S. P., *J. Chem. Ed.*, **31,** 560 (1954).
252. Nicora, C., G. Borsini, and L. Ratti, *J. Polymer Sci. B*, **4** (2), 151 (1966).
253. Nienburg, H. J., H. Böhm, and R. Herbeck, Ger. Pat. 1,105,167 (to Badische Anilin- u. Soda-Fabrik A. G.), April 20, 1961; *C.A.*, **55,** 26541 (1961).
254. Norman, V., and T. P. Whaley, U.S. Pat. 3,018,194 (to Ethyl Corp.), Jan. 23, 1962; *C.A.*, **56,** 8402 (1962).
255. Norrish, R. G. W., *Science*, **149,** 1470 (1965).
256. Nottes, E. G., and J. F. Cordes, *Erdöl Kohle*, **18,** 885 (1965).
257. Novak, L. J., U.S. Pat. 2,930,767 (to Commonwealth Engineering Co.), March 29, 1960; *C.A.*, **54,** 17752 (1960).
258. Nowakowska, M., *Wiadomosci Chem.*, **16,** 307 (1962); through *C.A.*, **57,** 11375 (1962).

259. Oosterhoff, L. J., *Rec. trav. chim.*, **59,** 811 (1940).
260. Osborn, S. W., and A. J. Yu, Brit. Pat. 973,318 (to Thiokol Chemical Corp.), Oct. 21, 1964; *C.A.*, **62,** 1817 (1965).
261. Osborn, S, W., and A. J. Yu, U.S. Pat. 3,201,376 (to Thiokol Chemical Corp.), Aug. 17, 1965; *C. A.*, **63,** 15055 (1965).
262. Overmars, H. G. J., and G. M. Van der Want, *Chimia*, **19,** 126 (1965).
263. Patnode, W., and F. Schmidt, *J. Am. Chem. Soc.*, **67,** 2272 (1945).
264. Perilstein, W. L., U.S. Pat. 3,287,265 (to Intern. Lead Zinc Res. Organ.), Nov. 22, 1966; *C.A.*, **66,** 48102z (1967).
265. Perry, R. H., Jr., C. J. DiPerna, and D. P. Heath, S.A.E Preprint, **207A,** 1960; *C.A.*, **57,** 8802 (1963).
266. Perry, R. H., Jr., P. L. Gerard, and D. P. Heath, *S.A.E. Journal*, **30,** 48 (1962).
267. Peski, A. J. van, U.S. Pat. 2,478,006 (to Shell Development Co.), Aug. 2, 1949; *C.A.*, **43,** 8666 (1949).
268. Peski, A. J. van, N. Max, and J. A. van Melsen, U.S. Pat. 2,228,039 (to Shell Development Co.) Jan. 7, 1941 ; *C.A.*, **35,** 2530 (1941).
269. Peters, R. A., U.S. Pat. 2,967,105 (to Stupa Corp.), Jan. 3, 1961; *C.A.*, **55,** 9129 (1961).
270. Petrochemicals, Ltd., Brit. Pat. 841,872, July 20, 1960.
271. Phillips Petroleum Co., Belg. Pat. 549,836, July 26, 1956.
272. Phillips Petroleum Co., Brit. Pat. 796,530, June 11, 1958; *C.A.*, **53,** 1836 (1959).
273. Phillips Petroleum Co., Brit. Pat. 796,539, 1958.
274. Phillips Petroleum Co., Brit. Pat. 931,313, July 17, 1963; *C.A.*, **59,** 10328 (1963).
275. Phillips Petroleum Co., Brit. Pat. 931,440, July 17, 1963; *C.A.*, **59,** 10328 (1963).
276. Phillips Petroleum Co., Ger. Pat. 1,124,699, March 1, 1962; *C.A.*, **58,** 3582 (1963).
277. Pipenberg, K. J., A. J. Pahnke, and R. H. Blaker, *Proc. Am. Pet. Inst., Sect. III,* **38,** 68 (1958).
278. Papova, Z. V., N. V. Tikhova, and N. S. Vyazankin, *Vysokomolekul. Soedin., Khim. Svoistva i Modifikatsiya Polimerov, Sb. Statei*, **1964,** 175; through *C.A.*, **62,** 14892 (1965).
279. Postell, D. L., and W. E. Morris, *S.A.E. Preprint, 207C*, 1960; *C.A.*, **57,** 8801 1962).
280. Preuss, H. P., *Metal Finishing*, **64** (3), 61 (1966); **64** (4), 63 (1966).
281. Ray, R. L., and T. O. Sistrunk, U.S. Pat. 2,868,772 (to Ethyl Corp.), Jan. 13, 1959; *C.A.*, **53,** 8711 (1959).
282. Razuvaev, G. A., N. S. Vyazankin, and Yu. I. Dergunov, *Dokl. Akad. Nauk, SSSR*, **132,** 364 (1960); *C.A.*, **54,** 20937 (1960).
283. Razuvaev G. A., N. S. Vyazankin, and Yu. I. Dergunov. *Zh. Obshch. Khim.*, **30,** 1310 (1960); *C.A.*, **55,** 362 (1961).
284. Reinhardt, R. C., U.S. Pat. 2,160,939 (to Dow Chemical Co.), June 6, 1939; *C.A.*, **33,** 7441 (1939).
285. Richard, W. R., U.S. Pat. 2,477,349, (to Monsanto Chemical Co.), July 26, 1949; *C.A.*, **44,** 4718 (1950).
286. Richardson, W. L., U.S. Pat. 3,010,980 (to California Research Corp.), Nov. 28, 1961; *C.A.*, **56,** 11620 (1962).
287. Richardson, W. L., U.S. Pat. 3,116,126 (to California Research Corp.), Dec. 31, 1963; *C.A.*, **60,** 6686 (1964).
288. Richardson, W. L., M. R. Barusch, G. J. Kautsky, and R. E. Steinke, *J. Chem. Eng. Data*, **6,** 305 (1961).
289. Richardson, W. L., M. R. Barusch, G. J. Kautsky, and R. E. Steinke, *Ind. Eng. Chem.*, **53,** 305 (1961).

290. Richardson, W. L., M. R. Barusch, W. T. Stewart, G. J. Kautsky, and R. K. Stone, *Ind. Eng. Chem.*, **53,** 306 (1961).
291. Richardson, W. L., P. R. Ryason, G. J. Kautsky, and M. R. Barusch, *Symp. (Intern.) Combust., 9th*, Ithaca, N.Y., **1962,** 1023 (pub. 1963); *C.A.*, **59,** 13744 (1963).
292. Riegert, J. and E. Siganos, U.S. Pat. 3,124,433 (to Esso Res. & Eng. Co.), March 10, 1964; *C.A.*, **60,** 14317 (1964).
293. Rifkin, E. B., Preprint, 23rd American Petroleum Institute Div. Refg., Midyear Meeting, Los Angeles, 1958.
294. Rifkin, E. B., and C. Walcutt, *Ind. Eng. Chem.*, **48,** 1532 (1956).
295. Rifkin, E. B., and C. Walcutt, *SAE Trans.*, **65,** 552 (1957).
296. Ritchie, W. Jiminez, and I. Joseph, *World Health Organization Bilharziasis Research Document, Mol. Info. Bull. No. 18, III* (1964).
297. Roedel, M. J., U.S. Pat. 2,409,996 (to E. I. du Pont de Nemours & Co.), Oct. 12, 1946; *C.A.*, **41,** 625 (1947).
298. Roedel, M. J., U.S. Pat. 2,439,528 (to E. I. du Pont de Nemours & Co.), April 12, 1948; *C.A.*, **42,** 2819 (1948).
299. Roedel, M. J., U.S. Pat. 2,462,678 (to E. I. du Pont de Nemours & Co.), Feb. 22, 1949; *C.A.*, **43,** 4515 (1949).
300. Rohm & Haas Co., Brit. Pat. 771,646, April 3, 1957; *C.A.*, **51,** 12542 (1957).
301. Rohm & Haas Co., Brit. Pat. 771,647, April 3, 1957; *C.A.*, **51,** 12542 (1957).
302. Rohm & Haas Co., French Pat. 1,130,443, Feb. 5, 1957.
303. Ross, A., and E. B. Rifkin, *Ind. Eng. Chem.*, **48,** 1528 (1956).
304. Sabina, J. R., paper presented to Lead Industries Association 35th Annual Meeting, April, 1963.
305. Salooja, K. C., *Combust. Flame*, **9,** 211 (1956); *C.A.*, **64,** 495 (1966).
306. Sandler, S. R., and K. C. Tsou, *Intern. J. Appl. Radiation Isotopes*, **15,** 419 (1964); through *C.A.*, **61,** 9139 (1964).
307. Sandler, S. R., and K. C. Tsou, *J. Phys. Chem.*, **68,** 300 (1964); through *C.A.*, **60,** 6988 (1964).
308. Sargent, D. E., U.S. Pat. 2,462,680 (to E. I. du Pont de Nemours & Co.), Feb. 22, 1949; *C.A.*, **43,** 4515 (1949).
309. Sargent, D. E., and W. E. Hanford, U.S. Pat. 2,467,234 (to E. I. du Pont de Nemours & Co.), April 12, 1949 *C.A.*, **43,** 6000 (1949).
310. Saunders, B. C., and G. J. Stacey, *J. Chem. Soc.*, **1949,** 919.
311. Scott, S. L., U.S. Pat. 2,407,181 (to E. I. du Pont de Nemours & Co.), Sept. 3, 1946; *C.A.*, **41,** 2931 (1947).
312. Scott, C. R., U.S. Pat. 2,994,189 (to Phillips Petroleum Co.), Aug. 1, 1961; *C.A.*, **55,** 27891 (1961).
313. Seyferth, D., and R. B. King, "E., Lead," in *Annual Survey of Organometallic Chemistry*, Vol. 1, 1964, Elsevier, Amsterdam, 1965.
314. Shapiro, H., "Preparation of Tetraalkyllead Compounds from Lead or Its Alloys," in *Metal-Organic Compounds*, Advances in Chemistry Series, Vol. 23, American Chemical Society, Washington, 1959.
315. Shapiro, H., and J. V. Capinjola, Ethyl Corporation, unpublished data.
316. Shapiro, H., and F. W. Frey, "Organolead Compounds," in *Kirk-Othmer Encyclopedia of Chemical Technology*, Interscience, New York, 1967.
317. Shapiro, H., V. F. Hnizda, and G. Calingaert, U.S. Pat. 2,674,610 (to Ethyl Corp.), Apr. 6 1954; *C.A.*, **49,** 4706 (1955).
318. Sheftel, V. O., *Gigiena i Sanit.*, **29** (10), 105 (1964); through *C.A.*, **62,** 4508 (1965).

319. Sidgwick, N. J., *Chemical Elements and Their Compounds*, Oxford Press, London, 1950.
320. Sijpesteijn, A. J., F. Rijkens, J. G. A. Luijten, and L. C. Willemsens, *Antonie van Leeuwenhoek J. Microbiol. Serol.*, **28,** 346 (1962); *C.A.*, **58,** 7308 (1962).
321. Smith, H. V., *Brit. Plastics*, **27,** 213 (1954); *C.A.*, **51,** 756 (1957).
322. Sobue, H., and Y. Saito, *J. Chem. Soc., Japan*, **62** (8), 1110 (1959); through *C.A.*, **55,** 13901 (1961).
323. Societa Edison S.p.A.-Settore Chimico, Neth. Appl. 6,503,797, Oct. 4, 1965; *C.A.*, **64,** 8340 (1966).
324. Solvay and Cie, Belg. Pat. 545,968, Sept. 10, 1956; *C.A.*, **54,** 16924 (1960).
325. Solvay and Cie, Belg. Pat. 547,618, May 7, 1956; *C.A.*, **54,** 16917 (1960).
326. Solvay and Cie, Belg. Pat. 553,839, Dec. 29, 1956; *C.A.*, **53,** 18550 (1959).
327. Solvay and Cie, Belg. Pat. 586,754, July 20, 1960; *C.A.*, **54,** 26003 (1960).
328. Sowa, F. J., and E. J. Kenney, U.S. Pat. 2,580,473, Jan. 1, 1952; *C.A.*, **46,** 4823 (1952).
329. Stormont, D. H., *Oil Gas J.*, **60,** 189 (1962).
330. Stuart, A. P., U.S. Pat. 2,921,060 (to Sun Oil Co.), Jan. 12, 1960; *C.A.*, **54,** 12657 (1960).
331. Sturgis, B. M., Preprint A51, Division of Petroleum Chemistry, A.C.S. Meeting, Chicago, Ill., Sept. 3–8, 1961.
332. Tanake, Y., and Y. Nagai, *J. Soc. Chem. Ind. Japan*, **29,** 266 (1926); through *C.A.*, **21,** 323 (1927).
333. Taylor, H. S., and W. H. Jones, *J. Am. Chem. Soc.*, **52,** 1111 (1930).
334. Thompson, R., U.S. Pat. 2,450,099 (to Universal Oil Products Co.), Sept. 28, 1948; *C.A.*, **43,** 1440 (1949).
335. Tisdale, W. H., Brit. Pat. 578,312 (to E. I. du Pont de Nemours & Co.), June 24, 1946.
336. Tsou, K. C., and S. R. Sandler, U.S. Pat. 3,244,637 (to Borden. Co.), April 5, 1966; *C.A.*, **65,** 1732 (1966).
337. Tsuruta, T., *Yuki Gosei Kagaku Kyokaishi*, **20,** 10 (1962); through *C.A.*, **56,** 7485 (1962).
338. Twiss, D. F., and F. A. Jones, Brit. Pat. 360,599 (to Dunlop Rubber Co.), Oct. 7, 1930; *C.A.*, **25,** 2878 (1931).
339. Union Carbide Corp., Belg. Pat. 581,026, July 24, 1959.
340. Union Carbide Corp., Brit. Pat. 808,956, Feb. 11, 1959; *C.A.*, **53,** 8593 (1959).
341. Union Carbide Corp., Brit. Pat. 873,498, Appl. March 25, 1958; through *C.A.*, **56,** 3654 (1962).
342. Vaughn, W. E., and F. F. Rust, *J. Org. Chem.*, **5,** 449 (1940).
343. Vaughn, W. E., and F. F. Rust, U.S. Pat, 2,299,441 (to Shell Development Co.), Oct. 20, 1942; *C.A.*, **37,** 1722 (1943).
344. Wall, H. H., Jr.; U.S. Pat. 3,158,636 (to Ethyl Corp.), Nov. 24, 1964; *C.A.*, **62,** 3870 (1965).
345. Wallen, L. L., *Iowa St. Coll. J. Sci.*, **29,** 526 (1955); *C.A.*, **49,** 10418 (1955).
346. Wheeler, R. W., D. Downs, and A. D. Walsh, *Nature*, **162,** 893 (1948).
347. Whelen, M. S., "Preparation of Alkyl Mercurials with Tetraethyllead," in *Metal-Organic Compounds*, Advances in Chemistry Series, Vol. 23, American Chemical Society, Washington, D.C., 1959, pp. 82–86.
348. Wiley, R. M., U.S. Pat. 2,160,932 (to Dow Chemical Co.), June 6, 1939; *C.A.*, **33,** 7443 (1939).

349. Wiley, R. M., U.S. Pat. 2,160,933 (to Dow Chemical Co.), June 6, 1939; *C.A.*, **33,** 7443 (1939).
350. Wiley, R. M., U.S. Pat. 2,160,935 (to Dow Chemical Co.), June 6, 1939; *C.A.*, **33,** 7443 (1939).
351. Willemsens, L. C., in *Organolead Chemistry*, Intern. Lead Zinc Res. Organ., New York, 1964, Chapt. 7, p. 64.
352. Willemsens, L. C., *Quarterly Report No. 16*, *Project No. LC-18*, to Intern. Lead Zinc Res. Organ., New York, March, 1964.
353. Willemsens, L. C., *Quarterly Report No. 20*, *Project No. LC-18*, to Intern. Lead Zinc Res. Organ., New York, Feb., 1965.
354. Willemsens, L. C., *Progress Report No. 24* (*1964 Annual Report*), *Project No. LC-18*, Intern. Lead Zinc Res. Organ., New York.
355. Willemsens, L. C., *A Fundamental Study of Organolead Chemistry*, *Report on Project LC-18*, to Intern. Lead Zinc Res. Organ., New York, July, 1964.
356. Wilson, T. P., U.S. Pat. 2,824,116 (to Union Carbide Corp.), Feb. 18, 1958; *C.A.*, **52,** 7668 (1958).
357. Wilzbach, K. E., F. R. Mayo, and R. Van Meter, *J. Am. Chem. Soc.*, **70,** 4069 (1948).
358. Woolford, R. G., *Dissertation Abstr.*, **20,** 104 (1959).
359. Wright, P. G., *Combust. Flame*, **5,** 205 (1961).
360. Wu, C. A., and C. L. Meaker, *PB Report A-3846*, U.S. Gov't. Res. Repts., 26, 1956.
361. Wyler, J. A., U.S. Pat. 2,486,773 (to Trojan Powder Co.), Nov. 1, 1949; *C.A.*, **44,** 1709 (1950).
362. Yngve, V., U.S. Pat. 2,267,777 (to Carbide & Carbon Chemicals Co.), Dec. 30, 1941; *C.A.*, **36,** 1405 (1942).
363. Yngve, V., U.S. Pat. 2,267,778 (to Carbide & Carbon Chemicals Co.), Dec. 30, 1941; *C.A.*, **36,** 2647 (1942).
364. Yngve, V., U.S. Pat. 2,307,090 (to Carbide & Carbon Chemicals Co.), Jan. 5, 1943; *C.A.*, **37,** 3532 (1943).
365. Young, C. O., and S. D. Douglas, U.S. Pat. 1,775,882 (to Carbide & Carbon Chemicals Corp.), Sept. 16, 1930; *C.A.*, **24,** 53084 (1930).
366. Zherdev, Yu. V., A. Ya. Korolev, and N. S. Lesnov, *Vysokomolekul. Soedin.*, *Khim. Svoista i Modifikatsiya Polimerov, Sb. Statei*, **1964,** 260; through *C.A.*, **62,** 2880 (1965).
367. Ziegler, K., French Pat. 1,137,459, May 29, 1957.
368. Ziegler, K., *Compt. rend. congr. intern. chim. ind. 31e*, *Liege*, *1958*, (*Ind. chim. belge*, *Suppl.*), **1,** 40 (1959); *C.A.*, **54,** 3182, (1960).

Author index

Numbers in parentheses are reference numbers and indicate that the author's work is referred to although his name is not mentioned in the text. Numbers in italics show the pages on which the complete references are listed.

B

C

D

E

F

G

H

I

J

K

L

M

N

S

T

U

V

W

Y

Z

Subject index

Practically all the compounds mentioned in the text are individually indexed but common compounds, such as tetraethyllead, are not indexed for trivial entries. The major compounds, such as tetraethyllead, tetraphenyllead, hexaethyldilead, hexaphenyldilead and triphenylplumbyllithium, are indexed to show typical physical and chemical properties, and all known compounds can be found in the tables associated with each chapter. *Classes* of compounds in which one or more of the bonds is not a lead-lead or lead-carbon bond are indexed by the name of the non-carbon bonded group, e.g., nitrates, organolead; amines, organolead; hydroxides, organolead. Finally, both the tetra-tetrakis, tri-tris and di-bis designations are used in the index so that individual compounds should be searched under both systems.

B

D

E

F

G

I

R

S

T

U